Landmarks in Organo-Transition Metal Chemistry

Profiles in Inorganic Chemistry

Series Editor:
John P. Fackler, Texas A & M University, College Station, Texas

Current Volumes in this Series:

Landmarks in Organo-Transition Metal Chemistry: A Personal View
Helmut Werner

From Coelo to Inorganic Chemistry: A Lifetime of Reactions
Fred Basolo

Helmut Werner

Landmarks in Organo-Transition Metal Chemistry

A Personal View

Helmut Werner
Institute of Inorganic Chemistry
University of Würzburg
Germany

ISSN: 1571-036X
ISBN: 978-1-4419-1892-5 e-ISBN: 978-0-387-09848-7
DOI 10.1007/978-0-387-09848-7

Printed on acid-free paper

springer.com

To Monika, Andreas, and Annemarie
And in Loving Memory of Helga

Foreword

Organometallic chemistry has witnessed an exponential growth in the past half decade, and today is represented at its frontiers by the second edition of a multi-volume text, two major journals and a plethora of monographs. Helmut Werner, a pioneer who has contributed extensively to the field, now offers us a personal view of important areas of transition metal chemistry. It is unusual in that it provides an historical perspective on some of the more significant developments in this area. He writes both with a great generosity of spirit and an obvious love of the subject. It is evident that both for him, and now his readers, it is not only the science, but also its protagonists, that are the focus of much attention.

The first two chapters provide interesting information on Helmut's family and scientific background, culminating in his Würzburg C4 professorship (since 1975); he has mentored 110 Ph.D. students and 40 postdoctoral and visiting scientists. He continues in Chap. 3 to provide an account of the birth of the subject and its development in the nineteenth century. Subsequent chapters deal with metal carbonyls and derived clusters, the discovery of "sandwich" compounds, triple-decker analogues, metal–ethene complexes and their congeners, metal carbenes and carbynes, and finally metal alkyls and aryls. Each chapter has ample references. Helmut's account is exceedingly modest; from around 800 citations, less than 20 are to his own contributions. The text is well illustrated with formulae, reaction schemes, biographies and photographs.

The work is of very high quality and the author is to be congratulated on having given us a very informative and eminently readable and enjoyable book. He clearly has a profound knowledge of the subject and, as one of its leading practitioners, offers his readers a unique overview. I commend it with confidence and much enthusiasm.

July 2008 Michael Lappert

Series Preface

A renaissance in the field of inorganic chemistry began in the middle of the twentieth century. In the years following the discoveries of A. Werner and S. M. Jørgensen at the turn of the century, the field was relatively inactive. The publication of Linus Pauling's *Nature of the Chemical Bond* in 1938 and World War II shortly thereafter launched this renaissance. The war effort required an understanding of the chemistry of uranium and the synthetic actinide elements that were essential to the production of the atom bomb. There was also a need for catalysts to produce rayon, nylon, synthetic rubber, and other new materials for the war effort. As a result, many gifted chemists applied their talents to inorganic chemistry. *Profiles in Inorganic Chemistry* explores the roles some of the key contributors played in the renaissance and development of the field.

Some of the early leaders in this reawakening are now deceased. Pioneers included John Bailar at the University of Illinois, W. Conard Fernelius, at Pennsylvania State University, and Harold Booth at Western Reserve University, who with some others, started the important series entitled *Inorganic Syntheses*. Several inorganic chemistry journals were born, as were various monograph series including the *Modern Inorganic Chemistry* series of Springer. Geoffrey Wilkinson, who along with E. O. Fischer was the first inorganic chemist since Werner to win the Nobel prize, started his career at Harvard in about 1950 but later that decade moved to the University of London's Imperial College. By then, Ron Nyholm already was building a strong inorganic program at the University of London's University College.

Physical and mathematical concepts including group theory gave inorganic chemists new tools to understand bonding, structure, and dynamics of inorganic molecules. Fischer, Wilkinson, and their contemporaries opened up a new subfield, organometallic chemistry, out of which many metal-based catalysts were developed. It was soon realized that many inorganic minerals play essential roles as catalysts in living systems. As a result, another subfield, bioinorganic chemistry, was born. The discipline of inorganic chemistry today includes persons of many different walks of life, some creating new material and catalysts, others studying living systems, many pondering environmental concerns with elements such as tin, mercury, or lead, but all focusing on questions outside the normal scope of organic chemistry.

Organic chemistry has enjoyed a long history as a great science, both in Europe and the United States. During the past 15 years or so, many of the U.S. contributors have produced interesting autobiographies as part of an American Chemical Society series entitled *Profiles in Inorganic Chemistry*. There is also, however, a need to have students and scientists of inorganic chemistry understand the motivating forces that lead prominent living inorganic chemists to formulate their ideas. I am grateful that Springer has undertaken to publish this series. These profiles in inorganic chemistry will portray the interesting and varied personalities of leaders who have contributed significantly to the renaissance of inorganic chemistry.

College Station, TX John P. Fackler, Jr.

Preface

In the short period between December 1951 and February 1952, two papers appeared which laid the roots for what a few years later was called by Sir Ronald Nyholm *The Renaissance of Inorganic Chemistry*. Two research groups, working in completely different fields, reported the isolation of a seemingly simple iron compound of the analytical composition $FeC_{10}H_{10}$ which quite soon became the flagship of a new chemical discipline. It was not the composition of the new compound but its surprising and absolutely unexpected molecular structure that stimulated both experimental and theoretical chemists. While in the nineteenth and even in the first half of the twentieth century, it usually took decades before an epoch-making idea such as the cyclic structure of benzene, the tetrahedral configuration of methane, or Alfred Werner's concept of coordination compounds has been accepted, the synthesis and structural elucidation of bis(cyclopentadienyl)iron $FeC_{10}H_{10}$ – later called *ferrocene* – initiated immediately a research avalanche for which there is almost no precedent. In less than 20 years, not only metal compounds containing planar three-, four-, five-, six-, seven- and eight-membered ring systems were prepared, but at the same time also the chemistry of compounds with metal–carbon double and triple bonds was brought to light. The synthetic techniques together with the newly emerging analytical tools, in particular IR and NMR spectroscopy, offered the opportunity to follow the course of a chemical reaction and thus to understand the mechanism of the process. This also led to the rebirth of the field of homogeneous catalysis, and it is only fair to say that without the pioneering work in the 1950s and early 1960s on transition metal organometallics a number of important industrial processes such as the oxidation of ethene to acetaldehyde by the Wacker reaction, the synthesis of L-Dopa by the Monsanto process or the stereoselective polymerisation of olefins with the Brintzinger-type *ansa*-metallocenes as catalysts would not have been developed.

When I started writing this book, it was exactly 50 years ago that I became acquainted with organo-transition metal chemistry. As an undergraduate at the University of Jena in the former *Deutsche Demokratische Republik* ("East Germany"), I attended a course in preparative inorganic chemistry and a junior colleague of Professor Franz Hein took care of the course. It was at this time, that Professor Ernst Otto Fischer visited Hein's laboratory to inform him that, based on his work at the *Technische Hochschule* in München, he was convinced that the unusual "polyphenylchromium compounds", reported by Hein mainly between 1919 and 1931, were indeed sandwich-type complexes. At first, Hein was irritated but after his coworkers proved that Fischer's proposal was correct, he accepted the new ideas. Since I had the fortune to work for my Diploma thesis with Hein and for my Ph.D. thesis with Fischer, I became automatically involved in the rapid and breath-taking development of modern organometallic chemistry, and I remained caught and fascinated by this subject ever since. The close personal contacts with Hein and Fischer, together with the fact that from 1968 to 1975 I taught coordination chemistry at the University of Zürich in the same building as Alfred Werner did, also awakened my interest in the history of "my" discipline, and it became a challenge to discover the links between the beginnings in the nineteenth and early twentieth century and our present research activities.

This book is neither a textbook nor an autobiography. If I take into account that in recent years organometallic chemistry has not only grown tremendously but also concerns the chemistry of the majority of the elements of the Periodic Table, it is nearly impossible to cover the historical development of the whole field. Therefore, I limited the content to compounds of the transition metals which also happen to be the most important components in homogeneous catalysis. I did my best to consider all the relevant literature and I apologize if I have missed some of the links. It is, of course, a personal view of the discipline and it may well be that some younger scientists in particular feel that I have over-emphasized what had happened in the past. Thus, I answer with a sentence written by the German author Bernhard Schlink in his novel *The Reader*: "Doing history means building bridges between the past and the present, observing both banks of the river, taking an active part on both sides."

Würzburg, Germany Helmut Werner

Acknowledgments

I owe the first and particular debt of gratitude to my friends and colleagues Lutz Gade and Adrian Parkins, who not only corrected and polished my written English but, equally important, also made countless valuable comments regarding the content of the Chaps. 3–9 and the list of references. They also pushed me ahead in those moments when I felt exhausted or were near to despair about the bulk of the literature. Moreover, I am grateful to numerous colleagues who provided biographical informations and offered useful hints for the manuscript. Taking the risk of being incomplete, I would like to name Anthony Arduengo, Didier Astruc, Wolfgang Beck, Susanne Becker, Martin Bennett, Robert Bergman, Friedrich Bickelhaupt, Marika Blondel-Mégrelis, Pierre Braunstein, Hans Brintzinger, Fausto Calderazzo, Ernesto Carmona, (the late) Albert Cotton, John Ellis, Christoph Elschenbroich, John Fackler, (the late) Ernst Otto Fischer, Helmut Fischer, Peter Gölitz, William Graham, Malcolm Green, Robert Grubbs, Max Herberhold, Wolfgang Herrmann, Volkan Kisakürek, Michael Lappert, Jack Lewis, Giuliano Longoni, Peter Maitlis, David Milstein, Ullrich Müller-Westerhoff, Luis Oro, Peter Pauson, Martyn Poliakoff, Philip Power, Warren Roper, (the late) Max Schmidt, Richard Schrock, Dietmar Seyferth, Gordon Stone, Rudolf Taube, Jim Turner, Egon Uhlig, Wolfgang Weygand, (the late) Nils Wiberg, and Günther Wilke. Most of the formulae and schemes were drawn with insight and proficiency by Sabine Timmroth. I give my sincere thanks to her as well as to Cornelia Walter and my former secretary Inge Bräunert, both of whom proved to be computer experts and helped me extensively. Finally, I am indebted to Kenneth Howard from Springer Publishers for the pleasant form of cooperation and his unlimited patience during the time of writing. Last but not least, I hope that the authors of present and future textbooks will not miss the past and tell their students about the roots on which the wonderful field of organo-transition metal chemistry rests.

Contents

1 Prologue 1
References 7

2 Biographical Sketch 9
2.1 The Years at Home 9
2.2 The First Move: From Mühlhausen to Jena 19
2.3 The Second Move: From Jena to Munich 24
2.4 The First Years at München 26
2.5 From München to Pasadena and Back 31
2.6 Crossing the Border: The Years at Zürich 41
2.7 Back to Germany 50
Biographies 64

3 The Nineteenth Century: A Sequence of Accidental Discoveries 69
3.1 The Beginnings of Organometallic Chemistry 69
3.2 Wilhelm Christoph Zeise and the First Transition Metal π-Complex 70
3.3 Edward Frankland's Pioneering Studies 71
3.4 Victor Grignard: The Father of "Organometallics for Organic Synthesis" 74
3.5 Paul Schützenberger and Ludwig Mond: The First Metal Carbonyls 75
Biographies 78
References 83

4 Transition Metal Carbonyls: From Small Molecules to Giant Clusters 85
4.1 A Class of "Peculiar Compounds" 85
4.2 The Giant Work of Walter Hieber 89
4.3 Hieber and his Followers 93
4.4 Surprisingly Stable: Multiply Charged Carbonyl Metallate Anions 98
4.5 Metal Carbonyl Cations: Not Incapable of Existence 100

4.6 Highly Labile Metal Carbonyls 102
4.7 The Exiting Chemistry of Metal Carbonyl Clusters........ 105
4.8 Otto Roelen and Walter Reppe: Industrial Applications of Metal Carbonyls.................................. 110
4.9 Biographies.. 114
References.. 119

5 A Scientific Revolution: The Discovery of the Sandwich Complexes 129
5.1 The Early Days: Ferrocene.............................. 129
5.2 The Rivalry of Fischer and Wilkinson.................. 135
5.3 Fischer's Star: Bis(benzene)chromium.................. 136
5.4 Hein's "Polyphenylchromium Compounds" 138
5.5 Zeiss and Tsutsui: Hein's Work Revisited 140
5.6 Wilkinson's Next Steps................................. 145
5.7 From Sandwich Complexes to Organometallic Dendrimers.. 146
5.8 The Taming of Cyclobutadiene: A Case of Theory before Experiment 150
5.9 The Smaller and Larger Ring Brothers of Ferrocene....... 152
5.10 Sandwiches with P_5 and Heterocycles as Ring Ligands..... 154
5.11 Two Highlights from the 21st Century.................. 157
5.12 Brintzinger's Sandwich-Type Catalysts 159
5.13 Woodward and the Nobel Prize.......................... 161
Biographies 163
References 169

6 One Deck More: The Chemical "Big Mac" 177
6.1 The Breakthrough: $[Ni_2(C_5H_5)_3]^+$ 177
6.2 The Iron and Ruthenium Counterparts.................... 182
6.3 Arene-bridged Triple-Decker Sandwiches 185
6.4 "Big Macs" with Bridging P_5, P_6 and Heterocycles as Ligands.. 186
6.5 Tetra-, Penta- and Hexa-Decker Sandwich Complexes..... 189
Notes.. 191
References 191

7 The Binding of Ethene and Its Congeners: Prototypical Metal π-Complexes 195
7.1 From 1827 to the 1930s: In the Footsteps of Zeise 195
7.2 Reihlen's Strange Butadiene Iron Tricarbonyl............ 199
7.3 Michael Dewar's "Landmark Contribution"............... 200
7.4 The Dewar–Chatt–Duncanson Model 202
7.5 An Exciting Branch: Mono- and Oligoolefin Metal Carbonyls.. 204

7.6 Schrauzer's Early Studies on Homoleptic Olefin Nickel(0) Complexes. 208
7.7 Wilke's Masterpieces and the "Naked Nickel". 209
7.8 Stone and the Family of Olefin Palladium(0) and Platinum(0) Compounds 214
7.9 Timms', Fischer's and Green's Distinctive Shares 216
7.10 A Recent Milestone: Jonas' Olefin Analogues of Hieber's Metal Carbonylates 219
Biographies. 221
References 228

8 Metal Carbenes and Carbynes: The Taming of "Non-existing" Molecules 235
8.1 The Search for Divalent Carbon Compounds 235
8.2 From Wanzlick's and Öfele's Work to Arduengo's Carbenes 237
8.3 The Breakthrough: Fischer's Metal Carbenes 238
8.4 The Next Highlight: Fischer's Metal Carbynes 241
8.5 Öfele's, Casey's and Chatt's Routes to Metal Carbenes 242
8.6 Lappert's Seminal Work on Bis(amino)carbene Complexes . 244
8.7 A Big Step: Schrock's Metal Carbenes and Carbynes 247
8.8 Fischer and His Followers 253
8.9 Using the Isolobal Analogy: Metal Complexes with Bridging Carbenes and Carbynes 256
8.10 The Seemingly Existing CCL_2 and Its Generation at Transition Metal Centers 259
8.11 The Congeners of Metal Carbynes with $M\equiv E$ Triple Bonds 263
8.12 The First and Second Generation of Grubbs' Ruthenium Carbenes 263
8.13 From Metal Carbenes to Open-Shell Metal Carbyne and Carbido Complexes 268
8.14 The Dötz Reaction and the Use of Metal Carbenes for Organic Synthesis 271
8.15 Olefin Metathesis: A Landmark in Applied Organometallic Chemistry 272
8.16 An Extension: Metal Complexes with Unsaturated Carbenes 274
Biographies. 276
References 284

9 Metal Alkyls and Metal Aryls: The "True" Transition Organometallics 297
9.1 The Extensions of Frankland's Pioneering Work 297
9.2 Heteroleptic Complexes with Metal–Alkyl and Metal–Aryl Bonds 299

9.3 Chatt and His Contemporaries 300
9.4 Lappert, Wilkinson and the Isolation of Stable Metal Alkyls und Aryls 304
9.5 An Apparent Conflict: Metal Alkyls and Aryls Containing σ- and π-Donor Ligands 310
9.6 Binary Metal Alkyls with M – M Multiple Bonds 314
9.7 The Recent Highlight: Power's RCrCrR and the Fivefold Cr – Cr Bonding 315
9.8 Novel Perspectives: Metal Alkyls and Aryls Formed by C – H and C – C Activation 317
9.9 Metal Alkyls and Aryls in Catalysis 324
References 325

10 Epilogue 337

Index 341

List of Abbreviations

acac	acetylacetonate
ACM	asymmetric cross-metathesis
AROCM	asymmetric ring-opening cross-metathesis
bpy	bipyridyl
*n*Bu	*n*-butyl
*t*Bu	*tert*-butyl
COD	1,5-cyclooctadiene
COT	cyclooctatetraene
Cy	cyclohexyl
DMSO	dimethyl sulfoxide
dme	1,2-dimethoxyethane
dmpe	1,2-dimethylphosphinoethane
dppe	1,2-diphenylphosphinoethane
dppm	1,2-diphenylphosphinomethane
DQ	duroquinone
en	1,2-ethylenediamin
Et	ethyl
HMPA	hexamethylphosphoric triamide
Me	methyl
Ment	mentyl
Mes	mesityl
Naph	naphthyl
NHC	*N*-heterocyclic carbene
Ph	phenyl
*i*Pr	isopropyl
py	pyridine
RCM	ring-closing metathesis
ROMP	ring-opening metathesis polymerization
tmeda	tetramethylethylenediamine
THF	tetrahydrofuran
Tol	tolyl
Xyl	xylyl

Synopsis

"Doing history means building bridges between the past and the present, observing both banks of the river, taking an active part on both sides." This sentence, cited from the novel *The Reader* by Bernhard Schlink, stands for the content of this book. Since the discovery of ferrocene and the sandwich-type complexes, the development of organometallic chemistry took its course like an avalanche and became one of the scientific success stories of the second half of the twentieth century. Based on this development, the traditional boundaries between inorganic and organic chemistry gradually disappeared and a rebirth of the nowadays highly important field of homogeneous catalysis occurred. It is fair to say that despite the fact that the key discovery, which sparked it all off, was made more than 50 years ago, organometallic chemistry remains a young and lively discipline.

The author of this book participated in the success story almost from the beginning. As an undergraduate student, he worked for his Diploma Thesis with Franz Hein, one of the key figures of coordination chemistry in Germany between 1920 and 1960, and obtained his Ph.D. in the laboratory of Ernst Otto Fischer, one of the great heroes of organo-transition metal chemistry in the latter half of the twentieth century. He prepared the first borazine metal complexes, isolated the chemical "Big Mac", promoted the concept of metal basicity, investigated the chemistry of metallacumulenes and, most recently, discovered a new bonding mode for tertiary phosphines, arsines and stibines. He held academic positions at the *Technische Hochschule* in Munich, the University of Zurich and the University of Würzburg, and from 1990 to 2001 was the Chairman of a collaborative research center in organometallic chemistry.

Chapter 1
Prologue

Schreib deine eigene Welt zu Ende, ehe das Ende dich abschreibt.[1]

Rose Ausländer, German Lyric Poet (1901–1988)

It happened on the 25th of November 1960. In the late afternoon of that day I was returning from Ludwigshafen to München, picked up my father-in-law and drove with him in my VW beetle to the beautiful town of Kempten, where my parents-in-law lived and where my wife was waiting for me. A few kilometers outside of Kempten, on a narrow road, we crashed into a big American car, my father-in-law was thrown out of the beetle, and I was trapped inside the car. In those days, German cars – even Mercedes and BMW – were not equipped with seat belts. An ambulance brought me to the main hospital in Kempten, where the doctors confirmed that I had broken my right leg, my right arm and a number of ribs but, much more seriously, also had pressure on the brain. Since I was unconscious, it took several days before I realized what had happened. Due to the brain damage, the doctors decided to avoid an operation and thus my leg and arm were fixed in a standard fashion. This continued for 4 months and only then was an operation carried out. I had to stay in the hospital for 6 additional weeks before I could return to our home.

The drive on that 25th of November to *the Badische Anilin und Sodafabrik* (BASF) at Ludwigshafen was for good reason. When I started my work at München in October 1958, my supervisor Professor Ernst Otto Fischer had suggested for my thesis on three different topics of which the preparation of palladocene seemed to me most challenging. After the serendipitous discovery of ferrocene by Peter Pauson and Samuel A. Miller and their coworkers in 1952 [1, 2], both Fischer's and Geoffrey Wilkinson's group had elaborated the chemistry of cyclopentadienyl complexes of nearly all the transition metals with the exception of the noble metals palladium and platinum. By taking into consideration that

[1] In English: "Write about your life to the end, before the end writes you off".

H. Werner, *Landmarks in Organo-Transition Metal Chemistry*,
Profiles in Inorganic Chemistry, DOI 10.1007/978-0-387-09848-7_1,

nickelocene $Ni(C_5H_5)_2$, although a 20-electron compound, was quite stable and accessible via different synthetic routes [3, 4], I presumed that the preparation of the heavier homologue $Pd(C_5H_5)_2$ should also be a realistic goal.

However, the great optimism which I had at the beginning of my work was not vindicated. The entries in my lab notebook between October and mid-December of 1958 showed that I carried out about 30 experiments aimed to generate palladocene from different precursors, but they all failed. In some cases, when I attempted to isolate the required product from the reaction mixture, I obtained nearly pure palladium black and occasionally even a palladium mirror, but this was not what I wanted. The decisive change of the direction of my work happened a week before Christmas 1958. At one of the regular meetings of the Chemical Society in München, Professor Rudolf Criegee from the *Technische Hochschule* at Karlsruhe talked about "New Insights into the Chemistry of Cyclobutadienes". In the first part of his talk, he mentioned that his coworker Gerhard Schröder had tried to abstract the two chloro substituents of the cyclobutene derivative **1** by nickel tetracarbonyl but instead of generating tetramethylcyclobutadiene **2** he isolated the corresponding nickel complex **3** (Scheme 1.1) [5]. Although at that time the complex could not be characterized crystallographically, there was no doubt that the proposed structure was correct [6].[2]

As the lecture continued and the auditorium became more and more fascinated by Professor Criegee's presentation, I was lost in my own thoughts. Since I was quite familiar with the organometallic chemistry of nickel, I realized that **3** was not only the first cyclobutadiene metal compound but also the first complex in which a diolefin was coordinated to the nickel center. Therefore, I concluded that if a diolefin, expected to be a good π-acceptor ligand, was able to form a stable bond to nickel(II), why should it not do the same to nickel(0). With the well-known complexes $Ni(CO)_4$, $Ni(CNPh)_4$, and $Ni(PPh_3)_4$ in mind, the target molecule should have the general composition $Ni(diolefin)_2$. The next day I started with my attempts to prepare a compound of this type. Since I was aiming to synthesize palladocene, I only had cyclopentadiene and no other

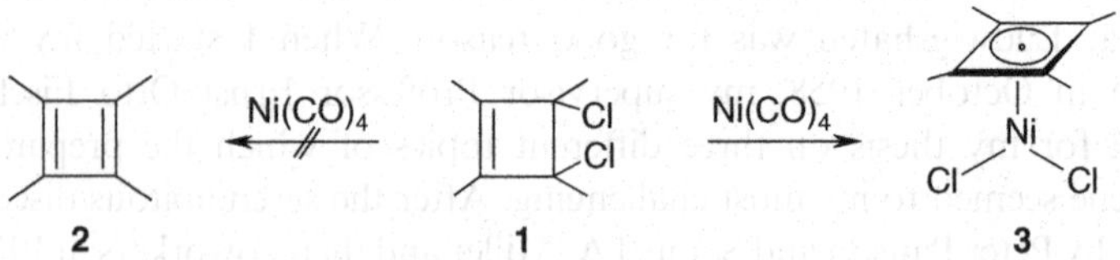

2 ← $Ni(CO)_4$ (crossed) — 1 (Cl, Cl) — $Ni(CO)_4$ → 3 (Ni, Cl, Cl)

Scheme 1.1 The reaction of nickel tetracarbonyl with 1,2-dichloro-1,2,3,4-tetramethylcyclobutene (**1**) which gave the tetramethylcyclobutadiene nickel complex **3** instead of tetramethylcyclobutadiene (**2**)

---

[2] Later it was shown by X-ray crystal structure analysis that in the crystalline state compound **3** is a dimer with two bridging chloro ligands between the metal centers.

diolefin such as 1,3-cyclohexadiene or 1,5-cyclooctadiene in the refrigerator, and thus I treated a solution of freshly distilled cyclopentadiene in hexane dropwise with $Ni(CO)_4$. I obtained a red air-sensitive solid which was highly volatile, soluble in all common organic solvents, and analyzed as $NiC_{10}H_{12}$. Based on these results, I was absolutely convinced to have an analogue of nickel tetracarbonyl with two cyclopentadiene units replacing the four CO ligands in my hand.

When I told Professor Fischer about the results, he became equally enthusiastic and after I repeated the synthesis four or five times, determined the molecular weight and proved that the compound was diamagnetic, we submitted the manuscript entitled *Di-cyclopentadien-nickel(0)* to the editor of *Chemische Berichte* on the 2nd of March 1959. In less than 2 weeks, it was accepted and published in the June issue 1959 [7]. In the meantime, I also prepared the supposed nickel(0) complex $Ni(C_5H_5Me)_2$ and began with attempts to generate a $Pd(diolefin)_2$ counterpart. Compared to the nickel compounds, this seemed to be a more ambitious goal since palladium tetracarbonyl – the homologue of $Ni(CO)_4$ – was unkown and no other palladium(0) complex, which could be used as starting material, was available. To circumvent this problem, I reacted the carbonyl palladium(II) compound $Pd(CO)Cl_2$ with a tenfold excess of 1,3-cyclohexadiene and isolated a yellow powder which, based on its elemental analysis, was assumed to be dimeric $[(C_6H_8)PdCl]_2$. Treatment of this compound with NaC_5H_5 gave a red crystalline solid, which by analogy with the above-mentioned nickel complex $NiC_{10}H_{12}$ was also very air-sensitive and had the analytical composition $PdC_{11}H_{14}$. Since we believed that owing to the properties, in particular the high volatility, it must be a palladium(0) derivative, Professor Fischer and I postulated in a second manuscript, submitted in April 1960, that the product of the stepwise reaction of $Pd(CO)Cl_2$ with C_6H_8 and NaC_5H_5 was the bis(diolefin) complex $(C_5H_6)Pd(C_6H_8)$ [8].

It was a mere coincidence that in those days when our paper appeared, Bernard Shaw reported the synthesis of the palladium(II) compound $(C_3H_5)Pd(C_5H_5)$ from $[(C_3H_5)PdCl]_2$ and NaC_5H_5 [9], which seemed to have similar properties as $PdC_{11}H_{14}$. On reading this report, we became suspicious whether the structures we had proposed for our nickel and palladium complexes were correct. We therefore asked Heinz Peter Fritz, who investigated the IR spectroscopic data of transition metal organometallics in his *Habilitation* thesis [10], to record the IR spectra of $NiC_{10}H_{12}$ and $PdC_{11}H_{14}$. The result was that *probably* both complexes contain a π-bonded cyclopentadienyl ring and thus should be formulated as $(C_5H_5)Ni(C_5H_7)$ and $(C_5H_5)Pd(C_6H_9)$ and not as $Ni(C_5H_6)_2$ and $(C_5H_6)Pd(C_6H_8)$, respectively. To confirm this proposal, both Heinz Peter Fritz and Walter Hafner (at that time completing the work on the Wacker process) recommended that we study the ^1H NMR spectra of $NiC_{10}H_{12}$ and $PdC_{11}H_{14}$ but in the autumn of 1960 there was no NMR spectrometer, either at the *Universität* or the *Technische Hochschule* in München.

In order to resolve the problem, Professor Fischer called Dr. Walter Brügel at BASF, who operated such an instrument, and asked him whether it would be possible to measure the NMR spectra of our nickel and palladium compounds. Since they were very air-sensitive, Dr. Brügel suggested that I should bring small samples of $NiC_{10}H_{12}$ and $PdC_{11}H_{14}$ to Ludwigshafen in a Dewar flask and he would then measure the ^1H NMR spectra immediately. The proposed date was Friday, the 25th of November 1960. After leaving München in early morning, I arrived at Ludwigshafen at around 10 a.m. and less than 2 h later I knew that the structures we had originally proposed for the two compounds were wrong. Instead of being diolefin nickel(0) and palladium(0) derivatives, $NiC_{10}H_{12}$ (see Scheme 1.2) and $PdC_{11}H_{14}$ were indeed cyclopentadienyl complexes with the metal in the oxidation state two.

To digest the embarrassing news, Dr. Georg Hummel – a member of the board of directors at BASF and a friend of Ernst Otto Fischer – invited me for lunch and thus it was not before 2.30 p.m. that I left Ludwigshafen. Back in München, I brought the Dewar and the NMR spectra to my lab, picked up my father-in-law and on the way to Kempten had the car crash. From the hospital I informed Professor Fischer about the new data and in March 1961 we reported in a short paper the correct structures of the nickel and palladium complexes, $NiC_{10}H_{12}$ and $PdC_{11}H_{14}$, respectively [11].

In mid-May 1961, almost 6 months after the accident, I was discharged from the hospital, learnt again to walk, to use my right arm and right leg properly, and – most importantly – to write up the results of my Ph.D. work. I handed the draft of my thesis to one of Professor Fischer's technicians and asked her to type the manuscript as quickly as possible. I submitted the thesis to the faculty in mid-June hoping that the oral examination could be fixed before the end of the summer semester. My wife, Helga, and I expected our first child in early August 1961 and we both desperately wanted to have a few days of rest before the birth. With the support of Professor Fischer, the Dean set July 21th as the date of the final examination.

But things turned out to be more complicated. Since I was used to (and I still do) work late in the day and, if necessary, at night, I made the final check of my

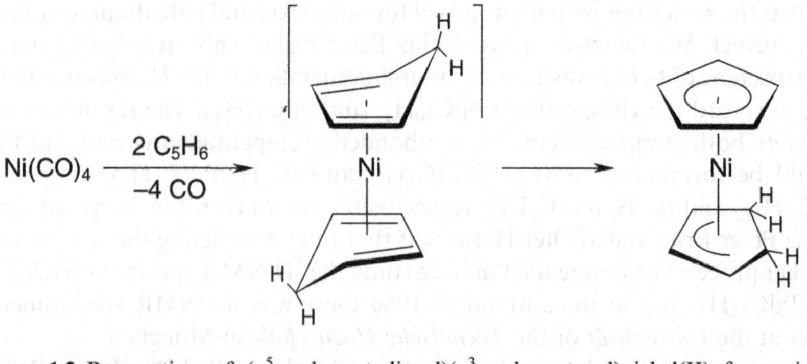

Scheme 1.2 Preparation of (η^5-cyclopentadienyl)(η^3-cyclopentenyl)nickel(II) from nickel tetracarbonyl via the nickel(0) complex $Ni(C_5H_6)_2$ as the supposed intermediate

notes in preparation of the exam at about midnight of July 20th. I went to bed at around 3 a.m. to find my wife awake, telling me that the labor pains had begun. I was of course extremely worried, could hardly sleep and, after the pains persisted, decided to take her to the clinic at 7 a.m. Without a telephone in our apartment and still handicapped by walking with crutches, it took me more than 30 min to catch a taxi and convince the driver to drive as careful as possible. When we arrived in the obstetrics ward of the clinic, the gynaecologist, seeing my ash pale face and trembling hands, instantly sent me home. I felt terrible since the oral examination was scheduled at 4 p.m. and I was unable to rest or sleep. All I could do was to revise my notes for the exam and to pray for a healthy child. It was the time of the Contergan ("thalidomide") case and, only a few months before, the first son of one of our friends was born with no arms, having his fingers directly connected to his shoulders.

However, despite my evident nervousness the examination proceeded quite well, the four examiners – Professor Fischer, Professor Walter Hieber, Professor Friedrich Weygand, and Professor Günter Scheibe – were satisfied and I received the best grade "summa cum laude". Less than 30 min later, I returned to the clinic but was told that it was still too early to initiate the birth. It took 6 more hours, and it was 10 min past midnight of July 22nd, that our daughter Monika was born. She was a healthy child (Fig. 1.1). The following week, when I returned to the department, I told Professor Fischer the news and I still remember right now – almost 50 years later – the incredulous expression in his face. After a while, he said *Meister Werner, Sie machen Sachen* but then he smiled and gave me a hug. This was not his usual type of behavior and even his secretary was surprised when she saw his emotional reaction.

Though pleased about the good news, Professor Fischer was also anxious about the progress of my career. Before the accident, I was supposed to spend a year as a postdoctoral fellow at Harvard with Professor Eugene G. Rochow, but in view of the latest events this idea had to be abandoned. Due to my

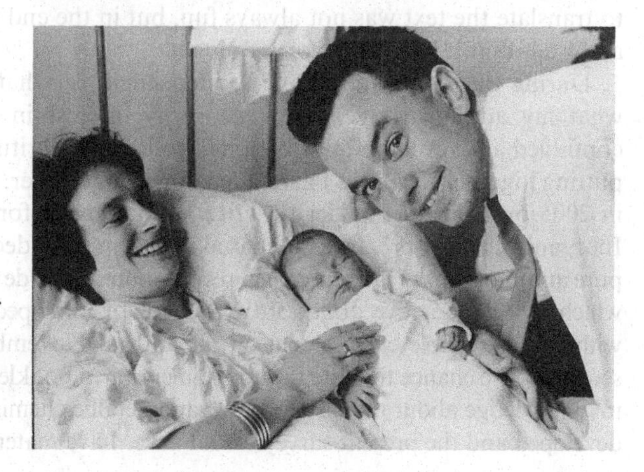

Fig. 1.1 My wife Helga, our daughter Monika and me in the gynaecology clinic in München, a few days after Monika was born (July 1961)

physical condition, I was also unable to work in the lab, and Professor Fischer therefore suggested that I review the literature on metal π-complexes with di- and oligoolefinic ligands and summarize the results in a booklet to be used internally for his research group. At the time when we still thought that the compounds $NiC_{10}H_{12}$ and $PdC_{11}H_{14}$ were nickel(0) and palladium(0) complexes with cyclopentadiene as ligand, Fischer was very keen to find out whether diolefin complexes of other transition metals in low oxidation states were accessible. The impetus to strengthen this type of research was not only the apparent novelty of the compounds but also the fact that, following the first report by Abel, Bennett and Wilkinson on the synthesis of cycloheptatriene molybdenum tricarbonyl [12], the chemistry of di- and oligoolefin metal complexes seemed to emerge as a highly promising new area in organometallic chemistry.

In August 1961, when I was asked to write the review, Fischer's appeal *Let the competitors not be alone* had already reached part of his research group and thus I was eager to get acquainted with this challenging new field. Having learned during my Ph.D. work about the serendipitous discovery of nickel tetracarbonyl by Langer and Mond and, earlier on at Jena, about the story on Hein's "polyphenylchromium compounds", my interest in the history of organometallic chemistry was kindled. Moreover, after reading the literature about the sharp controversy between Zeise and Liebig on the composition of Zeise's salt and the long time resistance by the chemical community to accept the view about the inherent ability of π-electrons to form dative bonds to transition metal ions, I became even more interested in historical events. Thus, writing the booklet seemed to be a good opportunity to study at least part of the history on organometallic chemistry in more detail and to learn more about the biographies of the pioneers in this field. The booklet was finished in summer 1962, one year later a somewhat extended version was published in German as a monograph by *Verlag Chemie* [13], a new and significantly extended edition in English was published by Elsevier in 1966 [14], and finally a Russian translation of the English version appeared in 1968 [15]. To write and to translate the text was not always fun, but in the end both Professor Fischer and I felt that it had been worth doing it.

During the following decades, independent of what I did in research and what my administrative duties were, my interest in the history of science continued and, even before I became Professor Emeritus, I had thought about putting together the notes I had taken during my career. Thus when I was asked in 2005 by John P. Fackler to write a monograph for the series "Profiles in Inorganic Chemistry", I was prepared to accept provided that it should not be a pure autobiography, but a personal recollection on the development of the field in which I became active about 50 years ago. In retrospect, I am convinced that without the terrible car accident on the 25th of November 1960 and, as a consequence, the chance to write the above-mentioned booklet on metal π-complexes, my knowledge about the history of organometallic chemistry would be much less developed and the present survey would never have materialized.

References

1. T. J. Kealy, and P. L. Pauson, A New Type of Organo-Iron Compound, *Nature* **168**, 1039–1040 (1951).
2. S. A. Miller, J. A. Tebboth, and J. F. Tremaine, Dicyclopentadienyliron, *J. Chem. Soc.* **1952**, 632–635.
3. E. O. Fischer, and R. Jira, Di-cyclopentadienyl-nickel, *Z. Naturforsch., Part B*, **8**, 217–219 (1953).
4. G. Wilkinson, P. L. Pauson, J. M. Birmingham, and F. A. Cotton, Bis-cyclopentadienyl Derivatives of some Transition Elements, *J. Am. Chem. Soc.* **75**, 1011–1012 (1953).
5. R. Criegee, and G. Schröder, Ein Nickel-Komplex des Tetramethyl-Cyclobutadiens, *Angew. Chem.* **71**, 70–71 (1959).
6. J. D. Dunitz, H. C. Mez, O. S. Mills, and H. M. M. Shearer, Kristallstruktur des Benzol-Addukts des 1,2,3,4-Tetramethylcyclobutadien-nickel(II)-chlorids, *Helv. Chim. Acta.* **45**, 847–665 (1962).
7. E. O. Fischer, and H. Werner, Di-cyclopentadien-nickel(0), *Chem. Ber.* **92**, 1423–1427 (1959).
8. E. O. Fischer, and H. Werner, Cyclohexadien-(1.3)-cyclopentadien-palladium(0), *Chem. Ber.* **93**, 2075–2082 (1960).
9. B. L. Shaw, Allyl(cyclopentadienyl)palladium(II), *Proc. Chem. Soc.* **1960**, 247.
10. H. P. Fritz, Infrared and Raman Spectral Studies of π-Complexes Formed between Metals and C_nH_n Rings, *Adv.Organomet. Chem.* **1**, 239–316 (1964).
11. E. O. Fischer, and H. Werner, Zur Struktur von $NiC_{10}H_{12}$ und $PdC_{11}H_{14}$, *Tetrahedron Lett.*, **1961**, 17–20.
12. E. W. Abel, M. A. Bennett, and G. Wilkinson, Cyclohepatriene Metal Complexes, *Proc. Chem. Soc.* **1958**, 152–153.
13. E. O. Fischer, and H. Werner, *Metall-π-Komplexe mit di- und oligoolefinischen Liganden* (Verlag Chemie, Weinheim, 1963).
14. E. O. Fischer, and H. Werner, *Metal π-Complexes* (Elsevier, Amsterdam, Vol. I, 1966).
15. E. O. Fischer, and H. Werner, *Metal π-Complexes* (Publishing Company Mir, Moscow, 1968).

Chapter 2
Biographical Sketch

Denn das Leben ist keine Rechnung und keine mathematische Figur, sondern ein Wunder.[1]
Herrmann Hesse, German Poet (1877–1962)

2.1 The Years at Home

I was born on 19th April 1934 in the town of Mühlhausen in Thüringen. Already as a child I was told that it is important to say "Mühlhausen *in Thüringen*" and not only "Mühlhausen" since there are at least a dozen towns or villages in Germany with that name. However, among those, Mühlhausen in Thüringen is not only the largest but – as everybody in my hometown believes – also the most beautiful and most important! The first document mentioning the name of the place dates back to 967 when Emperor Otto II donated the castle, built by the Carolingian dynasty in the ninth century, to his wife. At that time the former kingdom of Thuringia, having had its golden age in the fifth century, had ceased to exist and had been occupied since the sixth century by the Franconians. In the course of the political confusion following the decline of the ruling dynasties in the eleventh and twelfth centuries, around 1230 Mühlhausen became a "*Freie Reichsstadt*" (free city), which means that it was accountable only to the emperor but not to a regional prince. It maintained this privilege until 1802. An infamous "highlight" in the history of the city happened in 1525. Thomas Müntzer, a radical theologian, took over the city council, rallied troops formed by poor and badly armed peasants and led an uprising (called the "*Bauernkrieg*") against the local nobility. He lost the fight and was finally beheaded outside the city walls of Mühlhausen. The city was terribly punished and did not recover before the mid seventeenth century. After 1803, following the political agreement called "*Reichsdeputationshauptschluß*", Mühlhausen became part of Prussia and remained so until 1945 when the state of Prussia was erased by the Allies.

[1] In English "Life is not a calculation nor a mathematical figure, but a wonder."

H. Werner, *Landmarks in Organo-Transition Metal Chemistry*,
Profiles in Inorganic Chemistry, DOI 10.1007/978-0-387-09848-7_2,

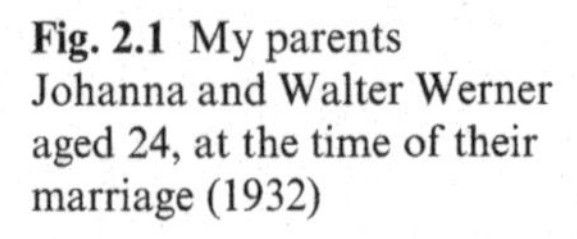

Fig. 2.1 My parents Johanna and Walter Werner aged 24, at the time of their marriage (1932)

My parents, Johanna (née Scharf) and Walter Werner (Fig. 2.1), were both born in 1907 and got married in 1932. My grandfather Hugo Scharf was a locksmith and extremely self-confident since he had not only his own workshop but also a trade shop selling nearly everything needed for the households of the people at Schlotheim, a small town 17 km north-east of Mühlhausen. Everybody called him "*Meister Scharf*" and he was very proud of this title. He was also the undisputed chief of the family clan and appeared to be particularly fond of my mother as well as me, his first grandson. I never met my other grandfather Christel Werner who was an umbrella-maker but had died in 1916 after being injured during World War I. His wife, my grandmother Ida Werner, was a highly ambitious and parsimonious person who was able in 1913 – with the financial support of her family – to buy a three storey house in the center of Mühlhausen and opened a shop in which umbrellas and walking-sticks were sold.

I was born in this house and lived there together with my parents and my grandmother. My father was an employee at the city council while my mother stayed at home, occasionally helping out my grandmother in her shop. I had no brothers or sisters and, since our house was in the main street, my mother was very anxious about me playing outside. Thus my playground was the small backyard of the house and it seems I was quite happy (Fig. 2.2).

In April 1940 I started elementary school (Fig. 2.3). Owing to the war, which had begun the year before, my teachers were relatively old which had its advantages and disadvantages. On the one hand, they were very experienced, knowing how to handle a class of nearly 40 boys, but on the other hand, they practised a style of education dating from the nineteenth to early twentieth century. I remember in particular one teacher we had in our second year who tried to hammer the times tables into our heads. Every morning when he entered the class-room we all had to stand up and the exercise began. The first one who could tell him how much is, for example, 12 times 16 was allowed to sit down

Fig. 2.2 Me at age 3 in the "Bavarian look", which was a common dress for children even outside Bavaria on special occasions such as church festivals, birthday parties, etc.

and this continued at least for the next 20 min. If one of those pupils, who remained standing at the end, was unable to answer the question, he would grasp the lobe of an ear, stretching it for about 10 s and, if the correct result was not given, would give him a hiding with a ruler. Although this certainly was not

Fig. 2.3 A picture of myself together with my classmates and our schoolmaster at the courtyard of our elementary school (1941). I am the fourth from left in the first row having both hands in my pockets

an appropriate educational method, it provided a strong incentive for me to learn the times tables and it thus frequently happened that I was among the first allowed to sit down. The approach certainly helped me to develop a good memory and even today some of my friends are rather surprised if I can tell them instantly how much, for example, 25 times 35 is.

Already in September 1939, at the beginning of Word War II, my father was drafted into the army. Fortunately, due to his former activity in the city administration, he was not sent into combat, but worked in an office in charge of the supply to the troops. In autumn 1944, when the Russian armies were approaching Germany's eastern border and the casualties of the German soldiers increased dramatically, my father was ordered to a place near Danzig (in Poland Gdansk) to be trained as a second lieutenant for the fighting troops. However, before he could take up the command of a unit, he was taken prisoner by the Russians and, subsequently, sent to the Ural mountains in Western Siberia to work in a mine.

Since I saw my father only twice a year when he returned for vacations, during the period of the war I was alone with my mother. She did everything to support me and gave me all the love usually provided by both parents. Nearly every weekend, we visited my grandparents at Schlotheim who had a large house including a workshop and a big garden. There I grew my own strawberries and, during the summer vacations, I helped my grandparents to harvest the potatoes and grapes grown on their fields. I also looked after the two or three pigs they had in the stable. The longer the war lasted, the more it became important to have at least partly your own resources for food, since otherwise it would have been difficult to survive.

In September 1944, after 4 years at elementary school, I moved to high school and, in addition to German grammar, maths, history, music and sports, had English, German literature, biology and geography as the new subjects. In contrast to places such as Jena, München or Würzburg, which suffered tremendously under the allied bombardments, my hometown was bombed only once by the Royal Air Force and thus, as far as I remember, teaching at school was never disrupted. My favorite subject in those days was English which was taught by Herr Ölgarte. I assume that he was at least 65 years old in those days and, like almost everybody of his generation, had never spent time in an English-speaking country. Due to this, we had no opportunity to practise conversation in English, but instead focussed on the intricacies of its grammar. Personally, I believe this clear and methodical approach was a useful mental training for a scientist to be, albeit at the expense of the ability to communicate in that language.

The time at school temporarily came to an end after American troops occupied Mühlhausen in the first week of April 1945. They approached the city from the West and, fortunately, were not seriously challenged by the remnants of the German Army. As a consequence, nothing – no building, no street, no bridge – was destroyed and for the population it took only a few days to continue with what seemed to be a "normal life". What I do remember well is that almost overnight all the Nazi emblems, in particular the photographs of Hitler and flags bearing the swastika, disappeared and – as I was told years later by my grandfather – many documents issued between 1933 and 1945 by the Nazi organizations were burnt.

Since initially all schools remained closed after the end of the war, my friends and I had plenty of time to devote to other things. We were curious to watch and possibly contact the American soldiers – mainly black people – who were either sitting around their local headquarters in the parks or playing baseball. The most courageous of us tried to sell them some remaining Nazi emblems or military medals and, if successful, got some chocolate or, what was even more popular, chewing gum. When I once attempted to do the same and to exchange my scout's knife for a bar of chocolate, I completely failed. The soldier who I contacted took the knife and gave me a slap in the face, thus ending the experiment. I felt very ashamed and did not tell my mother about the unsuccessful trade.

An event with terrible consequences for many citizens at my hometown happened at the beginning of July 1945. According to the agreement of Yalta, signed by Roosevelt, Churchill and Stalin in February 1945, the remaining part of Germany – West of the rivers Oder and Neisse – was divided into four Occupational Zones which were under the command of the American, British, French and Soviet military administrations. It was also decided by the allies to divide Berlin into four Sectors and to exchange Thuringia – which in April 1945 was completely occupied by the American army –for West Berlin. Therefore, Thuringia became part of the Soviet Zone and in the first week of July 1945 the Americans had to leave. Before the Soviet troops arrived at Mühlhausen, nearly everybody was nervous and tried to shut the doors of their house or flat as tightly as possible. Due to the expulsion of many millions of Germans from the former Eastern part of Germany, that now belongs to Russia, Poland or the Czech Republic, five to six thousand refugees stayed in my hometown, and it was the knowledge of their experience with the Soviet soldiers that frightened most of the citizens. As a consequence, thousands of people followed the American troops to the Western side of the border between the American and the Soviet Zone which was only 15 km to the west of Mühlhausen and in those days not barred by a wall or a barbed wire fence. Until at least 1948, it was not unusual that people from my hometown to ride west with their bicycles to exchange fruits, potatoes or other items mainly for cigarettes and nylon stockings. The Soviets only occasionally intervened since their main preoccupation was to establish a new administration strictly following their commands, to accommodate the officers and their families in buildings confiscated from upper-class citizens, and to provide food for the thousands of soldiers housing in the barracks previously used by the German army. In the American, British and French Zones (which in 1949 became the Federal Republic of Germany) the food supplies slowly improved after 1948, while in the Soviet Zone (since 1949 the German Democratic Republic or DDR) the situation remained strained and for several goods such as butter, meat, sausages and oil rationing lasted until 1956.

In September 1945, we were allowed to continue at our high school with most of the teachers we had before the end of the war. Some of them had disappeared, either because they had been dismissed due to their membership in the Nazi party or had decided to move to the West. In autumn 1946, the communist authorities decided that all children, independent of their talents and their educational background, would have to spend 8 years at the elementary school. Therefore, my friends and I were relegated from high school and mixed with other boys who had never been taught English, German literature, advanced maths, etc. before. Teaching such a heterogeneous pack of youngsters became quite difficult for our teachers and school became rather boring for me. As a type of compensation, I joined a soccer and a table-tennis club and, since those of my friends and classmates being neither interested in soccer nor in table-tennis founded a chess team, I also became a member of the latter. However, it is worth pointing out that despite playing chess quite well in those days, I always lost when playing against Professor Fischer during our group skiing vacations 12–15 years later.

The saddest day of my childhood was the 3rd of July 1947 when my mother died. In November 1946 my father had been released as a prisoner but when he returned from West Siberia to Mühlhausen he was very ill. In the following weeks both my mother and my grandmother did their best to help him regain his physical and mental composure. His experience as a Russian POW as well as his expulsion from the new communist dominated city council were a heavy burden for him. However, in early summer 1947 he seemed to have partly recovered and began to help my grandmother in her shop. Then at the end of June – over the course of a few days – my mother became seriously ill. She was hospitalized and operated upon instantly, but despite the efforts of the doctors and a blood-donation by my uncle, she passed away. Only my father was with her during the last hours, and when he returned home to tell me what happened, I broke down. I cried for hours, was unable to grasp the new situation, and – after the funeral service – vowed that I would not leave our house for 1 month. It was the decision of a 13-year old boy who had loved his mother so dearly. During that time at home I copied (with pencil) a book explaining some simple chemical experiments. A few weeks before my mother died I had borrowed this book from my classmate Hans-Karl Kobold, whose parents had a shop with plenty of space and allowed their youngest son to do some experiments in the laundry. Since we shared the way home from school, he invited me to see his "lab" and do something together. Thus I came in touch with chemistry at age 13. One of the first experiments we did was to react iron turnings with dilute sulfuric acid and to burn the hydrogen generated in the reaction. It worked as described in the book! However, when repeating the experiment, we did not wait until the air was completely replaced by hydrogen, and when we ignited the gas, an explosion occurred. Fortunately, our eyes (not protected by safety glasses) were not hit by the splinters but the acid spilt onto our shorts. It was hard to explain the result to our parents.

In the year following the death of my mother, the person who helped me cope emotionally with what had happened was our parson, Hans Falckner. He had already conducted the wedding of my parents, had baptized me and thus was well acquainted to my family. At the end of the war when my mother did not know whether my father had survived, he had supported her and, after July 1947, he attempted to do the same for me. He believed that doing sports in the spare time was not enough and convinced me to join the local Bach choir. Incidentally, Johann Sebastian Bach, the illustrious composer, served from 1707 to 1708 as the organist at Divi Blasii (Fig. 2.4), the same church in which I was baptized and later (1948) confirmed. It had been a challenge for all of his successors not only to keep up the tradition but also make the Bach choir the number one at Mühlhausen and its vicinity. It was certainly not due to me during the 2 years I belonged to it, that the choir maintained his high standards and it gave me great satisfaction to participate in it.

The boredom at school ended in late summer 1948, and in September of the same year I was allowed to re-enter high school. Despite the radical political change, dictated by the Soviet administration and readily adapted by the leaders of the East German communist party (since 1946 *Sozialistische Einheitspartei Deutschlands*, abbreviated SED) the high school of my home town provided a good general education, with an emphasis on the sciences during my final 4 years. In the humanities, the situation was delicate. Traditionally, Latin, followed by English, was the first foreign language, but it was decided by the communist authorities that it now should be replaced by Russian. This proved to be difficult since there were only a few people able to speak and teach

Fig. 2.4 Basilica Divi Blasii at Mühlhausen, built in the thirteenth century, the tower on the right completed somewhat later in the early Gothic style. Johann Sebastian Bach (1685–1750) was the organist of the church from 1707 to 1708 (photo by courtesy from Mrs. Karola Winkler)

Russian. For us this meant that we had Latin and English in the first year, only English in the second and third year, and both English and Russian in the final year. To enable us to pass the final examination in Russian, our class-master, Herr Preusse, decided that instead of having 3 h of English and 3 h of Russian per week the ratio should be 1:5 in favor of Russian. This recipe worked and helped me to earn some extra money by translating scientific articles from Russian to German later at München.

During my years at high school, from 1948 to 1952, two of my teachers made a lasting impression to me. The first was Miss Tappenbeck (nicknamed "Flora") who taught maths (Fig. 2.5). She was unmarried, about 50 years old, a strange person but as a teacher an absolute authority. After entering the classroom, from the first to the very last minute she was completely devoted to the subject and impressed upon us that maths is the cornerstone of all knowledge. In all these years I enjoyed doing maths exercises, and felt flattered when I was occasionally asked by Flora to help correcting the exercises of other pupils at her home. Although I assume that she was disappointed about my decision not to study maths later on, we maintained a good relationship, which continued after I left school.

The second inspiring teacher was Dr. Gerhard Hesse who taught chemistry. After finishing his Ph.D. in 1942, he had the good fortune of not becoming a soldier but to be assigned to a special research institute at Frankfurt, led by Professor Eugen Müller, where new polymers, eventually to be used for military purposes, were investigated. After World War II, this institute was closed and Dr. Hesse returned to Thuringia where his parents lived. With no specific background in teaching, he came to our school and, from scratch, installed a laboratory where we could perform some simple experiments usually in the afternoon when school was finished. With my "research" experience from the laundry, mentioned above, I was attracted from the very beginning and

Fig. 2.5 Miss Tappenbeck and Dr. Gerhard Hesse, the most important teachers I had at high school (1952)

Dr. Hesse did his best to foster this interest. Knowing that I enjoyed maths, he told me – apparently unintentionally – how exciting research could be and this got me hooked to the subject.

Since the piano lessons paid for by my grandfather seemed to have little effect on me, I got more and more drawn into sports in my spare time, in particular table-tennis. Pushed by a highly ambitious coach and in strong competition with a dozen of equally enthusiastic teammates, I often spent 4–5 h per day to exercise in the sports hall, and quite frequently was late for dinner. However, our efforts were rewarded and in 1950 we won the junior team championship and 2 years later, the senior team championship of Thuringia. Finally, on the 19th of April 1952 (which was my 18th birthday) we finished second in a contest with the top teams of the other five states in East Germany and rose to the premier league (Fig. 2.6). This success was instantly highlighted in the headlines of the Mühlhausen newspaper and made us local heroes for a few days. After enrolling at the university, I played one more year for Mühlhausen in the premier league but then decided to become a member of the (less ambitious) university team composed of students and assistants at Jena. Between 1953 and 1956 we actually made it first into the fourth and then into the third division. It was the perfect balance to the work in the lab and also created the lifelong friendship to Helmut Hanschmann, the table-tennis champion of East Germany in those days.

An extremely happy day was the 19th of August 1950 when my father remarried. About 6 months before, he had met a young woman named Ingeborg Leister who soon became one of my best friends. Inge (as we all call her) completely understood that at the bottom of my heart I was still attached to my mother and thus considered herself more like an elder sister. Already before the

Fig. 2.6 Table-tennis team from my club *Post Mühlhausen*, after rising to the premier league of East Germany (1952). I am to the left of the president of the club while on his right side is Heinz Schneider, the All-German table-tennis champion, 1952

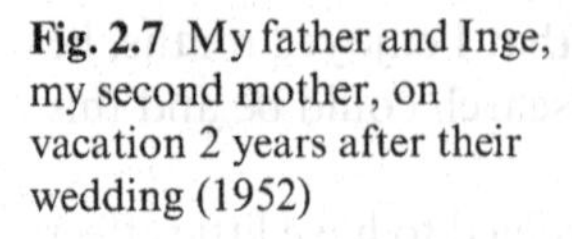

Fig. 2.7 My father and Inge, my second mother, on vacation 2 years after their wedding (1952)

wedding, my father and I saw her nearly every day and we instantly realized that she would provide sunshine and happiness to our home (Fig. 2.7). She also helped me to overcome a problem that weighed heavily on me in summer 1950. Like all of my friends, I had begun taking ballroom dancing lessons. However, it quickly became apparent that I had no talent for this, and thus felt like an outsider. After Inge became aware of this, she decided to exercise the basic steps with me at home and due to her efforts I finally even succeeded in dancing a Viennese waltz sufficiently well.

An important decision had to be made in autumn 1951 when my classmates and I were asked what we wanted to do after the final high school examinations. Despite the fact that no one in my family had attended a university, I was keen to break with this tradition and was strongly supported by my father and Inge. Nevertheless, it was rather uncertain whether my application to enter a university would be successful, since already at this early stage of the communist regime the ministry of education had decided that primarily children with a true working class background should be admitted at the universities. It was typical for the situation in those days that my father, who after the death of my grandmother ran the family shop without any employees, was considered as a "capitalist", a fact that could be held against me.

However, as several times before, I was lucky. During the final 2 years at school, I scored top grades in my exams which earned me the grade "with distinction" in my *Abitur* (Finals). Based on this, the headmaster of the school, Dr. Georg Möller, nominated me for the "Lessing Medal" which I received in September 1952 at a formal ceremony in the great hall of our school. Apart from the personal satisfaction, the important implication of the award was a guaranteed enrolment at a university despite my unpleasant "capitalist" background. I immediately applied to the University of Jena but was informed that lab space was limited and, therefore, my application was sent to the University of Greifswald, located in the North of the Baltic Sea. This was unfortunate, but at that moment political developments worked in my favor. At a conference at Leuna (a place near to Halle, where at the beginning of World War I the first plant for the Haber–Bosch process was built) Walter Ulbricht, the Secretary General of the SED, proclaimed that one of the prerequisites for establishing a socialist society in East Germany would be a strong chemical industry. Following this announcement, the universities were ordered to increase the intake of first year students majoring in chemistry significantly. Thus the authorities at Jena University, under the pressure of somc top communists, decided to accept instead of 30 (as in the years before) the fantastic number of 330 first year students! Everybody who had applied in June/July was now allowed to enrol and I was one of them. It was a sunny day when I received the respective letter, and I was looking forward to the near future with great expectations.

2.2 The First Move: From Mühlhausen to Jena

Although Mühlhausen was not a small town (in the early 1950s it had about 45,000 inhabitants), coming to Jena was like entering a new world. Jena was not only home to the well-known Carl-Zeiss and Otto-Schott companies but also, more important for me, the *Friedrich Schiller Universität*. This university was founded in 1558 by Prince Johann Friedrich of Saxony and became one of the most prestigious academic centers in Germany at the turn of the eighteenth to the nineteenth century. The upgrade was strongly supported by Johann Wolfgang von Goethe who lived in the neighboring city of Weimar. Brentano, Novalis, the Schlegels and Tieck founded the "Romantic School" in the literal arts, Fichte, Hegel and Schelling started a new way of thinking in philosophy, and – in the footsteps of the French Revolution – Friedrich Schiller taught modern history. Goethe's enthusiasm for the sciences was also most influential in establishing a chair of chemistry, and in 1810 Johann Wolfgang Döbereiner was appointed as the Professor of Chemistry and Technology. Following his earlier interests in commercially valuable products, he soon began to extract platinum metals from American platinum ores and in the course of his investigations not only found the platinum-catalyzed oxidation of ethanol to acetic acid (thereby designing a "vinegar lamp") but also constructed the pneumatic gas lighter that bears his name. After Döbereiner's death in 1849, inorganic chemistry remained strong at Jena and it was not a mere accident that before and after World War II the professors of inorganic chemistry, Adolf Sieverts and Franz Hein, were the respective Heads of the *Chemisches Institut*.

In September 1952, when I enrolled at the university, the situation for undergraduates was difficult indeed. Since the headquarters and most of the factories of the Zeiss and Schott companies were located at Jena, the city had been heavily bombed during the war. Despite some early efforts by the university to build dormitories, there was insufficient accommodation available. Numerous students, whose parents lived in the vicinity, traveled daily by train, bus or bicycle to Jena, and I remember that some of them even came from distant cities, such as Gotha, Gera or Erfurt. I myself was lucky since shortly before I moved to the city one of my cousins had left Jena. Thus I could rent his former room only a few days after my arrival. The landlady agreed that I could share the room with my friend from high school, Reinhard Wohlfarth, who initially had applied for enrolment as a biology student, but due to the political decision mentioned above became a student of chemistry. As was common in these days, our room was quite primitive, had no water or central heating and, when following the demands of nature, we had to walk to an outside toilet at the end of the back yard.

Nevertheless, I started at the university with great enthusiasm and so did most of my friends. Since the *Chemisches Institut*, formerly near to the main building of the Zeiss company, had been completely destroyed in 1944/45, the classes in inorganic and organic chemistry were held in the lecture theater of the Botanic Garden. Analytical chemistry was taught in the Institute of Pharmacy, which was just opposite the garden house where at the end of the eighteenth century Friedrich Schiller had lived. In the first year, laboratory facilities were extremely limited and therefore only a small number of students could participate in the practical courses. With the letter "W" at the beginning of my name, I had to wait nearly three semesters before doing the first analysis and even then got a lab place not in the provisional institute of inorganic chemistry but in the apprenticeship hall of the Schott company. However, thanks to my good luck in the analytical course I progressed quickly and in summer 1957 was among the first to start the research work for the Diploma thesis.

The decision as with whom I would like to work for this thesis was made much earlier since from the very beginning of my studies I was fascinated by the personality of Professor Franz Hein. He not only taught basic and advanced inorganic chemistry but also coordination chemistry. Even for me as an undergraduate, it was quite clear that the latter was his favorite subject. Although still influenced by the ideas of the former giants in this field, Alfred Werner and Paul Pfeiffer, Hein's teaching was not limited to classical coordination chemistry but also included the chemistry of metal carbonyls, metal isocyanides and sandwich compounds. Although he rarely gave hints to recent references, I got the feeling that it was cutting edge chemistry.

Organic chemistry was taught by Günther Drefahl who had become the Professor of Organic Chemistry at age 27. Compared with Hein, he was a completely different type of person with great ambitions not only in science, but also at the political level. Although not a member of the SED, he was to be the Rector of the *Friedrich Schiller Universität* from 1962 to 1967 and later the President of the "Peace Council" of the DDR, an institution formally supporting the "politics for peace" of the communist government. While I remained personally quite distant to him, his eloquence and charm attracted numerous students and in retrospect, I can understand why nearly 50% of those who enrolled with me in 1952 decided to work for the Diploma and eventually the Ph.D. thesis with him.

Teaching in physical chemistry seemed for me, at least at the beginning, very incoherent and was done by a *Privatdozent*, Dr. Bogislav Rackow. He belonged to the type of scientists who prefer to live in an ivory tower and avoid contacts with students. It was only at the end of 1954 that the Professor of Physical Chemistry, Horst Dunken, returned from the Soviet Union to Jena. After the Russian troops had occupied Thuringia including Jena in summer 1945, he, together with some other university professors and several top scientists from the Zeiss company, was "resettled" to a research center south of Moscow where they lived – as he told us later at a student's party – in a "golden cage" for nearly 10 years. In his classes he was not as brilliant as Hein but he taught physical chemistry in an absolutely up-to-date fashion.

My first steps in research occurred in the winter semester 1956/57. After passing the elementary lab courses and the preliminary exam (*Vordiplom*), we had to complete advanced courses both in organic and inorganic chemistry, each of them took about 2 months. Whereas I prepared some natural products using established procedures in the organic course, I had the chance to work in one of the research labs of Siegfried Herzog in the inorganic course. Herzog, a former student of Hein's, became a *Dozent* at Jena in the mid 1950s and later held chair positions in inorganic chemistry first at the University of Greifswald and then at the *Bergakademie* Freiberg in Saxonia. When I took the advanced course with him, he was mainly interested in 2,2'-dipyridyl complexes of metals in low oxidation states. After his group had prepared the corresponding chromium(II) and chromium(I) compounds $[Cr(dipy)_3]^{n+}$ (n = 1 or 2), the next challenge was to obtain the chromium(0) analogue. Although the experiments with which I was involved failed, I learned to isolate and handle air-sensitive substances and to work with a vacuum line.

Inspired by this experience, I wanted to work for my Diploma thesis in a related field, but when I approached Professor Hein, he suggested a completely different topic. He told me that chemists from the Academy of Science, which ran a large research institute at Jena, were looking for new methods for the resolution of racemic mixtures of chiral aminoalcohols and he thought that this could be achieved by coordinating the aminoalcohols to chiral transition metal centers. Hein's idea was to use Werner-type complexes such as $[Co(en)_2Cl_2]Cl$ and $[Cr(en)_2Cl_2]Cl$ as starting materials, to replace the two coordinated chlorides by a chelating (racemic) optically active aminoalcohol, and then to separate the expected pair of diastereoisomers by fractional crystallization. Since the people from the academy were primarily interested in separating ephedrine and pseudoephedrine, compounds of the general composition $C_6H_5CH(OH)CH(CH_3)NHCH_3$, I was supposed to explore the coordinating capabilities of these molecules.

Appreciating the importance of this project, I forgot my previous ideas and began my work enthusiastically. However, my great expectations were only partly fulfilled. While I completely failed to generate any cobalt(III) or chromium(III) compound with ephedrine as ligand, I was able to isolate the octahedral complex $[Co(en)_2\{\kappa(O,N)\text{-}OCH(C_6H_5)CH(CH_3)NHCH_3\}]Cl_2$ from $[Co(en)_2Cl_2]Cl$ and an excess of pseudoephedrine. Subsequent studies with $NH_2CH_2CH_2OH$ and aminoalcohols of the general composition $RNHCH_2CH_2OH$ and $R_2NCH_2CH_2OH$ (where R is an alkyl group) revealed that with increasing size of group R the rate of the reaction of the aminoalcohol with $[Co(en)_2Cl_2]Cl$ and $[Cr(en)_2Cl_2]Cl$ significantly decreased. Moreover, the alcohols with a tertiary amino group were more reluctant to coordinate than those with NHR as the substituent. In less than 6 months I prepared 16 new cobalt(III) and chromium(III) complexes with different aminoalcohols as ligands, carried out the elemental analyses myself (in the classical way) and determined the molar conductivity (Fig. 2.8). Even today, I am impressed by the accuracy of the analytical data which I obtained.

Although the original goal of the Diploma project was not achieved, Hein was quite satisfied and offered me the opportunity to carry out research towards the Ph.D. with him. Together with my friends Gerhard Grams and Frieder Löffler, I passed the orals in inorganic, organic, physical chemistry and chemical engineering on the 1st of July 1958, and we were highly satisfied to finish as the first of the 330 students, who enrolled with us in autumn 1952. Based on the results of the research project as well as the oral exams I was awarded the Diploma "with distinction" and was looking forward to the work to be carried out for my doctorate. Since I had told Hein that I would like to move from classical coordination to organometallic chemistry, he suggested that I should further develop the chemistry of oligophenylchromium compounds, which was a major research topic in his group. Richard Weiss, who had received his Ph.D.

Fig. 2.8 In the lab while working for the Diploma thesis (spring 1958). Since my starting materials as well as the new products were not air-sensitive, I usually worked in Erlenmeyer flasks with methanol as solvent

with Hein in 1957, had prepared the complex $Li_3Cr(C_6H_5)_6$ in his doctoral thesis and, as an extension of his work, it was suggested that I should try to generate a similar ate-complex of chromium(II). The interesting question was not only whether species such as $Li_3Cr(C_6H_5)_5$ or $Li_4Cr(C_6H_5)_6$ were accessible, but also whether they could be converted via reduction and possibly intramolecular C–C coupling to bis(arene)chromium derivatives.

Since the lab, in which I had previously worked, was not equipped with a vacuum line and the facilities for using Schlenk tubes, I moved to the main building of the inorganic institute and started immediately with the research for the Ph.D. Due to the fact that I was not a member of the SED or another political party, Hein suggested that I should not be funded by an East German government grant, but by a scholarship of the Saxonian Academy of Science of which he was a member. Although the monthly stipend was somewhat less compared to that of a regular assistantship, I agreed. I was unmarried and the salary (525 East German marks per month) was about three times of what I had before. In the first 3 weeks I prepared some chromium(II) precursors, learned to generate phenyllithium in its pure form and, inspired by Hein's chief assistant Erhard Kurras, whose lab was next to me, did the first experiments. Everything seemed to proceed well.

However, at the beginning of August 1958 I received a note from the SED secretary of the chemistry department to come to see him and some other party members the following evening. Being happily ignorant of what that was going to be about, I went to the meeting in my table-tennis outfit assuming that it would be all over soon. Most surprisingly, when I arrived at the party office I was surrounded by eight people, all of them graduate students or assistants from the department, who told me quite frankly that if I had any intentions to continue at the university and eventually start an academic career, I would have to play a more active role in the building of the socialist society. Noting that I was lost of words, they proposed that I should apply instantly to the *Elektrochemisches Kombinat Bitterfeld* (a big state-owned company near to Halle) and to participate in the establishment of the semi-military worker's brigades (*Kampfgruppen*). Provided I performed appropriately, I might be allowed, after an as yet unspecified period of time, to return to Jena and to take up again my studies for the Ph.D.

The next morning I informed Professor Hein about the unexpected proposal and asked him what to do. He realized my confusion and tried to ease my mind, by telling me that he would see the Secretary of State for Education and Science during the course of that week and would approach him about the matter. Given his reputation as a scientist, he was quite optimistic that he could do something about it. However, after he returned from Berlin he asked me immediately to come to his office where he told me that, unfortunately, there was nothing he could do to overturn the decision of the party. He himself was frustrated and before I left he said ambiguously: "I think you know what you ought to do".

2.3 The Second Move: From Jena to Munich

I had previously not considered leaving East Germany and it took me some days to appreciate that this was inevitable. Although I had no sympathy for the politics and the ideology of the communist government, I considered myself more or less an apolitical person and by no means involved in the illegal opposition against the system. However, after the meeting with the members of the SED, I had to admit to myself that I had been extremely naïve. The party secretary had filed a dossier of all my previous activities including what I had done on 17th June 1953 (the day when the workers at Berlin, Halle, Jena and several other East German cities protested against the increased production targets and with it the additional load of labor imposed by the government; the uprising was crushed by the Soviet military), what my comments were when Stalin died, and what I said after the Russian troops had suppressed the democratic movement in Hungary and murdered the non-communist leaders. Undoubtedly I was under observation, and had to realize that it was impossible to avoid that conclusion.

The following days belonged to the most thrilling of my life. I first went back to my hometown, informed my parents about the decision to leave, put a jacket, some trousers and several unsuspicious pieces of clothing into a suitcase and joined a sports delegation of *Motor Jena* which visited Berlin. In those days the only route to escape to the West was through Berlin, and remained so until the wall at Berlin was erected in 1961. However, it was not an easy route since public transport to and from East Berlin was closely monitored by the authorities, and East German citizens traveling by train to Berlin were scrutinized by the railway personnel. Being accompanied by my friend Helmut Hanschmann, whose achievement as a sportsman had earned him the title "*Meister des Sports*", I felt secure and arrived safely in Berlin. The following day, I succeeded in crossing the internal Berlin border and took my clothes and some important documents to my school friend Dieter Grubitzsch, who had to leave East Germany in 1955 and now lived at West Berlin. I did not stay in the West on that occasion since my disappearance would have had severe consequences for Helmut Hanschmann and the other members of the sports delegation. I therefore returned to Jena, brought the rest of my personal belongings to my parents in Mühlhausen and bade them farewell. My father had big tears in his eyes, but he attempted to smile and wished me all the best for my future. At Jena I bought a return train ticket to Greifswald to make the railway personnel believe that I was on a harmless journey to the Baltic seaside. When the train stopped in Berlin, I got off, waited for the morning rush hour and then joined the crowds of Berliners commuting to their work in Potsdam, which is geographically west of Berlin, but back in East German territory. I got onto a westbound train of the city railway at *Friedrich-strasse* (being in the East) and traveled to the West. In those days people living in East Berlin and working in Potsdam or its surroundings actually had to cross through West Berlin territory using the city railway system. This inner city railway system was run by the East and the police and border guards on the trains saw to it that nobody, who was not supposed to, got off at a "wrong" station.

I was lucky and, with a city railway ticket to Potsdam, I was able to convince the border guards that I had no intention to escape but planned to visit some relatives living at Potsdam. At the Zoo station in West Berlin I got off the train discreetly and brought my briefcase to Dieter Grubitzsch's room. I then went to see the authorities of the refugee camp at Marienfelde, a district of West Berlin, to report to them that I would not return to the East, but stay in the West. During the next 2 days, I was interviewed by British and US military intelligence personnel at Marienfelde who soon realized that what I had to say was already known to them. They allowed me to leave the refugee camp, after the debriefing was finished. Fortunately, Dieter's landlady agreed that I could stay in his room for the time being.

In spite of this opportunity, I was eager to leave West Berlin as soon as possible. Since I had been told in the camp that I could only continue in West Germany if I had a job offer or a guaranteed place as a graduate student at a university, I immediately wrote a letter to Professor Ernst Otto Fischer in München. I knew that he and Professor Hein highly respected each other and therefore I hoped that he would accept me as a Ph.D. student. However, I got no answer during the next 2 weeks and became more and more anxious, being unaware that at that time Professor Fischer was participating in an excursion to visit Troy (Greece) and other historic sites. Before running out of money, I contacted Dr. Karl Eisfeld, a former assistant of Hein and now working for BASF at Ludwigshafen, and he managed to get me out of West Berlin. This was accomplished by air in order to avoid crossing East German territory again. When I arrived at Ludwigshafen, I was very pleased that Dr. Eisfeld had arranged for me to participate in an international summer school organized by the BASF company. This was the best thing that could happen to me at this point, because it gave me accommodation and food for 4 weeks and, equally important, the opportunity to get information from other participants of the summer school about the situation at various chemistry departments in West Germany. Due to Fischer's trip to Greece and Turkey, I still did not know whether he would accept me and I therefore contacted Professor Georg Wittig in Heidelberg and Professor Walter Rüdorf in Tübingen. They both invited me for an interview and, after seeing my records, told me that I could do a Ph.D. with them.

While still trying to make up my mind, I received a telegram from Fischer in which he invited me to see him. He also sent a cheque for 100 Deutschmark to cover my travel expenses and, after the summer school was finished, I immediately traveled to München. He came to the railway station, invited me for lunch, showed me his labs and talked to me about his current research interests. I was absolutely fascinated and it took me only one night to decide that I would like to stay. Today I know that, from a professional point of view, it was the most decisive moment of my life and I am still grateful having had the opportunity to work with Fischer.

2.4 The First Years at München

At the time when I arrived at München, Ernst Otto Fischer was an associate professor at the *Ludwig Maximilians Universität* (abbreviated LMU) and had a group of about 15 graduate students and postdocs. The Director of the Institute of Inorganic Chemistry at the LMU was Professor Egon Wiberg (Fig. 2.9), who was well-known for his work on boron compounds and was also the Rector of the LMU from 1957 to 1959. He had gathered around him a team of young and highly talented inorganic chemists such as Max Schmidt, Heinrich Nöth, Hans Bock, Alfred Schmidpeter, Peter Pätzold and Nils Wiberg (his son), all of whom later had very successful careers in academia. Since their main research interests were in main group chemistry, teaching and research in transition metal chemistry was solely in the hands of Fischer. I noted quite soon that Wiberg and Fischer, although completely different personalities, got on well with each other. This created a pleasant working atmosphere in the institute.

My start in Fischer's group was a disaster. On the very first day, after I had enrolled at the university, I was cleaning my lab place and while doing this found some sealed pressure tubes made of thick glass in the hood which had been left by my predecessor. Since I had no experience in opening such tubes, I asked my labmate Siegfried Schreiner to tell me how to do it. He explained it to me, and when I opened the first tube nothing happened. However, the next attempt failed. The tube exploded and the top of the tube partly lacerated my right thumb. Siegfried and I were covered in blood and I was immediately taken to hospital. A few hours after the surgery had been performed, Fischer visited me. I was relieved indeed when he said that after leaving the hospital I should continue with my work as intended.

Despite my "bloody" start, Ernst Otto Fischer did his best in the following weeks to integrate me in his research group and to make me feel comfortable

Fig. 2.9 Egon Wiberg (1901–1976), the Director of the Institute of Inorganic Chemistry at the *Ludwig Maximilians Universität* in München (LMU) until 1970 (photo by courtesy of the late Professor Nils Wiberg)

in my new environment. Although he was a full blooded Bavarian with no relatives in East Germany, he seemed to be very interested in the political situation in the East and, particularly, in the attitude of the people who after 12 years of the Nazi regime now suffered again under the communist dictatorship. We had long and intense discussions on that topic and in retrospect I am sure that, our common interest in organometallic chemistry aside, this provided the basis for our wonderful lifelong personal relationship.

A few months later, these private conversations gained additional importance since in summer 1959 Fischer received an invitation from the University at Jena to succeed Franz Hein to the chair of inorganic chemistry. Our discussions started again and, as Fischer later mentioned to me, it was mainly the experience I had had, and my view of the political situation in East Germany, that made him refuse the call and stay in München. In the 1970s when I visited Franz Hein at his home in Jena and told him that I felt a bit guilty regarding Fischer's refusal to accept the invitation, he smiled at me and remarked: "It was the best thing he could do".

Apart from Fischer's efforts, two others helped me to settle at München. The first was related to table-tennis. Already before I arrived, I was aware of the top teams and since my student's hostel was not far away from the gymnasium where the Bavarian champion MTV 79 exercised, I tried to get into this team. However, I failed, not because of my playing skills but because my prospective teammates, all of them being significantly older than me, did not appreciate a "Prussian" newcomer. The next attempt was met with more success and, after 2 months of intense training, I became a member of the team of *Post München*. This proved to be fortunate for me. In particular the top player, Matthias Thurmaier, realized my difficult personal situation and that I suffered for being separated from my family and friends. He was the first at München who invited me to his home and, although I had great problems to understand the Bavarian dialect of his parents, I had the feeling that I was welcome. I tried to repay them with good results for the table-tennis team and in summer 1959 we became champion of the Upper Bavarian league.

The second aspect was completely different. It was a pure accident that in spring 1959 I met a group of young refugees from Jena who gathered once every month at a bar in München. At one of these gatherings I met Günter Rahm, who at Jena played in the premier soccer league and had left East Germany shortly after me. Günter was also a good lawn-tennis player and although I had no experience in this type of sports, I started playing tennis with him. In autumn 1959 and spring 1960 we played nearly every Saturday and after the match usually went to his flat and had breakfast or lunch. Sometimes we also picked up Günter's wife Gisela who worked in a pharmacy, and who I knew quite well from common courses in analytical chemistry at Jena. At one of these occasions Gisela told us that she had a new colleague, a young lady, who recently moved from Hamburg to München and would very much like to play tennis. We arranged for a mixed doubles on the 1st of May (Labor Day, a holiday in Germany) which, in the event, would completely change my life. The young lady with whom I played in the doubles against Gisela and Günter was Helga Schnadel who half a year later became my wife. It was – as we say in German – "love at first sight" and became a wonderful marriage for almost four decades.

Helga was a native of that part of Southern Bavaria which is called *Allgäu*. Her father was an active member of the *Alpenverein* (an organization taking care of the hiking routes and cottages in the Alps) and she herself was really attached to the area around the Allgäu mountains. Nevertheless, after she finished high school and a practical internship in a pharmacy, she moved to Hamburg in 1955 to study pharmacy. Helga's uncle was a Professor of Shipbuilding at the University of Hamburg and he invited her to live in his house. She was called by the other students "the girl from Austria", due to her South German dialect. She had a marvellous time, and also after we got married, we took every opportunity to visit her relatives and friends in the North.

In September 1960, before our wedding we flew from München to Berlin to see my parents who at that time – one year before the Wall was erected – were able to come to East Berlin by train and cross the internal Berlin border with the city railway system. Since my father and Inge had never been to West Berlin before, it was to them like entering a "new world". We showed them the city, the famous department stores and the *Funkturm* (the radio tower) where we had a fantastic view across both parts of Berlin (Fig. 2.10). My parents were extremely pleased with my choice and immediately took Helga to their hearts.

As already described in the Prologue, the work of my doctoral thesis proceeded quite well and within a relatively short period of time I became an established member of Fischer's research group. While I had only a stipend of 200 Deutschmarks per month during my first year (and thus was eager to

Fig. 2.10 Helga and me on the upper platform of the radio tower in Berlin (September 1960)

get some additional money from Russian to German translations), my financial situation improved significantly after September 1959 when I was fortunate to receive a government grant for scientists who were refugees from the East. From one day to the next I increased my salary from 200 to 630 Deutschmarks and felt like a millionaire. This allowed me to buy a new jacket and some modern shirts as well as a pair of skis plus the appropriate boots. It was a tradition in Fischer's group that after the end of the winter semester the *Chef* and his coworkers spend a week at a remote mountain cottage where everybody could relax. When I participated for the first time, in March 1959, I was an absolute beginner and was more off than on my skis. However, in winter 1959/60 I tried hard to improve my abilities and, with the support of Hubert Kögler and Karl Öfele, both members of Fischer's group, I succeeded on the second such event in March 1960 to finish the slalom competition (always scheduled on the last day of the skiing week) in 15th position out of 20 (Fig. 2.11). Later, when we returned from California, both Helga and I went with the Fischer group to the mountains, and since Helga grew up on the skis in the *Allgäu*, I was very ambitious to keep pace with her.

The highlight during my time as a Ph.D. student was the XVIIth IUPAC congress held at München from 30th August to 6th September 1959. It was the first IUPAC congress in Germany after World War II with inorganic chemistry, biochemistry and applied chemistry as the main topics. Due to the prestige of the IUPAC and the reputation of Professor Wiberg, who was the president of the organizing committee, even some scientists from East Germany were allowed to participate. Franz Hein was one of them and it was a particular pleasure for me to meet him in my lab and tell him about the progress of my work. The congress also offered me the chance to listen to such eminent

Fig. 2.11 Ernst Otto Fischer, my friend Frieder Löffler and me in the Alps during the skiing vacation with the Fischer group (March 1960). Frieder also worked for the Diploma degree with Hein and, after he escaped from East Germany, did his Ph.D. with Professor Georg Wittig at the University of Heidelberg

chemists as John Bailar, Herbert Brown, Anton Burg, Joseph Chatt, Harry Emeléus, Hans Meerwein, Eugene Rochow, Hermann Staudinger and Geoffrey Wilkinson, whose names I knew from the literature but who I had never seen before. Together with several other Ph.D. students from the institute of inorganic chemistry at LMU, I was involved in the preparation of the scientific program for the congress and thus Professor Wiberg invited me to attend the banquet at the famous *Mathäser* restaurant. As usual on such occasions, there was an after dinner speech which was given by Professor Georg Maria Schwab, the Director of the Institute of Physical Chemistry at the LMU. He was one of the most charming scientists I ever met and I think the timeless text of his speech is worth reproducing here:

We must never forget
The Ladies of the banquet.
And what they want most
Is the Ladies' toast.

And in every convention
Is old Adam to mention.
He was the first man in life,
Since he had the first wife.
But, it's not a good case:
He never went to a Congress.
He had no colleagues
So, he did not loose weeks.

The first conference round table
Was the famous tower of Babel,
And the languages used
Were already confused.
What the trouble all made, is:
They didn't bring ladies,
And so this Congress
Couldn't be a success.

Next, we have to deal
With the Church's Council
As International Congress,
And we know the success.
There were beyond doubt
No Ladies allowed
For the celibate's sake –
What tasteless a cake !

But the diet of Worms
Had much better forms:
Complexions and curls
Of nice noble girls.

At present one stresses
For all the Congresses
The female element
To be fifty percent.

To prove this conjecture:
You sleep in the lecture,
But during a festivity
You are full of activity.
Seize, gentlemen, your glass, and raise
It to the IUPAC ladies praise !

Ernst Otto Fischer was not only a professor at the LMU, but also held an external lectureship at the *Technische Hochschule* (abbreviated TH, since 1968 *Technische Universität*, abbreviated TU) in München. Therefore, in summer 1960, before finishing the work for my Ph.D., I had to decide at which of the two institutions I wanted to defend my doctoral thesis. I chose the TH because firstly it allowed me to avoid a non-scientific topic such as patent law or psychology (common at the LMU) as the second subject for the oral examination, and secondly because coordination chemistry including the chemistry of metal carbonyls had a long tradition at the TH but not at the LMU. The *Königliche Polytechnische Schule* (which in 1877 became the *Technische Hochschule*) was founded in 1868 with five departments (*Abteilungen*), one of which was for Pure and Applied Chemistry. It was headed by Emil Erlenmeyer (1825–1909), the inventor of the "Erlenmeyer flask", who was first succeeded by Wilhelm von Miller (1848–1899) and later by Wilhelm Manchot (1869–1945). The latter, appointed as "Professor of Inorganic and Analytical Chemistry with the additional duty to teach the basic principles of organic and physical chemistry", was a particularly able chemist whose interests ranged from the mechanisms of oxidation reactions, the superoxides of iron, and the preparation of metal silicides to the binding of oxygen in blood and the chemistry of nitrosyl and carbonyl metal compounds. Together with Rudolf Weinland at Tübingen he established the basis for an almost unexplored field of coordination chemistry, that went far beyond the limits set by Alfred Werner and his school. Manchot's successor Walter Hieber (see the biography in Chap. 4) became a giant in that field and made the TH to a worldwide renowned center of coordination chemistry. As the Professor of Inorganic Chemistry and the Director of the respective institute he became one of the examiners for my oral, and despite the rumors that he would be a "tough guy", I passed without any problems.

2.5 From München to Pasadena and Back

Although I had to abandon the initial arrangement with Professor Eugene G. Rochow for a postdoctoral position, I was still keen to spend a year abroad and, if possible, at a university in the United States. Ernst Otto Fischer strongly supported my intention and recommended, inter alia, Caltech at Pasadena as a top place. One of his former students, Klaus Plesske, was already there and worked as a postdoc with Professor John H. Richards.

In February 1962, as soon as I was able to move my right arm and right leg properly again, I wrote to Klaus and asked him for advice. Since I knew that

Richards was a Professor of Organic Chemistry with interests in biosciences, I thought initially that I should preferably join the group of an inorganic chemist. However, after Linus Pauling had left, research in inorganic chemistry at Caltech had been significantly reduced and in 1962 there were just a few of Pauling's former students working in that field. I therefore finally accepted an offer from Richards who, regarding the topic of my research, mentioned in the single letter, which I received from him, that it would be interesting to incorporate organometallic fragments as markers into bioorganic molecules.

Despite some doubts by my parents-in-law, Helga and I decided that we would go to California and take Monika, who at that time was 1-year old, with us. In mid-September 1962, the three of us crossed the Atlantic ocean on the "*Bremen*", the flagship of the Hapag–Lloyd line. We arrived at New York on September 25th, were absolutely fascinated by the skyline and walked for hours through Manhattan and Greenwich Village. The next day we flew on to Los Angeles. Klaus came to the airport, drove us to Pasadena, and invited us to his flat. After several unsuccessful attempts to find suitable accommodation, we were able to rent a small but charming summerhouse belonging to the chairlady of the Women's Club at Caltech. We were particularly pleased that it was near to the campus – at least by European standards – since after my accident I had not driven a car again and, before we arrived at Pasadena, had the naïve idea that we could stay without a car for a year. This was foolish of course, and it took us only a few days to realize that we had to have a car. At the end of October, with the first cheque from the university, I bought a 1954 Buick for $350, called by Helga the "*grüne Heinrich*" (green Henry), which was a real monster compared with the VW beetle, but served us fairly well for the next 12 months.

The start at Caltech was rather unusual. After Richards (Fig. 2.12) had given me a warm welcome, he immediately asked me what my research program was. With the experience from my time with Hein and Fischer, I expected that he would tell me what I should do in the next weeks or months but this was not the case. With the information he had from Fischer, that I might be a candidate for a successful academic career, he suggested that I should look around in the department and think about a promising research project. Since at that time there was no one like Robert Bergman, John Bercaw or Robert Grubbs, who in the 1970s and thereafter made Caltech a "Mecca" of organometallic chemistry, I first contacted some German and Swiss postdocs from John Robert's and George Hammond's groups. I was lucky that my lab was very near to George Hammond's office and since his door was always open, as is usual in the US but not in Germany, I asked him whether he would allow me to participate in his research seminars. He agreed (which I "paid back" later with some baby sittings) and I thus could listen to reports by people such as Peter Leermakers, Nicholas Turro, Angelo Lamola, Jack Saltiel, etc. During these discussions I heard about the controversy between Hammond and Günther Schenck from the Max Planck Institute at Mülheim in Germany, who both had completely different opinions about the mechanism of the photosensitized isomerization of stilbenes and other substituted olefins. Since ketones such

Fig. 2.12 John H. Richards is Professor of Organic Chemistry and Biochemistry at the California Institute of Technology. He received his B.Sc. from Oxford University in 1953 and his Ph.D. from the University of California at Berkeley in 1955. After 2 years at Harvard, he became Assistant Professor at Caltech and quite rapidly rose to the ranks. His present research is mainly aimed at gaining a molecular understanding of the mechanisms of protein function in such areas as biological catalysis, the transport of electrons, and the specific interactions between proteins and nucleic acids (reproduced with permission of Caltech)

as acetophenone or benzophenone were frequently used as sensitizers for these reactions, I thought that it would be interesting to find out how transition metal complexes containing those ketones as ligands would behave.

After I told Richards about the idea, he became really excited and suggested to start immediately. However, this was not so easy since my lab, located in the old Gates and Crellin building, was neither equipped with a vacuum line nor did it contain any facilities for working under inert gas. Fortunately, the glass-blower came from my German home state of Thuringia and, since I could talk to him in our mutual mother tongue, it took only a week to have the equipment ready to start my work. I prepared a variety of organometallic ketones such as $(RCOC_6H_5)Cr(CO)_3$, $(RCOC_5H_4)Fe(C_5H_5)$ and $(RCOC_5H_4)Mn(CO)_3$ ($R = CH_3$, C_6H_5) but none of them proved to be a good photosensitizer. While discussing the results with Hammond's group, we came to the conclusion that, due to the intramolecular quenching by the metal-containing fragment, the photo-excited triplet state of the ketone was probably extremely short-lived and, therefore, an energy transfer to the olefin did not occur.

Despite the disappointing results, Richards remained tolerant and let me think about a new research project. Being familiar with the chemistry of nickelocene, which is a paramagnetic molecule with two unpaired electrons, I thought it might interact with a carbene, in the most simple case with methylene

CH_2, which in the triplet state would have two unpaired electrons too. I therefore treated $Ni(C_5H_5)_2$ with diazomethane and, to my surprise, obtained polymethylene in a very fast reaction. Even in the presence of carbene acceptors, polymethylene was the only product formed. The catalytic effect of nickelocene, which was recovered unchanged, turned out to be unique since neither ferrocene nor chromocene behaved similarly. A series of experiments aimed to get some ideas about the mechanism of the unusual reaction of $Ni(C_5H_5)_2$ with CH_2N_2 indicated that no free methylene in the singlet or the triplet state was generated. Since the isolated polymethylene had a relatively high melting point, the properties of the polymer were studied in more detail at DuPont for which Richards acted as a consultant. This was the reason why the full paper describing our results was not published in the Journal of the American Chemical Society till 1968, which was 5 years after I left Caltech.

On the whole, Pasadena was a success. Professionally, I had broadened my scientific background, came in touch with new aspects of organometallic and physical chemistry, and privately we gained insight into a new world. Before our stay in the US, neither Helga nor I had been outside central Europe, had not even visited Great Britain or Scandinavia, and now from October 1962 to September 1963 we were privileged to spend 12 months in California! We were invited by members of the Rotary Club and representatives of religious communities several times, and thus got the opportunity to meet people both from academia and with other professional backgrounds. We noticed at such occasions that many Americans felt much less concerned about political events in the world than we did, and we realized this in particular in October 1962 when the Soviets attempted to install atomic missiles at Cuba. At first we were extremely nervous, having the pictures of the destroyed German cities at the end of World War II in mind, but finally we were also highly impressed by the courageous decision of President John F. Kennedy to block the Soviet armada and hence to avoid a military confrontation.

After Richards agreed that I could finish my work at Caltech in mid-September 1963, we had enough time to travel to New York. We decided to cross the continent stepwise, partly by train and partly by car, and made the first stop at Salt Lake City. We were fortunate to meet a young Mormon on the train who had performed his service in Southern Germany and was pleased to practise his spoken German with us. He also stopped at Salt Lake City, showed us around the town and the tabernacle and although we were not allowed to visit the interior of the Temple, we had the opportunity to listen to a rehearsal of the famous Mormon choir. We then drove to the Yellowstone Park in a rental car and were very impressed by the natural scenery. In some respects, it was just the opposite of what we had seen on our trips in spring and early summer 1962. These had taken us to Death Valley, the Arizona desert, the area around the Montezuma castle, the Indian reservations and the Petrified Forest and, of course, the Grand Canyon where we walked down in our hiking boots. But the Yellowstone and the neighboring Grand Teton National Park were totally different, and beside the natural beauty of the countryside I still remember the H_2S-rich smell surrounding the famous geyser Old Faithful and the hot-spring fields. From Pocatello (a place close to the Park) we boarded the Union Pacific Railroad in a Pullman cabin to Chicago, met relatives of Helga at Milwaukee, continued by car to Buffalo, visited the Niagara Falls at day and night, and finally returned to New York where we stayed for 3 more days. In contrast to late September 1962, the weather was wonderful and we enjoyed strolling through the Central Park, visiting the Metropolitan Museum and making a boat tour around Manhattan. On the eve of our return, Helga and I went up the Empire State Building, were once again fascinated by the breath-taking view of the city, and upon looking at each other we realized that we had spent an unforgettable year. With some tears in our eyes we said "good bye" to the United States.

On the 2nd of October 1963, we left New York again with the "*Bremen*" and after a stormy passage over the Atlantic arrived at Bremerhaven one week later. We continued to München, where less than an hour after our arrival, I got a phone call from Ernst Otto Fischer. He asked me whether we could meet immediately. The reason for the hurry was a vacancy at Professor Hieber's institute at the TH suitable for a candidate interested to do a *Habilitation.* Fischer indicated that the faculty at the TH had not yet decided who should succeed Hieber in the autumn of the following year, but that he would possibly receive the invitation.

Within a few hours I accepted the offer and the next day saw Professor Hieber who gave me a cordial welcome and showed me the new lab. He also asked me about the research project for my *Habilitation* and for a moment I hesitated. It was known that he considered kinetic and mechanistic studies as a type of "philosophy" which, in his opinion, it was not worth doing.[2] Nevertheless, he listened courteously and, after I told him that the class of compounds I was going to use as starting materials for my work were metal carbonyls, he wished me good luck.

However, before I could start my work in the lab another problem had to be solved. While still being at Caltech, Fischer had already informed me that Elsevier was most interested to publish an English edition of our monograph on metal π-complexes with di- and oligoolefinic ligands, possibly in an updated version. The publishers suggested that a translation office from London should first translate the German version, and then we should extend the original text by inserting some recent results. We agreed to this, and by the time I returned to München the first draft of the translation was sent to us. It was terrible! From a

[2] Fred Basolo (see Fig. 2.14) mentioned in his monograph "*From Coello to Inorganic Chemistry*" (being the first of the series "*Profiles in Inorganic Chemistry*") on p. 101 that, after he listened to the plenary lecture given by Professor Hieber on the reactivity of metal carbonyls at the IIIrd ICCC at Amsterdam in 1955, he asked him "You never told us how some of these reactions take place – in other words, observations on the reaction mechanisms". And Hieber replied: "Young man (Fred was 35!), we do real chemistry in my laboratory, not the philosophy of chemistry".

linguistic point of view it was correct, but the style was that of a novel rather than a scientific textbook. Smiling to me, Fischer asked "Are you ready to retranslate the prosaic English?" Although I was annoyed at first, there was no doubt that it had to be done, since the English edition of our monograph was expected to have a greater impact than the original German version. But it was hard work! Particularly, because in the meantime more than 100 new papers on the chemistry of complexes with di- and oligoolefinic ligands had appeared. Finally, I managed to finish it in about 8 months, but it would not have been possible without the unselfish support by Jeffery Leigh who was a postdoc in Fischer's group from 1963 to 1965. Jeff was not only extremely helpful but also very patient and, when I became upset about the waste of time retranslating the text, he usually proposed that we first should have a beer before continuing. In most cases a fresh *Paulaner* draft beer or a *Weissbier* helped to cool my spirit. With Jeff's assistance, I even succeeded in adding a chapter on metal π-complexes with allylic ligands to the English version, comprising 42 printed pages, which due to the results of my Ph.D. thesis I still considered as my "scientific pet". Two years later, I was very pleased when, at an annual meeting of the German Chemical Society, Professor Günther Wilke told me that he always recommended to new coworkers to read the allyl chapter of our monograph carefully.

In spring 1964, Ernst Otto Fischer received the invitation to succeed Professor Hieber, and after tough negotiations with the Bavarian ministry accepted the offer a few months later. I was greatly relieved and now put all my efforts into the work towards my *Habilitation*. After following the discussions with the Hammond group at Caltech and looking around for what might be an open field in organometallic chemistry, I decided to investigate the kinetics and mechanisms of substitution reactions of metal carbonyls. From the literature I knew that Fausto Calderazzo's group at Milan and Fred Basolo's group at Northwestern had already studied the kinetics of the exchange reaction of metal carbonyls with labeled carbon monoxide but, to my knowledge, there was no activity in this field in Germany.

Since I shared the lab and fume cupboard with Karl Öfele, one of Fischer's former students and later a research officer in Fischer's and Herrmann's group, I did not use nickel tetracarbonyl or iron pentacarbonyl, but the less toxic chromium, molybdenum and tungsten hexacarbonyls as the starting materials. I had no idea that at the same time Robert Angelici also started his work on this topic. It was only 2 years later, after our first papers appeared nearly simultaneously, that we became aware of this. At the beginning, Angelici was much more experienced than me since he had done his Ph.D. with Basolo and had already investigated the kinetics of substitution reactions of metal carbonyls and their derivatives in his doctoral thesis. It was purely by accident that we had not met in September 1962 at München, since he arrived to do postdoctoral work with Fischer 2 days before we left for Pasadena. For the kinetic studies of the substitution reactions of the hexacarbonyls, Angelici and our group used different ligands but, not surprisingly, our results agreed quite well. In collaboration with Wolfgang Beck (Fig. 2.13) we also investigated the kinetics and mechanism of the reactions of the hexacarbonyls with tetraethylammonium azide which

Fig. 2.13 Professor Wolfgang Beck and me, with our wives, at the VIIth International Conference on Organometallic Chemistry in Venice in 1975

afford, similarly to the Curtius degradation of acid azides, by stepwise addition and elimination, salts of the pentacarbonyl(isocyanato)metal anions.

Although Ernst Otto Fischer had much more teaching and administrative duties at the TU than previously at the LMU, he remained in close touch with his research group and was also eager to be up-to-date with the progress of my work. During one of our discussions he mentioned that he had recently received an invitation from Professor Klixbüll Jørgensen to participate in a workshop on "New Trends in Coordination Chemistry" to be held at Geneva in May 1965. Since he said that he was unable to leave for a whole week, he asked me whether I would be interested to participate. The spontaneous answer was "yes", firstly because I had not been to Geneva, not even to Switzerland before, and secondly, because the program with the names of the speakers appeared highly promising. The main topic of the workshop was Ralph Pearson's concept of "*Hard and Soft Acids and Bases*" and among the leading figures in coordination chemistry such as Kasimir Fajans, Gerold Schwarzenbach, Sten Ahrland and Klixbüll Jørgensen, a hot debate about the meaning of *hard* and *soft* developed. Ralph Pearson did his best to defend the concept, and though at the end several participants asked "what is really new?", I was impressed by the openness of the debate and the scholarly combativeness of the scientists involved. For the first time I also understood that if you presented a new concept to a general audience, you had to do this using simple terms in order to get the right amount of attention. Later I applied this recipe when I gave my inaugural lecture at Würzburg entitled "*Sandwiches in Chemistry*".

My second visit to Switzerland was to attend the IXth ICCC held in September 1966 at St. Moritz, with an opening session at the University of Zürich commemorating the 100th birthday of Alfred Werner. Although I did not know in those days that within 2 years I would move to Zürich, I was immediately caught by the charming atmosphere of the city. I well remember that we (Karl Öfele, Arnd Vogler – another former student of Fischer's – and I) arrived from München on a warm, sunny day, went to the lake and the *Bahnhofstrasse*, and strolled at night through the *Niederdorf* which is the Swiss version of New York's Greenwich Village. The next day we drove to St. Moritz where I presented a short lecture on the results of our kinetic and mechanistic investigations in metal carbonyl chemistry. At this meeting, I also met Fred Basolo for the first time (Fig. 2.14) who in those days was the leading figure in the field of reaction mechanisms in coordination chemistry. I was highly impressed by his frankness and, of course, very pleased when he said "you are on the right track". This first visit to St. Moritz is also unforgettable insofar, as I immediately felt in love with the Engadina, this beautiful valley in the South-eastern part of Switzerland. It was only recently that I celebrated with some friends the event of my 35th visit to this beloved place, where my family and I spent so many enjoyable weeks.

Fig. 2.14 Fred Basolo (1920–2007) was the Charles E. and Emma H. Morrison Professor of Chemistry at Northwestern University in Evanston in the US. He worked for his Ph.D. with one of the founders of coordination chemistry in the US, John C. Bailar, and received a doctorate from the University of Illinois in 1943. After working on then-classified projects for the war effort, he joined the chemistry department at Northwestern in 1946, where he was a force to be reckoned with for more than 60 years. Together with Ralph Pearson, he was one of the pioneers in the field of inorganic reaction mechanisms and one of the first studying the kinetics of substitution reactions of metal carbonyls. He coauthored two text books "*Mechanisms of Inorganic Reactions*" (with R. G. Pearson) and "*Coordination Chemistry*" (with R. C. Johnson). Fred was elected to the National Academy of Science in 1979, was the President of the American Chemical Society in 1983, and received the Priestley Medal, the highest award of the ACS, in 2001 (photo by courtesy from Professor Jim Ibers, Northwestern University)

While at the beginning of my work for the *Habilitation* I was only supported by Hermine Rascher, a very able technician, Richard Prinz joined me in spring 1965 and in the next 3 years proved to be a highly talented Ph.D. student. Richard did his undergraduate studies at the LMU and, like numerous other students at that time, was very impressed by Professor Egon Wiberg's work. Wiberg had obtained his Ph.D. with Alfred Stock, the *father of borane chemistry*, and in the 1930s and 1940s became well-known for his investigations on the formation and properties of borazines. Although in those days the aromaticity of these six-membered boron–nitrogen heterocycles had certainly been exaggerated, the formal analogy between the isoelectronic molecules $B_3N_3H_6$ and C_6H_6 gave the former the name *inorganic benzene*. Wiberg used this term quite frequently in his courses, so Richard was familiar with it and kept it at the back of his mind. After he had studied in the first part of his thesis the kinetics of the reactions of chromium and molybdenum hexacarbonyl with arenes leading to arene metal tricarbonyls, we both were interested to find out whether borazines behaved similarly and would yield the corresponding borazine metal tricarbonyls. Fortunately, we were unaware that several other groups (e.g., those of David Brown at Dublin, Joseph Lagowski at Austin and Kurt Niedenzu at Duke) had already tried to prepare compounds of the general composition $(R_3B_3N_3R'_3)M(CO)_3$ but failed.

Richard followed the course of the reactions of $Cr(CO)_6$ and $Mo(CO)_6$ with hexamethylborazine by UV–vis spectroscopy but although he observed that the intensity of the absorption maximum of the hexacarbonyls at around 290 nm decreased and a new band at around 350 nm appeared, he was unable to identify the new product. Experiments with other starting materials such as norbornadiene chromium and molybdenum tetracarbonyl or tris(aniline) molybdenum tricarbonyl which readily react with arenes by ligand exchange, also failed. The key to success was to use tris(acetonitrile) chromium tricarbonyl as the precursor, which in dioxan under reduced pressure afforded the desired hexamethylborazine chromium tricarbonyl as a stable crystalline solid in 90% yield. This was the breakthrough and, after we had communicated the synthesis and spectroscopic data of the complex in the January issue 1967 of *Angewandte Chemie*, Richard finished his work and defended his Ph.D. thesis in June 1967. Six months before, in December 1966, I defended my *Habilitation* thesis in front of the faculty and became *Privatdozent* (lecturer) on the 1st of January 1967.

The report of the synthesis of the first borazine–transition metal complex was important for me in two respects. Already in May 1967, only 4 months after our communication appeared, I was invited by David Brown to give a main lecture at the following annual meeting of the Royal Society of Chemistry to be held in spring 1968 at Dublin. When I received the letter, I could not believe it at first, because I still considered myself as a nobody in the organometallic community. At the conference, I was lucky that my lecture was scheduled for the afternoon session of the day in which we later had dinner in the Guinness brewery. There, I was able not only to relax but also to enjoy the strong dark beer. After the conference, I gave seminars at the University of Nottingham and University College London, which gave me the opportunity to meet Professor Clifford Addison and Professor Ronald Nyholm. Later when I heard about the untimely and tragic death of the latter, I was certainly not the only one to consider this as an irretrievable loss for the whole scientific community.

A few weeks after receiving the letter from David Brown, I was also invited to give a seminar in the autumn of 1967 at the University of Zürich. I gladly accepted, particularly because there were some rumors that a new tenure-track position in inorganic chemistry might be established in the near future. As for the RSC meeting in Dublin, the title of my talk was "*Kinetic Investigations as Forerunners for the Preparation of New Metal Carbonyl Complexes*". Arriving in Zürich, I was cordially welcomed by Hans Rudolf Oswald, at that time the only Professor of Inorganic Chemistry, and was briefly interviewed by the Professor of Organic Chemistry and Chairman of the Department Hans Schmid. Moreover, I was particularly pleased to see Professor Gerold Schwarzenbach from ETH Zürich in the auditorium (Fig. 2.15). The seminar started as usual but when, after some introductory remarks, I asked for the first slide to be shown, the projector broke down. Since it could not be replaced immediately, I

Fig. 2.15 Gerold Schwarzenbach (1904–1978) did his Ph.D. in 1928 at the ETH Zürich with Professor William D. Treadwell in analytical chemistry and, after a year with Sir Robert Robinson at Manchester and London, became *Oberassistent* at the *Chemische Institut* of the University of Zürich. In the 1930s he started his work on the coordination capabilities of a new class of polydentate ligands (*Komplexone*) which soon received world-wide recognition. In 1942, he was promoted as Associate Professor and in 1947 as Professor of Analytical Chemistry at the University of Zürich but returned in 1955 to the ETH, where he was the Director of the Laboratory of Inorganic Chemistry until his retirement in 1973 (photo from *Helv. Chim. Acta* **75**, 21–61 (1992); reproduced with permission of Dr. Kisakürek, Editor of Helvetica Chimica Acta)

continued my lecture on the black board using white and colored chalk. Although a new projector was available after about 15 min, I noticed that the audience seemed to be quite impressed by the "slideless" presentation. After finishing, Schwarzenbach came to me, congratulated me on my results and smilingly said: "Well done with the chalk". Since the faculty members obviously shared his opinion, I was placed number one on the shortlist for the new professorship. However, I did not receive the offer from the cantonal ministry for the tenure-track assistant professorship before February 1968. Having spent seven and a half years at Zürich, I would describe this as the "Swiss speed". In March and cantonal April 1968, I negotiated with the ministry about my salary and funding for our research and also came to an agreement with Oswald about the number of labs we could use. On the 1st of May 1968, I signed the contract in the office of the ministry and afterwards went with Helga to the "*Zunfthaus zum Rüden*", still one of the top restaurants at Zürich. Toasting with a glass of wine from a vineyard on the lake of Geneva, we were looking forward to a new – and hopefully pleasant – period of our life.

However, before I could start at the University of Zürich I had to pass a health test since otherwise I would not be accepted as a member of the *Pensionskasse* (pension fund). On the day of the medical test I was nervous, because I had to inform the physician about my previous brain pressure caused by the car accident, and was uncertain about his conclusion. Again I was lucky. When the physician looked at my documents, his first question was "*Are you a relative of Alfred Werner* ?" I was rather astonished but before I could reply he told me that at the beginning of his career he had met the widow of Alfred Werner and later was the family doctor of his daughter Charlotte. He became even more outspoken when he heard that I was born at Mühlhausen and, since Alfred Werner was born at Mülhausen (Mulhouse) in Alsace, he reasoned that this might not be mere accident. As Professor Franz Hein had already suggested a decade before, he encouraged me to study my family tree. However, a careful genealogical investigation did not establish any link of this kind. Finally, I passed the medical test, and had thus taken the last step to a professorship in Switzerland.

2.6 Crossing the Border: The Years at Zürich

In mid-July 1968, we moved from München to Zürich and took with us not only our private belongings, but also 30 big trunks containing all the glassware (Schlenk tubes, vacuum lines, sublimation flasks, etc.) needed for our research. Such equipment was not available in the Zürich department since nobody in the inorganic or organic institute had ever been working with air-sensitive compounds. It was of great help to be accompanied by two German coworkers, Edith and Karl Deckelmann (Fig. 2.16), who were not only familiar with our chemistry but, due to the generous start-up grant from the canton, succeeded in equipping the new labs within less than a month. Before the winter semester

Fig. 2.16 Edith and Karl Deckelmann, who both graduated at the TU in München and joined me in moving to Zürich, at my retirement party 2002

began, the first Swiss doctorate student, Viktor Harder, joined our group, and was followed by Wolfgang Kläui a few months later (Fig. 2.17).

While the group was in good shape from the very beginning, I had to accommodate to the imbalance between inorganic and organic chemistry in the Zürich department. In contrast to München, its resources in terms of lab space and assistantships were distributed in a ratio of 1:3 in favor of the organic section. The reason was that when Alfred Werner (Fig. 2.18) retired, he was succeeded by his former student Paul Karrer (Nobel laureate 1937), who became

Fig. 2.17 Helga together with the "native Swiss part" of my research group at Zürich, gathering at the party for my 60th birthday (from the left: Wolfgang Kläui, Heinrich Neukomm and Viktor Harder). Wolfgang received his Ph.D. in 1973 and did postdoctoral work with Professor Jack Lewis at Cambridge. After finishing his *Habilitation* in 1979, he became *Dozent* at the University of Würzburg and in 1982 Associate Professor at the *Technische Hochschule* in Aachen. Since 1991, he holds the Chair of Inorganic Chemistry at the University of Düsseldorf, Germany

Fig. 2.18 Alfred Werner (1866–1919) is usually described as the *founder of coordination chemistry*. Werner did his Ph.D. in 1889 with Professor Arthur Hantzsch and, after spending one semester with Marcellin Berthelot at the Collège de France at Paris, returned to the ETH at Zürich to finish his *Habilitation* in 1892. One year later, he became Associate Professor at the University of Zürich and was promoted as Professor of Chemistry in 1895. Remarkably, despite the widespread attention for his groundbreaking coordination theory, he was not permitted to give the basic lecture in inorganic chemistry before 1902. Werner attracted students from all over the world, supervised 230 Ph.D. theses and was the first Swiss to receive the Nobel Prize for Chemistry in 1913. In his famous book "*Valence and the Structure of Atoms and Molecules*", published in 1923, Gilbert N. Lewis wrote "...in attempting to clarify the fundamental ideas of valence, there is no work to which I feel so much personal indebtedness as to this of Werner's" (photo from *Helv. Chim. Acta* **75**, 21–61 (1992); reproduced with permission of Dr. Kisakürek, Editor of Helvetica Chimica Acta)

the Professor of Chemistry for 40 years and made organic chemistry the completely dominating discipline. Karrer even gave the classes in inorganic chemistry for the first and second year students, a fact which forced Gerold Schwarzenbach, at that time a pioneer in coordination chemistry, to move from the University to the ETH. Only after Karrer retired in 1959, at the age of 70, was a chair in inorganic chemistry established and first occupied by Ernst Schumacher. He had two labs for research allocated to him while the organic chemists had more than a dozen. Thus, not unexpectedly, Schumacher accepted an offer from industry 5 years later not without indicating to the cantonal administration that the situation for inorganic chemistry had to be improved. With some delay, the canton reacted, bought a private house behind the main building of the chemistry department and promised that a second staff position in inorganic chemistry would be established. Based on this, Hans Rudolf Oswald (formerly at the University of Bern) accepted in 1966 the invitation to Zürich and was followed by me 2 years later. In the meantime, the house in Schönleinstrasse 2, was restored and became the home for both the Oswald and the Werner research groups. Despite this substantial improvement, occasionally Oswald and I had to defend our position against our colleagues in organic chemistry, who only gradually accepted that state-of-the-art inorganic chemistry was not as old-fashioned as they imagined.

The year following our move to Switzerland was full of surprises. In early January 1969, Gerold Schwarzenbach phoned me and asked whether I would be interested in visiting Jena. He told me that as a member of the board of the Swiss Chemical Society he had received an invitation to participate in a symposium honouring the scientific achievements of Professor Franz Hein, but for lack of time was unable to attend. Since he knew about my relation to Hein, he suggested that I should replace him. At first I could not believe that it is possible because my name had been blacklisted and some of my friends as well as my father recommended me not to return. Therefore, I was reluctant to accept, but Schwarzenbach was confident that it would work out. A few weeks later, I received an official invitation to participate and even give a lecture at the symposium. This would be my first visit to Jena and East Germany after 11 years, and I was looking forward to meeting my friends and relatives, not knowing that this would be sooner than expected.

In July 1969, Helga and my daughter visited my parents in Mühlhausen and soon after they arrived my father fell ill.[3] He was immediately operated on in a hospital at Erfurt but before he awoke from anaesthesia, he passed away. Helga called me and told me that I was allowed to take part at the funeral service. The East German authorities had agreed that my entrance permit would be deposited at the railway station at the internal German border. Upon my arrival at this station the following day, I felt very confused. The East German railway personnel entered the carriages and ordered people like me having no entrance permit to leave the train. We were escorted by a squad of policemen to the building where we were supposed to receive the documents. I was the last in the line, became more and more nervous, and when I was asked to enter the commandant's office I already considered myself arrested. However, the reverse happened. The officer welcomed me, said that he was pleased to meet a professor from Zürich, and told me, inter alia, that the East German government would be interested in establishing diplomatic relations with Switzerland. Finally, he wished me all the best for my stay in Thuringia. Although he held my German passport in his hands, he obviously considered me as a Swiss. What a crazy world! Six weeks after the funeral I went back to East Germany, participated in the symposium dedicated to Franz Hein (Fig. 2.19), met most of my old

[3] There were no restrictions for Helga to visit my parents since she was born in Bavaria, had no family in East Germany and, unlike me, no political "black mark" in her CV. Although she was shocked when she passed the internal German border for the first time, seeing all the red flags and the numerous "bill boards" with the Communist slogans along the roads, she continuously traveled to my hometown at least once a year in the 1960s. Since East German citizens were not allowed to visit their relatives in the West, it was the sole opportunity for my parents to see their granddaughter. During these visits, Helga was regularly invited for coffee by the authorities at Mühlhausen, and was always told how superior life in East Germany was compared to that on the capitalist side. However, she knew the real situation and felt always extremely relieved after returning to the West.

Fig. 2.19 Ernst Otto Fischer and Franz Hein talking to each other at Hein's farewell symposium in Jena 1969. Standing behind Franz Hein is Rudolf Taube, at that time Professor of Inorganic Chemistry at the University of Greifswald, Germany

friends from the time at Jena, and finally visited my mother again for 2 days. It was strange: For 11 years I had not been allowed to return to my hometown, and now I could do it twice in less than 2 months.

In September 1969, a few days after returning from Jena, I received an invitation to the Chair of Inorganic Chemistry at the Eduard Zintl Institute of the TU Darmstadt, Germany. It was unexpected since I had not applied and not considered myself a candidate. I had mixed feelings from the beginning because although the offer seemed very attractive, the family was reluctant to leave the village of Herrliberg near the Lake of Zürich, where we had settled. Nevertheless, I went to Darmstadt, received a warm welcome from the predecessor Hans W. Kohlschütter, and was impressed by the size of the labs and the number of permanent positions linked to the chair. However, the spectroscopic equipment was rather scarce and there was no doubt that this had to be improved. While negotiations with the University at Darmstadt and the administration of the State of Hessen were under way, the faculty and the chairman of the chemistry department at Zürich reacted. They applied to the cantonal ministry for education to establish a new chair in inorganic chemistry and to promote me as full professor. I was really surprised by this decision since I had only arrived at Zürich 15 months previously and we had published only two papers about our research in this period of time. However, the letters of reference sent to the Dean were obviously in my favor and thus the cantonal ministry accepted the application from the faculty at the end of January 1970. As a consequence, I declined the offer from Darmstadt and the family was absolutely happy although they never attempted to influence my decision.

Fig. 2.20 Albrecht Salzer, who prepared in his doctoral thesis (finished in 1974) the first triple-decker sandwich, at my retirement party 2002. After a year as a postdoctoral fellow with me, he did his *Habilitation* under the mentorship of Hans Rudolph Oswald at the University of Zürich and became *Privatdozent* at this institution. In the 1980s, he published the textbook "*Organometallic Chemistry*" together with Christoph Elschenbroich, which became one of the standard texts in the field. In 1992, Albrecht accepted an invitation as Associate Professor of Inorganic Chemistry at the TH in Aachen where he still is

After I had decided to stay at Zürich, Albrecht Salzer joined me as a Ph.D. student and was assigned to investigate the reactivity of the electron-rich nickelocene towards electrophiles (Fig. 2.20). Viktor Harder had already studied the behavior of cobaltocene and nickelocene towards phosphines and phosphites and had prepared a series of compounds of the general composition $(C_5H_5)Co(L)_2$ and $Ni(L)_4$ by ring-ligand displacement. In a relatively short period of time, Albrecht succeeded in isolating several salts of the first triple-decker sandwich complex, the cation $[Ni_2(C_5H_5)_3]^+$, and together with Erich Dubler determined the molecular structure. This result had a considerable impact and became a conference highlight at the VIth International Conference on Organometallic Chemistry at Amherst in 1973. The report also initiated invitations to lecture at universities in Germany, Switzerland, Italy, France and even the DDR, where I was obviously no more considered as a "capitalist enemy". A grant from the British Council enabled me to go on a lecture tour to the UK and, despite the strain of 14 lectures in 4 weeks, I remember this visit with particular pleasure. In Scotland Peter Pauson strolled with me around Loch Ness, he and Evelyn Ebsworth made me familiar with Scottish malt whiskies, Mary and Bernard Shaw showed me the beauties of Yorkshire, Jack Lewis invited me to Sidney Sussex College at Cambridge and to a dinner party with Harry Emeléus and his wife, Jeffery Leigh and Michael Lappert went with me to the Royal Pavilion in Brighton, and Gerald Fowles informed me about the secrets of wine-making without using natural grapes. The last stop was at

Bristol in Gordon Stone's house, where we had a wonderful dinner party with Warren Roper and his wife. It was a giant tour but at the same time a great experience, and established friendships which have lasted until today.

Following the first report on the triple-decker sandwich complexes, several postdocs joined our research group, mainly from the UK. Trevor Court came from Nottingham, Graham Parker and John Clemens from Bristol, Kevin Turner from Oxford and Dave Tune from Brighton. Most of them stayed for 2 years. Dave Tune, who collaborated with me even after I had left Zürich, was particularly smart and prepared an unprecedented type of organometallic complexes in which metal–metal bonded Pd_2, Pt_2 or PdPt units were sandwiched between bridging cyclopentadienyl and allyl ligands.

An echo to our work also came from the Swiss chemical industry at Basel. In 1972/73 I was invited by Roche, Ciba/Geigy and Sandoz and lectured about up-to-date trends in organometallic chemistry. As a result I became a consultant for Roche for the following 25 years. In the early days, when I talked with chemists who had obtained their Ph.D. with Paul Karrer, Leopold Ruzicka, Tadeus Reichstein, Vladimir Prelog, Albert Eschenmoser or one of the other organic giants, it was not unusual that I first had to explain the 18-electron rule and why transition metal organometallics with a 16-electron configuration played an active role in catalysis. This changed in the 1980s and 1990s when students from Luigi Venanzi, Peter Kündig, Antonio Togni, Paul Pregosin, Hans-Jürgen Hansen and Heinz Berke joined the industry in Basel. As a result of the up-dated curriculae, organometallic chemistry became an integral part of training in chemistry and is currently well represented at the Swiss universities.

Apart from the discovery of the first triple-decker sandwich complex, 1972 was also the year of birth of our son Andreas. He was born on the same day as the first of my lectures within the new course in general chemistry took place, the content of which had been discussed in the department for more than a year. Occasionally, these discussions had been full of emotion, since for some members of the staff it was not easy to accept that the basic courses in experimental inorganic and organic chemistry, both having a long tradition, had to be abandoned in favor of a more integrated approach. During this first lecture, my mind was more focussed on what was happening in the maternity ward of the hospital than in the classroom, but everything worked out all right. Two months after Andreas' birth we moved to a larger and nicer flat across the street and at this stage we were convinced that this would be our home for a long period of time.

However, in January 1974 I received a phone call from Max Schmidt, Professor of Inorganic Chemistry at the University of Würzburg, asking me whether I would be interested to take a chair in the chemistry department. He told me that the predecessor, Hubert Schmidbaur, had accepted an invitation to München and that the faculty wanted to replace him by an organometallic chemist. When I went to Würzburg in February 1974, I was initially skeptical and told my group at Zürich that I am not intending to leave. However, the warm welcome which I received from Max Schmidt and the faculty, the guarantee to have lab space for at least 20 coworkers, and the promise of ten teaching assistantships changed my mind. Moreover, one nightlong discussion with Max Schmidt and Siegfried Hünig, the latter holding the Chair of Organic Chemistry, gave me the impression that in the department at Würzburg, unlike at Zürich, there was no subdiscipline dominating the other. Hünig also told me that he would do his best to keep that balance, and in retrospect I am pleased to say that he did.

After returning to Zürich, I was undecided in my mind and when asked by the coworkers hesitated to say yes or no. At home Helga supported me unconditionally, never saying that she would not like to go back to Germany, while Monika was rather upset. She was 13 years old, happy at school, had a lot of friends, and was a member of the team of child actors at the famous *Schauspielhaus* in Zürich. For her, Herrliberg and the surroundings at the Lake of Zürich were at home, and she not only spoke the Swiss dialect perfectly but also felt like a Swiss.

In July 1974, two weeks after I received the formal offer from the Bavarian Ministry of Science, the Conference on Organometallic Chemistry at Ettal, organized by Wolfgang Beck and his team on the occasion of the 80th birthday of Walter Hieber, took place. Eight months after receiving the Nobel Prize, Ernst Otto Fischer and Geoffrey Wilkinson were the special guests (Fig. 2.21) and, as Fred Basolo mentioned in his monograph "*From Coello to Inorganic Chemistry*", it was "a big success with excellent talks and exquisite Bavarian food, beer, wine, and fine organometallic chemists". At the meeting, I was asked repeatedly (in particular by people who were equally interested in the chair at Würzburg) whether I would leave Zürich or not, but I could not give a clear answer. I discussed the matter with Ernst Otto

Fig. 2.21 Ernst Otto Fischer and Geoffrey Wilkinson dancing at the final reception of the Conference on Organometallic Chemistry at Ettal, Germany, in July 1974 (photo by courtesy of Professor Wolfgang Beck)

Fischer and he was convinced that the conditions for basic research at Würzburg were excellent.

When I informed the faculty and the members of staff of the chemistry department at Zürich about the offer, the majority of my colleagues thought that I would decline. However, Hans Schmid, the chairman of the department and successor of Paul Karrer, noticed that I did not play around and that I considered the move as a real opportunity. He reacted by sending me the copy of a stage-play written by Paul Karrer in 1910, which was entitled "*Why He did not go to Würzburg*".[4] It referred to the fact that in spring 1910 Alfred Werner had received the invitation to the Chair of Chemistry at the University of Würzburg to follow his former supervisor Arthur Hantzsch, but then decided to stay at Zürich. The play was performed by Werner's students at the Christmas party of the institute in December 1910 and the main actors were *He* (Alfred), the first lover *Helvetia*, the second lover *Bavaria*, and several red and blue *Kobaltiake* (cobalt ammine complexes). In the play *Bavaria* tempted Alfred by saying:[5]

A future of glory I promise thee
Of fame and esteem, if you come to me.
To the Sciences your arrival will bring
A blossoming like a flower in spring.
Of Würzburg songs of praise will arise
Like the pious' chants of Paradise.

while *Helvetia*, taking the request of the *Kobaltiake* into consideration, replied:

Hold on, hold on! Oh Alfred, I plead thus
Appease the lot, or else, woe betide us.
Just one word from you to them by chance
And all will join hands and sing and dance.

and thus at the end of the play *He* said:

The whole affair has gripped me, it's true.
Well, so be it: I'll stay with you!

Despite an attractive counter offer by Zürich to increase my salary and the annual credits for teaching and research, I decided in May 1975 to accept the chair at Würzburg, following detailed discussions not with *Bavaria* but with the Bavarian Ministry for Science and Education. The colleagues in Zürich were fair enough to agree that those coworkers who had not finished their doctoral

[4] In German: "*Warum Er nicht nach Würzburg ging.*"

[5] In German: *Bavaria*: "*Eine Zukunft voll Ehre verspreche ich Dir, voll Ruhm und Ansehen kommst Du zu mir. Ich sehe bei uns die Wissenschaft, aufgehn wie die Knospe im Frühlingssaft. Von Würzburg wird man rühmen und reden, wie die frommen Leute vom Lande Eden.*"
Helvetia: "*Haltet ein, haltet ein! Alfred, ich flehe, beschwicht'ge das Volk, sonst wehe, wehe! Ein Wort von Dir zu diesem Schwarme und alle fliegen sich in die Arme.*"
Er: "*Die Sache geht mir wirklich zu nah, na meinetwegen, so bleib ich da.*"

theses or had contracts with the Swiss Science Foundation for one more year could stay in the laboratories until a successor would be elected. For several reasons, the latter took more than 3 years. Since 1988, Heinz Berke holds the Chair of Inorganic Chemistry at the University of Zürich and keeps organometallic chemistry awake.

2.7 Back to Germany

On the 1st of October 1975 I started in Würzburg and had about a dozen empty laboratories to fill. By the end of the winter semester, the group consisted of four Ph.D. students plus a Humboldt fellow, and in summer 1976 the first two postdocs joined us. With the support of the glassblower and the assistance of an excellent research associate, Konrad Leonhard, we succeeded in equipping all the labs with vacuum lines and the facilities for using the Schlenk techniques in a short period of time. At the end of 1976, research in organotransition metal chemistry was well established. Helga and Andreas came to Würzburg in January 1976 and Monika followed 3 months later after finishing high school in Zürich. In the first two years we rented a house near the famous pilgrim's church at Würzburg, but in autumn of 1977 were very fortunate to buy a piece of land within the walking distance of the campus. Here we built our home (Fig. 2.22) and I still live here.

Fig. 2.22 The Werner family at the terrace of our house in the late 1980s

Although in the initial stage my thoughts occasionally went back to Zürich, I noticed quite soon that Würzburg was a good place to be. Moreover, I realized that not only the university but also the department of chemistry was greatly esteemed. Having been originally founded in 1402, and after some turmoil, refounded by the powerful Prince-Bishop Julius Echter in 1582, the University of Würzburg had developed continuously to become a first-rate academic institution in Germany. A chair of chemistry was established in 1782 in the school of medicine and became part of the faculty of science in 1869. In the same year Adolf Friedrich Ludwig Strecker (well-known for the synthesis of amino acids via cyanohydrins) accepted the invitation to Würzburg and, after his sudden death in 1871, was succeeded by Johannes Wislicenus (1835–1902). Subsequently, two Nobel laureates, Emil Fischer (1852–1919) and Eduard Buchner (1860–1917), as well as Arthur Hantzsch (1857–1935), Alfred Werner's former supervisor, headed the *Chemisches Institut.* Wilhelm Manchot (from 1903 to 1914) and the young Walter Hieber (from 1919 to 1923), who later both held the Chair of Inorganic Chemistry at the *Technische Hochschule* in München, taught at Würzburg. The favorable economic situation in the 1960s enabled the university to establish a campus at the outskirts of the city and move the faculties of science, philosophy and social sciences to *Hubland.* Mainly due to the efforts of Siegfried Hünig, chemistry was the first to move and could double the space which it had occupied in the old building in the center of the city. Young members of staff, such as Gottfried Märkl, Fritz Vögtle, Egon Fahr, Theodor Eichler, Henning Hopf, Hubert Schmidbaur, Otto Scherer, Herbert Schumann, Walter Siebert, Peter Jutzi and Hans-Georg Kuball (all of whom later became full professors at other universities) had the chance to greatly expand their research activities, and did so very successfully.

Until 1995, I shared the running of the inorganic chemistry department with Max Schmidt (in the scientific community well-known as "*Sulfur Max*") and, in retrospect, I know that this was a fortunate constellation for both of us (Fig. 2.23). When I became the Dean of the Faculty in 1987, he was usually the first to be contacted for advice since his experience and his understanding of human behavior was unique. Max was a wonderful and inspiring colleague and we remained in close contact until his untimely death in 2002.

The increased laboratory space at Würzburg and the generous funding, in particular from the Deutsche Forschungsgemeinschaft and the *Fonds der Chemischen Industrie*, led to a rapid expansion of the research group which in the 1980s and 1990s generally consisted of about 20 Ph.D. students and postdocs. The expansion was accompanied by a diversification of the research activities which soon went beyond the chemistry of triple-deckers and bimetallic sandwiches. Following the conceptional idea of metal basicity and the use of electron-rich complex fragments to activate organic substrates, we found new coordination modes of CS and CSe, of carbon disulfide, carbon diselenide and the corresponding mixed species, prepared a series of complexes with thio-, seleno- and telluroformaldehyde as terminal and bridging ligands, and investigated cycloaddition reactions of 1,3-dipolarophiles with cobalt and rhodium

Fig. 2.23 Max Schmidt as Dean of the Faculty of Science presenting Professor Walter Hieber the Diploma of Doctor honoris causa. Left of Hieber is Ernst Otto Fischer who gave the laudatio at the ceremony held at Würzburg in 1969 (photo by courtesy from the late Professor Max Schmidt)

half-sandwich compounds. One particularly notable result of these studies, carried out by Justin Wolf and "Paco" Garcia Alonso (now a *Professore Titular* at the University of Oviedo in Spain), was the transformation of an acetylene rhodium complex to its vinylidene isomer in 1983. Following up this work during the next 15 years, metallacumulenes with up to five carbon atoms in the chain were prepared and completely characterized. The investigation of the reactivity of these complexes uncovered various new types of C–C, C–N and C–O coupling reactions. Justin Wolf (Fig. 2.24) played a leading role in most of these studies and remained associated with me until my retirement.

By the early 1990s, the investigation into the chemistry of the metallacumulenes with rhodium and iridium as the metal centers also provided the synthetic methodology which allowed filling a gap in the systematics of transition metal

Fig. 2.24 Justin Wolf did his Ph.D. with me in 1986 and later became a research associate in the group. He was the driving spirit for our work on carbene and vinylidene metal complexes and suggested several original research projects that I would not have conceived of myself

Fig. 2.25 Peter Schwab at the day of his doctoral exam with his fancy *Doktorhut* in February 1994. Every student who finished the Ph.D. with me received such a hat which was prepared and decorated by the other members of the group

carbenes. After some unsuccessful attempts to prepare and isolate four-coordinate carbene rhodium(I) complexes by reaction of labile rhodium(I) precursors with diazoalkanes, the serendipitous discovery that replacement of the usual phosphine coligands by stibines significantly changed the reactivity of the complexes led to a breakthrough in this field. Peter Schwab (Fig. 2.25) succeeded in making a series of simple rhodium carbenes of the type *trans*-[RhCl(=CRR')(L)$_2$] with L = PR_3, AsR_3, SbR_3, and after finishing his Ph.D. thesis brought the diazoalkane method to California, where he made the well-known metathesis catalyst $RuCl_2$(=CHPh)(PCy_3)$_2$ as a postdoc in Robert Grubbs' laboratory (see Chap. 8).

The use of stibines in coordination chemistry, which had been neglected for some time, also opened up a research project that preoccupied the group until I retired. It was again Peter Schwab who observed that upon warming the mononuclear compounds of the type *trans*-[RhCl(=CRR')(Sb*i*Pr_3)$_2$] to about 80°C stibine ligands were eliminated and dinuclear rhodium complexes with a bridging stibine were formed. The isolation and structural characterization of this exceptionally stable type of compounds provided the first example of an ER_3 (E = P, As, Sb) ligand in a bridging position, thus contradicting the received textbook wisdom. This finding could be extended to the lighter group XV derivatives and culminated at the turn of this century in the characterization of the first example of PMe_3 forming a bridge between two transition metal centers. Since then a whole series of such compounds also including $AsMe_3$ as a bridging ligand have been prepared (Fig. 2.26). It seems like a curiosity that my last Ph.D. student at Zürich, Heinrich Neukomm, uncovered the metal basicity of rhodium half-sandwich complexes (C_5H_5)Rh(L)$_2$ with L = PR_3 and P(OR)$_3$, and that my last Ph.D. student at Würzburg,

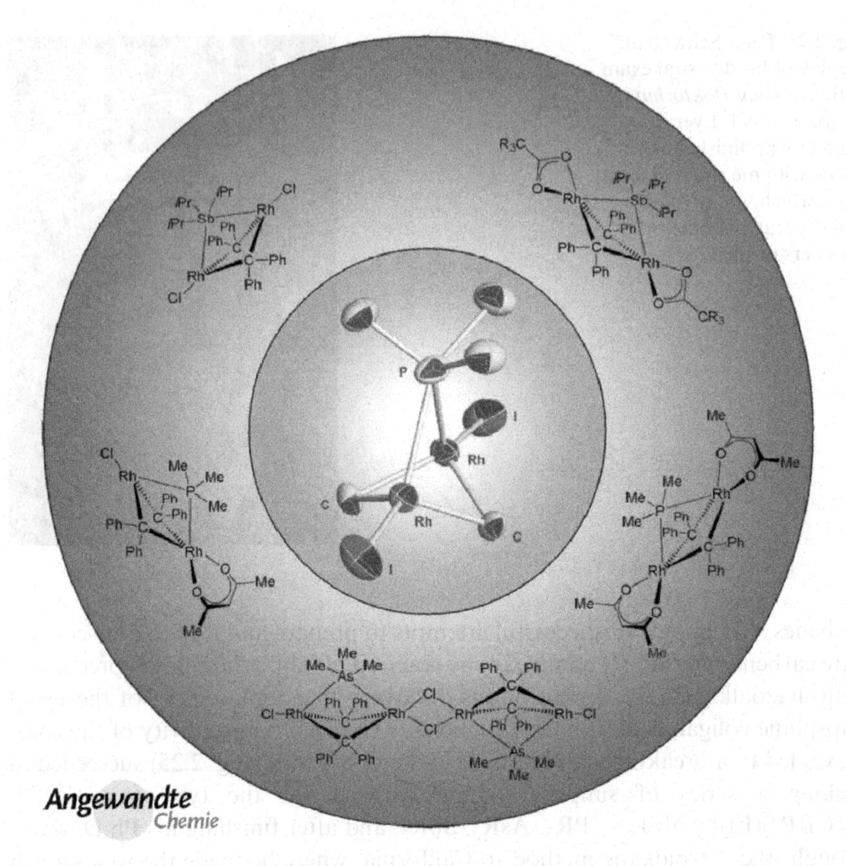

Fig. 2.26 The cover of the review summarizing our work on the chemistry of dinuclear rhodium complexes with bridging Phosphine, arsine and stibine ligands (from *Angew. Chem. Int. Ed.* **41**, 938-954 (2004), reproduced with permission of Wiley-VCH)

Thomas Pechmann, succeeded in transforming a tertiary phosphine from a terminal into a bridging position.

The fruits of the increasing activity in organometallic chemistry at Würzburg were twofold. Whereas before the mid 1970s the inorganic chemistry department was nationally and internationally recognized mainly as a center of research in main group chemistry, this changed during the next decade. Wolfgang Malisch and Wolfdieter Schenk, who were the youngest members of staff when I started, shifted their research interests more and more from phosphorus and sulfur chemistry to transition metal chemistry, and this change of direction continued in the 1980s when Ulrich Schubert and Gerhard Erker became associate professors in the department. Although Gerhard had a position formally "on the other side", i.e. in organic chemistry, we held joint seminars of our research groups and in 1987 we both organized the first international symposium on "*Organic*

Fig. 2.27 Gerhard Erker and me together with (from the left) Professor Günther Wilke, Dr. Walter Steinhardt from the *Volkswagen Stiftung*, and Professor Karl Schlögl from the *Universität Wien* before the opening session of the international symposium on "*Organic Synthesis via Organometallics*" held at Würzburg in 1987

Synthesis via Organometallics" sponsored by the *Volkswagen Stiftung* at Würzburg (Fig. 2.27). Due to the combined efforts, we succeeded in 1990 to establish a *Sonderforschungsbereich* ("SFB": collaborative research center) in the field of organometallic chemistry which was funded by the Deutsche Forschungsgemeinschaft and the University of Würzburg for the maximum of 12 years. I was elected as the Chairman of the SFB and, although Gerhard left in spring 1990 to take the Chair of Organic Chemistry at the University of Münster in Germany, we were able to bring 12–15 research groups from physics and inorganic, organic, physical and nutritional chemistry together to collaborate on projects of mutual interest. As part of the program of the SFB, international symposia on organometallic chemistry in all its aspects were held every 3 years. They attracted some of the key players from all over the world and were accompanied by wonderful wine tastings in the cellar of the *Residenz* (Palace) in Würzburg.

Due to the fact that the Würzburg chemistry department became one of the leading German institutions in organometallic chemistry, several young chemists decided to carry out their *Habilitation* under my mentorship. These include Wolfgang Kläui (see Fig. 2.17), Jörg Sundermeyer, Lutz Gade (see Fig. 2.28) and Martin Bröring who in the meantime occupy professorial positions at other universities. Their Ph.D. students often shared the laboratories with my coworkers and thus became part of the "Werner family" (Fig. 2.29). This "going together" made the second floor of the institute a stimulating and lively place in which every opportunity to party was gladly taken up.

Fig. 2.28 Lutz Gade (on the *left*) and Jörg Sundermeyer after Jörg had finished his *Habilitation* and became *Privatdozent* in 1995. Lutz now holds the Chair of Inorganic Chemistry at the University of Heidelberg and Jörg is the Associate Professor of Inorganic Chemistry at the University of Marburg

Particularly memorable are the Christmas parties, in the 1990s mainly organized by Lutz Gade, where group members staged various forms of entertainment, sometimes continuing throughout the night (Fig. 2.30). It was the ideal balance to the daily workload and impressed even those foreign students, postdocs and visiting scientists whose mother tongue was not German.

While the heavy teaching and administrative loads during the semesters left not much time for Helga and the children, we had fixed 2 weeks in March or early April for skiing vacations in the mountain area of the Engadina in Switzerland. This annual holiday with the family had absolute priority, and only twice in more

Fig. 2.29 The Werner family including the Ph.D. students of Lutz Gade and Jörg Sundermeyer, gathering on my 60th birthday. Right of me is my long-time secretary Gisela Sebald and next to her Jörg Sundermeyer and Lutz Gade

Fig. 2.30 Students from the Gade, Sundermeyer and Werner groups, led by Lutz, singing and marching at the Christmas party 1994

than 30 years did I leave the place for a few days to participate in a conference. After these winter breaks, I always felt that my "batteries" had been recharged and, because I still continue this tradition, this feeling is noticeable even today.

Summer vacations were usually abandoned and replaced by the participation in conferences in different parts of the world. In retrospect, I consider this as one of the most valuable privileges of a university professor and am most grateful for this. Since I moved to Würzburg, I spent several summer or autumn periods usually for 2–4 weeks in Japan, South Korea, Australia, New Zealand, South Africa, Brazil, Chile and various countries in Europe. Particularly memorable are a weekend with Akio Yamamoto in the mountains north of Tokyo where we stayed in a typical *ryokan*, several days spent with Michael Bruce and Martin Bennett in the winecellars around Adelaide and Canberra, a visit to Queenstown in the mountain part of New Zealand (according to my guidebook the "*St. Moritz of the Southern hemisphere*"), a ride with Mario Scotti – a former Ph.D. student of mine at Zürich and now an Emeritus Professor at the Universidad Católica de Chile at Santiago – to the Andes region at the border between Chile and Argentina, and a wonderful visit with Helga and Leentjie and Simon Lotz to the Kruger National Park in Transvaal. Simon was one of my first Humboldt fellows and later spent his first sabbatical together with his family at Würzburg. Since 1983, he is Professor of Inorganic Chemistry at the University of Pretoria in South Africa.

We also made many enjoyable visits to North America, in particular to California, New York and New England. In August 1982, the whole family went back to Pasadena to meet our landlord and landlady from the early 1960s and see John Richards and the new buildings for chemistry at Caltech. Six years later, I started the Pacific West Coast Inorganic Lecture tour at Pasadena which led me from Southern California via Santa Barbara, Davis, Eugene, Seattle, Victoria, Vancouver, Edmonton to Calgary and comprised 12 lectures in 4 weeks. In the same year, I was also invited to the Gordon conference on organometallic chemistry, and met Meta and Thomas Katz on this occasion. It was the beginning of a cordial friendship. Tom not only invited me for seminars at Columbia University, but Meta and he hosted Helga and me several times in their charming flat on Riverside Drive in Manhattan. We were shown the treasures of the Metropolitan Museum as well as the beauties of Greenwich Village, and we also learned where to find the best Chinese cuisine in New York. During our stays in the Northeastern part of the US, we always spent some time with our good friends Eda Easton and Ulrich Müller-Westerhoff in Connecticut. Their 40-year marriage proves that an artist and a scientist fit well together and it was (and still is) a real pleasure to be a guest in their unconventional house at Storrs (Fig. 2.31).

Although I was allowed to take a 6-month sabbatical leave every 4 years, I never used this opportunity to its full extent. The first long leave was in summer 1983 following an invitation by Jack Lewis to Cambridge (Fig. 2.32). I was hosted in a pleasant apartment at Robinson College and also became a Bye-Fellow of the college. David Ginsburg, a visiting professor from Haifa, stayed next door to me, and with him I walked nearly every day to the Chemical Laboratory at Lensfield Road. We rarely talked about chemistry but more about

Fig. 2.31 Eda Easton and Ulrich Müller-Westerhoff together with my sister-in-law (on the *right*) at the party for my 70th birthday. Ulli graduated at the LMU in München when I did my Ph.D. and later followed his supervisor Klaus Hafner to the University of Darmstadt, Germany. During his postdoctoral stay with Andrew Streitwieser at UC Berkeley he prepared the compound $U(C_8H_8)_2$, called uranocene, the first sandwich complex in which the metal center is coordinated to two planar eight-membered ring ligands (see Chap. 5). Since 1982, Ulli is Professor of Chemistry at the University of Connecticut in Storrs

Fig. 2.32 Professor Jack Lewis, now Lord Lewis of Newnham, and me at a dinner party in the Robinson College at Cambridge in 1994

Beethoven and Wagner and the misuse of these cultural giants by the Nazi regime. During these discussions, I got a proper understanding of how delicate and dangerous the situation for the state and the people in Israel is, and that it may take decades to come to an agreement with the Palestinians. When I was invited to the Weizmann Institute and other places in Israel including Haifa in spring 1988, I felt very sad that David had passed away.

During my time at Würzburg, 110 graduate students obtained their Ph. D. with me and about 40 postdocs and visiting scientists worked for different periods of time in my lab. I am grateful that most of them stayed in contact with me and also participated in regular group gatherings which began as early as 1981. In the meantime they have become family events which recently have taken place annually at Justin Wolf's "ranch" at Weikersheim in the Tauber valley. In June 2002, prior to the formal date of my retirement, I invited all the former coworkers to this place and together with the technicians, the secretaries, the Sundermeyer, Gade and Bröring students, as well as my children – we were about 180 people. It was a wonderful day and night, and even the result of the finals of the soccer World Cup (Germany lost 0:2 to Brazil) left us almost unaffected.

At occasions such as farewell parties or group gatherings it was always a great pleasure to meet some of the former research students and visiting scientists from abroad. Since the early 1980s, particularly closed ties had developed with several Spanish universities. It began at the XIth International Conference on Organometallic Chemistry at Callaway Gardens in Georgia, when Luis Oro (Fig. 2.33) and I agreed to initiate an exchange program for graduate students and postdocs

Fig. 2.33 Luis Oro and me while discussing joint research projects on the campus at Zaragoza. Luis received his Ph.D. from the University of Zaragoza and held academic positions at the universities of Madrid and Santander in the 1970s. Since 1982, he is Professor of Inorganic Chemistry at the University of Zaragoza and became Head of the *Instituto Universitario de Catálisis Homogénea* at the same city in 2004. Apart from his research activities, which center around the synthesis and catalytic activity of platinum group metals, he was the Director General of Scientific and Technical Research in Spain, the Vice-President of the European Science Foundation, the Coordinator of the Iberoamerican Network on Homogeneous Catalysis, and the President of the Spanish Royal Society of Chemistry. He is an elected member of the Academia Europaea, London, the European Academy of Science, the Académie des Sciences, France, and the German Academy of Sciences Leopoldina. He also received several major awards, among which the Humboldt Research Award, the Luigi Sacconi Medal, the Betancourt-Perronet Prize, the Catalán-Sabatier Prize, the King "Jaime I" Award and the Zaragoza Gold Medal deserve special attention

between Zaragoza and Würzburg. Miguel Angel Esteruelas was the first to join us in summer 1984, and was later followed by Ana López, Cristina Valero, Eduardo Sola, Marta Martín and some Ph.D. students. In my lab Miguel Esteruelas started the project on the reactivity of coordinatively unsaturated hydridoruthenium and hydridoosmium complexes containing triisopropylphosphine ligands, and after he returned to Spain followed up this work very successfully. In the 1990s, the universities at Murcia, Valencia and Sevilla were also involved in the exchange (Fig. 2.34) and this secured the existence of a small Spanish "colony" in my lab over the years.

The continuing efforts to strengthen the relations between German and Spanish universities and to deepen the collaborations with Luis Oro, Miguel Esteruelas and their research associates were recognized first by the J. C. Mutis/Alexander-von-Humboldt Prize, awarded to me in1995 by the Spanish Ministry of Science and Education in Madrid, and by an honorary doctorate of the University of Zaragoza in 2001 (Fig. 2.35). Among the other awards and honours which I received during my academic career are the Centenary Medal of the Royal Society of Chemistry, the Max Planck Research Award (which I received together with Martin Bennett, see Fig. 2.36), and the Alfred Stock Prize of the German Chemical Society. Alfred Stock was certainly one of

Fig. 2.34 José Vicente, who holds the Chair of Inorganic Chemistry at the University of Murcia, and me during a conference on organometallic chemistry at Sevilla in 1995. Two excellent Ph.D. students from José's group, Juan Gil-Rubio and Pablo González-Herrero, spent 3 years as postdocs with me and now hold academic positions at the University of Murcia

the masters of experimental chemistry in the twentieth century, and therefore this award is a particular honour for an inorganic chemist in Germany (Fig. 2.37). In November 2006, I received an honorary doctorate from my alma mater, the *Friedrich Schiller Universität* in Jena, for our research achievements as well as for the active role which I played in the reconstitution of the faculty at Jena after Germany's reunification (Fig. 2.38). In 1988, I became a member of the *Deutsche Akademie der Naturforscher Leopoldina*, which is the National Academy of Science since 2008.

The political event during the last decades which struck me most was undoubtedly the collapse of the communist regime in East Germany. Viewed from the outside, it began on the 18th of October in 1989 when Erich Honecker, the Secretary General of the SED, had to resign. On that day I finished a 2-week visit to Brazil and read at Sao Paulo airport on the front page of the newspapers

Fig. 2.35 Professor José Elguero and me waiting in front of the main hall at the University of Zaragoza, where we both received an honorary doctorate in 2001

Fig. 2.36 Martin A. Bennett is now Emeritus Professor at the Australian National University (ANU). He received his Ph.D. from the Imperial College London under the guidance of Geoffrey Wilkinson in 1960 and, after postdoctoral work with Arthur Adamson at the University of Southern California, became a Lecturer at the University College London in 1963. Four years later he moved "down under" and joined the Research School of Chemistry at ANU, where he was appointed Professorial Fellow in 1979 and Professor of Inorganic Chemistry in 1991. He is a Fellow of the Royal Australian Chemical Institute and the Royal Society of London, was awarded with the Nyholm Medal of the Royal Society of Chemistry in 1991 and with the Max Planck Research Award in 1994. Martin won fellowships from the Alexander-von-Humboldt Foundation in 1973 and 1980 and during the last time was a visiting professor in our department (photo by courtesy of M.B.)

Fig. 2.37 The President of the German Chemical Society, Professor Heinrich Nöth, presenting me with the certificate of the Alfred Stock Prize at Freiburg, Germany, in 1988

Fig. 2.38 The Rector of the *Friedrich Schiller Universität Jena* (on the *left*) and the Dean of the Faculty (on the *right*) presenting me with the certificate for the honorary doctorate at the main hall of the university in 2006

about the end of Honecker's dictatorship. At first I could not believe it and even when, back at home, I saw the television pictures of the continuing demonstrations of hundreds of thousands of people in Leipzig and elsewhere, I still considered this as a dream. However, on the 9th of November 1989 the wall at Berlin came down. Two days later one of my cousins, who had never been allowed to visit me or other relatives in the West for three decades, stood in front of our house and the city of Würzburg was overcrowded by several thousand visitors from the East. It was a sunny day and many people put tables and chairs on the street to celebrate the event which most of us had thought would not happen in our lifetime. A few weeks later, after one could pass through the internal German border without a special visa, I brought Inge from Mühlhausen to Würzburg and we had an unforgettable Christmas party with the family and some friends at home. We bought a nice flat for Inge near the city center, where she lived for 12 years before recently moving into a senior residence.

At the beginning of 1999 I had decided to accept no more Ph.D. students and when my successor, Holger Braunschweig, arrived from Imperial College, all the labs were at his disposal. In the mid 1990s, when I saw the collaborative research projects of the SFB on the right track, I had considered stepping aside to some degree and spend some more time with the family. Helga in particular was looking forward to travel with me not only to conferences, but also to places and countries where we could pursue our cultural interests. However, in spring 1995 she noticed that she had difficulties controlling the movements of her left hand and despite various therapies applied by the neurologists the illness continued steadily. It finally turned out to be an incurable disease, and from summer 2000 she was unable to use any muscles in her arms or legs. She remained absolutely conscious but died on the 18th of November 2000. It was a very painful experience for me, Monika, Andreas, Inge and our friends who all supported me as well as possible. In these sad days I was also extremely grateful for the sympathy I received from the research family and the members of the department. Nowadays, I greatly appreciate belonging to the faculty as an Emeritus Professor and observing the activities of those who had taken over the lead. I am pleased to see that organometallic chemistry is still alive at Würzburg, and hope that it can maintain this position in the future.

2.7.1 Biographies

Ernst Otto Fischer (referred to as "E. O." by his friends and students) made the most important decision of his life at the age of 33. Since his father Karl Tobias Fischer, a former Professor of Physics at the TH in München, had subscribed to the magazine *Nature*, E. O. had incidentally read the paper by Kealy and Pauson on dicyclopentadienyliron on one of the last days of December 1951. The unusual properties of this compound made him rather suspicious regarding the structure proposed by Pauson, and after he (exactly Reinhard Jira, in those days an undergraduate and later his second Ph.D. student) repeated the synthesis of $Fe(C_5H_5)_2$ the race began. It culminated in the Nobel Prize for Chemistry 1973, which he shared with Geoffrey Wilkinson, the strong competitor for nearly two decades. At the moment when he was informed about the decision of the Nobel Committee, he replied in his typical Bavarian dialect "*des glaub i nie*" (it's unbelievable), mainly because he thought that he and his coworkers could not assert themselves against the dominance of organometallic groups in the US and the UK. The most curious congratulation he received was from Wilkinson who sent a telegram containing only three words: "Everything is forgiven". While E. O. showed me the telegram, we both laughed, but at the same moment also remembered the time in the 1950 and 1960s when we – the *Chef* and the students – often worked 12 h per day in order to prevent "the others" from getting the lead (Fig. 2.39).

Ernst Otto Fischer, born on the 10th of November 1918 in München, finished grammar school at his hometown in summer 1937 and, following 2 years of *Arbeitsdienst* ("work service") and compulsory military service, was drafted into the German army at the beginning of World War II. While he was uncertain at the age of 19 whether he would choose chemistry or the history of art as his main subject at university, he decided in favor of the former in winter 1941/1942 during a 4 month study leave. He was taken prisoner of war and subsequently released by the Americans in autumn 1945. After the reopening of the TH in München, he continued his studies in spring 1946. He graduated in 1949 and, fascinated by the personality of Walter Hieber, decided to work with him for his Ph.D. Two years later he submitted his doctoral thesis entitled "*The Mechanism of the Reactions of Carbon Monoxide with Nickel(II) Salts in the Presence of Dithionites and Sulfoxylates*" to the faculty and received the Ph.D. in early 1952. Following the advice of Hieber, he stayed in academia and, due to his epoch-making studies on sandwich-type complexes, completed his

Fig. 2.39 Ernst Otto Fischer at the time when he received the Nobel Prize (photo by courtesy from the late Professor Max Schmidt)

Habilitation thesis already at the end of 1954. He became a *Privatdozent* at the TH in München in 1955, received an invitation to be Associate Professor at the LMU München in 1957, was promoted to Professor at the same university in 1959, and succeeded Walter Hieber to the Chair of Inorganic Chemistry at the TH in October 1964. He remained in this position until his retirement in summer 1984.

E. O. was a member of several academies and has received numerous awards among which, apart from the Nobel Prize, I believe he considered the *Bayerische Verdienstorden*, the *Bayerische Maximiliansorden für Wissenschaft und Kunst* and the Gold Medal *München leuchtet* to be the greatest honours. His dedication to scientific research as well as his personality attracted a great number of students both at the TH and the LMU in München. Between 1955 and 1985, he was supervisor to 130 doctoral students. Nineteen of them started an academic career and 13 (Henri Brunner, Karl Heinz Dötz, Alexander Filippou, Helmut Fischer, Rainer Dieter Fischer, (the late) Heinz Peter Fritz, Max Herberhold, Gerhard Herberich, Gottfried Huttner, Cornelius Kreiter, Jörn Müller, Ulrich Schubert and me) held or still hold chairs in inorganic or organic chemistry. E. O. described himself as "a conservative stubborn human-being" who constantly confessed "I really like basic research". Besides his life-long interests in the fine arts and the history of art, he enjoyed spending time in the Alps, where he had built a house, which was and still is mainly used by his former coworkers and their families. It was here that most of the Nobel Prize money was spent. Ernst Otto Fischer died on the 23rd of July 2007.

Franz Hein, born on the 30th of June 1892 in the *Badische Malerdorf* Grötzingen near to Karlsruhe, was in the best sense of the word a *grand-seigneur* (Fig. 2.40). His father was an artist and when he became Professor at the Academy of Arts in Leipzig in 1905, the family moved to this city. Hein finished high school in 1912 and enrolled at the University of Leipzig that

Fig. 2.40 Franz Hein at the age of 77, when he retired as the Director of the Research Laboratory of the Academy of Science (photo by courtesy from Professor Egon Uhlig)

same year. At that time Arthur Hantzsch (1857–1935) was the Director of the *Chemische Institut* and it was mere coincidence that Hantzsch guided the doctoral theses of both Alfred Werner in 1892/93 at Zürich and Hein in 1916/17 at Leipzig. However, Hein's doctoral thesis (co-directed by Hantzsch and Konrad Schäfer) entitled "*Investigations on Triphenylmethane Derivatives and Optical Studies on the Constitution of Bismuth Compounds*" apparently had nothing to do with coordination chemistry and was on the borderline between organic and inorganic chemistry. After Hein received the Ph.D. in 1917, he stayed at Leipzig as *Assistent* (after 1920 as *Oberassistent*) and began his groundbreaking work on the so-called "polyphenylchromium bases and their salts" for his *Habilitation*. In 1921 he finsished his *Habilitation* thesis which was well received by the faculty. One year later, Arthur Hantzsch and Max Le Blanc (the Dean of the Faculty) applied to the Ministry of Saxony to establish a "*planmässiges Extraordinariat für analytisch-anorganische Chemie*" and offer this position to Franz Hein. The ministry agreed and thus Hein was appointed *Extraordinarius* (Associate Professor) of Inorganic Chemistry in January 1923.

While Hein continued his research on polyphenylchromium compounds during the next decades, he also studied the properties of alkali-metal alkyls in solutions of zinc and aluminium alkyls and developed a method for the electrochemical preparation of alkyl derivatives of zinc, lead, aluminium, etc. The chemistry of metal carbonyls became another field of interest and he reported the first compounds with a covalent bond between a transition and a main-group metal, such as $[(C_2H_5)_3Sn]_2Fe(CO)_4$, $[(CH_3)_2PbFe(CO)_4]_2$ and $(C_6H_5)_3SnCo(CO)_4$. In November 1942, he accepted an invitation to be Associate Professor of Inorganic Chemistry and Head of the *Institut für Anorganische und Analytische Chemie* at the University of Jena. About 3 years later, when the American troops had to leave Thuringia including Jena, Hein was deported with more than 30 professors from the university to a camp near to Marburg in the American Zone, and lived there with his family for 9 months. In contrast to the majority of his companions in the camp who decided to stay in the West, Hein returned to Jena and, supported by young colleagues such as Georg Bähr, Kurt Issleib, Alfred Schubert and Helmut Zinner, put all his efforts into the reconstruction of the laboratories. Hein became *Ordinarius* (full professor) and Head of the *Chemische Institut* in June 1946 and held the Chair in Inorganic Chemistry until his retirement in autumn 1959. As a member of the Academy of Science, he could retain two laboratories in the institute until 1969, where he continued his studies on polyaryl transition metal compounds with a group of experienced coworkers. He retained a firm distance from the communist establishment in East Germany and when asked by the political authorities about his personal attitude he wrote in June 1946: "On account of my strong scientific attitude, I had neither inclination nor time to concern myself with politics or political organizations. Newspaper and radio have had not a great interest for me, in contrast to technical journals".[6] In spite of his attitude, he was twice awarded the "*Nationalpreis der DDR*" for his outstanding scientific achievements. Franz Hein died on the 26th of March 1976.

[6] In German: "*Ich hatte auf Grund meiner streng wissenschaftlichen Einstellung weder Neigung noch Zeit, mich um Politik und politische Organisationen zu kümmern. Zeitungen und Rundfunk haben für mich keine große Bedeutung gehabt, dagegen umso mehr Fachzeitschriften*". Taken by permission from a manuscript of a seminar held on the 15th of January 2004 at Jena by Professor Egon Uhlig entitled "*Geschichte der Chemie in Jena im 20 Jahrhundert – Die Ära Franz Hein (1942–1959)*".

Chapter 3
The Nineteenth Century: A Sequence of Accidental Discoveries

A journey of a thousand miles begins with a single step.
Confucius, Chinese Philosopher (551–479 B.C.)

3.1 The Beginnings of Organometallic Chemistry

At the beginning of the twenty-first century, we were told that we were entering the "century of bio-sciences". Some decades previously, we heard that we live in the "century of physics". Einstein was the giant and several others such as Planck, Rutherford, Millikan, Bohr, de Broglie, Schrödinger, Dirac, Heisenberg, Pauli and Fermi were the heroes who opened new horizons in their field. If one follows this type of labelling, one is tempted to ask: When was the "century of chemistry?" Was it the nineteenth century? Personally, I would say "Yes", since from John Dalton (who published in 1805 the famous "*Table of the Relative Weights of the Ultimate Particles of Gaseous and Other Bodies*") to Alfred Werner (who formulated in the 1890s the conceptual framework of modern coordination chemistry) a legion of eminent chemists such as Berzelius, Faraday, Gay-Lussac, Wöhler, Liebig, Bunsen, Kekulé, Mendeleev, Moissan, Pasteur, Ramsay, van't Hoff and Le Bel – to name only a few – set the pace in the discovery of a new chemical world and laid the cornerstones to what chemistry is nowadays.

But when does organometallic chemistry come in? Consulting current textbooks, the general impression is that the field of organometallic chemistry emerged fairly recently, with systematic studies only beginning in the 1950s. However, although this is not completely wrong, the roots of the field are much deeper and going back at least to the 1820s. Within a more generous interpretation, one might even argue that the birth of organometallic chemistry could be dated back to the 1750s and the discovery of the "*liqueur fumante de l'arsénic*" (better known as "*cacodyl*") by the French scientist Louis-Claude Cadet de Gassicourt. He was the "*apothicaire-major de l'Hotel Royal des Invalides*" in Paris and obtained a "slightly coloured liquid of an extremely penetrating garlic odour" on heating a mixture of arsenious oxide

H. Werner, *Landmarks in Organo-Transition Metal Chemistry*,
Profiles in Inorganic Chemistry, DOI 10.1007/978-0-387-09848-7_3,

and potassium acetate [1]. Cadet was unable to determine the exact composition of the new compound, and it took almost a 100 years until Adolf Baeyer (later Adolf von Baeyer) confirmed that cacodyl is tetramethyldiarsine $As_2(CH_3)_4$ [2]. The efforts undertaken in the interim between Cadet's and Baeyer's reports, particularly by Robert Bunsen, in elucidating the nature of cacodyl make for an interesting story by itself which has been covered in an essay by Dietmar Seyferth in 2001 [3].

Is tetramethyldiarsine an *organometallic* compound? By definition, arsenic is a so-called "metalloid" or "semi-metal", not a metal, and thus organoarsenic compounds like the organic compounds of boron or silicon are not per se organometallics. There is no doubt that the preparation of tetramethyldiarsine and the elucidation of its structure may in retrospect be linked to the development of organometallic chemistry, but it would be far beyond the scope of this book, if the chemistry of the organic compounds of the metalloids in its entirety was discussed.

3.2 Wilhelm Christoph Zeise and the First Transition Metal π-Complex

So let us return to the beginnings and to the isolation of the first *transition metal* compound with an organic ligand. It was the Danish pharmacist Wilhelm Christoph Zeise, who described the preparation of a lemon-yellow crystalline compound analysing as $KCl \cdot PtCl_2 \cdot C_2H_4 \cdot H_2O$ in 1827. He obtained this salt, which at that time was an entirely novel substance and today is known as "Zeise's salt" $K[(C_2H_4)PtCl_3] \cdot H_2O$, by boiling a solution of chloroplatinic acid in ethanol, and then adding KCl [4–6]. Incidentally, this material was also prepared by Magnus in Berzelius' laboratory and described as a double salt of "potassium chloride and platinum chloride with a peculiar ether-like substance" [7]. Initially the result remained unpublished and it was only after the appearance of Zeise's first report, that Berzelius realized that Magnus' and Zeise's compounds were identical [8].

The analytical composition of Zeise's salt became the object of a sharp controversy in the 1830s and 1840s, and it was an authority of the eminence of Justus Liebig (later Justus von Liebig) who criticized Zeise's work in an article [9, 10] which nowadays would not be permitted in a scientific journal as it was almost a personal attack. A flavour of Liebig's paper including its more extreme assertions can be appreciated from the selection of sentences which may be translated thus: "...apparently Zeise's formula is basically a mere supposition...Once an analysis has been carried out, the uncertainty about the way in which the elements of a compound are arranged makes it essential for a chemist, when postulating a theory, to develop reasons which will exclude all others...If in the following I seek to interpret Zeise's results then I wish to expressly emphasize that I desire thereby to provide a more thorough interpretation; undoubtedly, it is immaterial to science and all chemists which of our views is the correct one. I repeat, it matters not which one of us is right, provided, in conclusion, we learn the truth...As is well known, the deep-rooted fear of ghosts is not vitiated by reason, and despite my past and present achievements, I do not flatter myself that I can disperse the fantasy of the supporters of the ethereal theory of the existence of an oil-forming gas in the ethereal compounds. My writings are devoted exclusively to that generation which has been called to display the marvellous structure that is organic chemistry and not to those who seek to base this structure on air or cloud." Despite these objections, Zeise remained cool, repeated the synthesis and the analytical work, and reasserted his original formulation [11].[1] Nevertheless, he could not convince Liebig that the species in question, bonded to the metal, is ethene (in those days called "elalyl") but not ether, as Liebig believed. The conflict only died away after Griess and Martius confirmed in 1861 that ethene is evolved by heating Zeise's salt to 200°C [12], and when Birnbaum described in 1868 the preparation of $K[(C_2H_4)PtCl_3]\cdot H_2O$ by passing ethene into a solution of in situ generated $H_2[PtCl_4]$ under slightly increased pressure, and then adding KCl [13]. Birnbaum also prepared the analogous propene and 1-pentene platinum compounds, $K[(C_3H_6)PtCl_3]\cdot H_2O$ and $K[(C_5H_{10})PtCl_3]\cdot H_2O$, and came to the conclusion that "no further doubt can arise about the correctness of Zeise's views" [10]. In the context of Zeise's and Liebig's controversy, a comment by the famous German poet Johann Wolfgang von Goethe is worth mentioning. Goethe said of chemists [14]: "...they are learned, sensible, good men, each – regarded by himself – highly estimable, if they could only get on with each other! However, since this does not seem to be the case of mankind in general, we will also not expect it from this particular group".

3.3 Edward Frankland's Pioneering Studies

While the work of Zeise was not considered by most of his contemporaries as an epoch-making result, Edward Frankland's studies on organozinc compounds, carried out in Bunsen's laboratory at Marburg in Germany between 1848 and 1850 and during the following decades at Manchester and London, soon found worldwide recognition. Initially, on Bunsen's suggestion and in collaboration with Hermann Kolbe (in the 1840s Bunsen's assistant and since 1865 Professor of Chemistry at the University of Leipzig in Germany), Frankland attempted to prepare "radicals". In those days the term "radical" was the name for a "stable group of atoms that retains its integrity in its reactions and its formation of compounds with other atoms or groups of atoms (generally electronegative)"

[1] With regard to Zeise's analytical data, Berzelius made the following comment: "It would indeed be difficult to demand a more complete agreement between the theory and the actual results"; J. Berzelius, *Jahresber.* **18**, 445 (1839).

[15]. Ethyl was such a "radical" since it remained unchanged in the conversion of propionitrile C_2H_5CN to propionic acid $C_2H_5CO_2H$. To obtain the "ethyl radical", Frankland carried out the reaction of ethyl iodide with finely granulated zinc in a sealed thick-walled glass tube at ca. 200°C and obtained a white solid and a colourless, mobile liquid as well as some gaseous products [16].[2] When the sealed tube was opened over water with provision for collecting the gaseous products, a large volume of gas evolved and the mobile liquid disappeared. Frankland was convinced that among the gaseous products the component with the highest boiling point was the pure "ethyl radical" and thus entitled his paper "*Ueber die Isolirung der organischen Radicale*" [16]. As was realized later, the respective gas was not the "ethyl radical" but the dimer *n*-butane. Frankland could not exactly characterize the white solid, which probably was a mixture of C_2H_5ZnI and ZnI_2. Presumably, under the reaction conditions, diethylzinc was also produced by thermal disproportionation of ethylzinc iodide but had decomposed at 200°C by homolytic cleavage of the zinc–carbon bonds.

Although we presently know that C_2H_5ZnI and $Zn(C_2H_5)_2$ were the main components formed in this initial experiment, the first organozinc compound that Frankland prepared and clearly identified was dimethylzinc. Still working in Bunsen's laboratory, he treated zinc with methyl iodide and, after discharging the gases, isolated a solid residue that reacted violently with water to produce a *brilliant flame* [17]. Frankland confirmed that the gas generated on hydrolysis was "pure light carburetted hydrogen" (i.e. methane), and the "colorless pellucid liquid, possessing a peculiarly penetrating and exceedingly nauseous odour", obtained by distillation of the solid residue, was *zincmethylium* (i.e. dimethylzinc). He immediately recognized the importance of his discovery and was aware of the potential of the metal/RI reaction for the preparation of organic derivatives of other metals [17]. After he had moved to Manchester, he reported the synthesis of $Hg(CH_3)_2$ (*hydrargyromethylium*) and $Sn(C_2H_5)_4$ (*stanethylium*) from the metals and the corresponding alkyl iodides [18, 19] and published his full papers on dimethyl- and diethylzinc [18–20]. Moreover, following the preparation of $Hg(CH_3)_2$ and $Hg(C_2H_5)_2$ he also showed that heating metallic zinc with the appropriate mercury alkyl afforded pure samples of dimethyl- and diethylzinc without contamination by any side-products [21].

Soon after Frankland had isolated $Zn(CH_3)_2$, he noticed the potential utility of this compound for organic synthesis. His assertion [18, 19] that "the extraordinary affinity of zincmethylium for oxygen, its peculiar composition, and the facility with which it can be procured, cannot fail to cause its employment for a great variety of transformations in organic compounds; by its agency there is every probability that we shall be able to replace oxygen,

[2] Since it was not uncommon in those days, Frankland published his work also in English; see: *J. Chem. Soc.* **2**, 263–296 (1850).

chlorine etc., atom for atom by methyl, and thus produce entirely new series of organic compounds, and obtain clearer views of the rational constitution of others" was certainly an unusual prophecy and not generally accepted at that time. Nevertheless, before the Grignard reagents came into use, many of the reactions for which the latter are now commonly applied were first worked out with the zinc dialkyls, and it is very much to the credit of the chemists involved that so many successful reactions were carried out with these spontaneously inflammable and extremely air- and moisture-sensitive substances. It is worth remembering, that Frankland and his contemporaries had no inert atmosphere boxes, no Schlenk lines, no vacuum pumps, no efficient hoods and not the present-day glassware and thus one can only admire what they achieved at that time.

Yet the unambiguous characterization of the organozinc compounds by Frankland occupy a notable place in the history of chemistry also for another reason. The thorough studies on the vapour densities and the chemical reactions of the zinc dialkyls, which he carried out, led for the first time to a clearly defined theory of valency, suggesting that each element had a definite limited combining capacity [18, 19]. His statement that "...the compounds of nitrogen, phosphorus, antimony and arsenic especially exhibit the tendency of these elements to form compounds containing three or five equivalents of other elements, and it is in these proportions that their affinities are best satisfied..." not only provided the basis for much future research in inorganic and organic chemistry, but also led directly to the Periodic Table of the Elements.

Frankland's reports, that organic residues can be transferred intact to and from metals in a fashion quite similar to inorganic species, launched a widespread activity not only in England but also in Germany, France, Russia and elsewhere in Europe. As early as in 1850, Carl Jacob Löwig, a German professor at the ETH Zürich, described organic compounds of main-group metals, such as antimony, bismuth, tin and lead [22, 23], which he claimed were already prepared before 1850 but obviously, as Frankland indicated [20], not adequately characterized in those days. Later, Löwig continued his work at the University of Breslau in Germany (now Wroclaw in Poland) and, while studying the chemistry of antimony and bismuth alkyls, found that $Bi(C_2H_5)_3$ reacts with mercuric halides to form ethylmercuric halides [15]. In the 1860s, Wanklyn reported the synthesis of $Cd(C_2H_5)_2$ and $Mg(C_2H_5)_2$ [24], and also generated the first zincate, $NaZn(C_2H_5)_3$, from sodium and an excess of $Zn(C_2H_5)_2$ [25]. Buckton prepared $Pb(C_2H_5)_4$ by treatment of diethylzinc with an excess of $PbCl_2$ [26] and, together with Odling, applied Frankland's method of metal exchange to prepare trimethyl- and triethylaluminium from aluminium and dimethyl- or diethylmercury, respectively [27]. Buckton also noted the aggregation of $Al(CH_3)_3$, which posed a serious problem for valency theory at that time but also stimulated the study and theoretical interpretation of electron deficient molecules. Among the wealth of other main-group organometallic compounds, which were isolated and characterized in the late nineteenth century [28, 29],

$Ge(C_2H_5)_4$ deserves some prominence as it was prepared only one year after the discovery of elemental germanium by Clemens Winkler at the *Bergakademie* Freiberg in Germany [30].

3.4 Victor Grignard: The Father of "Organometallics for Organic Synthesis"

Until 1900, Frankland's organozinc compounds were the most useful sources of nucleophilic alkyl groups, but lost their predominant status when the more easily accessible and more reactive Grignard reagents came into use. The Grignard reagents, which only a few years after their discovery became a panacea in organic synthesis, are named after Victor Grignard, who began to develop their chemistry in 1900 and the following years. As mentioned above, some dialkyls of magnesium had been prepared previously, albeit with some difficulty, by Wanklyn [25] and Cahours [31], but unlike their zinc analogues had not been used for synthetic purposes.

Grignard started his work in the laboratory of Philippe Antoine Barbier at the University of Lyon, and it was one of his first aims to elaborate a generally applicable synthesis of tertiary alcohols from ketones. Somewhat earlier, Barbier had attempted to convert methylheptenone $(CH_3)_2C{=}CHCH_2CH_2C(O)CH_3$ (which occurs in nature) to dimethylheptenol $(CH_3)_2C{=}CHCH_2CH_2C(OH)(CH_3)_2$ by treatment with methyl iodide in the presence of zinc, in analogy to the Saytzeff reaction, but failed. When zinc was replaced by magnesium, more satisfactory results were obtained [32]. Barbier suggested to Grignard to continue this work and study the course of the reaction of ketones with CH_3I/Mg in more detail. Grignard discovered quite soon that, by using thoroughly dried diethyl ether as solvent, magnesium turnings react with alkyl bromides and iodides at room temperature and normal pressure to give organomagnesium halides (initially named "mixed organomagnesium compounds") which upon treatment with aldehydes or ketones gave alcohols in good to excellent yields. After he reported his initial results in a short note in 1900 [33], he continued his work with such vigour and diligence that he was able to publish seven additional papers in the following year [34, 35]. Grignard remained the leader in this area during the following decade, but was soon followed by legions of organic chemists who used the simple organomagnesium compounds for organic synthesis. As a measure of the wide-spread application, it is interesting to note that at the time of Grignard's death in 1935 more than 6,000 references to the use of Grignard's reagents were found in the literature [34, 35]. With regard to the behaviour of the organomagnesium halides in solution, Wilhelm Schlenk and his son Wilhelm Schlenk jr. recognized in the late 1920s that in ether solution an equilibrium between the mono- and di-alkyl or -aryl compounds (RMgX and MgR_2) existed which could be shifted to the side of the dialkyl or diaryl derivatives by change of solvent [36]. This observation provided the connection between the organomagnesium and organozinc compounds since it was known from Frankland's work in the 1850s that zinc dialkyls could be obtained from alkylzinc halides at elevated temperatures.

3.5 Paul Schützenberger and Ludwig Mond: The First Metal Carbonyls

At the same time, when research on organometallic compounds of zinc, mercury, aluminium, and other main-group metals was flourishing, the first transition metal carbonyl complexes appeared on the scene. Paul Schützenberger, who was born in Strasbourg in 1829 and had received both a doctorate in medicine and in chemistry, reported first in French in 1870 [37] and then in English in 1871 [38] that by passing a mixture of chlorine and carbon monoxide over platinum-black at 300–400°C in a glass tube, a "flocculent body containing platinum condenses in the cool part of the tube". After reducing the temperature to 240–250°C and treating the metallic platinum stepwise first with chlorine and then with CO, he obtained about 20 g of a solid compound in the course of 1 h, "partly in the form of bright yellow needles, partly as a yellow crystalline incrustation, and partly as yellow flocculi, lining the more remote parts of the tube". He noticed that "the entire product must be carefully excluded from moisture, as the least trace of water is sufficient to blacken it by causing a separation of metallic platinum" [38]. As the analysis and the melting point of the product indicated it to be a mixture, Schützenberger attempted to separate the components by fractional crystallization and finally isolated "diminishing quantities of crystals. . .having a composition which may be expressed either by $C_2O_2PtCl_2$ or by $C_4O_4Pt_2Cl_4$". Further experiments confirmed that the crude product consisted mainly of the monocarbonyl complex, described as $Pt(CO)Cl_2$, and the dicarbonyl complex, $Pt(CO)_2Cl_2$, which could both be purified by sublimation. Schützenberger noticed that they "are readily convertible into each other, the former into the latter by saturating it with carbonic oxide at 150°C, and the latter into the former by heating it to 250°C in a current of some indifferent gas" [38]. It took several decades to confirm that the monocarbonyl complex is a chloro-bridged dimer and the dicarbonyl complex contains the carbonyl and the chloro ligands in *cis* disposition [39]. Schützenberger did not continue his research on metal carbonyl compounds; instead, he turned to the properties and application of dyestuffs and produced cellulose acetate for the first time. In 1876, he became Professor of Chemistry at the Collège de France, and in the 1880s, the first director at the l'Ecole Supérieur de Physique et de Chimie Industrielles in Paris. He died in 1897 in Mézy-sur-Seine.

While Schützenberger had intended to study the action of CO on platinum and its compounds, the discovery of the first binary metal carbonyl $Ni(CO)_4$

occurred by accident. Ludwig Mond, an entrepreneur, who introduced Solvay's ammonia–soda process in England, was interested in making this process more profitable. To this end, he attempted to convert the ammonium chloride, formed as a by-product, to chlorine. His aim was to split the ammonium chloride into NH_3 and HCl, then – after removing the ammonia – to react HCl with metals or metal oxides to give metal chlorides, and finally to generate elemental chlorine from the metal chlorides upon heating, thereby recovering the respective metal. In collaboration with his long-time assistant Carl Langer, Mond decided to use iron vessels for the evaporation of ammonium chloride, which had to be rendered airtight with the aid of valves in order to prevent a large loss of ammonia. During his attempts, Langer found that "nickel was one of the few substances suitable for the construction of these valves, and that it was not at all attacked by ammonium chloride vapour. On the laboratory scale these nickel valves worked perfectly, but. . .on the manufacturing scale. . . they very soon became leaky. The faces became covered with a black crust, which, on examination, was found to contain carbon". While initially "the source of this carbon seemed mysterious", careful studies revealed that the CO_2, obtained from a lime kiln and used to sweep the ammonia out of the vessels, contained a few percent of CO. This observation "led us to study the action of CO on nickel" [40].

Another endeavour of Mond also led to a study of the interaction of CO with nickel. Earlier in the nineteenth century, John Grove in England had constructed the first hydrogen cell and, based on this knowledge, Mond wanted to convert the "Mond gas" (which was produced in his factory and contained hydrogen as the main component together with smaller amounts of CO_2 and CO) directly into electricity. Langer was also involved in this project but found that progress in his research was hindered by the presence of CO in the hydrogen feedstock. To overcome this inconvenience, he passed the gaseous mixture "containing hydrogen, CO and a certain quantity of steam over finely divided nickel at a temperature of 400°C. . .and could completely convert the CO into CO_2, obtaining its equivalent of hydrogen, which was just what we wanted". This formerly unknown affinity of CO to nickel was the second reason for Mond to find out "whether a definite compound of nickel and carbon was formed" [40].

Although in retrospect this work initiated the era of metal carbonyl chemistry, it is rather curious that several versions of the history leading to the discovery of nickel tetracarbonyl exist. Perhaps the most concise is given by Ludwig Mond's biographer, J. M. Cohen, who described it as follows: "In the course of these experiments (done by Langer, who was joined by a young German, Friedrich Quincke) a combustion tube was set up, in which finely divided nickel was treated with pure carbon monoxide made by the action of sulfuric acid on a formate. To keep the poisonous carbon monoxide out of the atmosphere of the laboratory, the tail-gases from the combustion tube were passed through a glass jet and there burned. – The experiment was started each morning and shut down each night, the nickel being allowed to cool off in the stream of carbon monoxide before the whole apparatus was shut down. – One evening in 1889, however, the assistant (Quincke) who usually remained behind to close down the experiment, had gone home early, and Langer himself was waiting for the flame to die down and the last of the carbon monoxide to be burnt before locking up. It was the first time that he had done so, and he was quite unprepared for what he saw. – For as the apparatus cooled he saw, to his great surprise, the flame of the jet grow luminous and increase in brightness until the temperature fell below the boiling point of water, when the flame faded again. Furthermore, from its usual lambent blue the flame had turned a sickly green. – Ludwig Mond was called at once to witness the strange phenomenon, and together they stared in silent wonder. At first they both thought of the arsenic that might be present in the sulphuric acid. ... But that was readily tested – a cold porcelain tile was thrust into the flame and was immediately coated with a shiny mirror, not unlike, but significantly different from the spots left by arsenic in Marsh's test; and when the neck of the combustion tube was heated, a bright mirror formed on the glass and the luminosity of the flame disappeared. The mirrors were analysed and appeared to be nickel, but such was the improbability of so heavy a metal as nickel forming a volatile compound that Ludwig would not believe his own tests, and postulated an unknown element in the nickel. But the carbon monoxide and the nickel were then purified as carefully as possible, but still the phenomenon occurred. It was not, one feels, until the gaseous nickel carbonyl had actually been frozen to a mass of needle-shaped crystals that Ludwig could really believe that he had – in Lord Kelvin's words – given wings to a heavy metal" [41].

The discovery was made in October 1889 and published by Mond, Langer and Quincke in 1890 [42]. The nature and properties of nickel tetracarbonyl (which at room temperature is a colourless liquid boiling at 43°C) were so novel that they initiated the quest for the carbonyls of other metals. The next member of the family, iron pentacarbonyl, was reported almost simultaneously by Marcellin Berthelot [43] and Ludwig Mond [44]. At a meeting of the *Academie des Sciences* in Paris on the 15th of June 1891, Moissan was admitted to the membership vacated on the death of Cahours, and among the papers read on that day was "*Sur une combinaison volatile de fer et d'oxyde de carbone, le fer carbonyl*" by Berthelot. Three days later, on the 18th of June 1891, the handwritten minutes of the meeting of the Chemical Society recorded the London lecture "*Note on a Volatile Compound of Iron and Carbonic Oxide*" by Mond and Quincke. Although initially they misformulated their product as $Fe(CO)_4$, they announced the compound to be liquid iron pentacarbonyl later in the year [44]. By subjecting this liquid to the action of light, they obtained crystalline $Fe_2(CO)_9$. However, at the time Mond and Quincke did not elucidate the nature of this compound, and it was Dewar and Jones [45] who determined the exact composition of this second iron carbonyl in the early twentieth century.

While Mond was aware of the unusual properties of nickel tetracarbonyl, he had "for a long time...no suspicion that this substance...should ever become available for industrial purposes". But the longer he and his collaborators "went on preparing it for our investigations, the more easy we found it to prepare it in quantity, after we once knew exactly the best conditions for so doing. After that I came to the conclusion that it ought to be possible to make use of the ease with which nickel is converted into a volatile gas by CO, while practically all other metals, and notably cobalt. . .was not acted upon by this gas, for separating nickel from cobalt and other metals on a manufacturing scale, and for obtaining it in a pure state" [40]. Mond first built a large scale laboratory plant and in 1892, after he found it to be sufficiently promising, a pilot plant at Smethwick in South Wales. In 1895, the Mond nickel process was commercially viable and produced purest nickel "at the rate of a ton and a half per week from the Canadian nickel copper matte imported into England" [40]. When Mond presented his results at the meeting of the Society of Chemical Industry in November 1895, he made some comments which, considering the present discussions about the conservation of energy, appear topical even today: "What gives me the greatest satisfaction. . .is that I believe I have succeeded in working out a purely chemical process for extracting nickel from its ores, which will be cheaper and simpler than any electrolytic process that can be used for the same purpose. . . I know there are many chemical operations which will always be carried out to much greater advantage by the old chemical methods, and I have no doubt that newer methods will be found, of which nobody thinks at present, based upon purely chemical reactions, such as the process which I have brought before you to-night, which will effect the chemical changes we want to produce at a smaller expense of energy than can be done by electrolysis" [40].

Although Ludwig Mond's work was not very much appreciated by most of his contemporaries working in academia, he had nevertheless made one of the truly seminal discoveries in science [46]. Moreover, he had also founded an essential part of organometallic chemistry that began to blossom during the course of the second quarter of the twentieth century.

Biographies

Wilhelm Christoph Zeise was born on the 15th of October 1789 in Slagelse, Denmark, and first trained as a pharmacist in his father's pharmacy. After finishing school he went to Copenhagen, where he lived with the family of H. C. Oersted, who was Professor of Physics and Chemistry at the University of Copenhagen and later discovered electromagnetism. Zeise (Fig. 3.1) enrolled as a student of medicine, chemistry and physics, and soon became Oersted's lecture assistant. After he obtained his degree in pharmacy in 1815, he set up a small laboratory in his father's pharmacy and carried out independent research on the action of alkalis on organic substances. On this basis he wrote a doctoral thesis, as required at that time in Latin, entitled "*De vi corporum alcalinorum materias regno organico peculiaris transformandis*" and received his Ph.D. degree in 1817. He

Fig. 3.1 Wilhelm Christoph Zeise (1789–1847) as Professor of Chemistry at Copenhagen (photo reproduced by permission of the History of Technology Division, Technical University of Denmark)

spent some time abroad, particularly at Göttingen and Paris, and then returned to Copenhagen where he began his studies on organosulphur compounds. In 1822 he was appointed Associate Professor of Chemistry at the University of Copenhagen and in 1829, on Oersted's recommendation, Professor of Organic Chemistry at the newly founded Polytechnic Institute of Copenhagen. In the course of his research on organosulphur compounds, he discovered the xanthates and thiols, naming the latter mercaptans, and later was one of the earliest investigators of carotenes. His careful studies, including those on the composition of the platinum complexes, were undoubtedly landmarks and the beginnings of chemical research in Denmark. Zeise held his position at the Polytechnic Institute until his death on the 12th of November 1847.

Edward Frankland was born on the 18th of January 1825 near Garstang in Lancashire in the UK and, after an apprenticeship to a druggist in Lancaster, began his career in London as a lecture assistant of Lyon Playfair in the government's Museum of Economic Geology in 1845. Between 1847 and 1849, he worked with Robert Wilhelm Bunsen at Marburg and also spent 3 months in Justus Liebig's laboratory at Giessen in Germany. Bunsen was a giant at that time or – in the words of John Tyndall – "*. . . a man whose superior as a chemist is not to be found within a radius of 8000 miles from the Piece Hall in Halifax*" [47].[3] Frankland (Fig. 3.2) received his Ph.D. with Bunsen in 1851 and in the same year became the first Professor of Chemistry at the newly founded Owens College in Manchester, which later became the University of Manchester. Here he developed an interest in applied chemistry and, based on his analytical skills, became a consultant for agriculturists, railway companies, gas works, chemical manufacturers and other industries. He returned to London in 1857, obtained a lectureship at St. Bartholomew's Hospital, and also

[3] J. Tyndall in a letter dated the 2nd of July 1849 to T. A. Hirst (see [47]). Halifax is a town near Leeds in England.

Fig. 3.2 Edward Frankland (1825–1899) at the time when we was the Director and Professor of Chemistry at the Royal College of Chemistry in London (photo reproduced with permission of the Science Museum, London)

taught at the Military College at Addiscombe. As successor to Faraday, he held the prestigious position of Professor of Chemistry at the Royal Institution from 1863 to 1869, and in 1865 also succeeded August Wilhelm Hofmann (later August Wilhelm von Hofmann) as Director and Professor of Chemistry at the Royal College of Chemistry (today Imperial College). There he stayed until his retirement in 1885. Frankland was the first to set up courses specifically designed to train teachers of chemistry, and to encourage them to perform demonstration lectures and offer their pupils the opportunity to carry out some laboratory work of their own. For his manifold achievements, he was widely regarded as one of Europe's leading chemist in the second half of the nineteenth century, probably the most distinguished in Victorian Britain. He was knighted by Queen Victoria in 1897 and died on the 9th of August 1899 while on vacation in Norway.

Victor Grignard was born on the 6th of May 1871 in Cherbourg in France and, after attending local schools, won a scholarship to study at the *École Normale Spéciale* at Cluny in 1889. Owing to a serious dispute between the teachers, the school was closed 2 years later and the pupils were transferred to other establishments. Grignard (Fig. 3.3) was fortunate to join the University of Lyon and received in 1894 the degree *Licencié ès Sciences Mathématiques.* Although he had at that time "a poor opinion of this science (chemistry)...and regarded it, in comparison with mathematics, as an inferior, purely empirical and mnemotechnical discipline" [34], he changed his mind after doing some laboratory work with Louis Bouveault at the same place. Promoted to *Préparateur*, he came into contact with Philippe Antoine Barbier, who was the head of the department and at whose side he worked for 14 years. In 1898, he became *Chef des travaux pratiques* and also wrote his first joint paper with Barbier. He submitted his ground-breaking doctoral thesis entitled "*Sur les combinaisons organomagnésiennes mixtes et leur application à des syntheses d'acides, d'alcools et d'hydrocarbures*" to the faculty in 1901 and received the degree *Docteur ès Sciences Physiques de Lyons* with "*Mention très honorable*". Following an interim period at the University of Besançon, Grignard returned

Fig. 3.3 Victor Grignard (1871–1935) as Professor of Chemistry at the University of Nancy (photo from *Nobel Lectures, Chemistry 1901–1921*, Elsevier, Amsterdam, 1966; reproduced with permission of Elsevier)

to Lyon in 1906 and 3 years later took charge of the Department of Organic Chemistry at Nancy, succeeding Emile Edmond Blaise. Together with Paul Sabatier, he was awarded the Nobel Prize for Chemistry in 1912 for "the discovery of the so-called Grignard reagent, which in recent years has greatly advanced the progress of organic chemistry". With regard to the Nobel Prize, Grignard felt that there was an injustice and that he should have received it jointly with Barbier.

After World War I, during which he was commissioned to study the cracking of alkylated benzenes with a view to increasing supplies of toluene, and also to work on problems of chemical warfare in Paris, he returned to Nancy and in 1919 succeeded Barbier as Professor of General Chemistry at Lyon. Two years later, he took an additional position as Director of the *École de Chimie Industrielle de Lyons*, became a member of the University Council, and in 1929 Dean of the Faculty of Science. In the 1930s he was heavily involved in publishing a comprehensive chemical handbook in the French language entitled "*Traité de Chimie Organique*", which was published in six volumes and became a standard text in France. Besides the Nobel Prize, Grignard received several prestigious awards and two honorary doctorates from the Universities of Brussels and Louvain. In 1920, he was appointed *Officer* and in 1933 *Commandeur de Légion d'Honneur*. Victor Grignard died on the 13th December 1935.

Ludwig Mond was born on the 7th of March 1839 in Ziegenhain near the city of Kassel in Germany. It was his maternal grandfather Aaron Levinsohn, who inspired in him an interest in scientific and cultural reading. Already at the age of 14, Mond (Fig. 3.4) enrolled at the Polytechnic School in Kassel, where he was taught the basics of chemistry known in 1853. He continued his studies at Marburg, similarly to Frankland, collaborating with Hermann Kolbe, and then moved to Heidelberg where Robert Bunsen had held the chair of chemistry since 1852. Although Mond was highly impressed by Bunsen's teaching and

Fig. 3.4 Ludwig Mond (1839–1909) at around 1900 (photo reproduced with permission of the Science Museum, London)

research, he enjoyed his time at Heidelberg in more senses than one. He was continuously in debt and, despite the financial support by his family, he left the university in 1858 before he took his degree.

Subsequently, he worked for several companies in Germany and Holland, and made his first visit to England in 1862. There he developed a process for the recovery of sulphur from the waste of alkali manufacture for which he received a patent. Although this process was not a particularly successful enterprise, it was used in several plants and eventually provided Mond with reasonable royalties. He became a partner at the John Hutchinson's company in Widnes and became a British citizen in 1866. He met Ernest Solvay on a visit to Belgium in 1872 and heard about the success in converting sodium chloride to sodium carbonate. Mond returned to England with a licence for the Solvay process and with his close friend John T. Brunner, also a former employee at Hutchinson's, built a factory at Winnington near Liverpool. The company Brunner, Mond & Co developed extremely well, and at the end of the nineteenth century became the biggest producer of soda in the world.

In 1889, the Mond family moved to London and bought a house near Regents Park. Mond extended the stables of that house and converted them to a well-equipped laboratory, where he continued to do research. Mond's activity is illustrated by the 40 patents from this laboratory which were registered over the years. He became, in the words of Edward Abel [48], "*an industrial pioneer and one of the most influential figures in the founding of the UK's chemical industry*". Mond acquired great wealth and supported a wide range of institutions such as the Royal Society, the Royal Institution, the British Academy, and many universities and museums in Britain and central Europe. His scientific and industrial achievements were recognized inter alia by an honorary doctorate from the University of Heidelberg, which he had left without a degree. Ludwig Mond died on the 11th of December 1909.

References

1. L. C. Cadet de Gassicourt, Qui peuvent servir à l'analyse du Cobolt; et Histoire d'une liqueur fumante, tirée de l'Arsenic, *Mem. Math. Phys.* **3**, 623 (1760).
2. A. Baeyer, Ueber die Verbindungen des Arsens mit dem Methyle, *Ann. Chem. Pharm.* **107**, 257–293 (1858).
3. D. Seyferth, Cadet's Fuming Arsenical Liquid and the Cacodyl Compounds of Bunsen, *Organometallics* **20**, 1488–1498 (2001).
4. W. C. Zeise, Eine besondere Platinverbindung, *Pogg. Ann. Phys. Chem.* **9**, 632 (1827).
5. W. C. Zeise, Von der Wirkung zwischen Platinchlorid und Alkohol, und von den dabei entstehenden neuen Substanzen, *Pogg. Ann. Phys. Chem.* **21**, 497–541 (1831).
6. W. C. Zeise, Gekohlenwasserstofftes Chlorplatin-Ammoniak, *Pogg. Ann. Phys. Chem.* **21**, 542–549 (1831).
7. J. Berzelius, Eine eigene Art von Platinsalzen, *Jahresber.* **9**, 162–163 (1830).
8. J. Berzelius, Sauerstoffäther, Aethersalze, *Jahresber.* **12**, 300–303 (1833).
9. J. Liebig, Ueber die Aethertheorie, in besonderer Rücksicht auf die vorhergehende Abhandlung Zeise's, *Ann. Pharm.* **23**, 12–42 (1837).
10. D. Seyferth, $[(C_2H_4)PtCl_3]^-$, the Anion of Zeise's Salt, $K[(C_2H_4)PtCl_3]\cdot H_2O$, *Organometallics* **20**, 2–6 (2001).
11. W. C. Zeise, Neue Untersuchungen über das entzündliche Platinchlorür, *Ann. Pharm.* **23**, 1–11 (1837).
12. P. Griess, and C. A. Martius, Notiz über Aethylenplatinchlorid, *Ann. Chem. Pharm.* **120**, 324–327 (1861).
13. K. Birnbaum, Ueber die Verbindungen des Aethylens und seiner Homologen mit dem Platinchlorür, *Ann. Chem. Pharm.* **145**, 67–77 (1868).
14. W. Mohr, *Naturwissenschaft und Naturphilosophie bei Goethe* (Verlag Scheiner, Würzburg, 1932, p. 17).
15. D. Seyferth, Zinc Alkyls, Edward Frankland, and the Beginnings of Main-Group Organometallic Chemistry, *Organometallics* **20**, 2940–2955 (2001).
16. E. Frankland, Ueber die Isolirung der organischen Radicale, *Ann. Chem. Pharm.* **71**, 171–213 (1849).
17. E. Frankland, Notiz über eine neue Reihe organischer Körper, welche Metalle, Phosphor u. s.w. enthalten, *Ann. Chem. Pharm.* **71**, 213–215 (1849) (*J. Chem. Soc.* **2**, 297–299 (1850)).
18. E. Frankland, On a New Series of Organic Bodies Containing Metals, *Phil. Trans. Roy. Soc.* **142**, 417–444 (1852).
19. E. Frankland, Ueber eine neue Reihe organischer Körper, welche Metalle enthalten, *Ann. Chem. Pharm.* **85**, 329–376 (1853).
20. E. Frankland, Ueber organische Verbindungen, welche Metalle enthalten, *Ann. Chem. Pharm.* **95**, 28–54 (1855).
21. E. Frankland, and B. F. Duppa, Ueber ein neues Verfahren zur Herstellung der Zinkverbindungen der Alkoholradicale, *Ann. Chem. Pharm.* **130**, 117–126 (1864) (*J. Chem. Soc.* **17**, 29–38 (1864)).
22. C. J. Löwig, and E. Schweizer, Ueber Stibäthyl, ein neues antimonhaltiges organisches Radical, *Ann. Chem. Pharm.* **75**, 315–355 (1850).
23. C. J. Löwig, Ueber Zinnäthyle, *Ann. Chem. Pharm.* **84**, 308–333 (1852).
24. J. A. Wanklyn, Ueber ein neues Verfahren zur Bildung organometallischer Verbindungen, *Ann. Chem. Pharm.* **140**, 353–356 (1866).
25. J. A. Wanklyn, Ueber einige neue Aethylverbindungen, welche Alkalimetalle enthalten, *Ann. Chem. Pharm.* **108**, 67–79 (1858).
26. G. B. Buckton, Untersuchungen über organische Metallverbindungen, *Ann. Chem. Pharm.* **109**, 218–227 (1859) and **112**, 220–227 (1859).
27. G. B. Buckton, and W. Odling, Preliminary Note on some Aluminium Compounds, *Proc. Roy. Soc.* **14**, 19–21 (1865).

28. E. Krause, and A. von Grosse, *Die Chemie der metall-organischen Verbindungen* (Verlag Gebrüder Bornträger, Berlin, 1937).
29. J. S. Thayer, Organometallic Chemistry: A Historical Perspective, *Adv. Organomet. Chem.* **13**, 1–45 (1975).
30. C. Winkler, Mittheilungen über das Germanium, *J. Prakt. Chem.* **36**, 177–209 (1887).
31. A. Cahours, Untersuchungen über die metallhaltigen organischen Radikale,*Ann. Chem. Pharm.* **114**, 227–255 (Part 1) and 354–382 (Part 2) (1860).
32. P. A. Barbier, Synthese du dimethylheptenol, *Compt. rend. Acad. Sci.* **128**, 110–111 (1899).
33. V. Grignard, Sur quelques nouvelles combinaisons organométalliques du magnésium et leur application à des syntheses d'alcools et d'hydrocarbures, *Compt. rend. Acad. Sci.* **130**, 1322–1323 (1900).
34. H. Rheinboldt, Fifty Years of the Grignard Reaction, *J. Chem. Educ.* **27**, 476–488 (1950).
35. M. S. Kharasch, and O. Rheinmuth, *Grignard Reactions of Nonmetallic Substances* (Constable & Co, London, 1954).
36. W. Schlenk, and W. Schlenk jun., Über die Konstitution der Grignard'schen Magnesiumverbindungen, *Ber. dtsch. chem. Ges.* **62**, 920–924 (1929).
37. P. Schützenberger, Mémoire sur une nouvelle classe de composés platiniques, *Bull. Soc. Chim. Paris* **14**, 17–27 (1870).
38. P. Schützenberger, On a New Class of Platinum-compounds, *J. Chem. Soc.* **24**, 1009–1014 (1871).
39. F. Calderazzo, R. Ercoli, and G. Natta, Metal Carbonyls: Preparation, Structure, and Properties, in: *Organic Synthesis via Metal Carbonyls*, (Eds. I. Wender and P. Pino, Interscience Publishers, New York, 1968, Vol. I, p. 216).
40. L. Mond, The History of my Process of Nickel Extraction, *J. Soc. Chem. Ind.* **14**, 945–946 (1895).
41. J. M. Cohen, *The Life of Ludwig Mond* (Methuen, London, 1956).
42. L. Mond, C. Langer, and F. Quincke, Action of Carbon Monoxide on Nickel, *J. Chem. Soc.* **57**, 749–753 (1890).
43. M. Berthelot, Sur une combinaison volatile de fer et d'oxyde de carbone, le fer carbonyl, et sur le nickel-carbonyle, *Compt. rend Acad. Sci.* **112**, 1343–1348 (1891).
44. L. Mond, and F. Quincke, Note on a Volatile Compound of Iron with Carbonic Oxide, *J. Chem. Soc.* **59**, 604–607 (1891) (*Ber. dtsch. chem. Ges.* **24**, 2248–2250 (1891)).
45. J. Dewar, and H. O. Jones, On a New Iron Carbonyl, and on the Action of Light and of Heat on the Iron Carbonyls, *Proc. Roy. Soc.* **A79**, 66–80 (1907).
46. E. Abel, Ludwig Mond – Father of Metal Carbonyls – and so Much More, *J. Organomet. Chem.* **383**, 11–20 (1990).
47. C. Meinel, *Die Chemie an der Universität Marburg seit Beginn des 19. Jahrhunderts* (Verlag N. G. Elwert, Marburg, 1978).
48. E. Abel, The Mond Connection, *Chemistry in Britain* **1989**, 1014–1016.

Chapter 4
Transition Metal Carbonyls: From Small Molecules to Giant Clusters

Do not go where the path may lead. Go instead where there is no path and leave a trail.

Ralph Waldo Emerson, American Poet (1803–1882)

Although nickel tetracarbonyl, iron pentacarbonyl, and diiron enneacarbonyl were already prepared in the 1890s, more than three decades passed before the chemistry of transition metal carbonyls took off. Undoubtedly, some parts of the chemical community had recognized that compounds such as $Ni(CO)_4$ and $Fe(CO)_5$ deserved special attention, in particular due to the use of $Ni(CO)_4$ for the production of pure metallic nickel. However, since the structure of those compounds was unknown, transition metal carbonyls remained, more or less, a curiosity.

4.1 A Class of "Peculiar Compounds"

To get a sense of the situation in the early stages of the twentieth century, it is interesting to refer to the chemical literature of that period. In his famous textbook "*Neuere Anschauungen auf dem Gebiet der Anorganischen Chemie*", first published in 1905, Alfred Werner described Schützenberger's platinum carbonyl complexes $Pt(CO)_2Cl_2$ and $Pt(CO)Cl_2$ as "compounds in which the central atom is connected to two different radicals (CO and Cl)" [1]. In the equally well-known textbook "*Chemie der Kohlenstoffverbindungen oder Organische Chemie*", published in 1928, Richard Anschütz discussed the properties and reactions of carbon monoxide and noticed that "it seems rather strange that it (CO) combines directly with some metals, for instance with nickel to form nickel carbonyl $(CO)_4Ni$" [2]. No comment about the possible bonding mode of CO to nickel! And even the organometallic standard text of that time "*Die Chemie der metall-organischen Verbindungen*", published in 1937, Erich Krause and Aristid von Grosse stated that "carbides, cyanides and metal

H. Werner, *Landmarks in Organo-Transition Metal Chemistry*,
Profiles in Inorganic Chemistry, DOI 10.1007/978-0-387-09848-7_4,

carbonyls. . .do not belong to the class of organometallic compounds, because they contain no hydrocarbon radical" [3].

Thus, before we discuss the rise of metal carbonyl chemistry and the pioneering work of Walter Hieber, let us first summarize what was known in this area when Hieber began his research in the late 1920s. It had already been mentioned in Chap. 3, that the discovery of $Ni(CO)_4$ and $Fe(CO)_5$ in 1890/91 was followed by the serendipitous generation of $Fe_2(CO)_9$ from $Fe(CO)_5$ and sun light [4]. About 15 years later, James Dewar and Humphrey Owen Jones at Cambridge University determined the exact composition of this compound [5] and also observed, that by heating $Fe(CO)_5$ or $Fe_2(CO)_9$ a third iron carbonyl, assumed to be $Fe(CO)_4$ and "exhibiting some very striking properties", was formed in moderate yield [6]. In contrast to orange-yellow $Fe_2(CO)_9$, the third iron carbonyl crystallized in "dark green, lustrous prisms" which were thermally significantly more stable than $Fe_2(CO)_9$, decomposing at 140–150°C. While Dewar and Jones thought that "the molecule would appear to be of the order of $[Fe(CO)_4]_{20}$, or, at any rate, the compound is a polymer consisting of many $Fe(CO)_4$ units" [6], it was later shown by Hieber and Becker on the basis of molecular weight determinations in iron pentacarbonyl as the solvent (!) that the formula $Fe_3(CO)_{12}$ is correct [7].

Encouraged by the success in preparing $Ni(CO)_4$ and $Fe(CO)_5$, Ludwig Mond (see Fig. 3.4) and his coworkers also made several attempts to synthesize carbonyls from all other metals which were at their disposal. Owing to its presence in the Canadian nickel ores, cobalt was of particular interest but when treated with CO under various conditions it "was not acted upon by this gas" [8]. Ludwig Mond continued his research in the field of metal carbonyls until his death and in a posthumous paper claimed, without providing direct evidence, to have obtained cobalt, ruthenium, and molybdenum carbonyls [9]. His elder son Robert L. Mond followed his footsteps and reported in 1910 the preparation of "cobalt tetracarbonyl" from metallic cobalt and carbon monoxide with "a minimum pressure. . .between 30 and 40 atmospheres at a temperature of about 150°C" [10]. The compound was described as "cobalt tetracarbonyl", but shown by a molecular weight determination to have the formula $Co_2(CO)_8$. By heating $Co_2(CO)_8$, "cobalt tricarbonyl" was obtained, which in fact was the tetramer $Co_4(CO)_{12}$. Using high pressure and temperature techniques, Robert L. Mond also prepared $Mo(CO)_6$ from freshly reduced molybdenum and CO, though in low yield [10]. Attempts to obtain the hexacarbonyls of chromium and tungsten under similar conditions failed.

A different procedure to prepare metal carbonyls was developed by André Job and his group in the 1920s. In the course of studying the reactivity of Grignard reagents toward small multiply bonded molecules, they observed that the slow reaction of organomagnesium halides with CO was accelerated by salts of the transition metals, in particular by anhydrous chromium trichloride. The organic products of this reaction were complex, but an ether-soluble compound of chromium could be isolated which was $Cr(CO)_6$ [11]. Subsequently, $Mo(CO)_6$ and $W(CO)_6$ were prepared in a similar way, using molybdenum pentachloride and tungsten hexachloride as the starting materials [12].

At the time these experiments were done, the mechanism of this process was not understood and it was not before Walter Herwig and Harold Zeiss showed the existence of chromium(III) aryls in the late 1950s [13], that the formation of organometallic compounds as intermediates in the reaction of metal halides with Grignard reagents had been accepted. Wilhelm Manchot and his son Wilhelm J. Manchot first demonstrated in their preparation of ruthenium pentacarbonyl from ruthenium triiodide, that not only metal aryls but also metal carbonyl halides were potential intermediates in the formation of zero-valent metal carbonyls [14]. At ambient pressure, RuI_3 reacted with CO to give $Ru(CO)_2I_2$, which in the presence of an acceptor for halogen, such as finely divided silver, afforded $Ru(CO)_5$ at 170°C under a CO atmosphere. The Manchots also showed that ruthenium pentacarbonyl is very sensitive to light, releasing CO and forming $Ru_2(CO)_9$ analogous to $Fe_2(CO)_9$ [14].

Not less interesting than the early developments in the preparation of metal carbonyls were the speculations about the possible generation of unknown species of the general composition $M_x(CO)_y$. The familiar observation that copper appears to be carried along the hot tube, when hydrogen or carbon monoxide is part of the gas stream, suggested the possibility of a carbonyl of copper for a long time. However, all the attempts to isolate this compound failed [15]. Considerable efforts, beginning as early as in 1898, were made to confirm the existence of carbonyls of zerovalent platinum and palladium [16]. As it was found that these metals absorb, respectively, 60 and 36 times their volume of carbon monoxide and that, in the case of platinum, the absorption product was stable liberating CO at 250°C, the formation of a carbonyl, possibly with the composition $Pt(CO)_4$, seemed reasonable. Since initial studies to extract a compound of this type from the absorption product had remained unsuccessful, Robert Mond attempted to prepare a platinum carbonyl by passing carbon monoxide under high pressure over platinum sponge at elevated temperatures but equally failed [10].

The fact that research in the field of transition metal carbonyls was not widely recognized until the 1930s, was not only due to the numerous failures to isolate complexes of the general composition $M_x(CO)_y$ but also to the lack of knowledge about the structures of these *peculiar compounds*. Taking into consideration, that the ideas of Frankland about the valencies of the elements were generally accepted at the beginning of the twentieth century, it was not surprising that Ludwig Mond and his contemporaries followed the trend. In accordance with the position of nickel in group VIII of the Periodic Table, they considered nickel tetracarbonyl to be a compound of octavalent nickel and, owing to its similarities to organic compounds (e.g., its high-molecular refraction and solubility in hydrocarbon solvents), assumed to have the structure shown in Scheme 4.1 [17]. An analogous formula was proposed for $Fe(CO)_5$. At the time of his death, Ludwig Mond appeared to have discarded the ring formulae, but his son Robert still used his father's expressions as late as in 1930 [18]. Some strange molecular formulae for $Ni(CO)_4$, $Fe(CO)_5$, and $Mo(CO)_6$ were also proposed by Reihlen and coworkers [19, 20], who considered the CO groups to be constituents of hypothetical organic *pseudo-acids* with the oxygen atoms of the terminal carbonyls *neutralizing* the

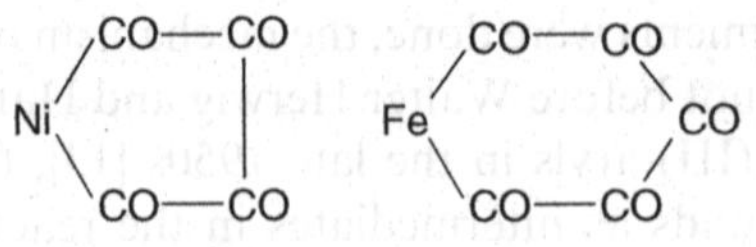

Scheme 4.1 Formulae for $Ni(CO)_4$ and $Fe(CO)_5$ proposed by Ludwig Mond in 1892

positive charge of the bivalent metals. Equally strange with respect to the structure and bonding of the mononuclear metal carbonyls of nickel, iron, and molydenum were the geometrical diagrams drawn by Blanchard and Gilliland [21], in which each corner of the squares represented an electron pair (see Scheme 4.2). Even after Brill had determined the crystal structure of $Fe_2(CO)_9$, with the kind of accuracy which was possible in 1927, and suggested the formula shown in Scheme 4.3 [22], Sidgwick and Bailey proposed in 1934 a linear arrangement Fe—C≡O—Fe for the central part of this molecule [23]. In retrospect it is interesting to note that it was not until 1976 that the existence of a linear M–C–O–M carbonyl bridge between two transition metal atoms was confirmed by Erwin Weiss at the University of Hamburg, Germany, for the V(II) and V(–I) centers of the trinuclear vanadium complex $[V(THF)_4][V(CO)_6]_2$ [24].

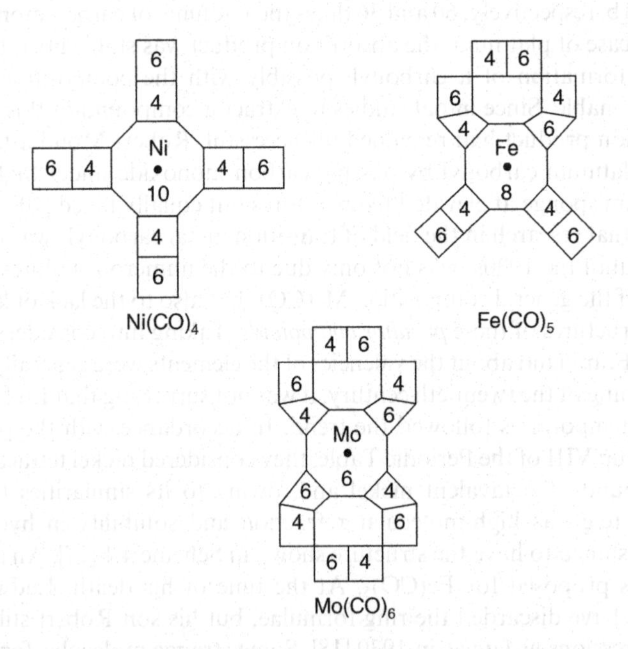

Scheme 4.2 Geometrical diagrams, representing the formulae for $Ni(CO)_4$, $Fe(CO)_5$, and $Mo(CO)_6$, proposed by Blanchard and Gilliland in 1926. The *dot* in the center of the octagons should be the projection of the pair of electrons in the polar axis while the numerals are the nuclear charges

Scheme 4.3 Formulae for $Fe_2(CO)_9$ proposed by Brill, and for $Fe_3(CO)_{12}$ proposed by Hieber and Becker, where the superscripts indicate the coordination number of the iron atoms

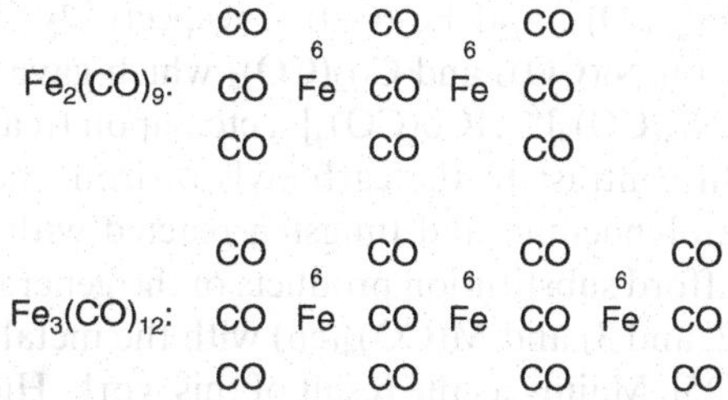

$$Fe_2(CO)_9:\quad \begin{array}{ccccc} CO & & CO & & CO \\ CO & \overset{6}{Fe} & CO & \overset{6}{Fe} & CO \\ CO & & CO & & CO \end{array}$$

$$Fe_3(CO)_{12}:\quad \begin{array}{ccccccc} CO & & CO & & CO & & CO \\ CO & \overset{6}{Fe} & CO & \overset{6}{Fe} & CO & \overset{6}{Fe} & CO \\ CO & & CO & & CO & & CO \end{array}$$

4.2 The Giant Work of Walter Hieber

From a historical perspective, it is remarkable that despite the widespread interest in the structure and bonding of the metal carbonyls and the early statement by Dewar and Jones, that "the peculiarities exhibited by this compound ($Fe(CO)_5$) call for further attention" [5], systematic studies on the reactivity of $Fe(CO)_5$ and other metal carbonyls were not carried out before the late 1920s. It required the genius of Walter Hieber to appreciate the challenge, and he soon became the leader in this field. In a personal account, published a few years before his death, he described in detail how he became acquainted with the chemistry of metal carbonyls: "...it was only in the autumn of 1927 at the Institute of Chemistry of the University of Heidelberg that I took up research experiments with iron pentacarbonyl, which was kindly provided by Dr. A. Mittasch of BASF... On the basis of his own experience with nickel carbonyl he warned me emphatically of the danger inherent in the use of these highly toxic substances, coupling his warning with the comment that in this field one could only expect a great deal of trouble and results of little scientific value!" But Hieber did not feel discouraged and continued in his account: "Apart from certain cyano complexes, transition metal compounds with direct metal-carbon bonds had scarcely been examined systematically. In contrast to many well-known scientists of that period, I believed metal carbonyls to be organometallics, and thereby anticipated the now known special position of the transition metals in forming bonds to organic groups" [25].

Following his previous work on the preparation of coordination compounds of tin(IV), thallium(III), and silver(II) with amines and oximes, Hieber started his work in the field of metal carbonyls by investigating the reactivity of $Fe(CO)_5$ toward pyridine and ethylenediamine [26]. He found that these reactions resulted in the partial elimination of CO and the formation of deep red solutions from which iron carbonyl complexes containing amines as ligands could be isolated. While these products were initially assumed to be molecular compounds, it was later shown that they are ionic in nature containing cations such as $[Fe(py)_6]^{2+}$ or $[Fe(en)_3]^{2+}$ and mono- and oligonuclear carbonylferrates as counter ions [27, 28]. Hieber's conclusion was that basic reagents could initiate a valency disproportionation of the metal which in the case if iron led to the complex anions $[Fe(CO)_4]^{2-}$, $[Fe_2(CO)_8]^{2-}$, and $[Fe_3(CO)_{11}]^{2-}$ that have the same number of valence electrons as the neutral iron carbonyls $Fe(CO)_5$,

$Fe_2(CO)_9$, and $Fe_3(CO)_{12}$, respectively. Similar disproportionations occurred with $Ni(CO)_4$ and $Co_2(CO)_8$ which gave anionic species such as $[Ni_2(CO)_6]^{2-}$, $[Ni_3(CO)_8]^{2-}$, $[Co(CO)_4]^-$, etc., upon treatment with ammonia or other amines. In contrast to the carbonyls of iron, nickel and cobalt, those of chromium, molybdenum and tungsten reacted with pyridine and 1,2-ethylenediamine to afford substitution products of the general composition $M(CO)_{6-n}(py)_n$ ($n = 1$, 2, and 3) and $M(CO)_4(en)$ with the metal remaining in the oxidation state zero [25]. Mainly as the result of this work, Hieber became convinced that the metal carbonyls should be regarded as true coordination compounds, and the coordinated CO should not be considered a radical but a monodentate ligand like NH_3, pyridine, etc. He held this view despite the criticism by several of his contemporaries [3, 19] and was very pleased to see that in most textbooks published after 1940 this view had been accepted.

However, the most spectacular result of Hieber's work in the 1930s was undoubtedly the isolation and characterization of iron and cobalt carbonyl hydrides. Hieber described in his account that he could "clearly remember the day when I, together with my co-workers Leutert and Vetter, at the Heidelberg Institute, was able to freeze out a volatile water-clear liquid from the decomposition of the ethylenediamine containing iron carbonyls, identifying it as $H_2Fe(CO)_4$. To our considerable advantage, we found that we could prepare this absolutely novel compound – the first complex metal hydride having a formally negative oxidation number on a metal atom – via the reaction of iron pentacarbonyl with alkali with subsequent acidification of the solution" [25]. A few years later, through superb experimental work, Hieber and Schulten also succeeded in isolating the corresponding cobalt carbonyl hydride $HCo(CO)_4$, which, similar to the iron counterpart, was stable only at low temperature [29]. Owing to some similarities in the physical properties of $H_2Fe(CO)_4$, $HCo(CO)_4$, and $Ni(CO)_4$, Hieber postulated a "drawing-in of the hydrogen atom into the electron shell of the metal" (see Fig. 4.1) [30] and, in agreement with the effective atomic number rule formulated by Sidgwick in 1927 [31], considered the H_2Fe and HCo units as *pseudo-nickel atoms*. In the context of the preparation of pure $HCo(CO)_4$ by Hieber and Schulten in 1937, the interesting aspect is that 4 years

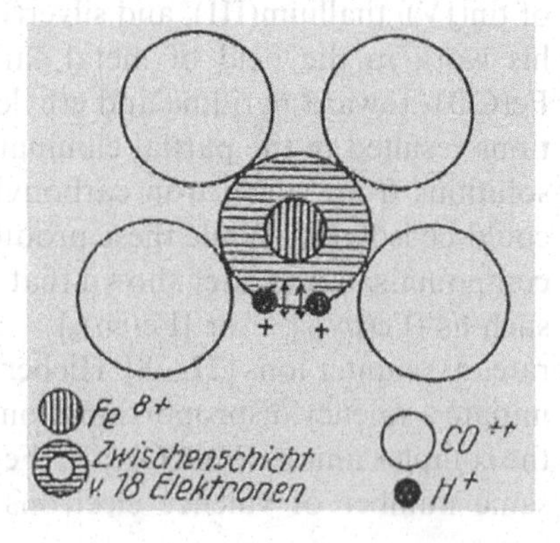

Fig. 4.1 Hieber's schematic representation of the structure and bonding in $H_2Fe(CO)_4$ (from ref. 30; reproduced with permission of Wiley–VCH)

earlier, Maxwell Schubert at the Rockefeller Institute for Medical Research in New York had observed that strongly alkaline aqueous solutions of a cobalt(II) cysteine complex efficiently absorbed carbon monoxide and that upon acidification of the resulting solutions with aqueous HCl "some gas is liberated and an extremely choking odor is produced strongly suggestive of impure acetylene" [32]. The substance with "an evil odor" was assumed to be $HCo(CO)_4$ on the basis of the isolation of the mercury and silver derivatives, $Hg[Co(CO)_4]_2$ and $AgCo(CO)_4$, and Hieber's initial report on the action of oxygen and nitrogen bases on $Co_2(CO)_8$ [33]. With his "bioinorganic approach", Schubert was indeed the first to generate a binary metal carbonyl anion by reductive carbonylation.

Since the IR or NMR spectra of the iron and cobalt carbonyl hydrides could not be measured at the time these compounds were prepared, Hieber's structural proposals were not generally accepted. Several scientists believed that the more electropositive hydrogen atoms should be linked to the most electronegative part of the molecules which were the oxygen atoms of the CO groups. These arguments were apparently supported by the fact, that the iron and cobalt carbonyl hydrides behaved as acids in aqueous or alcoholic solutions, the acidity of $HCo(CO)_4$ being comparable with that of HNO_3 and the acidity of $H_2Fe(CO)_4$ with that of acetic acid [25]. Despite this conceptual criticism, Hieber remained firm and was strongly supported by Hein who stated in his textbook "*Chemische Koordinationslehre*" (which at the time I was a student in Jena was called "the bible") that in the carbonyl hydrides "the hydrogen must be in some way directly involved in the coordination sphere of the respective metal which, inter alia, would explain the low stability as well as the great tendency to give binary carbonyls such as $[Co(CO)_4]_2$ by elimination of elemental hydrogen" [34]. It was only 30 years after the preparation of the first metal carbonyl hydrides $H_2Fe(CO)_4$ and $HCo(CO)_4$, that the structures proposed by Hieber (including $HMn(CO)_5$) were confirmed by electron, neutron and X-ray diffraction analyses [35]. These studies proved that the hydrido ligands do indeed occupy definite coordination sites in the metal carbonyl hydrides, and that in polynuclear compounds such as $[HFe_3(CO)_{11}]^-$ the hydride can be coordinated, similar to CO, not only in a terminal but also in a bridging position.

A more favorable synthesis of salts of the iron, cobalt and nickel carbonyl anions, which were initially prepared by disproportionation reactions of $Fe(CO)_5$, $Co_2(CO)_8$, and $Ni(CO)_4$ with pyridine and other amines, was found by treatment of the neutral carbonyls with alkali in aqueous or alcoholic solutions. Careful studies by Hieber revealed that $Fe(CO)_5$ as well as $Fe_3(CO)_{12}$ reacted with exactly four equivalents of hydroxide ions to give the corresponding dianionic iron carbonylates (Scheme 4.4). These dianions are relatively strong bases and readily accept a proton from a water molecule to give the monoanionic hydrido carbonylates $[HFe(CO)_4]^-$ and $[HFe_3(CO)_{11}]^-$, respectively [36]. The related carbonylates of cobalt and manganese, $[Co(CO)_4]^-$ and $[Mn(CO)_5]^-$, were obtained by a similar way as $[Fe(CO)_4]^{2-}$ [25]. With regard to the mechanism of Hieber's "*Basenreaktion*", the most plausible explanation is based on an initial nucleophilic attack by the hydroxide ion at the carbon atom of a CO

$$Fe(CO)_5 + 4\,OH^- \longrightarrow [Fe(CO)_4]^{2-} + CO_3^{2-} + 2\,H_2O$$

$$Fe_3(CO)_{12} + 4\,OH^- \longrightarrow [Fe_3(CO)_{11}]^{2-} + CO_3^{2-} + 2\,H_2O$$

$$[Fe(CO)_4]^{2-} + H_2O \rightleftharpoons [HFe(CO)_4]^- + OH^-$$

$$[Fe_3(CO)_{11}]^{2-} + H_2O \rightleftharpoons [HFe_3(CO)_{11}]^- + OH^-$$

Scheme 4.4 Generation of iron carbonylates and hydrido carbonylates from neutral iron carbonyls according to Hieber's "*Basenreaktion*"

ligand, leading to a labile M–CO_2H unit, which in the presence of OH^- eliminates CO_2, and gives one equivalent of carbonate [37].

Apart from the metal carbonylates and hydrido metal carbonylates, the "*Basenreaktion*" also provided an access to *organometallic carbonyls* such as $CH_3Co(CO)_4$, $CH_3COCo(CO)_4$, $CH_3Re(CO)_5$, etc. [25]. A remarkable observation was that, in contrast to $CH_3Co(CO)_4$, which is highly air-sensitive and decomposes above –35°C [38], the trifluoromethyl counterpart $CF_3Co(CO)_4$ was significantly more stable and other derivatives such as $C_6F_5Co(CO)_4$ could even be handled in air [39]. Moreover, using the metal carbonylates as starting materials, it was possible to prepare metal carbonyl compounds containing covalent bonds between main group metal and transition metal centers. Typical representatives such as $(CH_3Hg)_2Fe(CO)_4$ and $[(C_6H_5)_3Pb]_2Fe(CO)_4$ were reported by Hein in the 1940s [40, 41], while Hieber obtained the related cobalt–tin compounds $(C_4H_9)_3SnCo(CO)_4$ and $(CH_3)_2Sn[Co(CO)_4]_2$ in 1957 [42]. In this context it is worth mentioning, that as early as in 1928 Hock and Stuhlmann studied the reaction of iron pentacarbonyl with mercuric salts which resulted inter alia in the "*Anlagerungsprodukt*" $Fe(CO)_5(HgCl_2)$, a non-ionic Lewis acid–base adduct of $HgCl_2$ with the metal carbonyl [43]. The same authors also isolated an insoluble oligo- or polymeric compound of the analytical composition $[Fe(CO)_4Hg]_n$, the structure of which is still unknown. More recently, the analogous cadmium compound was characterized crystallographically by Jim Ibers' group at Northwestern University in Evanston and found to be a cyclic tetramer [44].

The chemistry of the metal carbonyl hydrides and metal carbonylates remained the principal research topic for Hieber until the 1960s. He mentioned in his account [25], that it was a particular pleasure for him that in his laboratory the first hydrido carbonyl complexes of the manganese group, $HMn(CO)_5$ and $HRe(CO)_5$, were prepared by careful addition of concentrated phosphoric acid to solid samples of the sodium salts of the $[M(CO)_5]^-$ anions, giving the highly volatile hydrido derivatives in nearly quantitative yield [45, 46]. In contrast to $HCo(CO)_4$ and its rhodium and iridium analogues, the pentacarbonyl hydrido compounds of manganese and rhenium are thermally remarkably stable, and in aqueous or alcoholic solutions practically not acidic at all. Hieber also succeeded in isolating the hydrido vanadium complex $HV(CO)_5(PPh_3)$ [47] and illustrated with this and other examples that the stability of intrinsically labile metal carbonyl hydrides such as $HV(CO)_6$ could be significantly improved if one (or two) CO group(s) are substituted by phosphine or phosphite ligands [25].

4.3 Hieber and his Followers

Since the late 1930s, also new synthetic methods for neutral mono- and dinuclear metal carbonyls were developed. With the high pressure and temperature techniques, introduced by Mond [10] and extended by Hieber [25], it was possible to prepare the iron, cobalt, molybdenum, and tungsten carbonyls from the corresponding metal halides in the presence of Cu, Ag, or Zn as a halogen acceptor in high yield and to obtain for the first time $Re_2(CO)_{10}$ and $Os(CO)_5$, respectively. While the reaction of $ReCl_5$ with CO and Cu gave the mononuclear rhenium(I) compound $Re(CO)_5Cl$ instead of $Re_2(CO)_{10}$, treatment of Re_2O_7 with CO at 250 bar and 250°C in the absence of a solvent afforded the desired dirhenium decacarbonyl [48]. Later, the same type of reductive carbonylation was applied by Fausto Calderazzo (Fig. 4.2) to prepare $Os(CO)_5$ from OsO_4 and CO at 300 bar and 300°C [49]. Under slightly different conditions, the highly stable trinuclear carbonyl $Os_3(CO)_{12}$ could be obtained. Already in the 1940s, Hieber and his school prepared the tetranuclear rhodium and iridium carbonyls $M_4(CO)_{12}$ from the metal trichlorides MCl_3 as starting materials, via the respective carbonyl halides $[Rh(CO)_2Cl]_2$ and $Ir(CO)_3Cl$ as intermediates [25]. Another oligonuclear rhodium carbonyl, initially formulated as $Rh_4(CO)_{11}$ [50], was subsequently shown by Larry Dahl at the University of Wisconsin in Madison to be $Rh_6(CO)_{16}$ [51]. This structural study represented a landmark insofar as it revealed for the first time that carbonyl ligands could adopt a bridging position between three metal centers. Soon after it had been published, the drawing of the molecular structure of $Rh_6(CO)_{16}$ appeared on the cover of a well-known textbook (Fig. 4.3) [52] and since then it is to be found in virtually all monographs on metal carbonyl or organometallic chemistry. With regard to the above mentioned rhodium(I) and iridium(I) carbonyl halides, it is worth mentioning that a derivative of $Ir(CO)_3Cl$ with the composition *trans*-$IrCl(CO)(PPh_3)_2$, named "Vaska's compound", received special attention since the 1960s, because it readily reacts with numerous electrophilic substrates A–B to afford hexa-coordinate iridium(III) complexes $IrCl(A)(B)(CO)(PPh_3)_2$, mostly in excellent yields. Based on Vaska's pioneering research, *trans*-$IrCl(CO)(PPh_3)_2$ subsequently became the prototype of transition metal complexes undergoing "oxidative addition reactions" [53, 54].

Walter Hieber's lifelong endeavor to explore the chemistry of metal carbonyls as completely as possible is best illustrated by the synthesis of $Tc_2(CO)_{10}$ from Tc_2O_7 and CO at elevated pressure and temperature, for which specially

Fig. 4.2 Fausto Calderazzo (born 1930 in Parma) studied chemistry at the University of Florence where he graduated in 1952 under the supervision of Luigi Sacconi. After his military service he joined the research group of Giulio Natta at the Polytechnic Institute in Milano, working at the beginning on the cobalt-catalyzed hydroformylation reaction. In 1960 he received an A. P. Sloan Foundation Fellowship and worked for 14 months with Al Cotton at MIT. Back in Milano for one more year, he accepted an invitation from the Cyanamid European Research Institute near to Geneva as a research associate in 1963 and 2 years later became the head of the synthetic inorganic chemistry group. He stayed in Switzerland until the end of 1968 and then moved to the University of Pisa, where he was appointed to the Chair of General and Inorganic Chemistry. His research interests comprise the preparation and reactivity of metal carbonyls, in particular of the group IV, V, and VI elements, the characterization of the carbonyl(chloro) complexes of palladium, platinum, and gold, mechanistic studies of carbon monoxide insertion reactions, and more recently the chemistry of dialkylcarbamato derivatives of transition and main group metals. A recipient of several major awards, Fausto is an honorary member of the Société Royale de Chimie and the Società Chimica Italiana, and a corresponding member of the Accademia Nazionale dei Lincei (photo by courtesy from F. C.)

designed equipment to avoid radioactive contamination had to be used [55, 56].[1] The corresponding dimanganese decacarbonyl $Mn_2(CO)_{10}$ was first prepared by Eugene Brimm (see Chap. 5) in 1954, who treated a mixture of magnesium powder, manganese iodide, copper, and copper iodide, suspended in ether, with carbon monoxide at 200 bar for 15–17 h at room temperature [57]. The isolated yield was 1%! Closson as well as Podall attempted to improve the synthesis by using sodium benzophenone ketyl or aluminum alkyls as reducing agents but the procedure remained rather tedious. The best method, which is still in use, was developed by Calderazzo and gave pure $Mn_2(CO)_{10}$ in 20 g quantities and 48% yield [58, 59]. Larry Dahl together with Robert Rundle determined the structure of the dinuclear manganese carbonyl and confirmed that, in contrast to $Co_2(CO)_8$ and $Fe_2(CO)_9$, the two $Mn(CO)_5$ halves are not connected via CO bridges, but just by a direct metal–metal bond [60, 61]. The same structural

[1] $Tc_2(CO)_{10}$ was independently prepared by Herb Kaesz and his group.

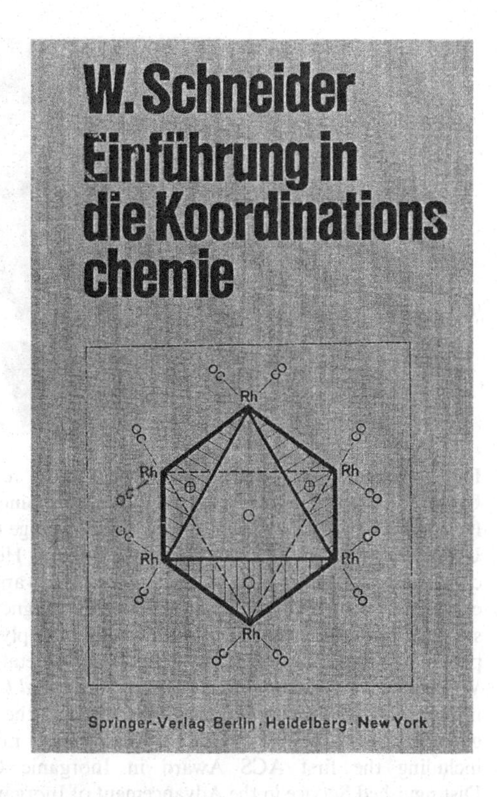

Fig. 4.3 Cover of the textbook "*Einführung in die Koordinationschemie*", published by Professor Walter Schneider from ETH Zurich, which was very popular in Switzerland in the 1970s (from ref. 52; reproduced with permission of Springer)

motif was found for $Tc_2(CO)_{10}$ and $Re_2(CO)_{10}$, the metal–metal bond in those molecules being ca. 0.1 Å longer than in the manganese counterpart [62].

The elucidation of the molecular structure of $Mn_2(CO)_{10}$, published in 1957, undoubtedly marked a breakthrough in the chemistry of metal–metal bonding and initiated the impressive growth of a field for which Albert Cotton (Fig. 4.4) became one of the most active advocates [63]. Based on detailed crystallographic studies, mainly on dinuclear iron and vanadium carbonyl complexes, Cotton also established the concept of "semibridging carbonyl groups" which "may occur singly or in sets, and connect nonequivalent metal atoms" [64]. The most prominent example for this type of bonding is found in $Fe_3(CO)_{12}$, where the bond lengths between the two bridging CO groups and the two corresponding iron centers differ by ca. 0.2 Å at 320 K [65].

Apart from the work on binary metal carbonyls and metal carbonyl hydrides, Hieber and his school also greatly extended the field of carbonyl(nitrosyl) metal complexes. The first compound of the general composition $M(CO)_x(NO)_y$ was obtained by Robert L. Mond and Albert Wallis as early as in 1922 [67]. While studying the reactivity of $Co_2(CO)_8$ toward various substrates, they observed that "slowly at room temperature, but almost instantaneously at 40°C, nitric oxide gas reacts with cobalt tetracarbonyl to form a cherry-red liquid, with the evolution of carbon monoxide". This liquid was

Fig. 4.4 Frank Albert (Al) Cotton (1930–2007) received his Ph.D. in 1955 from Harvard University with Geoffrey Wilkinson and then joined the faculty at MIT, where he became Professor of Inorganic Chemistry in 1961 at an age of 31. In 1972, he moved to Texas A&M University to accept a Welch Foundation Chair. He actively took part in the renaissance of coordination chemistry and was one of the first to apply crystal field theory to transition metal complexes with unusual spin states and magnetic properties. Moreover, he explored systematically the chemistry of metal–metal multiply bonded systems which culminated in the proposal of the bond multiplicity of four for the metal–metal bond in the $[Re_2Cl_8]^{2-}$ anion. With Wilkinson, he coauthored the textbook "*Advanced Inorganic Chemistry*" which was translated into at least 15 foreign languages and became the leading source in the field of inorganic chemistry in the 1960s–1990s. Cotton received numerous honorary degrees and awards, including the first ACS Award in Inorganic Chemistry, the first ACS Award for Distinguished Service in the Advancement of Inorganic Chemistry, the US National Medal of Science, the ACS Priestley Medal, the Robert A. Welch Award and the highly prestigious Wolf Prize for Chemistry. As Stephen Lippard wrote in his obituary, "he was a champion of fundamental research and .... ranks with the very best twentieth-century chemists" [66] (photo by courtesy of Professor John P. Fackler, Jr., Texas A&M University)

analyzed as *cobalt nitrosotricarbonyl* $Co(CO)_3(NO)$. Mond and Wallis also investigated the reaction of $Fe_2(CO)_9$ with NO and isolated a red liquid which on distillation gave $Fe(CO)_5$. "Owing to the difficulties of manipulation, with the very small amounts of liquid available" [67], the analysis indicated a composition "FeNO, $3Fe(CO)_5$", which of course was obscure. Hieber solved this problem and remembered in his account "with great satisfaction my collaboration with J. S. Anderson (Hieber's first postdoc)…who made the volatile, previously unrecognized as such, dinitrosyldicarbonyliron by the action of pure nitric oxide on a solution of $Fe_3(CO)_{12}$ in iron pentacarbonyl" [25]. In retrospect, by taking the pronounced air-sensitivity of the product and the use of $Fe(CO)_5$ as solvent into consideration, the preparation of $Fe(CO)_2(NO)_2$ must be regarded as an experimental masterpiece, carried out in 1932 [68]. The formal analogy between $Ni(CO)_4$, $Co(CO)_3(NO)$, and $Fe(CO)_2(NO)_2$, which was later confirmed by structural analyses [69], led Friedrich Seel (a former student of Hieber, who later held the Chair of Inorganic Chemistry at the University of Saarbrücken in Germany) to propose the "nitrosyl shift relationship" [70] which is still valuable as a heuristic principle. The series of isoelectronic tetrahedral molecules $M(CO)_{4-x}(NO)_x$ ($x = 0$–4) was completed by Jack Lewis at Cambridge and Max Herberhold (one of E. O. Fischer's Ph.D. students and from 1978 until 2002 Professor of Inorganic Chemistry at the University of Bayreuth) who prepared the manganese and chromium compounds $Mn(CO)(NO)_3$ [71] and $Cr(NO)_4$ [72] from the respective starting materials $Mn(CO)_5I$ and $Cr(CO)_6$ under thermal or photochemical conditions (Scheme 4.5). The nitrosyl manganese carbonyl analogue of iron pentacarbonyl, $Mn(CO)_4(NO)$, is also known [73] while other complexes of the general composition $M(CO)_{5-x}(NO)_x$ with $x = 2, 3, 4$, and 5 [e.g., $Cr(CO)_3(NO)_2$, $V(CO)_2(NO)_3$, $Ti(CO)(NO)_4$, and $Sc(NO)_5$] still remain to be discovered. With regard to the chemistry of compounds with both CO and NO as ligands, it is worth mentioning that the first organotransition metal complex containing a chiral metal center was $[(C_5H_5)Mn(CO)(NO)(PPh_3)]^+$, the PF_6^- salt of which was prepared by Henri Brunner (also an E. O. Fischer student, who held the Chair of Inorganic Chemistry at the University of Regensburg from 1970 to 2004) from $[(C_5H_5)Mn(CO)_2(NO)]PF_6$ and PPh_3. The racemic cation $[(C_5H_5)Mn(CO)(NO)(PPh_3)]^+$ was treated with the L-mentholate anion

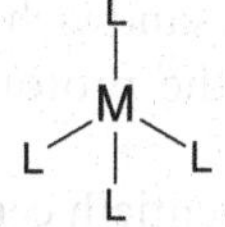

L = CO			L = CO/NO		
$[Cu(CO)_4]^+$	S. H. Strauss	1999			
$Ni(CO)_4$	L. Mond	1890	$Ni(CO)_4$	L. Mond	1890
$[Co(CO)_4]^-$	W. Hieber	1934	$Co(CO)_3(NO)$	R. L. Mond	1922
$[Fe(CO)_4]^{2-}$	W. Hieber	1932	$Fe(CO)_2(NO)_2$	W. Hieber	1932
$[Mn(CO)_4]^{3-}$	J. E. Ellis	1975	$Mn(CO)(NO)_3$	J. Lewis	1960
$[Cr(CO)_4]^{4-}$	J. E. Ellis	1978	$Cr(NO)_4$	M. Herberhold	1972

Scheme 4.5 Two isoelectronic series of tetrahedral metal carbonyl and metal carbonyl nitrosyl compounds, prepared in the course of one century from 1890 to 1990 (with the names of the principal investigators and the date of discovery)

yielding a pair of diastereoisomers $(C_5H_5)Mn[C(O)OC_{10}H_{19}](NO)(PPh_3)$, which could easily be separated by fractional crystallization [74, 75].

Soon after their discovery, the unusual nature of the metal carbonyls and their derivatives prompted some crystallographers to investigate the structure of this novel type of compounds. In 1927, Rudolf Brill was the first to carry out an X-ray structure analysis of $Fe_2(CO)_9$ and, despite the limited instrumentation available at that time, came to the conclusion that three of the CO ligands were in a bridging position and that the coordination spheres of the two iron centers corresponded to an octahedron [22]. With the aid of more sophisticated equipment, Powell and Ewens confirmed this result in 1939 [76], and the most accurate refinement of the structure of $Fe_2(CO)_9$ was reported by Cotton and Troup in 1974 [77]. Owing to the development of high-quality instruments in the 1960s and 1970s, the molecular structures of $Ni(CO)_4$, $Fe(CO)_5$ and the hexacarbonyls of chromium, molybdenum, and tungsten were also determined as were those of several di- and oligonuclear metal carbonyls [62]. The precise determination of the M–C and C–O bond lengths led to the concept of a synergistic interaction between σ donation from CO to the metal and π back donation from the metal to CO. This bonding scheme, which implied a bond order for the metal–carbonyl bond of greater than one, was supported by MO and ab initio calculations and found to be in good agreement with both spectroscopic data and dissociation energies [78]. While the synergistic concept was also applied to isonitrile, olefin and sandwich-type metal compounds, there is still no doubt that CO represents the prototype of the so-called π-acceptor ligands.

In the last three decades of the twentieth century, following Walter Hieber's retirement, four aspects of the research on mono- and polynuclear metal carbonyl complexes found particular attention. These were the preparation of highly reduced carbonyl metallate anions, the generation of stable metal carbonyl cations, the matrix isolation of uncharged metal carbonyls obeying or not the 18-electron rule and, last but not least, the giant metal carbonyl clusters.

4.4 Surprisingly Stable: Multiply Charged Carbonyl Metallate Anions

The main protagonist for the development of the field of multiply charged carbonyl metallate anions has been John Ellis at the University of Minnesota in the US (Fig. 4.5), who began his work in the early 1970s. At that time the knowledge about complexes of the general composition $[M(CO)_n]^{x-}$ was limited to those with $x = 1$ ($n = 4$, M = Co, Rh, Ir; $n = 5$, M = Mn, Tc, Re) and $x = 2$ ($n = 4$, M = Fe, Ru, Os; $n = 5$, M = Cr, Mo, W), of which the majority were first prepared by Hieber [25] and Helmut Behrens [79] and their schools. While comparing the reactivity pattern of the respective carbonyl metallate anions with formally analogous main group anions, Ellis realized that "several derivatives of

Fig. 4.5 John Emmett Ellis (born in 1943), a native of California, received his Ph.D. in 1971 from MIT under the direction of Alan Davison who "opened his eyes to the challenge and wonder of inorganic syntheses" (J. E. E.). In 1971, John joined the faculty at the University of Minnesota, where he is Professor of Chemistry since 1984. His recent honors include a Humboldt Senior Scientist Award, spent in Wolfgang Beck's group in München, and the F. Albert Cotton Award in Synthetic Inorganic Chemistry for his studies on the synthesis and characterization of compounds containing d-block elements in their lowest known formal oxidation states. John's research also includes investigations on the reactions of polyarene radical anions with high-valent transition metal precursors as a route to new classes of homoleptic polyarene metal species, in particular anionic ones. Those species can function as highly reactive sources of "naked" metal anions which previously had only been detected in the gas phase (photo by courtesy of J. E. E.)

carbonylmetallate trianions existed, but the parent compounds were unknown" [80]. He first showed, mainly on the basis of derivative chemistry and infrared spectra, that in the unusually effective reducing medium of Na/HMPA the monoanion $[Mn(CO)_5]^-$ underwent facile reduction to form a yellowish solution containing as the major soluble component the trisanion $[Mn(CO)_4]^{3-}$ [81]. Subsequent studies revealed, that under slightly different conditions the reduction of both $Mn_2(CO)_{10}$ and $Re_2(CO)_{10}$ with Na/HMPA afforded the unsolvated and exceedingly thermally stable complexes $Na_3[M(CO)_4]$ (M = Mn, Re) in nearly quantitative yield [80, 82]. Moreover, in what was described by Wolfgang Beck as "a series of impressive experimentally demanding and systematically planned studies" [83], Ellis and coworkers succeeded in the generation of the trisanions $[M(CO)_5]^{3-}$ and $[M'(CO)_3]^{3-}$ of the group V (M = V, Nb, Ta) and group IX (M' = Co, Rh, Ir) elements as well as in that of the tetraanions $[M(CO)_4]^{4-}$ of chromium, molybdenum, and tungsten [80, 82]. Who would have thought of the existence of those anions with a formal oxidation number of −4 for the group VI elements before 1978? As for the preparation of $H_2Fe(CO)_4$ and $HCo(CO)_4$ by Hieber 40–45 years previously, the isolation of the sodium salts $Na_4[M(CO)_4]$ (M = Cr, Mo, W) was an experimental adventure since these salts were explosive even under argon with little or no provocation.

In the 1990s, the research of the Ellis group was further highlighted by the isolation of the $[K(cryptand)]^+$ salts of the hexacarbonylmetallate dianions

$[M(CO)_6]^{2-}$ of titanium, zirconium and hafnium which were the first stable homoleptic carbonyl compounds of the group IV elements. Their synthesis was achieved, inter alia by reductive carbonylation of the seven-coordinate chelate complexes $[(triphos)M(CO)_4]$ (triphos = $MeC(CH_2PMe_2)_3$) with potassium naphthalenide as reducing agent in the presence of cryptands [83]. The existence of the dianions $[(Ph_3Sn)_2Ti(CO)_5]^{2-}$ containing titanium in the oxidation state –4 and $[(Ph_3Sn)_4M(CO)_4]^{2-}$ (M = Zr, Hf) containing zirconium and hafnium in the oxidation state –6 (!) could also be established [82]. What, according to Ellis, appear to be the next promising candidates for synthesis, are the group X tricarbonyl anions, in particular $[Ni(CO)_3]^{2-}$, the isoelectronic copper compound $[Cu(CO)_3]^-$ and also the platinum and gold counterparts. And following Ellis again, with regard to the preparation of a stable triscyclopentadienyl uranium carbonyl by Ernesto Carmona (see Fig. 5.11) [84], even the possibility of isolating salts of the uranium anions $[U(CO)_5]^{2-}$ and $[U(CO)_6]^{2-}$ "may not be total fantasy" [82].

4.5 Metal Carbonyl Cations: Not Incapable of Existence

In contrast to metal carbonyl anions, of which typical representatives such as $[Fe(CO)_4]^{2-}$ and $[Co(CO)_4]^-$ had been prepared by Hieber as early as in the 1930s [25, 36], related metal carbonyl cations were considered for some decades as incapable of existence. The first clue that this might not be true was found by Irving Wender's group at the Pittsburgh Coal Research Center in 1955, who observed a dramatic change in the infrared spectrum of iron pentacarbonyl upon addition of organic nitrogen bases [85]. They attributed this change to the presence of the iron hexacarbonyl dication formed by disproportionation of two molecules of $Fe(CO)_5$ to give $[Fe(CO)_6]^{2+}$ and $[Fe(CO)_4]^{2-}$, respectively. Although this view was not generally accepted, Sacco and Freni [86] as well as Hieber and Freyer [87] reported in 1958 that the reaction of $Co_2(CO)_8$ with tertiary phosphines gave apart from $[Co(CO)_4]^-$ the stable cations $[Co(CO)_3(PR_3)_2]^+$, being the phosphine-substituted analogues of $[Co(CO)_5]^+$. The first unsubstituted metal carbonyl cation $[Mn(CO)_6]^+$, isoelectronic to $Cr(CO)_6$ and $[V(CO)_6]^-$, with $AlCl_4^-$ as the counterion was prepared by Fischer and Öfele in 1961 from $Mn(CO)_5Cl$ and CO in the presence of $AlCl_3$ as the halide acceptor [88]. In the same year, Hieber and Kruck reported the isolation of the corresponding rhenium cation $[Re(CO)_6]^+$, which was found to be considerably more stable than the manganese counterpart [89]. Later, the technetium cation $[Tc(CO)_6]^+$ was also obtained [25]. While subsequently to this work several substituted metal carbonyl cations such as $[M(CO)_2(PR_3)_2]^+$ (M = Rh, Ir) [25] and $[Fe(CO)_5X]^+$ (X = H, Cl, Br) [90] were described, the generation of binary carbonyl cations such as $[Co(CO)_5]^+$ or $[M(CO)_6]^{2+}$ (M = Fe, Ru, Os) remained a challenging problem which was not solved before the end of the twentieth century.

As it often happens in science and in organometallic chemistry in particular, the breakthrough occurred by accident. In attempting to prepare the elusive HCO^+ cation by protonation of CO with the conjugate superacid HSO_3F/ $Au(SO_3F)_3$, Helge Willner (since 1998 Professor of Inorganic Chemistry at the University of Wuppertal, Germany) in collaboration with Friedhelm Aubke at the University of British Columbia in Vancouver, Canada, generated the solvated gold(I) dicarbonyl cation $[Au(CO)_2]^+$, which was the first binary metal carbonyl complex with the coordination number two of the metal center [91]. In the presence of HSO_3F, this cation gave the neutral compound $Au(CO)SO_3F$, which reacted with CO in pure SbF_5 as solvent to afford the stable crystalline complex $[Au(CO)_2](Sb_2F_{11})$ [92]. This salt decomposed at 156°C ! Parallel to this work, Steven Strauss at Colorado State University in the US prepared the thermally labile, extremely hygroscopic silver(I) complex $[Ag(CO)_2][B(OTeF_5)_4]$ and determined its structure by single-crystal X-ray crystallography at low temperature [93]. Somewhat later, the same group generated mono-, di-, tri- and tetracarbonyls of copper(I) and characterized the tetrahedral $[Cu(CO)_4]^+$ cation, stabilized by the *superweak* anion $[1\text{-Et-}CB_{11}F_{11}]^-$, crystallographically [94].

Initiated by the above mentioned reports, research in the field of cationic metal carbonyls developed rapidly and since the mid 1990s was dominated by Willner and Aubke and their schools [95, 96]. One particular highlight was the preparation of the stable mercury carbonyl compounds $[Hg(CO)_2](Sb_2F_{11})_2$ and $[Hg_2(CO)_2](Sb_2F_{11})_2$, which were described as the first "post-transition metal carbonyls". The key to success for preparing these compounds and several other cationic metal carbonyls was the use of highly acidic reaction media like liquid SbF_5 or the superacid HF/SbF_5, previously not belonging to the experimental tools of organometallic chemists. Up to 2003, cations of the general composition $[M(CO)_n]^{x+}$, with the metal M ranging from group VI (Mo, W) to group XII (Hg), had been generated and most of them characterized by X-ray structural analyses. The principal coordination numbers (and geometries) are $n = 2$ (linear), 4 (square planar), and 6 (octahedral), of which four (examples are $[Pd(CO)_4]^{2+}$ and $[Pt(CO)_4]^{2+}$) deserves special mention insofar as presently no neutral or anionic metal carbonyl is known in which the metal center is surrounded by four CO ligands in a square planar coordination mode. The ionic charge x is not limited to 1 and 2 but can also be 3, as in the unusual compound $[Ir(CO)_6](SbF_6)_3 \cdot 4HF$, where one of the HF molecules is coordinated in a tridentate fashion to three carbon atoms of the particular cation (Fig. 4.6) [97]. The isolation of this iridium(III) complex also corrected the view that binary metal carbonyl cations with a positive charge of +2 or +3 should be incapable of existence [25].

In contrast to the carbonyl metallate anions, the $[M(CO)_n]^{x+}$ cations are in general substitutionally labile and readily undergo ligand exchange in solution. Even the complex $[Au(CO)_2](Sb_2F_{11})_2$, which is stable as a solid, loses the CO ligands in acetonitrile and affords the gold(I) complex $[Au(NCMe)_2](SbF_6)$ [92]. The reason for the lability is presumably the lack (or low degree) of π back-bonding from the positively charged metal atom to the carbonyl ligands, which

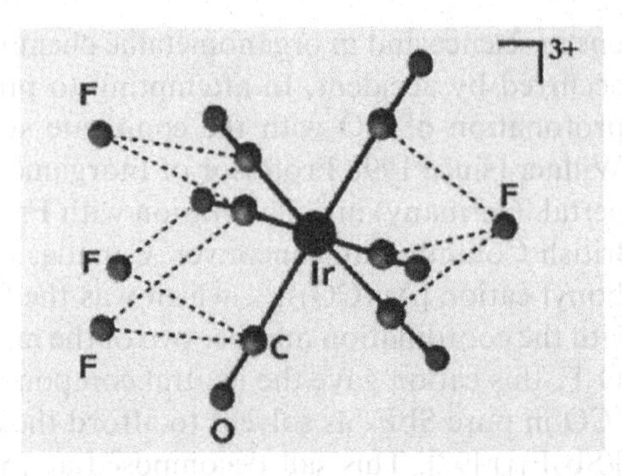

Fig. 4.6 Molecular structure of the solvated cation in the iridium(III) carbonyl complex $[Ir(CO)_6](SbF_6)_3 \cdot 4HF$ (from ref. 97; reproduced with permission of the American Chemical Society)

leads to a significant decrease of the M–CO bond order compared with neutral or anionic metal carbonyls. Based on this fact, Willner and Aubke suggested the name "σ-bonded metal carbonyls" for the cationic complexes [96], while Strauss preferred the term "nonclassical metal carbonyls", in particular for the carbonyl cations of Ag(I) and Cu(I) [98]. Apart from the strength of the M–CO bond, the cationic metal carbonyls also differ from the neutral and anionic compounds insofar as the latter obey the 18-electron rule (with the exception of $V(CO)_6$), whereas some of the former do not. Typical examples are $[M(CO)_4]^{n+}$ ($n = 1$, M = Rh; $n = 2$, M = Pd, Pt) and $[M(CO)_2]^{n+}$ ($n = 1$, M = Ag, Au; $n = 2$, M = Hg), all of which are electron deficient and have, respectively, 16 and 14 electrons in the valence shell. The lability of electron-deficient cations such as $[M(CO)_2]^+$ (M = Ag, Au) and $[Pt_2(CO)_6]^{2+}$ has recently been used also for catalytic processes, for example for the carbonylation of mixtures of olefins and alcohols, which in concentrated sulfuric acid as solvent and at ambient pressure of CO gave tertiary carboxylic acids in excellent yield. The "precious" gold(I) cation $[Au(CO)_2]^+$ was found to be particularly active [99].

4.6 Highly Labile Metal Carbonyls

The metal carbonyl family, however, not only consists of stable mono- and oligonuclear compounds, either uncharged or ionic, but also of a range of highly labile, though mainly well characterized molecules that exist in frozen matrices or for a very short time in solution. The work in this field was initiated by Walter Strohmeier at the University of Würzburg in the late 1950s and further developed by Raymond Sheline at Florida State University and Ernst Koerner von Gustorf at the Max Planck Institute in Mülheim in the 1960s [100–102]. These research groups not only prepared a great number of substituted metal carbonyls $M(CO)_{x-y}(L)_y$ by photolysis of closed-shell $M(CO)_x$ precursors in the presence of ligands L, but also investigated the mechanism of the photochemical substitution reactions. The conclusion was that the initial step of the reaction is the generation of an excited state $[M(CO)_x]^*$ which eliminates a CO ligand to yield a short-lived, coordinatively unsaturated intermediate $M(CO)_{x-1}$ [103–105]. In some cases, these short-lived species $M(CO)_{x-1}$ could be temporarily stabilized in donor solvents such as tetrahydrofuran or diethyl ether forming labile intermediates $M(CO)_{x-1}(THF)$ or $M(CO)_{x-1}(OEt_2)$, which upon addition of appropriate σ- or π-donors L afforded the substitution products $M(CO)_{x-1}(L)$. In addition to mononuclear compounds $M(CO)_x$, also dinuclear and oligonuclear metal carbonyls as well as substituted carbonyl complexes such as $(C_5H_5)M(CO)_4$ (M = V, Nb, Ta), (arene)$M(CO)_3$ (M = Cr, Mo), $(C_5H_5)M(CO)_3X$ (M = Mo, W; X = Cl, CH_3), $M(CO)_5X$ (M = Mn, Re; X = H, Cl, Br, I, CH_3, CF_3), $(C_5H_5)Mn(CO)_3$, $[(C_5H_5)M(CO)_2]_2$ (M = Fe, Ru, Os) and $(C_5H_5)M(CO)_2$ (M = Co, Rh) were used as precursors and their photoreactions were investigated in much detail. Cleavage of the metal–metal bond(s) was observed for di- and oligonuclear metal carbonyls apart from CO substitution, resulting in the formation of mononuclear products or decomposition of the intermediates [103–105].

The matrix isolation studies, intended not only to generate but also to characterize coordinatively unsaturated metal carbonyls, started around 1970 and were pioneered by Jim Turner and his group (Fig. 4.7). In their experiments, they usually trapped a stable metal carbonyl in a large excess of an inert frozen solid matrix such as methane, dinitrogen or a noble gas at low temperature, and then photolyzed the material by generating unstable fragments [106–109]. In most cases, the open-shell species [e.g., $Cr(CO)_5$, $Cr(CO)_4$, $Fe(CO)_4$, $Ni(CO)_3$] thus formed were identified by IR or UV/vis spectroscopy. The important features, which emerged from this work, were (1) the low symmetry of several of these fragments [e.g., both $Cr(CO)_4$ and $Fe(CO)_4$ have C_{2v} symmetry in the matrix], and (2) the enormous reactivity of the frozen complex species. A typical example is $Fe(CO)_4$, which has been thoroughly investigated, in particular by Martyn Poliakoff, for more than two decades [110–113]. It reacted with N_2 reversibly in the matrix yielding $Fe(CO)_4(N_2)$ and, under similar conditions, with methane and xenon producing $Fe(CO)_4(CH_4)$ and $Fe(CO)_4(Xe)$, respectively. Using $Cr(CO)_6$ as the precursor, $Cr(CO)_5(CH_4)$ and $Cr(CO)_5(Xe)$ were formed via $Cr(CO)_5$ as the transient species [114–116].

The remarkable xenon complex $Cr(CO)_5(Xe)$ was also obtained by UV photolysis of $Cr(CO)_6$ in liquefied xenon and found to have a lifetime of ca. 2 s at −98°C [117]. Later, the corresponding molybdenum and tungsten complexes $Mo(CO)_5(Xe)$ and $W(CO)_5(Xe)$ were generated in a similar way and characterized by time-resolved IR spectroscopy [118]. The bond energies Cr–Xe, Mo–Xe, and W–Xe were not as low as previously anticipated and determined to be 8–9 kcal/mol, independent of the group VI metal.

Matrix studies, however, were not only performed to generate transient, open-shell $M(CO)_{x-y}$ fragments from closed-shell $M(CO)_x$ precursors but also to produce new, mainly labile metal carbonyls which in several cases obey the 18-electron (or if they are dinuclear, the 34-electron) rule. The leading work in this field was carried out by Steven Ogden at the Inorganic Chemistry

Fig. 4.7 James (Jim) J. Turner was born 1935 in Darwen, Lancashire, UK, and read chemistry at King's College in Cambridge. He received his Ph.D. in 1960 on High-Resolution NMR Spectroscopy under the supervision of Norman Sheppard. Following election to a Research Fellowship at King's College, he spent a year studying theoretical chemistry with Christopher Longuet-Higgins. He then obtained a Harkness Fellowship and spent 2 years working with George Pimentel in Berkeley, where he acquired his interest in matrix isolation. In 1963, he returned to King's College and became a Demonstrator/Lecturer in Chemistry in the University of Cambridge. In 1972, he was appointed to the Chair of Inorganic Chemistry at the University of Newcastle-upon-Tyne and, 7 years later, moved to the same position at the University of Nottingham. Since 1997, he has continued to work as Emeritus Professor at this institution. Jim's principal research interests have centered on organometallic photochemistry and the structure and reactivity of reaction intermediates and excited states. This has involved conventional and time-resolved IR spectroscopy coupled with matrix isolation and the use of liquid noble gases as solvents. Much of this work has been done in collaboration with Martyn Poliakoff. Jim has been awarded the Tilden and Liversidge Medals of the Royal Society of Chemistry and was elected to the Royal Society in 1992. He was the Pro-Vice-Chancellor at the University of Nottingham from 1986 to 1990 and the President of the Dalton Council from 1994 to 1996 (photo by courtesy from J.J.T.)

Laboratory in Oxford, and Geoffrey Ozin at the University of Toronto in the early 1970s. With the metal atom cocondensation technique (which as described in Chaps. 6 and 7 was also used to prepare a series of zerovalent arene and olefin metal complexes), they reported simultaneously that the elusive palladium and platinum tetracarbonyls, $Pd(CO)_4$ and $Pt(CO)_4$, as well as the coordinatively unsaturated fragments $M(CO)_3$, $M(CO)_2$, and M(CO) (M = Pd, Pt) were formed by cocondensation reactions of Pd and Pt atoms with CO in inert gas matrices at 4–10 K [119–122]. The comparison of the CO bond stretching force constants for $Pd(CO)_n$ and $Pt(CO)_n$ (n = 1–4) revealed that, in analogy to $Ni(CO)_n$, the most stable compounds were the tetracarbonyls. In a xenon matrix, $Pd(CO)_4$ existed up to about 80 K [120]. Ozin's group as well as others also reported the formation of paramagnetic $V(CO)_6$ and diamagnetic $V_2(CO)_{12}$ [123, 124], open-shell $Ti(CO)_6$ [125], $Ir_2(CO)_8$ (the existence of which had been previously claimed but not substantiated) [126], $Au(CO)_n$ ($n = 1$ or 2) [127] and the copper and silver carbonyls $M(CO)_n$ ($n = 1, 2$, or 3) and $M_2(CO)_6$ (M = Cu, Ag) [128–130]. In analogy to $M_2(CO)_{10}$ (M = Mn, Tc, Re) and $M_2(CO)_8$ (M = Co, Rh, Ir), the composition of the dicopper and disilver hexacarbonyls is consistent with the 34-electron rule. To explain why, in contrast to the neighboring iron, cobalt and nickel carbonyls, $Cu_2(CO)_6$ is only known as a *cryomolecule*, John Ellis supposed that "the valence d electrons are at sufficiently low energies that the metal may be incapable of effectively participating in $d\pi$–π^* back-bonding, due to its poor π-donor ability" [82]. In agreement with the IR data of $M_2(CO)_6$ (M = Cu, Ag) (and of $Pd(CO)_4$ and $Pt(CO)_4$ as well), it seems that in these compounds the M–CO bonds are indeed simply too weak to prevent facile dissociation of CO and subsequent formation of bulk metal. With regard to the divanadium complex $V_2(CO)_{12}$, it is worth mentioning that its formation from $V(CO)_6$ had been previously postulated but later definitely refuted on the basis of temperature-dependent magnetic measurements [131].

4.7 The Exiting Chemistry of Metal Carbonyl Clusters

Apart from the plethora of neutral, anionic, and cationic mono- and dinuclear metal carbonyls, the last three decades of the twentieth century also saw the impressive growth of the field of metal carbonyl clusters. Although various smaller representatives, such as trinuclear $Fe_3(CO)_{12}$, $Ru_3(CO)_{12}$, and $Os_3(CO)_{12}$, tetranuclear $Co_4(CO)_{12}$, $Rh_4(CO)_{12}$, and $Ir_4(CO)_{12}$, and hexanuclear $Rh_6(CO)_{16}$, had been known and their bonding and structures well understood [25, 52], before 1970 almost nobody working in the field would have imagined that molecular species, either neutral or anionic, containing 30 metal atoms or more would be stable and could be isolated. It was initially Paolo Chini and somewhat later Jack Lewis and their schools, who opened up the field of high-nuclearity metal carbonyl clusters and thus created a link between molecular aggregates and metallic crystallites.

Paolo Chini began his work in the late 1950s with the characterization of cobalt carbonyl species involved in the hydroformylation of olefins with cobalt catalysts, and in the course of these studies developed improved synthetic methods for the known cobalt carbonyls $Co_2(CO)_8$ and $Co_4(CO)_{12}$ [132]. His next steps were the preparation of the heterometallic hydrido complex $HFeCo_3(CO)_{12}$ (isoelectronic to $Co_4(CO)_{12}$) and the corresponding anion $[FeCo_3(CO)_{12}]^-$, both a novelty at that time, and of the new hexanuclear cobalt clusters $[Co_6(CO)_{15}]^{2-}$, $[Co_6(CO)_{14}]^{4-}$, and $Co_6(CO)_{16}$ [133–139]. This work was followed by the synthesis of carbido carbonyl cluster anions $[Co_6(CO)_{14}C]^-$, $[Co_6(CO)_{15}C]^{2-}$ and $[Co_8(CO)_{18}C]^{2-}$, containing an interstitial

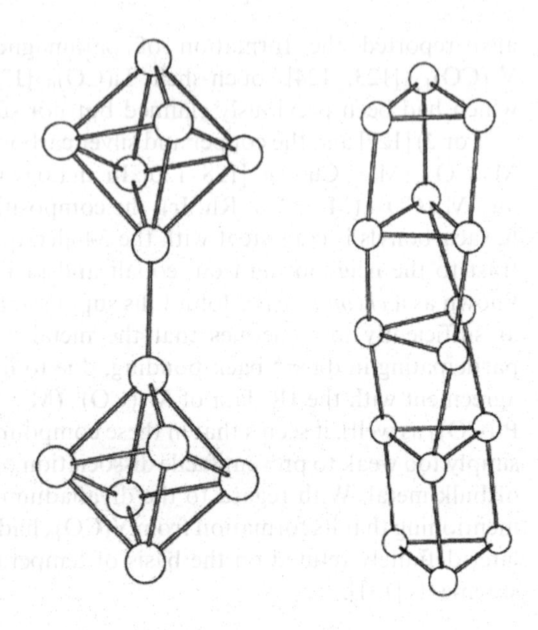

Fig. 4.8 The solid state structures of the metal core of $[Rh_{12}(CO)_{30}]^{2-}$ (*left*) and $[Pt_{15}(CO)_{30}]^{2-}$ (*right*). The stacked $Pt_3(CO)_6$ units of the Pt_{15} assembly possesses a helical twist (from ref. 143; reproduced with permission of Oxford University Press)

carbido ligand generated by scission of one CO group, and of higher nuclearity rhodium clusters such as $[Rh_{12}(CO)_{30}]^{2-}$ (see Fig. 4.8), $[Rh_{13}(CO)_{24}H_{5-n}]^{n-}$ ($n =$ 2 or 3), and $[Rh_{15}(CO)_{28}(C)_2]^-$ [136–140]. In collaboration with Brian Heaton at the University of Kent in Canterbury it was shown by NMR spectroscopy, that the hydrogens in the hydridocarbonyl Rh_{13} anions migrate rapidly around the inside of the hexagonal close-packed unit, and that nine of the 12 edge-bridging CO groups undergo exchange with the 12 terminal carbonyl ligands [141]. Brian Johnson, in those days collaborating with Jack Lewis at the University of Cambridge, confirmed that those fluxional polytopal rearrangements were a wide-spread phenomenon in carbonyl cluster chemistry and also intimately related to the fragmentation of carbonyl clusters [142].

The most spectacular result of Chini's work, however, was the synthesis and characterization of the platinum carbonyl dianions $[Pt_3(CO)_6]_n^{2-}$ (n = 1, 2, 3, 4, 5, 6, and ~10), representing a series of new inorganic oligomers with a formerly unknown chemical architecture [144–146]. These anionic species, usually isolated as the tetraalkyl ammonium or tetraphenyl phosphonium salts, were prepared either by reduction of $Pt(CO)_2Cl_2$ with alkali metals in the presence of CO or, more conveniently, by reduction of $[PtCl_6]^{2-}$ with carbon monoxide and NaOH in methanol at atmospheric pressure and room temperature. As for the procedure for obtaining polynuclear rhodium carbonyls [147], the composition of the products is strongly dependent upon the amount of NaOH used. Scheme 4.6 illustrates the sequence of the individual steps along which the reductive carbonylation of $[PtCl_6]^{2-}$ in basic media occurs [144–146]. Crystallographic studies revealed that the structures of the $[Pt_6(CO)_{12}]^{2-}$, $[Pt_9(CO)_{18}]^{2-}$ and $[Pt_{15}(CO)_{30}]^{2-}$ oligomers result from the stacking of $Pt_3(CO)_3(\mu_2\text{-}CO)_3$ triangular units along

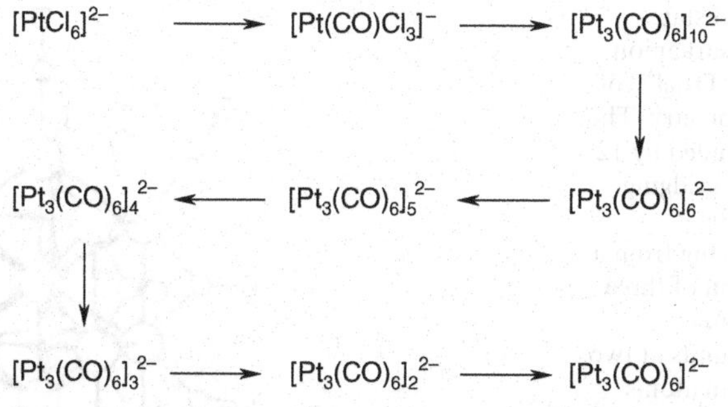

$[PtCl_6]^{2-}$ ⟶ $[Pt(CO)Cl_3]^-$ ⟶ $[Pt_3(CO)_6]_{10}^{2-}$ ⟶ $[Pt_3(CO)_6]_6^{2-}$ ⟶ $[Pt_3(CO)_6]_5^{2-}$ ⟶ $[Pt_3(CO)_6]_4^{2-}$ ⟶ $[Pt_3(CO)_6]_3^{2-}$ ⟶ $[Pt_3(CO)_6]_2^{2-}$ ⟶ $[Pt_3(CO)_6]^{2-}$

Scheme 4.6 The sequence of reaction steps passed along the reductive carbonylation of $[PtCl_6]^{2-}$ with alkali hydroxides in methanol

the pseudo-threefold axis (see Fig. 4.8). To explain this feature, Larry Dahl (whose group did the structural work) assumed that the "observed distortions from a regular prismatic stacking of platinum atoms represent a compromise between steric effects (imposed by nonbonding repulsions mainly between the carbonyl ligands of adjacent layers) and electronic effects which appear to favor a regular trigonal-eclipsed metal geometry" [144]. In contrast to trigonal-prismatic $[Pt_6(CO)_{12}]^{2-}$, the congener nickel analog $[Ni_6(CO)_{12}]^{2-}$ has a trigonal-antiprismatic (or distorted octahedral) metal framework [148]. Chini's group prepared salts of the dianions $[Ni_6(CO)_{12}]^{2-}$ and $[Ni_5(CO)_{12}]^{2-}$ by reduction of $Ni(CO)_4$ with alkali hydroxides in methanol and Dahl confirmed the trigonal-bipyramidal structure of the Ni_5 cluster by X-ray diffraction analysis [149]. A trigonal-bipyramidal arrangement of the metal core also exists in the heterometallic dianions $[M_2Ni_3(CO)_{13}(\mu_2\text{-}CO)_3]^{2-}$ (M = Cr, Mo, W), in which the two $M(CO)_5$ units occupy the apical positions [150].

Among the last achievements of Chini's research, published before his untimely death in 1980, was the synthesis of the rhodium tri- and tetraanions $[Rh_{15}(CO)_{13}(\mu\text{-}CO)_{14}]^{3-}$ and $[Rh_{14}(CO)_9(\mu\text{-}CO)_{16}]^{4-}$ [151], and the platinum tetraanion $[Pt_{19}(CO)_{12}(\mu\text{-}CO)_{10}]^{4-}$ [152]. The structure of the latter corresponds to a bicapped triple-decker all-metal sandwich of idealized fivefold geometry (Fig. 4.9). Paolo's former coworkers Secondo Martinengo and Gianfranco Ciani continued the rhodium work by preparing Rh_{17} and Rh_{22} carbonyl clusters, of which the closed-packed $[Rh_{22}(CO)_{37}]^{4-}$ supported the view that high-nuclearity metal carbonyls are useful models for small metallic crystallites covered by ligands on the surface [153, 154]. Giuliano Longoni, another former coworker of Chini, recently succeeded in isolating salts of the dianion $[Pt_{24}(CO)_{48}]^{2-}$, which were obtained by the reaction of $[Pt_6(CO)_{12}]^{2-}$ with the unusual oxidant $SbCl_3$. The remarkable feature of this cluster is that in the crystal the $[Pt_{24}(CO)_{48}]^{2-}$ molecular ions are arranged in infinite chains composed of alternating $Pt_{21}(CO)_{42}$ moieties and single $Pt_3(CO)_6$ units. The chains resemble morphologically CO-insulated platinum wires with the interesting consequence that pellets of the tetraethyl ammonium salt behave as electric

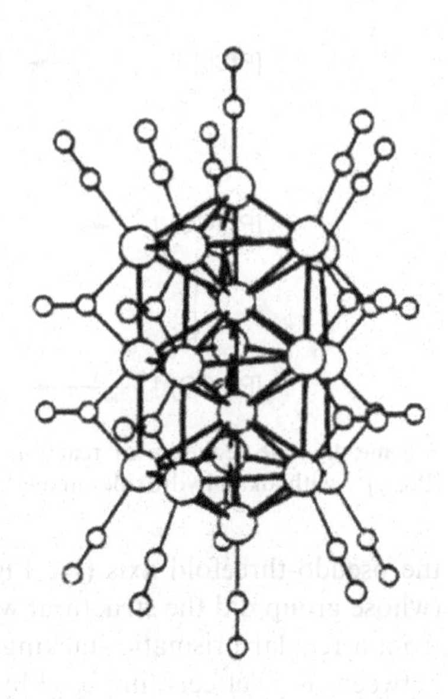

Fig. 4.9 The solid state structure of the tetraanion $[Pt_{19}(CO)_{12}(\mu\text{—}CO)_{10}]^{4-}$ of idealized D_{5h} geometry. The Pt_{19} core, surrounded by 12 terminal and ten bridging CO ligands, can be envisioned as arising from a head-to-tail fusion of three eclipsed pentagonal-bipyramidal Pt_7 units at two common platinum atoms (from ref. 152; reproduced with permission of the American Chemical Society)

conductors [155–158]. In this context it is interesting to note that palladium, unlike its nickel and platinum congeners, does not form stable oligo- or poly-nuclear metal carbonyls. However, extensive studies by Larry Dahl's group and others revealed that a variety of large substituted palladium carbonyl phosphine clusters exist with a truly remarkable diversity of geometries around the metal cores [159]. The largest known crystallographically determined discrete palladium cluster contains a capped three-shell Pd_{145} core, the geometry of which closely conforms to I_h icosahedral symmetry [160].

Simultaneous with Chini's work in the 1970s, the chemistry of high-nuclearity carbonyl clusters of ruthenium and, in particular, osmium was developed in Jack Lewis' group at the University of Cambridge. This field was effectively opened up in 1972 by the isolation of a series of osmium carbonyl compounds from the vacuum pyrolysis of $Os_3(CO)_{12}$, with nuclearities ranging from four to eight [161, 162]. The most prominent species in the product mixtures generated in these reactions was $Os_6(CO)_{18}$, obtained in up to 80 % yield under the appropriate reaction conditions. In the original publication by Eady, Johnson, and Lewis a regular octahedral structure was assigned to this cluster on the basis of the Effective Atomic Number (EAN) rule. However, the X-ray structure analysis of the compound, carried out in Ron Masons group at the University of Sussex in Brighton, revealed a bicapped tetrahedral core geometry [163], a result which was explained in terms of the newly developed Polyhedral Skeletal Electron Pair Theory [164]. Based on this, the required electron count of 86 cluster valence electrons for a closo M_6-structure formally necessitated a two-electron reduction of the neutral hexaosmium cluster, which proved to be a readily achieved synthetic goal. Even weak reducing agents like zinc and iodide were found to generate the octahedral dianion $[Os_6(CO)_{18}]^{2-}$, while mild oxidants converted this species back to the neutral starting material [165]. The elucidation of the structures of $[Os_6(CO)_{18}]^{2-}$ as well as the conjugate acids

$[HOs_6(CO)_{18}]^-$ and $H_2Os_6(CO)_{18}$ marks one of the highlights of the early period of osmium carbonyl cluster chemistry [162]. While both the dianionic cluster and the monoanionic system have the expected octahedral metal core, a capped square based pyramidal structure was found for $H_2Os_6(CO)_{18}$. This turned out to be the archetypal example for the "capping rule", a concept which proved to be very successful in the analysis of the structures of metal carbonyl clusters of the iron and cobalt triads [166].

The second phase in the chemistry associated with the pyrolytic generation of osmium clusters was initiated with the synthesis and structural characterization of $[Os_{10}C(CO)_{24}]^{2-}$ [167]. This carbido cluster dianion was obtained in good yields from the vacuum pyrolysis of $Os_3(CO)_{11}(py)$ at 260°C over a period of nearly 3 days. The same cluster could be identified in the "metallic residues" obtained in the reactions with the unsubstituted carbonyl $Os_3(CO)_{12}$ in a reinvestigation of this process. The structure of $[Os_{10}C(CO)_{24}]^{2-}$ (Fig. 4.10) was found to consist of a tetrahedral metal core of ten cubic close-packed metal atoms with a carbido ligand situated in the central octahedral cavity. Due to its stability and accessibility in high yields, $[Os_{10}C(CO)_{24}]^{2-}$ proved to be the ideal starting material for a rich high-nuclearity cluster chemistry of osmium [168, 169]. It complemented another useful strategy, particularly for hydrido carbonyl clusters, developed by Lewis' group in the early 1980s. A pertinent example was the tetrahydrido decaosmium dianion $[H_4Os_{10}(CO)_{24}]^{2-}$, which was first isolated from the thermolysis of $Os_3(CO)_{12}$ in refluxing isobutanol [170]. Its metal core and CO-ligand shell is isostructural to that of the Os_{10} carbido cluster.

Osmium clusters with larger metal cores than Os_{10} began to be characterized in Lewis' and Johnson's laboratory by the end of the 1980s. The first such example was $[Os_{17}(CO)_{36}]^{2-}$[171] and, finally, $[Os_{20}(CO)_{40}]^{2-}$ which like the Os_{17} dianion was isolated from the pyrolysis reaction of $Os_3(CO)_{10}(NCMe)_2)$ [172]. The Os_{20} cluster contains a highly symmetrical tetrahedral metal core, the symmetry of which is matched by the 40 terminally bonded carbonyl ligands that model the CO chemisorption on the (111)-surface of cubic close-packed metals.

The isolation and characterization of $[Os_{20}(CO)_{40}]^{2-}$ and its homologous relationship to $[Os_{10}C(CO)_{24}]^{2-}$ raised the question of the preferred cluster geometries and nuclearities generated under the thermodynamically controlled

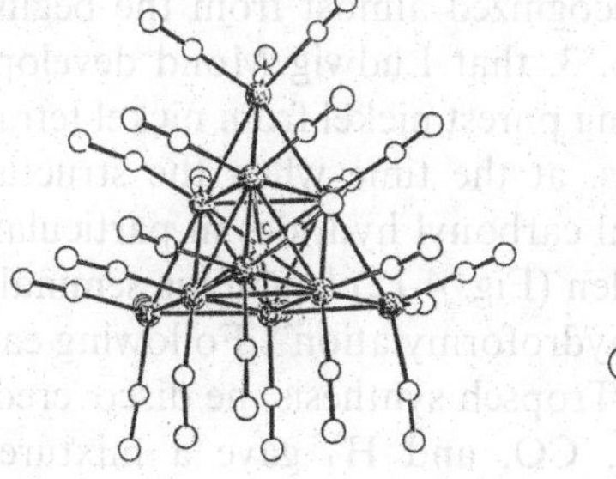
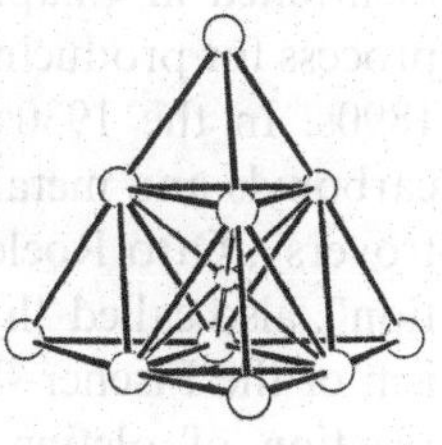

Fig. 4.10 The molecular structure of $[Os_{10}C(CO)_{24}]^{2-}$ (from ref. 167; reproduced with permission of the Royal Society of Chemistry)

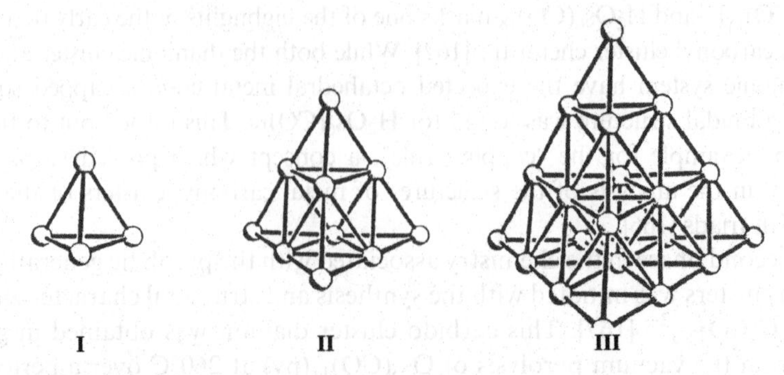

Fig. 4.11 The metal cores of a series of structurally characterized tetrahedral osmium carbonyl clusters (**I**: $[H_2Os_4(CO)_{12}]^{2-}$; **II**: $[Os_{10}C(CO)_{24}]^{2-}$; **III**: $[Os_{20}(CO)_{40}]^{2-}$; from ref. 173; reproduced with permission of the Royal Society of Chemistry)

conditions of most pyrolysis reactions of osmium carbonyl derivatives [173]. In these thermolyses, certain nuclearities are found to dominate the product distributions (thereby allowing highly selective syntheses of some species under the appropriate conditions) while others appear only as minor fractions. For the osmium carbonyl clusters, the generation of metal cores containing 4, 6, 10, and 20 metal atoms seemed to be particularly favored. With the exception of the Os_6 compounds, these nuclearities correspond to the series of the "magic numbers" 4, 10, 20,... associated with molecular tetrahedra of edge lengths two (atoms), three, and four (see Fig. 4.11). They represent a gradual build up of cubic close-packed atoms with overall tetrahedral symmetry, which is not found in any other transition metal cluster system.

4.8 Otto Roelen and Walter Reppe: Industrial Applications of Metal Carbonyls

While Hieber's work as well as that of the next generation, including those who pioneered the research on highly reduced metal carbonyl anions, stable metal carbonyl cations, labile metal carbonyls, and large metal carbonyl clusters, was curiosity-driven basic research, the usefulness of metal carbonyls for industrial applications was recognized almost from the beginning. It has already been mentioned in Chap. 3, that Ludwig Mond developed a commercially viable process for producing purest nickel from nickel tetracarbonyl as early as in the 1890s. In the 1930s, at the time when the structure and bonding of metal carbonyls and metal carbonyl hydrides in particular were still subject to controversy, Otto Roelen (Fig. 4.12) began his seminal studies on the "oxo reaction", also called "hydroformylation". Following earlier work on the mechanism of the Fischer–Tropsch synthesis, he discovered that the cobalt-catalyzed reaction of olefins, CO, and H_2 gave a mixture of linear and branched

Fig. 4.12 Otto Roelen (1897–1993) was born in Mülheim, 17 years before the prestigious *Kaiser-Wilhelm-Institut für Kohlenforschung* (later Max-Planck-Institut) was founded at this place. In summer 1915, he studied chemistry at the *Technische Hochschule* in München for one semester and, after being conscripted into the army and injured during the fighting in Northern France, continued his studies at the *Technische Hochschule* in Berlin and the *Technische Hochschule* in Stuttgart. He obtained his Ph.D. with William Küster at Stuttgart in 1924 and joined the research group of Franz Fischer, the director of the *Kaiser-Wilhelm-Institut für Kohlenforschung* in Mülheim, in that same year. He participated in the discovery of the Fischer–Tropsch (F–T) synthesis, was promoted to group leader (*Abteilungsvorsteher*) in 1927, and directed the technical development of the F–T synthesis since 1929. Five years later, he accepted an offer from *Ruhrchemie AG* in Oberhausen, which had acquired the license for the commercialization of the F–T process from the Mülheim institute. In the course of these studies, he discovered the "oxo reaction". After World War II, the British Ministry of Fuel and Power ordered Roelen's transfer to Wimbledon, where he was interned at the Beltane School for 10 months and repeatedly interviewed by several allied commissions of inquiry. Upon returning to Oberhausen, he was initially engaged in literature work for *Ruhrchemie AG* as well as for the military government and pursued no research of his own. In 1949, he became Assistant Head and on the 1st of October 1955 Head of the Ruhrchemie Research Laboratories. After retiring in 1962, he was awarded the *Adolf von Bayer Denkmünze* of the German Chemical Society and received a honorary doctorate of the *Technische Hochschule* in Aachen for "his accomplishments in fundamental processes of coal and petroleum chemistry" (photo from ref. 174; reproduced with permission of Wiley–VCH)

aldehydes. Ketones were formed as byproducts if ethene was used as the starting material [174]. The first patent application for the oxo reaction with Otto Roelen as the main investigator was filed by *Ruhrchemie AG* on the 20th of September 1938, and the first test plant was built at Oberhausen–Holten in the Western part of Germany in 1941. Until the late 1950s, the industrial capacity for "oxo compounds" was rather small but grew to about six million tons per year worldwide in the following three decades. Already in 1945, Roelen postulated that volatile "cobalt hydrocarbonyls" could be the active species in the cobalt-catalyzed process [174], but it was not before the landmark paper by Heck and Breslow appeared in

1961, that the mechanism of the reaction (with the formation of $HCo(CO)_3$, $HCo(CO)_3$(olefin), alkyl and acyl cobalt carbonyls as intermediates) was understood [175]. Given the commercial scope of the oxo reaction, intensive investigations on the catalytic potential of other transition metal compounds have been carried out with the result, that for various applications rhodium carbonyl derivatives, in particular Wilkinson's $HRh(CO)(PPh_3)_3$, were superior compared with cobalt compounds [176]. Since in several cases the rhodium catalysts ensure higher chemo- and regioselectivity, they have even been used more recently for the preparation of pharmaceuticals and molecules of biological importance [177].

Almost at the same time as Roelen discovered the cobalt-catalyzed oxo reaction, Walter Reppe (Fig. 4.13) began his fundamental research on the

Fig. 4.13 Walter Reppe (1892–1969) studied mathematics, physics, and chemistry at the Universities of Jena and München from 1911 to 1914 and 1918–1920 and received his Ph.D. at the *Ludwig-Maximilans-Universität* in München with Kurt H. Meyer in 1920. The title of his doctorate thesis was "*On the Reduced Aryl Derivatives of Nitric Acid*". He stayed for one year as *Assistent* at the University in München and joined BASF at Ludwigshafen in 1921. After working in some production plants with the main duty to improve the applied process technologies, he became the leader of a research group on fine chemicals in 1934 and, 4 years later, the director of the highly reputated Central Research Laboratory of BASF. It was at that time that the *Reppe Chemie*, involving the use of acetylene at high pressures in the presence of suitable catalysts, was mainly developed. Immediately after the end of World War II, he strongly supported the restoration of the research laboratories, most of which were completely destroyed by bombardment in 1944/45. He was promoted to Head of Research of BASF in 1950, and he became a member of the board of directors of the company in 1952. Reppe's achievements were honored by several awards among which the *Adolf von Bayer Denkmünze* of the German Chemical Society, the *Werner-von-Siemens Ring*, and the *Großes Verdienstkreuz mit Stern des Verdienstordens der Bundesrepublik Deutschland* deserve special attention. Apart from his memberships in several academies in Europe, Japan, and the US, he was an honorary professor at the University of Mainz and the Technical University of Darmstadt. After he retired in 1957, he was allowed to keep his *Privatlaboratorium* at BASF, where he remained active in research until a few weeks before his death (photo reproduced with permission of BASF)

carbonylation of unsaturated hydrocarbons, particularly of alkynes, at BASF [178, 179]. Apart from the preparation of propionic acid (from ethene, CO, and H_2O) and tetrahydrofuran (from acetylene, formaldehyde, and H_2), the *Reppe Chemie* received great industrial importance for the synthesis of acrylic acid from acetylene, carbon monoxide, and water, and of acrylic acid esters from acetylene, CO, and primary alcohols. For both processes, $Ni(CO)_4$ acted as the catalyst. Although acetylene was later replaced by ethene and other olefins as feedstocks, the process technology developed for the carbonylation reaction is still used for the synthesis of fine chemicals and intermediates for the plastic industry. As part of his work with alkynes, Reppe also achieved the cyclotetramerization of acetylene to cyclooctatetraene with modified nickel carbonyls as catalysts, a goal which was on the agenda of organic chemists since Richard Willstätter's pioneering studies on C_8H_8 in the first decade of the twentieth century [180, 181]. Although the cyclooligomerization reactions of acetylene and other alkynes with $Ni(CO)_4$ and $Fe(CO)_5$ [182, 183] have not found industrial applications as yet, they have nevertheless played a decisive role in the chemistry of cyclobutadiene metal carbonyls and quinone and cyclooctatetraene metal complexes as well (see Chaps. 5 and 7).

When Walter Hieber finished his research in the 1960s and was looking back [25], the general view from the time he started his work that metal carbonyls were "far from being firmly categorized as a class of compounds and were considered by many as curiosities" had changed completely [184]. Hieber himself ascertained that "in collaboration with outstanding co-workers. . .it was granted to me to obtain novel results for some decades before any serious competition arose from other researchers! I can even remember the occasional allusions to the strange nature of the work which was directed upon a field where success was hardly to be expected. ... However, starting from modest beginnings it has been shown that the metal carbonyls in no way represent an isolated area of chemistry" [25]. In addition it is fair to say that nowadays the metal carbonyls and their relatives represent one of the big cornerstones of modern organometallic chemistry [185]. Helmut Behrens (a former student of Hieber's, who later held the Chair of Inorganic Chemistry at the University of Erlangen-Nürnberg) pointed out in a personal address on the occasion of Hieber's 80th birthday that not only "at present (1975) more than 1500 papers on carbonyl complexes are published yearly" but also that "the fascinating and exciting development of metal carbonyl chemistry provides an impressive example of the fact that real progress in science often results from discoveries, such as that of nickel carbonyl, which at first are difficult or even impossible to understand. The development of metal carbonyl chemistry has fundamentally influenced the whole of chemical thought. It has opened up the organometallic chemistry of the transition metals; it has played a most significant role in the development of bonding theories; it has presented new problems and new answers in structural chemistry; it has important influence on catalytic and industrial processes and even has a relevance to biochemical research" [186]. Now, more than three decades later, this statement is still true.

Biographies

Biography of Ludwig Mond – see Chap. 3

Walter Hieber is generally regarded by the chemical community as the *father of metal carbonyl chemistry*. This title reflects in the best possible way the role Hieber played in the development of a field that bridges the domains of coordination and organometallic chemistry. Educated at a time when the phrase "organometallic chemistry" was still unknown, Hieber considered himself mainly as a coordination chemist and he was always eager to convince students and his co-workers that the principles of coordination chemistry as implemented by Alfred Werner at the turn of the nineteenth to the twentieth century be equally valid for the chemistry of metal carbonyls. I remember quite well that when I had just started the experimental work for my *Habilitation* in summer 1964 (some months before E. O. Fischer took the Chair of Inorganic Chemistry at the *Technische Hochschule* in München) Hieber came into my lab with Professor Wilhelm Klemm, in those days one of the giants in solid state chemistry. Hieber introduced me to his guest, mentioned that I did my first degree with Franz Hein at Jena, and then said to Klemm: "You know there are two well known coordination chemists in Germany, and the other is Professor Hein."

Walter Hieber, born on 18th December 1895 at Waldhausen near to Stuttgart in Germany, did his undergraduate and graduate studies at the University of Tübingen. There he also received his Ph.D. with Rudolf Weinland in 1919 after completing the doctoral thesis entitled *Iron(III) Complexes Formed from Phosphinic Acid*. In that same year he moved to the University of Würzburg and while working for his *Habilitation* (in the course of which he lost his right hand by an accident in the lab) he became more or less incidentally acquainted with the chemistry of metal carbonyls. In a personal account published in 1970, he mentioned that "my unforgettable teacher R. F. Weinland encouraged me to set up a demonstration of a synthesis of nickel carbonyl by passing carbon monoxide over activated metal..... Since that time I have always been interested in experimental studies on metal carbonyls. In the early years little was known of the chemical behavior of these compounds in spite of the important development of the Mond-Langer process for the processing of Canadian magnetic pyrites ore" [25]. He finished his *Habilitation* in 1924 and became a Lecturer at the University of Jena in 1925. One year later, he accepted an invitation to be Associate Professor of Inorganic Chemistry (*Extraordinarius*) at the University of Heidelberg, where Karl Ziegler held a similar position at that time. At Heidelberg he began his fundamental research on metal carbonyls which he continued for almost four decades.

Following a short stay as the acting Head of the Institute of Inorganic and Analytical Chemistry at the *Technische Hochschule* Stuttgart, Walter Hieber (Fig. 4.14) took the Chair of Inorganic Chemistry at the *Technische Hochschule* in München in 1935 at the age of 39. With the fate of his father

Fig. 4.14 Walter Hieber at the time when he was the Director of the *Anorganisch-chemische Laboratorium* at the *Technische Hochschule* in München (photo by courtesy from Professor Wolfgang Beck)

in mind (who was the President of the State of Württemberg until 1933, before he was dismissed by the Nazi government), he never became a member of the Nazi party and even during World War II resisted the political pressure to do so [187]. After the war, Hieber put all his efforts into the restoration of the laboratories of the chemistry department of the *Technische Hochschule* [187, 188], being strongly supported by his younger colleagues and coworkers Fritz Seel, Reinhard Nast, Helmut Behrens, Ernst Otto Fischer, and Erwin Weiss (all of whom became professors of inorganic chemistry in the 1950s and 1960s). When Hieber received the Alfred Stock Prize of the German Chemical Society in 1951, he was probably very pleased to read in the laudatio that "the chemistry of carbon monoxide, like that of the nitric oxide complexes, has contributed in a special way to the refinement and deepening of our valence concepts and the growth of the theory of coordination chemistry" [189].

Although Walter Hieber came across as a rather shy person, avoiding any publicity, he had a particular sense of humor. When Egon Wiberg (his rival as the Director of the Institute of Inorganic Chemistry at the University in München, see Fig. 2.9) showed him the newly built main lecture hall of the chemistry department at the university and pointed to the huge illuminated periodic table of the elements at the front wall of the hall, Hieber replied: "Your students obviously need that table at the wall, while ours have it in their head" [188]. Despite the fact that he spoke only a few words of English and published all his papers in German, his outstanding contributions to coordination and organometallic chemistry were recognized world-wide. It was a particular pleasure for him to meet the *crème de la crème* of the field at a conference in Ettal in July 1974, which was organized by his former student Wolfgang Beck in his honor. He held the Chair of Inorganic Chemistry at the *Technische*

Hochschule in München until his retirement in 1964 and died on the 29th of November 1976.

Paolo Chini was most influential in the development of the chemistry of large metal carbonyl clusters but could not witness the impressive and continuing growth in this field owing to his sudden premature death on the 2nd of February 1980. To characterize him, Fausto Calderazzo recalled in a biographical memoir that "he was a straightforward and frank person. It was easy to see what he thought and meant, both in science and in his relationships with other people; .... he regarded personal ambition as beneficial as long as it did not become predominant and therefore detrimental to the progress of science" [190].

Born on the 14th of March 1928 in Florence, Chini (Fig. 4.15) was very proud of his birthplace during his whole life. It is thus not surprising that from the beginning of his education he had firmly established a preference for certain canonical features of beauty and harmony in nature and art. He obtained a degree in chemistry (*laurea*) from the University of Florence in 1951, and in the following years started industrially funded research, supported by the company Montecatini, at the Polytechnic Institute in Milan under the supervision of Giulio Natta. In 1953, he spent 6 months at the Max Planck Institute in Mülheim with Karl Ziegler and Günther Wilke, where he prepared various organometallic compounds and used these to polymerize olefins and aldehydes. Back in Milan, he played an important role in the initial stages of the research on the stereoregular polymerization of olefins. Moreover, he also began his collaboration with Raffaele Ercoli on the preparation and characterization of cobalt carbonyls. It was definitely in those days that his interest in the chemistry of metal carbonyls and metal clusters emerged.

Fig. 4.15 Paolo Chini at the time when he became Professor of Inorganic Chemistry at the University of Milan (photo by courtesy from Professor Giuliano Longoni)

Paolo started his academic career in 1965 with a NATO scholarship, working as a postdoctoral research associate with Joseph Chatt at the University of Sussex. There he took up his work on transition metal carbonyl clusters, which he had begun in Milan in the late 1950s. At the end of 1965, he became a Lecturer in Organometallic Chemistry at the University of Milan and worked at the Institute of General and Inorganic Chemistry directed by Lamberto Malatesta. Already at that time, prior to his promotion to full professor in 1973, he began his intensive and fruitful collaboration with two well-known X-ray crystallographers, Vincenzo Albano and Larry Dahl, as well as with the NMR expert Brian Heaton, without whom much of the cluster characterization would have been impossible. Between the late 1960s and 1980, the Chini school spear-headed the staggering growth of cluster chemistry, which soon was acclaimed world-wide. After Paolo's death, his pioneering work was continued in Milan by his former coworkers Giuliano Longoni, Gianfranco Ciani, and Secondo Martinengo, and also by Larry Dahl and others. The scientific community will always remember Paolo as a man of experimental ingenuity and the highest integrity who had a good sense of humor. His modesty is best typified by a remark made at a conference, where he was once introduced as *king of the clusters*. Chini's reply was: "In Italy we do not have kings – we are a republic".

Jack Lewis, now Lord Lewis of Newnham, was born on the 13th of February 1928 at Barrow-in-Furness on the edge of the Lake District, one of the most beautiful parts of the UK. He graduated in 1949 with a bachelor's degree in chemistry from the University of London and then moved to the University of Nottingham to work in the group of the eminent inorganic chemist Clifford Addison. Under his supervision he obtained his Ph.D. in 1952 and 2 years later, at the age of 26, was appointed Lecturer at the University of Sheffield. In 1956, he returned to London and held lectureships and readerships at Imperial and University College. It was mainly at these institutions that he turned his attention to the newly emerging field of physical inorganic chemistry, in particular magnetochemistry. It was the time when magnetochemists still wound their own electromagnets, and thus Jack not only studied a multitude of new types of coordination compounds but equally contributed to the development of the respective analytical equipment. This work, partly in collaboration with Brian Figgis, rapidly established his reputation as one of the leading magnetochemists world-wide and also led to the publication of the monograph *Modern Coordination Chemistry*, jointly edited with Ralph Wilkins, which soon became "the Bible" of physical coordination chemistry in the English speaking world.

In 1961, Lewis (Fig. 4.16) accepted the Chair of Inorganic Chemistry at the University of Manchester and stayed there for 6 years. At that time, following investigations of magnetic interactions in polynuclear complexes, he turned his attention to the area of metal–metal bonding where he soon became a leading contributor right up to his retirement in 1995. At Manchester as well as at University College London, where he was Professor of Chemistry from 1967 to

Fig. 4.16 Jack Lewis (now Lord Lewis of Newnham) at the time when he became the Warden of Robinson College at Cambridge (photo by courtesy from J. L.)

1970, he began to develop the field of polynuclear metal carbonyl clusters, mainly of ruthenium and osmium as the core elements. After moving to Cambridge to the Chair of Inorganic Chemistry in 1970, these activities expanded considerably and made the Lewis group to an internationally acclaimed center in metal clusters. During the next decade, a paper on high-nuclearity, neutral and anionic ruthenium and osmium carbonyl clusters appeared almost every 2 weeks and made this area a borderline discipline between molecular and solid state chemistry. In the late 1980s, Jack's continuing interest in the combination of preparative and physical methods also led him to start another research project, in which he again played a leading role. It has been mainly concerned to organometallic acetylide polymers, which have fascinating liquid crystalline, conducting and non-linear optical properties and for which the Lewis group, in collaboration with Paul Raithby, developed new methods of synthesis. This field is still expanding world-wide.

It is characteristic for Lewis' personality, that apart from his scientific interests he was also active in the development of Cambridge University and as an advisor in governmental policy issues related to scientific research and education of science. He was primarily responsible for establishing Robinson College, the "youngest" of the Cambridge colleges, for which he served as the Warden (Rector) from its foundation in 1975 until 2001. He was knighted in 1982, was President of the Royal Society of Chemistry from 1986 to 1988, and elevated to a life peerage in 1989. He is presently a member of various select committees on science and technology in the House of Lords. Jack received manifold recognition, became an Honorary Fellow of the Royal Society of Chemistry and the Royal Australian Chemical Institute, was awarded a large number of honorary doctoral degrees and numerous international prizes. Quite remarkably, he is also the recipient of the Gold Medal of the Apothecaries Society!

References

1. A. Werner, *Neuere Anschauungen auf dem Gebiete der Anorganischen Chemie, 4. Aufl.* (Vieweg & Sohn, Braunschweig, 1920, p. 192).
2. R. Anschütz, *Chemie der Kohlenstoffverbindungen oder Organische Chemie, 12. Aufl.* (Akademische Verlagsgesellschaft, Leipzig, 1928, Vol. I, p. 319).
3. E. Krause and A. von Grosse, *Die Chemie der metall-organischen Verbindungen* (Gebrüder Bornträger, Berlin, 1937, p. 1).
4. L. Mond, and F. Quincke, Note on a Volatile Compound of Iron with Carbonic Oxide, *J. Chem. Soc.* **1891**, 604–607 (*Ber. dtsch. chem. Ges.* **24**, 2248–2250 (1891)).
5. J. Dewar, and H. O. Jones, The Physical and Chemical Properties of Iron Carbonyl, *Proc. Roy. Soc.* **A76**, 558–577 (1905).
6. J. Dewar, and H. O. Jones, On a New Iron Carbonyl, and on the Action of Light and of Heat on the Iron Carbonyls, *Proc. Roy. Soc.* **A79**, 66–80 (1907).
7. W. Hieber, and E. Becker, Über Eisentetracarbonyl und sein chemisches Verhalten, *Ber. dtsch. chem. Ges.* **63**, 1405–1417 (1930).
8. L. Mond, The History of my Process of Nickel Extraction, *J. Soc. Chem. Ind.* **14**, 945–946 (1895).
9. E. Abel, Ludwig Mond – Father of Metal Carbonyls – and so Much More, *J. Organomet. Chem.* **383**, 11–20 (1990).
10. R. L. Mond, H. Hirtz, and M. D. Cowap, Some New Metallic Carbonyls, *J. Chem. Soc.* **97**, 798–810 (1910).
11. A. Job, and A. Cassal, Le chrome-carbonyle, *Bull. Soc. Chim. Fr.* **41**, 1041–1046 (1927).
12. A. Job, and J. Rouvillois, Préparation d'un tungstène-carbonyle par l'intermédiaire d'un magnésien, *Compt. rend.* **187**, 564–565 (1928).
13. W. Herwig, and H. Zeiss, The Preparation and Reactions of Triphenylchromium(III), *J. Am. Chem. Soc.* **81**, 4798–4801 (1959).
14. W. Manchot, and W. J. Manchot, Darstellung von Rutheniumcarbonylen und –nitrosylen, *Z. anorg. allg. Chem.* **226**, 385–415 (1936).
15. W. E. Trout jr., The Metal Carbonyls. I. History. II. Preparation, *J. Chem. Educ.* **14**, 453–459 (1937).
16. E. Harbeck, and G. Lunge, Über die Einwirkung von Kohlenoxyd auf Platin und Palladium, *Z. anorg. allg. Chem.* **16**, 50–66 (1898).
17. W. E. Trout jr., The Metal Carbonyls. III. Constitution. IV. Properties, *J. Chem. Educ.* **14**, 575–581 (1937).
18. R. L. Mond, Metal Carbonyls, *J. Soc. Chem. Ind.* **49**, 271–278 (1930).
19. H. Reihlen, A. von Friedolsheim, and W. Oswald, Über Stickoxyd- und Kohlenoxydverbindungen des scheinbar einwertigen Eisens und Nickels, *Liebigs Ann. Chem.* **465**, 72–96 (1928).
20. H. Reihlen, A. Gruhl, and G. von Hessling, Über den photochemischen und oxydativen Abbau von Carbonylen, *Liebigs Ann. Chem.* **472**, 268–287 (1929).
21. A. A. Blanchard, and W. L. Gilliland, The Constitution of Nickel Carbonyl and the Nature of Secondary Valence, *J. Am. Chem. Soc.* **48**, 872–882 (1926).
22. C. Brill, Röntgenographische Untersuchung des Eisennonacarbonyls $Fe_2(CO)_9$, *Z. Krist.* **65**, 85–93 (1927).
23. N. V. Sidgwick, and R. W. Bailey, Structures of the Metallic Carbonyl and Nitrosyl Compounds, *Proc. Roy. Soc.* **A144**, 521–537 (1934).
24. M. Schneider, and E. Weiss, Kristall- und Molekülstruktur von Tetrakis(tetrahydrofuran)vanadium(II)-bis(hexacarbonylvanadat(–I), ein Beispiel für eine lineare Carbonylbrücke, *J. Organomet. Chem.* **121**, 365–371 (1976).
25. W. Hieber, Metal Carbonyls, Forty Years of Research, *Adv. Organomet. Chem.* **8**, 1–28 (1970).
26. W. Hieber, and F. Sonnekalb, Reaktionen und Derivate des Eisencarbonyls, *Ber. dtsch. chem. Ges.* **61**, 558–565 (1928).

27. F. Feigl, and P. Krumholz, Über Salze des Eisencarbonylwasserstoffs, *Z. anorg. allg. Chem.* **215**, 242–248 (1933).
28. W. Hieber, J. Sedlmeier, and R. Werner, Neuere Anschauungen über Entstehung und Konstitution Äthylendiamin-haltiger Eisencarbonyle, *Chem. Ber.* **90**, 278–286 (1957).
29. W. Hieber, and H. Schulten, Darstellung und Eigenschaften des freien Kobaltcarbonylwasserstoffs, *Z. anorg. allg. Chem.* **232**, 29–38 (1937).
30. W. Hieber, and F. Leutert, Die Basenreaktion des Eisenpentacarbonyls und die Bildung des Eisencarbonylwasserstoffs, *Z. anorg. allg. Chem.* **204**, 145–164 (1932).
31. N. V. Sidgwick, *The Electronic Theory of Valence* (Oxford University Press, London-New York, 1927).
32. M. P. Schubert, The Action of Carbon Monoxide on Iron and Cobalt Complexes of Cysteine, *J. Am. Chem. Soc.* **55**, 4563–4570 (1933).
33. W. Hieber, F. Mühlbauer, and E. A. Ehmann, Derivate des Kobalt- und Nickelcarbonyls, *Ber. dtsch. chem. Ges.* **65**, 1090–1101 (1932).
34. F. Hein, *Chemische Koordinationslehre* (S. Hirzel Verlag, Leipzig, 1950, p. 339).
35. P. B. Armentrout, and L. S. Sunderlin, Gas-Phase Organometallic Chemistry of Transition Metal Hydrides, in: *Transition Metal Hydrides* (Ed. A. Dedieu, VCH Publishers, New York, 1992, Chap. 1.4).
36. W. Hieber, W. Beck, and G. Braun, Anionische Kohlenoxyd-Komplexe, *Angew. Chem.* **72**, 795–801 (1960).
37. T. Kruck, M. Höfler, and M. Noack, Reaktionsweisen von Rhenium(I)-Kohlenoxidkomplexen und neue Anschauungen über den Mechanismus der Basenreaktion von Metallcarbonylen, *Chem. Ber.* **99**, 1153–1167 (1966).
38. W. Hieber, O. Vohler, and G. Braun, Über Methylkobalttetracarbonyl, *Z. Naturforsch., Part B*, **13**, 192–193 (1958).
39. P. M. Treichel, and F. G. A. Stone, Fluorocarbon Derivatives of Metals, *Adv. Organomet. Chem.* **1**, 143–220 (1964).
40. F. Hein, and H. Pobloth, Umsetzungen von Metallorganoverbindungen mit Eisenpentacarbonyl und Eisencarbonylwasserstoff, *Z. anorg. allg. Chem.* **248**, 84–104 (1941).
41. F. Hein, and E. Heuser, Über Organoblei-Eisentetracarbonyle, *Z. anorg. allg. Chem.* **254**, 138–150 (1947).
42. W. Hieber, and R. Breu, Über Organometall-Kobaltcarbonyle, *Chem. Ber.* **90**, 1270–1274 (1957).
43. H. Hock, and H. Stuhlmann, Über die Einwirkung von Quecksilbersalzen auf Eisenpentacarbonyl, *Ber. dtsch. chem. Ges.* **61**, 2097–2101 (1928).
44. R. D. Ernst, T. J. Marks, and J. A. Ibers, Metal-Metal Bond Cleavage Reactions. The Crystallization and Solid State Structural Characterization of Cadmium Tetracarbonyliron, $CdFe(CO)_4$, *J. Am. Chem. Soc.* **99**, 2090–2098 (1977).
45. W. Hieber, and G. Wagner, Über "Manganpentacarbonylwasserstoff", $HMn(CO)_5$, *Z. Naturforsch., Part B*, **13**, 339–347 (1958).
46. W. Hieber, and G. Braun, "Rheniumcarbonylwasserstoff" und Methylpentacarbonylrhenium, *Z. Naturforsch., Part B*, **14**, 132–133 (1959).
47. W. Hieber, E. Winter, and E. Schubert, Reaktionen des Vanadinhexacarbonyls mit verschiedenartigen Basen und die Säurefunktion von Vanadincarbonylwasserstoff-Verbindungen, *Chem. Ber.* **95**, 3070–3076 (1962).
48. W. Hieber, and H. Fuchs, Über Rheniumpentacarbonyl, *Z. anorg. allg. Chem.* **248**, 256–268 (1941).
49. F. Calderazzo, and F. L'Eplattenier, The Pentacarbonyls of Ruthenium and Osmium. I. Infrared Spectra and Reactivity, *Inorg. Chem.* **6**, 1220–1224 (1967).
50. H. Lagally, Das Rhodium im System der Metallcarbonyle, *Z. anorg. allg. Chem.* **251**, 96–113 (1943).
51. E. R. Corey, L. F. Dahl, and W. Beck, $Rh_6(CO)_{16}$ and its Identity with Previously Reported $Rh_4(CO)_{11}$, *J. Am. Chem. Soc.* **85**, 1202–1203 (1963).

52. W. Schneider, *Einführung in die Koordinationschemie* (Springer Verlag, Berlin-Heidelberg-New York, 1968).
53. L. Vaska, Reversible Activation of Covalent Molecules by Transition Metal Complexes. The Role of the Covalent Molecule, *Acc. Chem. Res.* **1**, 335–344 (1968).
54. J. P. Collman, L. S. Hegedus, J. R. Norton, and R. G. Finke, *Principles and Applications of Organotransition Metal Chemistry* (University Science Books, Mill Valley, California, 1987, Chap. 5).
55. W. Hieber, and C. Herget, Technetiumcarbonyl, *Angew. Chem.* **73**, 579–580 (1961).
56. J. C. Hileman, D. K. Huggins, and H. D. Kaesz, Technetium Carbonyl, *J. Am. Chem. Soc.* **83**, 2953–2954 (1961).
57. E. O. Brimm, M. A. Lynch, and W. J. Sesny, Preparation and Properties of Manganese Carbonyl, *J. Am. Chem. Soc.* **76**, 3831–3835 (1954).
58. F. Calderazzo, The Manganese-Catalyzed Carbonylation of Amines, *Inorg. Chem.* **4**, 293–296 (1965).
59. F. Calderazzo, R. Ercoli, and G. Natta, Metal Carbonyls: Preparation, Structure, and Properties, in: *Organic Synthesis via Metal Carbonyls* (Eds. I. Wender and P. Pino, Wiley Interscience, New York, 1968, Vol. I, p. 27).
60. L. F. Dahl, E. Ishishi, and R. E. Rundle, Structures of $Mn_2(CO)_{10}$ and $Re_2(CO)_{10}$, *J. Chem. Phys.* **26**, 1750–1751 (1957).
61. L. F. Dahl, and R. E. Rundle, The Crystal Structure of Dimanganese Decacarbonyl, $Mn_2(CO)_{10}$, *Acta Cryst.* **16**, 419–426 (1963).
62. C. M. Lukehart, *Fundamental Transition Metal Organometallic Chemistry* (Brooks/Cole Publishing Company, Monterey, 1985, p. 25).
63. F. A. Cotton, and M. H. Chisholm, Bonds between Metal Atoms. A New Mode of Transition Metal Chemistry, *Chem. Eng. News* **60**, 40–54 (1982).
64. F. A. Cotton, Metal Carbonyls: Some New Observations in an Old Field, *Progr. Inorg. Chem.* **21**, 1–28 (1976).
65. D. Braga, F. Grepioni, L. J. Farrugia, and B. F. G. Johnson, Effect of Temperature on the Solid-state Molecular Structure of $[Fe_3(CO)_{12}]$, *J. Chem. Soc., Dalton Trans.* **1994**, 2911–2918.
66. S. J. Lippard, Frank Albert Cotton (1930–2007), *Nature* **446**, 626 (2007).
67. R. L. Mond, and A. E. Wallis, The Action of Nitric Oxide on the Metallic Carbonyls, *J. Chem. Soc.* **121**, 32–35 (1922).
68. J. S. Anderson, Über ein flüchtiges Eisen-Nitrosocarbonyl $Fe(CO)_2(NO)_2$, *Z. anorg. allg. Chem.* **208**, 238–248 (1932).
69. G. B. Richter-Addo, and P. Legzdins, *Metal Nitrosyls* (Oxford University Press, New - York-Oxford, 1992, p. 81).
70. F. Seel, Struktur- und Valenztheorie anorganischer Stickoxydkomplexe, *Z. anorg. allg. Chem.* **249**, 308–324 (1942).
71. C. G. Barraclough, and J. Lewis, Trinitrosylcarbonylmanganese, *J. Chem. Soc.* **1960**, 4842–4846.
72. M. Herberhold, and A. Razavi, Tetranitrosylchromium $[Cr(NO)_4]$, *Angew. Chem. Int. Ed. Engl.* **11**, 1092–1094 (1972).
73. P. M. Treichel, E. Pitcher, R. B. King, and F. G. A. Stone, Tetracarbonylnitrosylmanganese, *J. Am. Chem. Soc.* **83**, 2593–2594 (1961).
74. H. Brunner, Optical Activity at an Asymmetric Manganese Atom, *Angew. Chem. Int. Ed. Engl.* **8**, 382–383 (1969).
75. H. Brunner, Optical Activity from Asymmetric Transition Metal Atoms, *Angew. Chem. Int. Ed. Engl.* **10**, 249–260 (1971).
76. H. M. Powell, and R. V. G. Ewens, The Crystal Structure of Iron Enneacarbonyl, *J. Chem. Soc.* **1939**, 286–292.

77. F. A. Cotton, and J. M. Troup, Accurate Determination of a Classic Structure in the Metal Carbonyl Field: Nonacarbonyldi-iron, *J. Chem. Soc., Dalton Trans.* **1974**, 800–802.
78. E. R. Davidson, K. L. Kunze, F. B. C. Machado, and S. J. Chakravorty, The Transition Metal–Carbon Bond, *Acc. Chem. Res.* **26**, 628–635 (1993).
79. H. Behrens, Four Decades of Metal Carbonyl Chemistry in Liquid Ammonia: Aspects and Prospects, *Adv. Organomet. Chem.* **18**, 1–53 (1980).
80. J. E. Ellis, Highly Reduced Metal Carbonyl Anions: Synthesis, Characterization, and Chemical Properties, *Adv. Organomet. Chem.* **31**, 1–51 (1990).
81. J. E. Ellis, and R. A. Faltynek, The Tetracarbonyl Trianions of Manganese and Rhenium, $[M(CO)_4]^{3-}$, *J. Chem. Soc., Chem. Comm.* **1975**, 966–967.
82. J. E. Ellis, Metal Carbonyl Anions: From $[Fe(CO)_4]^{2-}$ to $[Hf(CO)_6]^{2-}$ and Beyond, *Organometallics* **22**, 3322–3338 (2003).
83. W. Beck, Highly-Reduced Metal Carbonyls, *Angew. Chem. Int. Ed. Engl.* **30**, 168–169 (1991).
84. J. Parry, E. Carmona, S. Coles, and M. Hursthouse, Synthesis and Single Crystal X-ray Diffraction Study on the First Isolable Carbonyl Complex of an Actinide, $(C_5Me_4H)_3U(CO)$, *J. Am. Chem. Soc.* **117**, 2649–2650 (1995).
85. H. W. Sternberg, R. A. Friedel, S. L. Shufler, and I. Wender, The Dissociation of Iron Pentacarbonyl in Certain Amines, *J. Am. Chem. Soc.* **77**, 2675–2677 (1955).
86. A. Sacco, and M. Freni, Salts of Bis(triphenylphosphine)tricarbonylcobalt(I), *Ann. Chim. (Rome)* **48**, 218–224 (1958).
87. W. Hieber, and W. Freyer, Über triphenylphosphinhaltige Kobaltcarbonyle, *Chem. Ber.* **91**, 1230–1234 (1958).
88. E. O. Fischer, and K. Öfele, Mangan(I)-pentacarbonyl-äthylen-Kation, *Angew. Chem.* **73**, 581 (1961).
89. W. Hieber, and T. Kruck, Über kationische Kohlenoxyd-Komplexe, I. Hexacarbonylrhenium(I)-Salze, *Z. Naturforsch., Part B*, **16**, 709–713 (1961).
90. Z. Iqbal, and T. C. Waddington, Liquid Hydrogen Chloride as an Ionizing Solvent. Protonation and Oxidation Reactions of Pentacarbonyliron, *J. Chem. Soc. A*, **1968**, 2958–2961.
91. H. Willner, and F. Aubke, Reaction of Carbon Monoxide in the Superacid HSO_3F-Au$(SO_3F)_3$ and the Gold(I) Bis(carbonyl) Cation $[Au(CO)_2]^+$. Isolation and Characterization of Gold(I) Carbonyl Fluorosulfate, $Au(CO)SO_3F$, *Inorg. Chem.* **29**, 2195–2200 (1990).
92. H. Willner, J. Schaebs, G. Hwang, F. Mistry, R. Jones, J. Trotter, and F. Aubke, Bis(carbonyl)gold(I) Undecafluorodiantimonate(V), $[Au(CO)_2][Sb_2F_{11}]$: Synthesis, Vibrational and ^{13}C NMR Study, and the Molecular Structure of Bis(acetonitrile)gold(I) Hexafluoroantimonate(V), $[Au(NCCH_3)_2][SbF_6]$, *J. Am. Chem. Soc.* **114**, 8972–8980 (1992).
93. P. K. Hurlburt, J. J. Rack, S. F. Dec, O. P. Anderson, and S. H. Strauss, $[Ag(CO)_2][B(OTeF_5)_4]$: The First Structurally Characterized $M(CO)_2$ Complex, *Inorg. Chem.* **32**, 373–374 (1993).
94. S. M. Ivanova, S. V. Ivanov, S. M. Miller, O. P. Anderson, K. A. Solntsev, and S. H. Strauss, Mono-, Di-, Tri-, and Tetracarbonyls of Copper(I), Including the Structure of $Cu(CO)_2(1\text{-}Bn\text{-}CB_{11}F_{11})$ and $[Cu(CO)_4][1\text{-}Et\text{-}CB_{11}F_{11}]$, *Inorg. Chem.* **38**, 3756–3757 (1999).
95. H. Willner, and F. Aubke, Homoleptic Metal Carbonyl Cations of the Electron-rich Metals: Their Generation in Superacidic Media Together with Their Spectroscopic and Structural Characterization, *Angew. Chem. Int. Ed. Engl.* **36**, 2402–2425 (1997).
96. H. Willner, and F. Aubke, σ-Bonded Metal Carbonyl Cations and Their Derivatives: Syntheses and Structural, Spectroscopic, and Bonding Principles, *Organometallics* **22**, 3612–3633 (2003).

97. B. von Ahsen, M. Berkei, G. Henkel, H. Willner, and F. Aubke, The Synthesis, Vibrational Spectra, and Molecular Structure of $[Ir(CO)_6][SbF_6]_3 \cdot 4HF$ – The First Structurally Characterized Salt with a Tripositive, Homoleptic Metal Carbonyl Cation and the First Example of a Tetrahedral Hydrogen-Bonded $(HF)_4$ Cluster, *J. Am. Chem. Soc.* **124**, 8371–8379 (2002).
98. S. H. Strauss, Copper(I) and Silver(I) Carbonyls. To Be or not to Be Classical, *J. Chem. Soc., Dalton Trans.* **2000**, 1–6.
99. Q. Xu, Metal Carbonyl Cations: Generation, Characterization and Catalytic Application, *J. Organomet. Chem.* **231**, 83–108 (2002).
100. W. Strohmeier, and K. Gerlach, Die Photochemische Darstellung von Pentacarbonyl-pyridin-chrom(0) und Pentacarbonyl-anilin-chrom(0), *Z. Naturforsch., Part B,* **15**, 413–414 (1960).
101. I. W. Stolz, G. R. Dobson, and R. K. Sheline, The Infrared Spectrum and Evidence for the Structure of a New Metal Carbonyl, *J. Am. Chem. Soc.* **84**, 3589–3590 (1962).
102. G. O. Schenck, E. Koerner von Gustorf, and M.-J. Jun, Photochemische Umsetzungen von Eisenpentacarbonyl mit Philodienen, *Tetrahedron Lett.* **1962**, 1059–1064.
103. W. Strohmeier, Photochemical Substitution of Metal Carbonyls and Their Derivatives, *Angew. Chem. Int. Ed. Engl.* **3**, 730–738 (1964).
104. E. A. Koerner von Gustorf, and F.-W. Grevels, Photochemistry of Metal Carbonyls, Metallocenes, and Olefin Complexes, *Fortschr. Chem. Forsch.* **13**, 366–450 (1969).
105. M. S. Wrighton, The Photochemistry of Metal Carbonyls, *Chem. Rev.* **74**, 401–430 (1974).
106. J. K. Burdett, Matrix Isolation Studies on Transition Metal Carbonyls and Related Species, *Coord. Chem. Rev.* **27**, 1–58 (1978).
107. R. B. Hitam, K. A. Mahmoud, and A. J. Rest, Matrix Isolation Studies of Organometallic Intermediates, *Coord. Chem. Rev.* **55**, 1–29 (1984).
108. M. Poliakoff, and E. Weitz, Detection of Transient Organometallic Species by Fast Time-Resolved IR Spectroscopy, *Adv. Organomet. Chem.* **25**, 277–316 (1986).
109. R. N. Perutz, Organometallic Intermediates: Ultimate Reagents, *Chem. Soc. Rev.* **22**, 361–369 (1993).
110. M. Poliakoff, and J. J. Turner, Structure and Reactions of Matrix-isolated Tetracarbonyliron(0), *J. Chem. Soc., Dalton Trans.* **1974**, 2276–2285.
111. M. Poliakoff, $Fe(CO)_4$, *Chem. Soc. Rev.* **7**, 527–540 (1978).
112. M. Poliakoff, and E. Weitz, Shedding Light on Organometallic Reactions: The Characterization of $Fe(CO)_4$, a Prototypical Reaction Intermediate, *Acc. Chem. Res.* **20**, 408–414 (1987).
113. M. Poliakoff, and J. J. Turner, The Structure of $[Fe(CO)_4]$ – An Important New Chapter in a Long-Running Story, *Angew. Chem. Int. Ed.* **2001**, 2809–2812 (2001).
114. M. A. Graham, M. Poliakoff, and J. J. Turner, The Pentacarbonyls of Chromium, Molybdenum, and Tungsten, *J. Chem. Soc. A* **1971**, 2939–2948 (Part I).
115. R. N. Perutz, and J. J. Turner, Infrared Spectra and Structures of ^{13}CO-Enriched Hexacarbonyls and Pentacarbonyls of Chromium, Molybdenum, and Tungsten, *Inorg. Chem.* **14**, 262–270 (1975) (Part II).
116. R. N. Perutz, and J. J. Turner, Interaction of the Pentacarbonyls with Noble Gases and Other Matrices, *J. Am. Chem. Soc.* **97**, 4791–4800 (1975) (Part III).
117. M. B. Simpson, M. Poliakoff, J. J. Turner, W. B. Maier II, and J. G. McLaughlin, $[Cr(CO)_5Xe]$ in Solution; the First Spectroscopic Evidence, *J. Chem. Soc., Chem. Comm.* **1983**, 1355–1357.
118. J. R. Wells, and E. Weitz, Rare Gas-Metal Carbonyl Complexes: Bonding of Rare Gas Atoms to the Group VI Pentacarbonyls, *J. Am. Chem. Soc.* **114**, 2783–2787 (1992).
119. J. H. Darling, and J. S. Ogden, Infrared Spectroscopic Evidence for Palladium Tetracarbonyl, *Inorg. Chem.* **11**, 666–667 (1972).

120. J. H. Darling, and J. S. Ogden, Spectroscopic Studies on Matrix-isolated Metal Carbonyls. Infrared Spectra and Structures of $Pd(CO)_4$, $Pd(CO)_3$, $Pd(CO)_2$, and PdCO, *J. Chem. Soc., Dalton Trans.* **1973**, 1079–1085.
121. E. P. Kündig, M. Moskovits, and G. A. Ozin, Intermediate Binary Carbonyls of Palladium $Pd(CO)_n$ where n = 1–3; Preparation, Identification, and Diffusion Kinetics by Matrix Isolation Infrared Spectroscopy, *Canad. J. Chem.* **50**, 3587–3593 (1972).
122. E. P. Kündig, D. McIntosh, M. Moskovits, and G. A. Ozin, Binary Carbonyls of Platinum, $Pt(CO)_n$ (Where n = 1–4). A Comparative Study of the Chemical and Physical Properties of $M(CO)_n$ (Where M = Ni, Pd, or Pt; n = 1–4), *J. Am. Chem. Soc.* **95**, 7234–7241 (1973).
123. T. C. DeVore, and H. F. Franzen, Synthesis of Dodecacarbonyldivanadium in Low-Temperature Matrices, *Inorg. Chem.* **15**, 1318–1321 (1976).
124. T. A. Ford, H. Huber, W. Klotzbücher, M. Moskovits, and G. A. Ozin, Direct Synthesis with Vanadium Atoms. Synthesis of Hexacarbonylvanadium and Dodecacarbonyldivanadium, *Inorg. Chem.* **15**, 1666–1669 (1976).
125. R. Busby, W. Klotzbücher, and G. A. Ozin, Titanium Hexacarbonyl, $Ti(CO)_6$, and Titanium Hexadinitrogen, $Ti(N_2)_6$. Synthesis Using Titanium Atoms and Characterization by Matrix Infrared and Ultraviolet-Visible Spectroscopy, *Inorg. Chem.* **16**, 822–828 (1977).
126. A. J. L. Hanlan, and G. A. Ozin, Iridium Atom Chemistry: A Reappraisal of the Matrix Synthesis of Diiridium Octacarbonyl, $Ir_2(CO)_8$, *J. Organomet. Chem.* **179**, 57–64 (1979).
127. D. McIntosh, and G. A. Ozin, Synthesis of Binary Gold Carbonyls, $Au(CO)_n$ (n = 1 or 2). Spectroscopic Evidence for Isocarbonyl(carbonyl)gold, a Linkage Isomer of Bis (carbonyl)gold, *Inorg. Chem.* **16**, 51–59 (1977).
128. J. S. Ogden, Infrared Spectroscopic Evidence for Copper and Silver Carbonyls, *Chem. Comm.* **1971**, 978–979.
129. H. Huber, E. P. Kündig, M. Moskovits, and G. A. Ozin, Binary Copper Carbonyls. Synthesis and Characterization of $Cu(CO)_3$, $Cu(CO)_2$, CuCO, and $Cu_2(CO)_6$, *J. Am. Chem. Soc.* **97**, 2097–2106 (1975).
130. D. McIntosh, and G. A. Ozin, Synthesis Using Metal Vapors. Silver Carbonyls. Matrix Infrared, Ultraviolet-Visible, and Electron Spin Resonance Spactra, Structures, and Bonding of $Ag(CO)_3$, $Ag(CO)_2$, AgCO, and $Ag_2(CO)_6$, *J. Am. Chem. Soc.* **98**, 3167–3175 (1976).
131. J. C. Bernier, and O. Kahn, Magnetic Behaviour of Vanadium Hexacarbonyl, *Chem. Phys. Lett.* **19**, 414–417 (1973).
132. R. Ercoli, P. Chini, and M. Massi-Mauri, Sintesi del tetracobalto dodecacarbonile, per riduzione del cobalto cationico con idrogeno e dicobalto ottacarbonile, *Chim. Ind. (Milan)* **41**, 132–135 (1959).
133. P. Chini, L. Colli, and M. Peraldo, Preparazione a proprietà dell'idrocarbonile $HFeCo_3(CO)_{12}$ e di alcuni composti derivati dall'anione $[FeCo_3(CO)_{12}]^-$, *Gazz. Chim. Ital.* **90**, 1005–1020 (1960).
134. P. Chini, A New Cluster Carbonylcobaltate, *Chem. Comm.* **1967**, 29.
135. P. Chini, Hexacobalt Hexadecacarbonyl and its Derivatives, *Chem. Comm.* **1967**, 440–441.
136. P. Chini, The Closed Metal Carbonyl Clusters, *Inorg. Chim. Acta Rev.* **2**, 31–51 (1968).
137. P. Chini, G. Longoni, and V. G. Albano, High Nuclearity Metal Carbonyl Clusters, *Adv. Organomet. Chem.* **14**, 285–344 (1976).
138. P. Chini, Synthesis of Large Anionic Carbonyl Clusters as Models for Small Metallic Crystallites, *Gazz. Chim. Ital.* **109**, 225–240 (1979).
139. P. Chini, Large Metal Carbonyl Clusters (LMCC), *J. Organomet. Chem.* **200**, 37–61 (1980).
140. M. D. Vargas, and J. N. Nicholls, High-Nuclearity Carbonyl Clusters: Their Synthesis and Reactivity, *Adv. Inorg. Chem. Radiochem.* **30**, 123–222 (1986).

141. S. Martinengo, B. T. Heaton, R. J. Goodfellow, and P. Chini, Hydrogen and Carbonyl Scrambling in $[Rh_{13}(CO)_{24}H_{5-n}]^{n-}$ (n = 2 and 3): A Unique Example of Hydrogen Tunneling, *J. Chem. Soc., Chem. Comm.* **1977**, 39–40.
142. B. F. G. Johnson, and A. Rodgers, Polyhedral Rearrangements and Fragmentation Reactions in Cluster Complexes, in: *The Chemistry of Metal Cluster Complexes* (Eds. D.F. Shriver, H. D. Kaesz, and R. D. Adams, VCH Publishers, New York, 1990, Chap. 6).
143. C. E. Housecroft, *Metal–Metal Bonded Carbonyl Dimers and Clusters* (Oxford University Press, Oxford-New York-Tokyo, 1996, pp. 22 and 23).
144. J. C. Calabrese, L. F. Dahl, P. Chini, G. Longoni, and S. Martinengo, Synthesis and Structural Characterization of Platinum Carbonyl Cluster Dianions, $[Pt_3(CO)_3(\mu_2\text{-}CO)_3]_n^{2-}$ (n = 2, 3, 4, 5). A New Series of Inorganic Oligomers, *J. Am. Chem. Soc.* **96**, 2614–2616 (1974).
145. G. Longoni, and S. Martinengo, Synthesis and Chemical Characterization of Platinum Carbonyl Dianions, $[Pt_3(CO)_3(\mu_2\text{-}CO)_3]_n^{2-}$ (n = ~10, 6, 5, 4, 3, 2, 1). A New Series of Inorganic Oligomers, *J. Am. Chem. Soc.* **98**, 7225–7231 (1976).
146. C. Brown, B. T. Heaton, A. D. C. Towl, P. Chini, A. Fumagalli, and G. Longoni, Stereochemical Non-rigidity of a Metal Polyhedron; Carbon-13 and Platinum-195 Fourier Transform Nuclear Magnetic Resonance Spectra of $[Pt_n(CO)_{2n}]^{2-}$ (n = 3, 6, 9, 12 or 15), *J. Organomet. Chem.* **181**, 233–254 (1979).
147. V. G. Albano, A. Ceriotti, P. Chini, G. Ciani, S. Martinengo, and W. M. Anker, Hexagonal Close Packing of Metal Atoms in the New Polynuclear Anions $[Rh_{13}(CO)_{24}H_{5-n}]^{n-}$ (n = 2 or 3); X-Ray Structure of $[(Ph_3P)_2N]_2[Rh_{13}(CO)_{24}H_3]$, *J. Chem. Soc., Chem. Comm.* **1975**, 859–860.
148. J. C. Calabrese, L. F. Dahl, A. Cavalieri, P. Chini, G. Longoni, and S. Martinengo, Synthesis and Structure of a Hexanuclear Nickel Carbonyl Dianion, $[Ni_3(CO)_3(\mu_2\text{-}CO)_3]_2^{2-}$, and Comparison with the $[Pt_3(CO)_3(\mu_2\text{-}CO)_3]_2^{2-}$ Dianion. An Unprecedented Case of a Metal Cluster System Possessing Different Metal Architectures for Congener Transition Metals, *J. Am. Chem. Soc.* **96**, 2616–2618 (1974).
149. G. Longoni, P. Chini, L. D. Lower, and L. F. Dahl, Synthesis and Structural Characterization of a New Type of Homonuclear Metal Carbonyl, $[Ni_5(CO)_9(\mu_2\text{-}CO)_3]^{2-}$. A Trigonal Bipyramidal Metal Cluster System, *J. Am. Chem. Soc.* **97**, 5034–5036 (1975).
150. J. K. Ruff, R. P.White, and L. F. Dahl, Preparation, Structure, and Bonding of a New Type of Metal Cluster System, $[M_2Ni_3(CO)_{16}]^n$ (M = Cr, Mo, W; n = –2): A Noncorformist to the Nobel Gas Metal Family, *J. Am. Chem. Soc.* **93**, 2159–2176 (1971).
151. S. Martinengo, G. Ciani, A. Sironi, and P. Chini, Analogues of Metallic Lattices in Rhodium Carbonyl Cluster Chemistry. Synthesis and X-ray Structure of the $[Rh_{15}(\mu\text{-}CO)_{14}(CO)_{13}]^{3-}$ and $[Rh_{14}(\mu\text{-}CO)_{16}(CO)_9]^{4-}$ Anions Showing a Stepwise Hexagonal Close-Packed/Body-Centered Cubic Interconversion, *J. Am. Chem. Soc.* **100**, 7096–7098 (1979).
152. D. M. Washecheck, E. J. Wucherer, L. F. Dahl, A. Ceriotti, G. Longoni, M. Manassero, M. Sansoni, and P. Chini, Synthesis, Structure, and Stereochemical Implications of the $[Pt_{19}(CO)_{12}(\mu_2\text{-}CO)_{10}]^{4-}$ Tetraanion: A Bicapped Triple-Decker All-Metal Sandwich of Idealized Fivefold (D_{5h}) Geometry, *J. Am. Chem. Soc.* **101**, 6110–6112 (1979).
153. S. Martinengo, G. Ciani, and A. Sironi, Synthesis and X-ray Structural Characterization of the $[Rh_{22}(\mu_3\text{-}CO)_7(\mu\text{-}CO)_{18}(CO)_{12}]^{4-}$ Anion Containing a Large Close-Packed Cluster with an ABAC Sequence of Compact Layers, *J. Am. Chem. Soc.* **102**, 7564–7565 (1980).
154. G. Ciani, A. Magni, A. Sironi, and S. Martinengo, Synthesis and X-Ray Characterization of the High-nuclearity $[Rh_{17}(\mu_3\text{-}CO)_3(\mu\text{-}CO)_{15}(CO)_{12}]^{3-}$ Anion containing a Tetra-capped Twinned Cuboctahedral Cluster, *J. Chem. Soc., Chem. Comm.* **1981**, 1280–1281.
155. C. Femoni, F. Kaswalder, M. C. Iapalucci, G. Longoni, M. Mehlstäubl, S. Zacchini, and A. Ceriotti, Synthesis and Crystal Structure of $[NBu_4]_2[Pt_{24}(CO)_{48}]$: An Infinite 1D Stack of $[Pt_3(CO)_6]$ Units Morphologically Resembling a CO-Insulated Platinum Cable, *Angew. Chem. Int. Ed.* **45**, 2060–2062 (2006).

156. C. Femoni, F. Kaswalder, M. C. Iapalucci, G. Longoni, and S. Zacchini, Infinite Molecular $\{[Pt_{3n}(CO)_{6n}]^{2-}\}_\infty$ Conductor Wires by Self-Assembly of $[Pt_{3n}(CO)_{6n}]^{2-}$ (n = 5–8) Cluster Dianions Formally Resembling CO-Sheathed Three-Platinum Cables, *Eur. J. Inorg. Chem.* **2007**, 1483–1486.
157. G. Longoni, C. Femoni, M. C. Iapalucci, and P. Zanello, Electron-sink Features of Homoleptic Transition-metal Carbonyl Clusters, in: *Metal Clusters in Chemistry* (Eds. P. Braunstein, L. A. Oro, and P. R. Raithby, Wiley-VCH, Weinheim, 1999, Vol. 2, Chap. 3.9).
158. C. Femoni, M. C. Iapalucci, F. Kaswalder, G. Longoni, and S. Zacchini, The Possible Role of Metal Carbonyl Clusters in Nanoscience and Nanotechnologies, *Coord. Chem. Rev.* **250**, 1580–1604 (2006).
159. E. G. Mednikov, S. A. Ivanov, I. V. Slovokhotova, and L. F. Dahl, Nanosized $[Pd_{52}(CO)_{36}(PEt_3)_{14}]$ and $[Pd_{66}(CO)_{45}(PEt_3)_{16}]$ Clusters Based on a Hypothetical Pd_{38} Vertex-Truncated ν_3 Octahedron, *Angew. Chem. Int. Ed.* **44**, 6848–6854 (2005).
160. N. T. Tran, D. R. Powell, and L. F. Dahl, Nanosized $Pd_{145}(CO)_x(PEt_3)_{30}$ Containing a Capped Three-Shell 145-Atom Metal-Core Geometry of Pseudo Icosahedral Symmetry, *Angew. Chem. Int. Ed.* **39**, 4121–4125 (2000).
161. C. R. Eady, B. F. G. Johnson, and J. Lewis, Products from the Pyrolysis of $Ru_3(CO)_{12}$ and $Os_3(CO)_{12}$, *J. Organomet. Chem.* **37**, C39–C40 (1972).
162. B. F. G. Johnson, and J. Lewis, Transition-Metal Molecular Clusters, *Adv. Inorg. Chem. Radiochem.* **24**, 225–355 (1981).
163. R. Mason, K. M. Thomas, and D. M. P. Mingos, Stereochemistry of Octadecacarbonylhexaosmium(0). A Novel Hexanuclear Complex Based on a Bicapped Tetrahedron of Metal Atoms, *J. Am. Chem. Soc.* **95**, 3802–3804 (1973).
164. K. Wade, The Structural Significance of the Number of Skeletal Bonding Electron-pairs in Carboranes, the Higher Borane Anions, and Various Transition-metal Carbonyl Cluster Compounds, *Chem. Comm.* **1971**, 792–793.
165. C. R. Eady, B. F. G. Johnson, and J. Lewis, Synthesis and Carbon-13 Nuclear Magnetic Resonance Studies of the Hexanuclear Osmium Clusters $[H_2Os_6(CO)_{18}]$, $[HOs_6(CO)_{18}]^-$, and $[Os_6(CO)_{18}]^{2-}$, *J. Chem. Soc., Chem. Comm.* **1976**, 302–303.
166. D. M. P. Mingos, and M. I. Forsyth, Molecular-orbital Calculations on Transition-metal Cluster Compounds containing Six Metal Atoms, *J. Chem. Soc., Dalton Trans.* **1977**, 611–616.
167. P. F. Jackson, B. F. G. Johnson, J. Lewis, M. McPartlin, and W. J. H. Nelson, Synthesis of the Carbido Cluster $[Os_{10}(CO)_{24}C]^{2-}$ and the X-Ray Structure of $[Os_{10}(CO)_{24}C][(Ph_3P)_2N]$, *J. Chem. Soc., Chem. Comm.* **1980**, 224–226.
168. L. H. Gade, B. F. G. Johnson, J. Lewis, M. McPartlin, T. Kotch, and A. J. Lees, Photochemical Core Manipulation in High-Nuclearity Os–Hg Clusters, *J. Am. Chem. Soc.* **113**, 8698–8704 (1991).
169. J. Lewis, and P. R. Raithby, Reflections on Osmium and Ruthenium Carbonyl Compounds, *J. Organomet. Chem.* **500**, 227–237 (1995).
170. D. Braga, J. Lewis, B. F. G. Johnson, M. McPartlin, W. J. H. Nelson, and M. D. Vargas, Synthesis and X-Ray Analysis of the Tetrahydrido-dianion $[H_4Os_{10}(CO)_{24}]^{2-}$, the First Non-carbido Decaosmium Cluster, *J. Chem. Soc.. Chem. Comm.* **1983**, 241–243.
171. E. Charalambous, L. H. Gade, B. F. G. Johnson, J. Lewis, M. McPartlin, and H. R. Powell, Synthesis and X-Ray Structure Analysis of the Largest Binary Osmium Carbonyl Cluster: $[Os_{17}(CO)_{36}]^{2-}$, *J. Chem. Soc. Chem. Comm.* **1990**, 688–690.
172. A. J. Amoroso, L. H. Gade, B. F. G. Johnson, J. Lewis, P. R. Raithby, and W.-T. Wong, $(nBu_4N)_2[Os_{20}(CO)_{40}]$, a Thermally Generated Polynuclear Cluster Compound with a Tetrahedral Cubic Thickly Packed Cluster Nucleus, *Angew. Chem. Int. Ed. Engl.* **30**, 107–109 (1991).
173. L. H. Gade, B. F. G. Johnson, J. Lewis, M. McPartlin, H. R. Powell, P. R. Raithby, and W.-T. Wong, Synthesis and Structural Characterisation of the Osmium Cluster

Dianions $[Os_{17}(CO)_{36}]^{2-}$ and $[Os_{20}(CO)_{40}]^{2-}$, *J. Chem. Soc., Dalton Trans.* **1994**, 521–532.
174. B. Cornils, W. A. Herrmann, and M. Rasch, Otto Roelen, Pioneer in Industrial Homogeneous Catalysis, *Angew. Chem. Int. Ed. Engl.* **33**, 2144–2163 (1994).
175. R. F. Heck, and D. S. Breslow, The Reaction of Cobalt Hydrotetracarbonyl with Olefins, *J. Am. Chem. Soc.* **83**, 4023–4027 (1961).
176. P. M. Maitlis, and A. Haynes, Hydroformylation of Olefins, in: *Metal-catalysis in Industrial Organic Processes* (Eds. G. P. Chiusoli, and P. M. Maitlis, RSC Publishing, Cambridge, 2006, Chap. 4.6).
177. M. Beller, and K. Kumar, Hydroformylation: Applications in the Synthesis of Pharmaceuticals and Fine Chemicals, in: *Transition Metals for Organic Synthesis, Second Ed.* (Eds. M. Beller and C. Bolm, Wiley-VCH, Weinheim, 2004, Vol. 1, Chap. 2.1).
178. W. Reppe, *Neue Entwicklungen auf dem Gebiete der Chemie des Acetylens und des Kohlenoxyds* (Springer, Berlin, 1949).
179. W. Reppe, *Chemie und Technik der Acetylen-Druck-Reaktionen, 2. Aufl.* (Verlag Chemie, Weinheim, 1952).
180. W. Reppe, O. Schlichting, K. Klager, and T. Toepel, Cyclisierende Polymerisation von Acetylen I. Über Cyclooctatetraen, *Liebigs Ann. Chem.* **560**, 1–92 (1948).
181. R. E. Colborn, and K. P. C. Vollhardt, On the Mechanism of the Cyclooctatetraene Synthesis from Ethyne Employing Nickel Catalysts, *J. Am. Chem. Soc.* **108**, 5470–5477 (1986).
182. C. Hoogzand, and W. Hübel, Cyclic Polymerization of Acetylenes by Metal Carbonyl Compounds, in: *Organic Synthesis via Metal Carbonyls* (Eds. I. Wender and P. Pino, Wiley Interscience, New York, 1968, Vol. I, pp. 343–371).
183. H. Bönnemann, and W. Brijoux, Cyclomerization of Alkynes, in: *Transition Metal for Organic Synthesis, Second Ed.* (Eds. M. Beller and C. Bolm, Wiley-VCH, Weinheim, 2004, Vol. 1, Chap. 2.8).
184. F. A. Cotton, A Millenial Overview of Transition Metal Chemistry, *J. Chem. Soc., Dalton Trans.* **2000**, 1961–1968.
185. H. Werner, Complexes of Carbon Monoxide and its Relatives: An Organometallic Family Celebrates Its Birthday, *Angew. Chem. Int. Ed. Engl.* **29**, 1077–1089 (1990).
186. H. Behrens, The Chemistry of Metal Carbonyls: "The Life Work of Walter Hieber", *J. Organomet. Chem.* **94**, 139–159 (1975).
187. H. Behrens, *Wissenschaft in Turbulenter Zeit* (Münchner Universitätsschriften, Heft 25, Institut für Geschichte der Naturwissenschaften, München, 1998).
188. E. O. Fischer, In memoriam Walter Hieber, *Chem. Ber.* **112**, XXI–XXXIX (1979).
189. R. E. Oesper, Walter Hieber, *J. Chem. Educ.* **31**, 140–141 (1954).
190. Calderazzo, Prof. Paolo Chini: Biographical Memoir, *J. Organomet. Chem.* **213**, ix–xii (1981).

Chapter 5
A Scientific Revolution: The Discovery of the Sandwich Complexes

Gepriesen sei mir der Zufall; er hat größere Dinge getan als klügelnde Vernunft.[1]
Friedrich Schiller, German Poet (1759–1805)

5.1 The Early Days: Ferrocene

It was in the mid 1950s that some eminent scientists such as Ron Nyholm were convinced that they witnessed a "renaissance of inorganic chemistry" [1]. What had happened? In December 1951 and February 1952 two papers appeared in *Nature* and the *Journal of the Chemical Society*, which the majority of the readers of those journals probably did not take notice [2].[2] The communication in *Nature* was entitled "A New Type of Organo-iron Compound" [3] and that in the *Journal of the Chemical Society* "Dicyclopentadienyliron" [4]. Independently, two groups, one led by Peter Pauson, who at that time was at the Duquesne University in Pittsburgh, and the other by Samuel Miller, who was working for the British Oxygen Company, reported the preparation of an organoiron compound which they had obtained by accident. Peter Pauson's original aim was to prove or disprove a suggestion, made by Robert D. Brown in 1950, that the conjugated system $C_{10}H_8$, called "fulvalene", might show aromatic properties. Pauson was rather sceptical but thought that the respective compound **2** (see Scheme 5.1) could be accessible in just two steps: The coupling of two cyclopentadienyl radicals, formed from the cyclopentadienyl anion by oxidation, should afford the dihydro species **1**, which by dehydrogenation should give the fulvalene. He assumed that $FeCl_3$ might be a useful reagent, both for the coupling of the Grignard reagent to generate **1**, and for the oxidation of this intermediate to form **2** [5].

---

[1] In English: "Praise be to my luck; it has done greater things than thoughtful reasoning."

[2] In a historical account, Luigi Venanzi wrote: "I wonder how many readers have actually seen Kealy and Pauson's 1951 Nature article?"

H. Werner, *Landmarks in Organo-Transition Metal Chemistry*,
Profiles in Inorganic Chemistry, DOI 10.1007/978-0-387-09848-7_5,

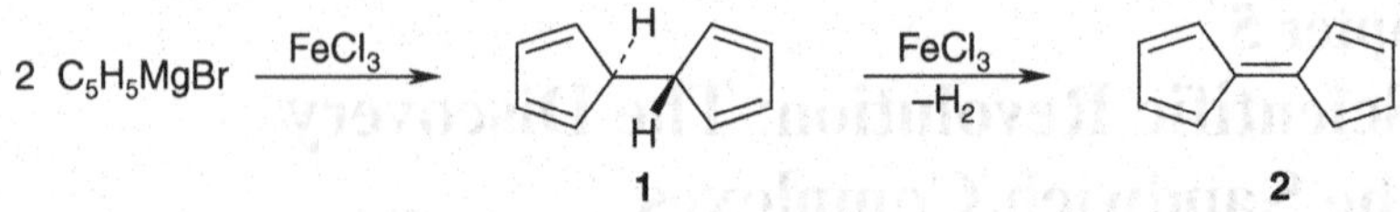

Scheme 5.1 The proposed course of the reaction of the cyclopentadienyl Grignard reagent with $FeCl_3$

However, the outcome of the experiment, which Peter Pauson and his graduate student Tom Kealy did together in July 1951, was very surprising. After standard workup of the reaction mixture, obtained from treatment of a solution of C_5H_5MgBr in diethyl ether with $FeCl_3$, pretty orange-yellow crystals were isolated, which at first sight were thought to be the desired fulvalene. However, this proved to be wrong, since the C, H microanalysis did not fit the composition $C_{10}H_{10}$; instead it seemed to be consistent with the composition $FeC_{10}H_{10}$. As Peter Pauson recalled [5], they initially had difficulties to get a quantitative analysis for iron, since the unknown compound, after it was dissolved in concentrated sulphuric acid, was largely recovered unchanged upon dilution with water. Only after evaporating a solution of the compound in perchloric acid to dryness, was an accurate iron analysis obtained. With this result in hand, Pauson and Kealy submitted a note to *Nature* which was accepted within a few weeks [3].

The aim of the second paper leading to $FeC_{10}H_{10}$, which was submitted 4 weeks prior to Kealy and Pauson's note but appeared 2 months later [4], was completely different. Samuel Miller and his coworkers had attempted to prepare new catalysts for the formation of ammonia and amines, using olefins and nitrogen as the substrates in the latter case [6]. For this reason, they treated reduced iron in the form of the well known "doubly promoted synthetic ammonia catalyst" with cyclopentadiene at 300°C in nitrogen at atmospheric pressure. Instead of the desired amine they isolated a yellow solid. This "remarkable substance" sublimed above 100°C without decomposition, was volatile in steam, and soluble in organic solvents. Although the crude product obtained from iron and cyclopentadiene adsorbed nitrogen at comparatively low temperatures, the Miller team found that "the compound was not capable of acting as a vapour-phase catalyst for ammonia synthesis" [4].

While there is no doubt that Peter Pauson and Samuel Miller and their coworkers have the priority regarding the isolation and characterization of $FeC_{10}H_{10}$, they were probably not the first to have prepared this novel organometallic compound. Eugene O. Brimm from Linde Air Products, nowadays well known for the first synthesis of $Mn_2(CO)_{10}$ (see Chap. 4), was one of several industrial chemists who, after reading the communication by Kealy and Pauson, wanted to prepare a sample of the new substance. For this reason, he asked a colleague at Union Carbide (at that time Linde's parent company) whether they had any cyclopentadiene. The reply, that they not longer did so, was accompanied by the statement that, some years previously, they had terminated work on the cracking of dicyclopentadiene because of a "yellow sludge" which clogged the iron pipes they used. They had not attempted to isolate or analyse this material but had kept a bottle of it: It was $FeC_{10}H_{10}$! The preparative method used by Miller's group was, in fact, very similar as it involved passing cyclopentadiene over a heated iron-containing ammonia catalyst.

However, whilst its formulation was thus established, the structure of the new organoiron compound remained obscure. Peter Pauson, who was aware of the common belief that bonds between transition metals and hydrocarbon groups were unstable [5], nevertheless suggested in his note to *Nature* the linear arrangement **I** with two planar cyclopentadienyl rings linked to the metal as shown in Fig. 5.1. He attributed the remarkable stability of $FeC_{10}H_{10}$ "to the tendency of the cyclopentadienyl group to become 'aromatic' by acquisition of a negative charge, resulting in important contributions from the resonance form **II** and intermediate forms" [3]. Miller made only a short comment regarding the structure of $FeC_{10}H_{10}$ and argued that "by analogy with the well known cyclopentadienylpotassium, it is believed that substitution has occurred in the methylene group" [4].

While most of Pauson's contemporaries took no notice of his structural proposal, it immediately attracted the interest of at least three people: Ernst Otto Fischer in München, and both Geoffrey Wilkinson and Robert Burns Woodward at Harvard. They were convinced that the linear arrangement was incorrect, and with the investigations of the real structure of bis(cyclopentadienyl)iron, the age of modern organometallic chemistry began. The reason for not believing in the proposed structure of $Fe(C_5H_5)_2$ was similar on both sides of the Atlantic. Ernst Otto Fischer had investigated new synthetic routes to metal carbonyls at ambient pressure and temperature in his doctoral thesis, which he finished at the end of 1951. He had attempted, inter alia, to obtain $Fe(CO)_5$ by reductive carbonylation of iron(II) salts but failed, in contrast to analogous experiments with cobalt(II) and nickel(II) salts. Given this previous experience, he was curious about the properties of the new iron compound, reported by Kealy and Pauson, and asked Reinhard Jira, an undergraduate student working with him, to repeat the synthesis of $Fe(C_5H_5)_2$. This was done and the stability of the product towards air and moisture confirmed. Jira then treated the orange solid with CO in an autoclave at 200 bar and ca. 150°C and found that it remained unchanged. Due to this unusual behavior, completely unexpected for an organoiron compound with Fe–C σ-bonds, Fischer concluded that the whole set of the six π-electron pairs of the two $C_5H_5^-$ anions would

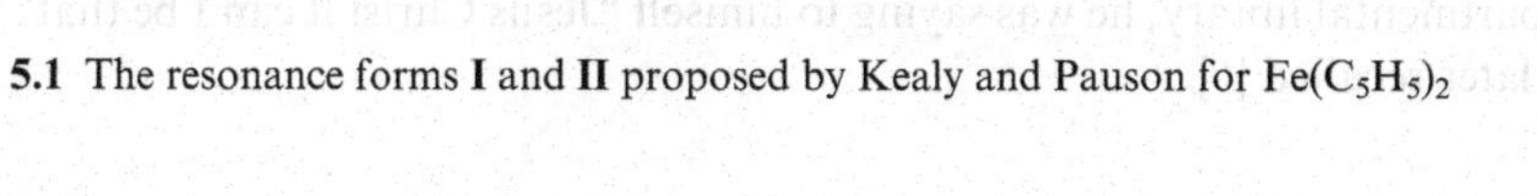

Fig. 5.1 The resonance forms **I** and **II** proposed by Kealy and Pauson for $Fe(C_5H_5)_2$

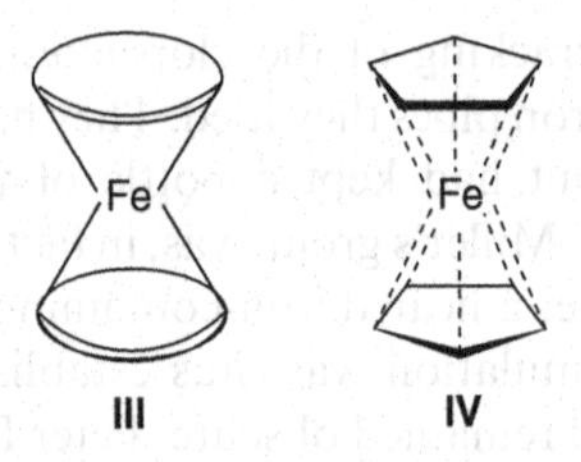

Fig. 5.2 The structural proposals for $Fe(C_5H_5)_2$, offered by Fischer (**III**) and Woodward and Wilkinson (**IV**)

participate in the bonding to the metal atom, "with each cyclopentadienyl ring functioning as a tridentate ligand and filling the iron orbitals up to 18 electrons – the krypton configuration" [7]. In analogy to the well-known iron(II) complex $[Fe(CN)_6]^{4-}$, Fischer proposed an octahedral structure for $Fe(C_5H_5)_2$, with two parallel planar cyclopentadienyl rings surrounding the central iron atom. This *Doppelkegel* (double-cone) arrangement **III** (Fig. 5.2) was supported by a crystallographic study carried out by Wolfgang Pfab who, like Fischer, was working with Hieber and "activated some X-ray equipment that had been stored unopened at the institute" [7]. Further evidence for the proposal of a *Durchdringungskomplex* (the terminology for kinetically inert low-spin complexes introduced by Alfred Werner was still used in Germany in the early 1950s) was the diamagnetism, determined by Fischer with a Gouy balance. Moreover, after Reinhard Jira found that the neutral bis(cyclopentadienyl)iron could be oxidized by bromine to give the cation $[Fe(C_5H_5)_2]^+$ (which was precipitated as the Reinecke salt), the first short communication from Fischer's group was submitted to the *Zeitschrift für Naturforschung* in June 1952 [8].

The first paper by the Harvard team had been submitted to the *Journal of the American Chemical Society* 3 months previously [9]. The two main protagonists involved not only differed in age, but also in their academic position. Geoffrey Wilkinson was a young assistant professor, who came to Harvard in September 1951, following a year as research associate at MIT. The other main actor was Robert Burns Woodward, a full professor at Harvard since 1950 and already in those days a giant in synthetic organic chemistry. Although Wilkinson owed his appointment largely to his *nuclear background* [6], he began his teaching career with a course in inorganic chemistry. In terms of his research, he first continued his MIT work on phosphine substitution reactions but began "to look at reactions of the carbonyls with unsaturated compounds" at the same time [6]. He knew from the books he used for his teaching (in particular Sidgwick's *The Chemical Elements and Their Compounds* and Emeléus and Anderson's *Modern Aspects of Inorganic Chemistry*) that metal phenyls and alkyls were unstable and hence, when he saw the note by Kealy and Pauson in January 1952 in Harvard's departmental library, he was saying to himself "Jesus Christ it can't be that", as he later recalled [6].

Despite some discrepancies in the recollections with regard to the sequence of the events [6, 10, 11], there is no doubt that Woodward had seen the note in *Nature*, as well as the title of the paper by Miller, Tebboth and Tremaine, prior to Wilkinson. Woodward assumed at first sight that the linear arrangement was wrong and, when talking to his graduate student Myron Rosenblum, finally drew out "in architecturally exact Woodwardian form" the structure with "the iron atom nested neatly between two cyclopentadienyl rings" [11]. To prove this proposal, Woodward asked Rosenblum to repeat the synthesis of $Fe(C_5H_5)_2$ and try to obtain the ruthenium counterpart as well. However, Rosenblum had neither $RuCl_3$ nor any other ruthenium salt in his lab and therefore contacted Wilkinson posing his request. Wilkinson's reaction was rather harsh [6, 11], because he also had the idea to prepare analogues of $Fe(C_5H_5)_2$. The outcome was, that he and Woodward went for lunch at the Harvard Faculty Club and sorted things out. As Wilkinson recalled [6], "the possibility that the C_5H_5 ring in the iron compound could possibly undergo Friedel-Crafts or other aromatic reactions simply had not dawned on me, but other than the structure, this seemed to be Bob's main interest, whereas mine was to go on to other transition metals". Although Rosenblum offers a somewhat different perspective [11], he gave the ruthenium work to Wilkinson and carried out the reaction of bis (cyclopentadienyl)iron with acetyl chloride. He isolated a "beautifully crystalline red" product, being the diacetyl derivative of the starting material [12]. He also measured the IR spectrum of $Fe(C_5H_5)_2$, which showed only one C–H stretching mode, and determined the dipole moment, which was effectively zero. Wilkinson showed, at the same time as Fischer, that $Fe(C_5H_5)_2$ was diamagnetic, and that it could be readily oxidized, for example, by aqueous silver sulphate, to give the paramagnetic cation $[Fe(C_5H_5)_2]^+$. The results of these studies were summarized in the first Harvard paper, and the pentagonal antiprismatic structure **IV**, shown in Fig. 5.2, was suggested [9]. It was favored over the related prismatic structure although the possibility of the latter was not excluded. In the second paper from Harvard, submitted in June 1952, Woodward, Rosenblum and Whiting described the aromatic properties of the "unique iron compound" and, due to these properties, the felicitous name *ferrocene*, the *ene* ending implying aromaticity, was proposed in analogy to benzene [12].[3]

However, despite the convincing results obtained at München and Harvard, there was also a certain amount of scepticism, for example, by Jack Dunitz as well as by Marshall Gates, the assistant editor of the *Journal of the American Chemical Society*. In a letter to Woodward he wrote: "We have dispatched your communication to the printers but I cannot help feeling that you have been at the hashish again. 'Remarkable' seems a pallid word with which to describe this substance" [13]. The sceptical voices disappeared when Philip Frank Eiland and Ray Pepinsky reported a complete X-ray crystal structure analysis of $Fe(C_5H_5)_2$

[3] In [10] it is mentioned in lit. cit. 12, that the name "ferrocene" is due to Mark Whiting (a postdoc in Woodward's group), while "ferrozene" was an early favorite.

in October 1952, confirming the structural proposal offered by Fischer, Woodward and Wilkinson [14]. Independently, Jack Dunitz and Leslie Orgel also determined the structure of $Fe(C_5H_5)_2$ and came to the same result [15]. To characterize the molecular architecture of the compound, they used the name *molecular sandwich*, while at the same time Wilkinson mentioned the *sandwich structure* of bis(cyclopentadienyl)iron [16]. So the term *sandwich compounds* or *sandwich complexes* was created.[4] Dunitz and Orgel also provided an explanation for the stability of the sandwich structure in terms of molecular orbital theory and, based on this, they predicted that "manganese and chromium, with sufficient donor and acceptor orbitals available, should form compounds of the same order of stability as that of iron". While they expected that for nickel (having "no orbital available for covalent bonding") and cobalt the corresponding compounds would be "much less stable than for iron" [15], both Wilkinson, in collaboration with Pauson [16, 17], and Fischer [18, 19] had already begun to prepare bis(cyclopentadienyl) compounds of these metals. Both groups reported the synthesis of stable salts of the cation $[Co(C_5H_5)_2]^+$, which is isoelectronic to ferrocene, and of neutral $Ni(C_5H_5)_2$ between December 1952 and May 1953.

Before we summarize the breath-taking race between Fischer and Wilkinson, it should be noted that, in retrospect, Peter Pauson felt he should have come to the same conclusion about the structure of ferrocene somewhat earlier than Fischer, Woodward and Wilkinson. In his personal account [5], he remembered that when he visited Columbia University in New York in autumn of 1951 and told William von E. Doering about the new organoiron compound, Doering suggested a measurement of the magnetic susceptibility. If the compound were paramagnetic, the linear structure could be correct; but if it were diamagnetic, it had to be wrong. Peter confessed that he "was too reluctant to expose my ignorance and my lack of understanding and was content to await the definite answer expected from X-ray work" [5]. The fact was, that some weeks before he had handed crystals of ferrocene to J. Monteath Robertson, the distinguished crystallographer from Glasgow University, who unfortunately was too busy to determine the structure immediately. Thus, this opportunity passed and the credit for the correct structural proposal went to the groups in München and Harvard. With regard to the original aim of his work, Pauson later reasoned that "we (Kealy and he) had indeed prepared a novel, non-benzenoid aromatic, but a very unexpected addition to this class" [5].

[4] N. A. M. Rodger describes in his book *The Insatiable Earl: A Life of John Montagnu, Fourth Earl of Sandwich* (Harper Collins, London, 1993), that the Earl did not like to intercept his card play, not even for the meals, and ordered his kitchen staff to prepare double slices of bread filled with sausage or cheese, which he could eat while playing. This passion made him immortal for all time and the word *sandwich* became a familiar term since the late eighteenth century.

5.2 The Rivalry of Fischer and Wilkinson

With the elucidation of the structure of $Fe(C_5H_5)_2$, the development of the chemistry of sandwich-type complexes began at an unprecedented rate. In less than one decade, Fischer's and Wilkinson's groups prepared bis(cyclopentadienyl) metal and cyclopentadienyl metal carbonyl compounds of almost all of the transition metals. It was inevitable that they were in constant competition as to who would be first with the next, fairly obvious target. To get a feeling of the hard work carried out at Harvard and in München, it is instructive to see the list of publications which appeared between December 1952 and December 1954 (Table 5.1). From the synthetic point of view, the greatest obstacle at the beginning was the insolubility of most of the metal halides, which prevented their reaction with the ethereal solution of the cyclopentadienyl Grignard reagent. Fischer overcame this problem by turning to the hexammine complexes of the corresponding metals, and Wilkinson by using the acetylacetonates. The key to further success in this field was using sodium cyclopentadienide instead of C_5H_5MgBr and tetrahydrofuran, instead of diethyl ether as the solvent [20, 21]. In retrospect, it is curious that NaC_5H_5 had not been used prior to this finding since the potassium counterpart KC_5H_5 was known from Thiele's work since 1901 [22].

Table 5.1 Summary of the race between Fischer's and Wilkinson's groups between December 1952 and December 1954 (the dates are those of the appearance in print of the respective papers; abbreviations used: Cp=C_5H_5, Ind=C_9H_7)

		Fischer	Wilkinson
1952:	December		$Ru(Cp)_2$; $[M(Cp)_2]^+$ (M=Ru, Co)
1953:	January	$[Co(Cp)_2]^+$	
	February		$Ni(Cp)_2$; $[Ni(Cp)_2]^+$ $MX_2(Cp)_2$ (M=Ti, Zr, V)
	May	$Ni(Cp)_2$; $[Ni(Cp)_2]^+$	
	June	$Co(Cp)_2$	
	July		$[M(Cp)_2]^+$ (M=Rh, Ir)
	August	$Cr(Cp)_2$	
	November	$M(Ind)_2$ (M=Fe, Co, Ni) $[Co(Ind)_2]^+$	
1954:	January		$Cr(Cp)_2$; $[CpM(CO)_3]_2$ (M=Mo, W)
	March		$M(Cp)_2$ (M=Mn, Mg)
	April		$Co(Cp)_2$, $Fe(Ind)_2$, $[Co(Ind)_2]^+$
	June		$MoCl_2(Cp)_2$
	July	$V(Cp)_2$; $CpV(CO)_4$	
	September	$M(Cp)_2$ (M=Mn, Mg); $CpMn(CO)_3$	$MBr_3(Cp)_2$ (M=Nb, Ta) $[M(Cp)_2]^+$ (M=Ti, V)
	December		$M(Cp)_3$ (M=Sc, Y, La, Ce, Pr, Nd, Sm)

5.3 Fischer's Star: Bis(benzene)chromium

When counting the number of cyclopentadienyl metal compounds and the number of publications by Fischer and Wilkinson, which appeared between April 1952 and late 1955, it is fair to say that the two groups were roughly on par [23, 24]. A big leap ahead, which temporarily gave Fischer the lead, occurred in December 1955, when Fischer and his Ph.D. student Walter Hafner reported the synthesis of bis(benzene)chromium [25]. Dietmar Seyferth recalled, that even a year later "it caused quite a stir" [26] when Fischer at the Gordon Research Conference on Inorganic Chemistry in August 1956 described the preparation and surprising properties of this novel transition metal compound. The idea that a "sandwich complex built up of two neutral benzene molecules and a zerovalent chromium atom that was thermally stable to 300°C in the absence of air" [26] could exist, was obviously beyond the expectations of the chemical community at that time.

Fischer's belief that a compound such as bis(benzene)chromium could be stable, was based on the hypothesis that in bis(cyclopentadienyl) metal compounds such as ferrocene or nickelocene each cyclopentadienyl anion contributes six electrons to the ring–metal bond and that the neutral benzene molecule could do the same. Provided that, as in ferrocene, a compound with an 18-electron configuration would lead to the most stable system, chromium(0) should be the metal of choice. Fischer discussed this idea as early as in summer 1952 with Walter Hafner, who in those days worked in the advanced inorganic chemistry laboratory of Hieber's institute at the *Technische Hochschule* in München. About 1 year later, Hafner decided to do research for his Diploma and, later, for his doctoral thesis with Fischer and to study organochromium chemistry. In less than 15 months, he prepared $Cr(C_5H_5)_2$, $[(C_5H_5)Cr(CO)_3]_2$, $(C_5H_5)Cr(CO)_3H$ and $(C_5H_5)Cr(CO)_2(NO)$ as well as $V(C_5H_5)_2$ and $(C_5H_5)V(CO)_4$ [7], but despite this success did not forget the challenge to prepare the unknown bis(benzene)chromium. In January 1954, he carried out an exploratory experiment and treated $CrCl_3$ with *m*-xylene in the presence of aluminium and $AlCl_3$ under reflux. He obtained a yellow solid which, however, did not sublime and thus could not be the desired neutral molecule. Although a second experiment with benzene under autogeneous pressure at 150°C led to a similar result, Walter Hafner did not thoroughly investigate the products at that time. Instead, following Fischer's advice, he repeated the reaction of $CrCl_3$, Al, $AlCl_3$ and *m*-xylene under a pressure of CO and obtained chromium hexacarbonyl in high yield [27, 28].

In July 1955, Harold Zeiss (Fig. 5.3) visited the chemistry department of the University (LMU) in München and gave a seminar, in which he discussed the results of Minoru Tsutsui's doctoral thesis. In this context, he mentioned that Hein's "polyphenylchromium compounds" might be in fact π-arene chromium(I) complexes. Walter Hafner attended the seminar and instantly remembered the outcome of his previous experiments. A few days later, after having repeated the reaction of $CrCl_3$, Al, $AlCl_3$ and

Fig. 5.3. Harold H. Zeiss (1917–1995) studied chemistry at the University of Indiana, where he received his B.S. in 1938. After having done research for 1 year in Germany, working with the Nobel laureate Hans Fischer in München and with the eminent organic chemist Karl Freudenberg in Heidelberg, he spent 6 years in industry in the US. He obtained his Ph.D. at Columbia University with William von E. Doering in 1949, and in the same year was appointed Instructor at Yale University. He was promoted to Assistant Professor in 1951, but in 1955 left Yale and joined the Monsanto Chemical Company, first as a Research Associate in the Central Research Laboratory in Dayton, Ohio, and then in 1961 as President and Director of Monsanto Research S.A. in Zürich, Switzerland. This laboratory, with an outstanding international staff, made significant contributions to basic research in organometallic, organophosphorus and organosilicon chemistry. After its closure in 1975, Zeiss spent 2 more years with Monsanto in St. Louis, before he retired in 1978 (photo from ref. 32; reproduced with permission of the American Chemical Society)

benzene in the absence of CO, he reduced the yellow product which he now anticipated to be a salt. Zinc dust in aqueous HCl did not work, but sodium dithionite in dilute aqueous NaOH was effective. He isolated a dark brown crystalline solid, which was sublimable, soluble in benzene, and diamagnetic. The dipole moment was zero and the elemental analysis (C, H and Cr) confirmed the composition $Cr(C_6H_6)_2$ [25]. Erwin Weiss (who was a Ph.D. student of Hieber's and later held the Chair of Inorganic Chemistry at the University of Hamburg, Germany) carried out a preliminary X-ray crystal structure analysis and showed that the molecule was centrosymmetric, being in agreement with the expected sandwich structure [29]. Walter Hafner prepared several salts of the bis(benzene)-chromium cation by air oxidation of $Cr(C_6H_6)_2$ in water and found that they were light-sensitive but stable in neutral or basic solution.[5]

[5] In my opinion (and I am convinced that Ernst Otto Fischer would have agreed), Walter Hafner was one of the best Ph.D. students, probably the best Fischer ever had. After he finished his doctoral research, he joined the *Consortium für Elektrochemische Industrie*, a research subsidiary of Wacker-Chemie in Munich, where he laid the foundations for the palladium-catalyzed oxidation of ethene to acetaldehyde, the Wacker process. This process has been licenced to various chemical companies all over the world and initiated an ever-growing area of synthetic organic chemistry. Walter Hafner retired in 1992 and died in 2004.

The reason why Fischer succeeded with the preparation of $Cr(C_6H_6)_2$ and Wilkinson did not, is a noteworthy facet of the bis(benzene)chromium story. In his *Recollections of the First Four Months*, Wilkinson recalled that shortly after William (Bill) Moffit, a young theoretician, came to Harvard in autumn of 1952, he asked him "what was the chance of benzene binding to a transition metal?" [6]. After a few days, Moffit came back arguing that such an arrangement should not be stable since it would readily decompose to solid metal, e.g., chromium and stable benzene molecules. As Wilkinson had, in his own words, "implicit faith in theoretical chemists at that time", he forgot all about it and went on with cyclopentadienyl metal chemistry. It was, as he ironically mentioned in 1975, "the first of more than one interesting experience with theoretical chemists" [6].

5.4 Hein's "Polyphenylchromium Compounds"

But despite the breakthrough by Fischer and Hafner with their synthesis of bis(benzene)chromium, the history of this unusual compound dates back to 1919. In January of that year, Franz Hein (see Fig. 2.40), in those days a research assistant at the University of Leipzig in Germany, reported in a short communication that the reaction of phenylmagnesium bromide with chromium(III) trichloride in diethyl ether, undertaken with the intention to prepare triphenylchromium, gave a mixture of products. The major component was an amorphous orange powder which could not be recrystallized but formed a 1:1 adduct with $HgCl_2$. Based on the elemental analysis of this adduct, Hein assumed that the orange powder was with "reasonable certainty" pentaphenylchromium bromide [30]. Apart from the colour, being typical for chromium(VI), the fact that the product could also be obtained from CrO_2Cl_2 and C_6H_5MgBr, seemed to support the proposal. Since Hein was convinced that it would be worthwhile to extend the work on the reactivity of chromium compounds with organometallic reagents, his first paper ended with the sentence: "I therefore direct to all my esteemed colleagues in this area the request to leave to me the organochromium compounds for my further investigation" [30].

While today this request appears rather quaint, it was respected not only because it was quite common at that time to stake a claim of one's own, but also because the first full paper by Hein, published in 1921, clearly illustrated the great experimental difficulties he experienced in obtaining his results [31]. The yield of the orange product (which Hein described as the *raw bromide*) was "not excellent, scarcely 20% of the chromium trichloride used, and became even lower if care is not taken to insure good cooling and mixing. Otherwise the reaction may proceed explosively with volatilization of all the ether within a few minutes. Then of course, the yield is zero". But even if care was taken, the isolation of the raw bromide was rather complicated and thus one can understand Dietmar Seyferth's comment in his essay on bis(benzene)chromium [32], "that after having read this paper, none of Hein's 'esteemed colleagues' would have felt any desire to intrude in this area of organometallic chemistry".

After he had finished his *Habilitation* in 1921, Hein continued his work on organochromium compounds, and was able to convert the raw bromide into pure pentaphenylchromium hydroxide $(C_6H_5)_5CrOH$. Treatment of this compound with KI or HI surprisingly gave red–brown crystalline $(C_6H_5)_4CrI$ instead of the expected chromium(VI) iodide $(C_6H_5)_5CrI$ [33]. In searching for the fate of the fifth phenyl group, Hein found a substantial amount of phenol and assumed that a "free" phenyl group was generated as an intermediate, and then oxidized. With $(C_6H_5)_5CrOH$ as the starting material, a Reineckate salt with the analytical composition $(C_6H_5)_3Cr[Cr(NH_3)_2(SCN)_4]$ was also isolated, which was quite labile and on exposure to air formed a dark solid that smelled intensely of biphenyl and benzene. In his further research, Hein was able to reduce $(C_6H_5)_4CrI$ electrochemically in liquid ammonia under purified nitrogen to iodine-free, orange–red tetraphenylchromium which was extremely reactive and thermally unstable at room temperature [34]. Similarly to other polyphenylchromium compounds, it decomposed to give biphenyl. Electrolysis of $(C_6H_5)_3CrI$ under similar conditions resulted in the formation of air-sensitive brown–yellow triphenylchromium, which was even less stable than tetraphenylchromium and upon treatment with ethanol gave triphenylchromium hydroxide. Triphenylchromium was also obtained by chemical reduction of $(C_6H_5)_3CrI$ with sodium in liquid ammonia [35].[6]

The formation of $(C_6H_5)_3CrOH$ from $(C_6H_5)_3Cr$ and C_2H_5OH was only one in a series of unusual results, that made the chemistry of the polyphenylchromium compounds mysterious. When Hein summarized his work in 1932 [36], he had to confess that many experiments had been done, a large amount of information had been gathered, but an understanding of the nature of the neutral and ionic arylchromium derivatives still had not been achieved. The fact was that all the salts, independent of whether three, four or five aryl groups were assumed to be linked to the metal centre, had nearly the same colour, all were moderately stable in water, all showed an absorption in the UV spectrum at around 350 nm, and all could be reduced to neutral species. The dilemma of defining oxidation states for the tri-, tetra- and pentaarylchromium compounds became even more evident when Wilhelm Klemm and Anna Neuber determined the magnetic properties of the $(C_6H_5)_5CrX$, $(C_6H_5)_4CrX$ and $(C_6H_5)_3CrX$ derivatives and found that all had a magnetic susceptibility of about 1.73 Bohr Magnetons [37]. This value, which is consistent with the presence of one unpaired electron, was not in agreement with Hein's suggestion of oxidation states VI, V and IV, respectively, for the penta-, tetra- and triphenylchromium compounds and required further explanation. Based on the then current

[6] The reader should take in mind that Hein's formulations of the "polyphenylchromium compounds" were incorrect and had to be revised after Fischer and Zeiss had published their work.

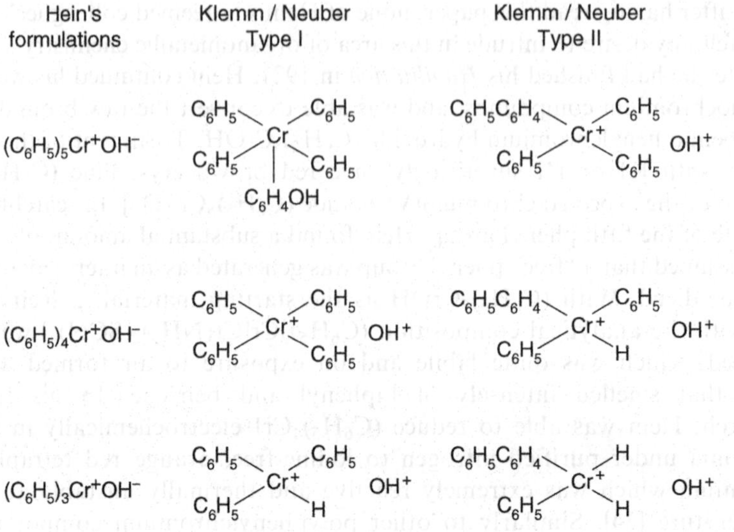

Hein's formulations | Klemm / Neuber Type I | Klemm / Neuber Type II

$(C_6H_5)_5Cr^+OH^-$

$(C_6H_5)_4Cr^+OH^-$

$(C_6H_5)_3Cr^+OH^-$

Fig. 5.4 Hein's formulations and the proposed structures (Type I and II) by Klemm and Neuber for the so-called penta-, tetra- and triphenylchromium derivatives

knowledge of inorganic chemistry, Klemm and Neuber suggested that all three phenylchromium families contained the rare Cr(V) valence state. Two types of structures were proposed (see Fig. 5.4), in which the organic groups were all phenyl (Type I) or in which one of the organic groups was biphenylyl (Type II). Although these proposals seemed to be in agreement with some experimental facts, the presence of hydride ligands and of biphenylyl units in the Type II formulas created a problem. Hein, like Klemm, was not especially happy with the proposed structures, particularly because the hydride ligands were mere speculation. He therefore suggested to react $CrCl_3$ with a mixture of C_6H_5MgBr and $C_6H_5C_6H_4MgBr$ [38], but there is no evidence that this experiment has ever been done.

5.5 Zeiss and Tsutsui: Hein's Work Revisited

Despite the fact that even 20 years after Hein's first paper had appeared the "polyphenylchromium" chemistry was full of anomalies and questions, it was according to Al Cotton "one of the most fascinating and perplexing phases of organometallic chemistry" [39]. While some chemists doubted the validity of Hein's results in the late 1930s and 1940s, the scepticism disappeared after Minoru Tsutsui, a Ph.D. student in Harold Zeiss' group at Yale, successfully repeated Hein's work on the reactivity of $CrCl_3$ towards C_6H_5MgBr. Despite inordinate experimental difficulties, Tsutsui was able to obtain small amounts of the "raw bromide", from which he could prepare the key compounds of Hein's series, $(C_6H_5)_5CrOH$, $(C_6H_5)_4CrI$ and $(C_6H_5)_3CrI$, respectively. Without having the chance to determine the structure of these compounds by X-ray crystallography or mass spectrometry, he decided to get some insight into the bonding mode by using degradation methods. He found that diphenylmercury reacted with lithium aluminium hydride (a compound not available to Hein) to give benzene as the sole organic product, and so treated the tri-, tetra- and pentaphenylchromium compounds with this reagent in diethyl ether. The reaction of $(C_6H_5)_4CrI$ with $LiAlH_4$ produced two equivalents of biphenyl but no benzene, while similar reductive cleavage of $(C_6H_5)_3CrI$ gave biphenyl and benzene in a ratio of 1:1. The hydroxide $(C_6H_5)_5CrOH$ was reduced by $LiAlH_4$ to afford a mixture of biphenyl and phenol. Also in this case no benzene was detected and it thus became clear that Hein's "pentaphenylchromium hydroxide" was in fact a member of the tetraphenylchromium family with its anion not being hydroxide but phenoxide [40]. Moreover, these results disproved the original view that the tetra- and pentaphenylchromium derivatives contained $Cr–C_6H_5$ bonds.

Whereas some anomalies associated with Hein's polyphenylchromium compounds had been resolved on the basis of these experiments, they did not lead directly to the correct structure. It was Lars Onsager of Yale University (who received the 1968 Nobel Prize for Chemistry for his work on irreversible thermodynamic processes) who in early 1954 became interested in the structural problem and made the connection to ferrocene during discussions with Tsutsui and Zeiss. Onsager proposed a *biconoidal* (e.g., sandwich) structure for the tetraphenylchromium cation and pointed to the isoelectronic resemblance between this cation and the known $[Fe(C_5H_5)_2]^+$ ion. The bis(biphenyl)chromium(I) structure, in which chromium is one electron short of the krypton configuration, should be paramagnetic with a magnetic moment of 1.73 Bohr Magnetons, in agreement with Klemm and Neuber's results. The final conclusion was that the electrolytic reduction of Hein's $(C_6H_5)_4CrI$ had not led to $(C_6H_5)_4Cr$ but resulted in the formation of bis(biphenyl)chromium(0). Hein's $(C_6H_5)_3Cr$, obtained upon reduction of $(C_6H_5)_3CrI$, could then possibly be benzene(biphenyl)chromium(0).

Onsager's proposal, that Hein's polyphenylchromium compounds were analogues of ferrocene, was without any doubt a revolutionary idea, and one can understand that initially Zeiss and Tsutsui had trouble getting their results published. At the 126th meeting of the American Chemical Society in September 1954, they presented a paper concerning the structure and bonding of Hein's polyphenylchromium compounds, and at the same time also submitted a preliminary communication to the *Journal of the American Chemical Society*. This manuscript, however, was rejected by the referees "for lack of conclusive evidence", particularly because an X-ray crystal structure analysis was missing [41]. In early 1955, a portion of the results was published in abbreviated form [42, 43] but it was only after the Fischer–Hafner paper appeared at the end of 1955, that the real nature of Hein's compounds as π-arene metal complexes was accepted.[7]

Fig. 5.5 Hein's formulations of the penta-, tetra- and triphenylchromium compounds and the correct structures elucidated by Zeiss and Tsutsui

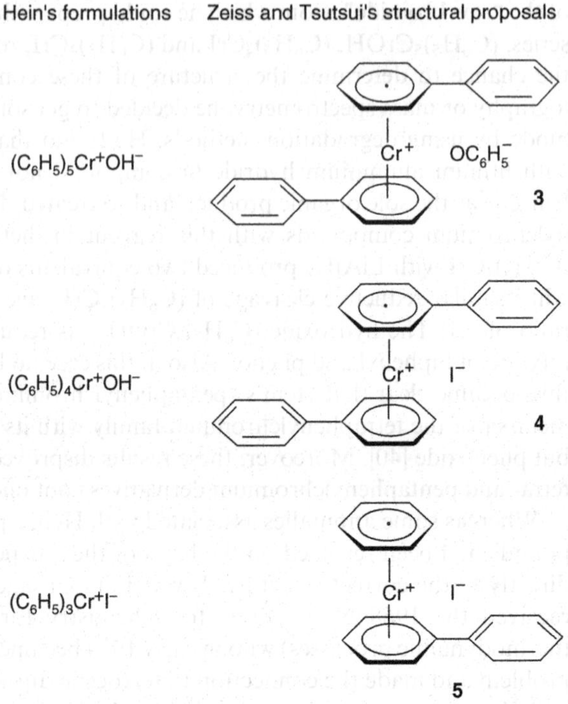

Hein's formulations | Zeiss and Tsutsui's structural proposals

$(C_6H_5)_5Cr^+OH^-$ — Cr^+ $OC_6H_5^-$ **3**

$(C_6H_5)_4Cr^+OH^-$ — Cr^+ I^- **4**

$(C_6H_5)_3Cr^+I^-$ — Cr^+ I^- **5**

In retrospect, there is no doubt that bis(arene)chromium compounds had been prepared by Hein more than 30 years prior to bis(benzene)chromium, but the sandwich structure was not recognized, until the reinvestigation of Hein's work by Zeiss and Tsutsui led to their correct identification. After Fischer and Seus reported the synthesis of bis(biphenyl)chromium(0) and the corresponding chromium(I) cation in August 1956 [44], the reviewers of the *Journal of the American Chemical Society* recommended acceptance of the full paper by Zeiss and Tsutsui, which finally appeared in June 1957 [41]. In this article, Zeiss and Tsutsui drew the structures **3**, **4** and **5** (see Fig. 5.5) for Hein's penta-, tetra- and triphenylchromium compounds, which were in agreement with Fischer's work. Also in 1957, Hein and Eisfeld isolated the iodide of the benzene(biphenyl) chromium(I) cation, by employing a mixture of benzene and biphenyl as the arene components in the Fischer–Hafner synthesis, and found it to be indistinguishable from the original triphenylchromium iodide [45].

[7] In his monograph "*Leaving No Stone Unturned*" (American Chemical Society, Washington, 1993, p. 24) Gordon Stone probably expressed the opinion of most of his contemporaries by saying that "the idea of a metal atom hexahapto-coordinated to a benzene ring... was evidently alien to some referees of that era. Since ferrocene was by then well known they should have had more imagination".

Analysing Hein's work on the reaction of $CrCl_3$ with C_6H_5MgBr and the repetition of these studies by Tsutsui, Harold Zeiss was puzzled by the failure to find any "diphenylchromium" salts, which in fact should be salts of the bis (benzene)chromium cation [40]. He assumed that the corresponding hydroxide $[Cr(C_6H_6)_2]OH$ had been missed by Hein because of its high water solubility, and thus suggested to his coworker Walter Herwig (a former Ph.D. student of Georg Wittig's at Heidelberg) to look for it in the aqueous phases left after isolation of the $[(C_6H_6)Cr(C_6H_5C_6H_5)]^+$ and $[Cr(C_6H_5C_6H_5)_2]^+$ products. After addition of $NaBPh_4$ to the aqueous solution, Herwig could indeed precipitate the tetraphenyloborate of the $[Cr(C_6H_6)_2]^+$ cation, which on reduction gave neutral bis(benzene)chromium that was identical in all respects with the compound described by Fischer and Hafner [46]. Thus it turned out that Hein had the first bis(arene) metal cation right from the start and that the history of bis(arene) metal complexes did indeed begin in 1919 and not in the 1950s when the Fischer–Hafner and the Zeiss–Tsutsui papers appeared.

Before the next steps in the historical development of sandwich complexes, and of bis(arene) metal compounds in particular, will be reviewed, it is worth briefly commenting on Hein's view about the results reported by Fischer and Zeiss. On the one hand, he was of course shocked to realize that for more than three decades he had been convinced that he had prepared the first genuine polyaryl transition metal compounds which proved to be erroneous. On the other hand, he was relieved that the long lasting mystery about the structural chemistry of these compounds had been solved. When I visited him at his home in Jena in the early 1970s and told him about the preparation of our triple-decker sandwiches, he said that he was very pleased to see the progress in the field of sandwich-type complexes. However, he emphasized again (as he had already done in a paper published in 1956 [47][8]) that had he considered a sandwich structure for the polyphenylchromium compounds in the 1920s and 1930s, he would not have been taken seriously. The accepted wisdom about the reactivity of metal salts MX_n towards Grignard reagents RMgBr at that time definitely predicted that the substitution of X for R would occur with no change in the nature and composition of the organic group. At the end, Hein was very satisfied that Herwig and Zeiss [48] and, independently, he and his coworker Richard Weiss [49] could confirm that σ-bonded polyphenylchromium compounds did in fact exist (see Chap. 9) and could rearrange intramolecularly to π-arene chromium complexes [50].[9]

[8] Hein wrote (translation in English): "It should be mentioned, and briefly explained, why I considered the organochromium compounds, at the time of their discovery, to be 'tetraphenylchromium' salts, and not formulations of the type now suggested by H. Zeiss and E. O. Fischer. The main reason was that at that time, there were no precedents for such a view, and I firmly believe that if I had brought such a formulation of this type into consideration, I would not have taken seriously in the context of the state of knowledge at the time".

[9] For a detailed discussion of the σ-to-π conversion of oligoarylchromium compounds to π-arene chromium complexes, see [50].

With the existence of bis(benzene)chromium and its analogues being established, the Fischer–Hafner synthesis was extended to numerous other transition metals. Besides $Mo(C_6H_6)_2$ and $W(C_6H_6)_2$, a whole series of neutral and cationic bis(arene) complexes of vanadium, rhenium, iron, ruthenium, osmium, cobalt, rhodium, iridium and even technetium were prepared, most of them in Fischer's laboratory [51]. While initial studies on the functionalization of the arene ligands in $Cr(C_6H_6)_2$ met with limited success, ring metalation with excess *n*-amylsodium under carefully chosen conditions gave $(C_6H_6)Cr(C_6H_5Na)$ that could be converted to $(C_6H_6)Cr(C_6H_5CH_2OH)$, $(C_6H_6)Cr(C_6H_5CH(OH)CH_3)$, $(C_6H_6)Cr(C_6H_5CO_2CH_3)$, etc. [52, 53]. By using *n*-butyllithium instead of *n*-amylsodium, in the presence of tmeda, Christoph Elschenbroich (one of E. O. Fischer's Ph.D. students, who later became Professor of Inorganic Chemistry at the University of Marburg) succeeded in obtaining pure bismetalated $Cr(C_6H_5Li)_2$, which upon treatment with $(CH_3)_3SiCl$ and other electrophiles gave a variety of ring-substituted complexes $Cr(C_6H_5R)_2$ [54–56]. With $Cr(C_6H_5Li)_2$ as the starting material, the silylene-bridged and germylene-bridged compounds, $Cr(C_6H_5SiR_2C_6H_5)$ and $Cr(C_6H_5GeR_2C_6H_5)$ ($R{=}CH_3$, C_6H_5) [57, 58] as well as phosphino derivatives $Cr(\eta^6\text{-}C_6H_5PR_2)_2$ (formally bis(phosphine) chromium(0) complexes) could be prepared [59]. Elschenbroich's group also made use of the metal atom/arene vapour cocondensation technique, pioneered by Philip Skell at the Pennsylvania State University and Peter Timms at the University of Bristol in the late 1960s [60, 61], to obtain disubstituted bis(benzene)chromium compounds. Timms' method of metal atom/ligand vapour cocondensation [62] had a significant advantage over the reductive Friedel–Crafts and Grignard procedures insofar as it could be used to prepare bis(arene) chromium complexes of all kinds of substituted arenes, polynuclear aromatic hydrocarbons and heteroarenes such as pyridine, phospha- and arsabenzene.[10] Malcolm Green at Oxford (see Fig. 7.12) extended this work and obtained the first bis(arene) metal compounds of titanium, zirconium, hafnium, niobium, tantalum and even the lanthanides [63]. In the meantime, substantial progress had been made in this field and, by using a modified version of Timms' apparatus, the metal atom/ligand vapour cocondensation technique was introduced in many advanced inorganic laboratory courses.

Apart from bis(benzene)chromium and other bis(arene) metal compounds, a huge number of mono(arene) transition metal complexes were also prepared in the "golden 1950s". In 1957, Fischer and Öfele reported the synthesis of $(C_6H_6)Cr(CO)_3$, the first representative of the family of arene metal carbonyls [64]. The original preparative procedure (a sealed-tube reaction of $Cr(CO)_6$ and $Cr(C_6H_6)_2$ at 220°C) was soon replaced by a more convenient synthesis, based

[10] For a summary of this work see D. Seyferth, Bis(benzene)chromium. 2. Its Discovery by E. O. Fischer and W. Hafner and Subsequent Work by the Research Groups of E. O. Fischer, H. H. Zeiss, F. Hein, C. Elschenbroich, and Others, *Organometallics* **21**, 2800–2820 (2002).

on the direct reaction of $Cr(CO)_6$ with the arene. This procedure, which was simultaneously developed by three research groups [65–67], was found to be much more versatile and applicable to various substituted arenes, such as chlorobenzene, aniline, phenol, etc., as well as heteroarenes such as thiophene or methylpyridine. As nicely illustrated by Martin Semmelhack, Peter Kündig and others [68–70], the arene chromium tricarbonyl complexes have a well-defined chemistry and in the latter part of the twentieth century have become valuable reagents and/or intermediates in organic synthesis.

5.6 Wilkinson's Next Steps

Although Fischer's work on bis(benzene)chromium and benzene chromium tricarbonyl made a big splash in scientific journals for several years, Wilkinson ("the other") did not give up and intensified his research on novel transition metal organometallics with great vigour. As Cotton recalled [71], for "him to stay abreast continued to be a challenge". The next highlights from Wilkinson's lab were the characterization of $(C_5H_5)_2ReH$, being not only the first bis(cyclopentadienyl) metal hydride but also the first molecular transition metal hydride to be studied by ^1H NMR spectroscopy [72],[11] the preparation of the first cyclopentadienyl actinide compounds $(C_5H_5)_3MCl$ (M=Th, U) [73], and the synthesis of chromium, molybdenum and iron complexes having alkyl groups attached to $(C_5H_5)M(CO)_n$ moieties [74]. At the time when these studies were done, it was still uncertain whether transition metal–alkyl bonds were stable and corresponding complexes isolable under normal conditions. The work on $(C_5H_5)Fe(CO)_2R$ derivatives also included the synthesis of $(\eta^5\text{-}C_5H_5)Fe(CO)_2(\eta^1\text{-}C_5H_5)$, a compound that had independently been prepared in Peter Pauson's laboratory [75]. The interpretation of the ^1H NMR spectrum, at room temperature, of this compound which, based on the 18-electron rule, was believed to contain one π-bonded and one σ-bonded cyclopentadienyl ring, initially created a problem since it not only showed a single resonance for the $\eta^5\text{-}C_5H_5$ ligand (as expected), but also a single signal (slightly broader) for the five hydrogen atoms of the $\eta^1\text{-}C_5H_5$ ring, instead of the anticipated A_2B_2X pattern [76].[12] Wilkinson explained this observation by assuming that "the metal is executing a 1,2 rearrangement at a rate greater than the expected chemical shift (difference)... and the cyclopentadienyl group may thus be regarded as rotating... and all of the protons thus become equivalent" [74]. Although this was a clever idea, it was not until a decade later that it was shown to be correct [77]. Nowadays, the bis(cyclopentadienyl) iron dicarbonyl complex $(\eta^5\text{-}C_5H_5)Fe(CO)_2(\eta^1\text{-}C_5H_5)$ and the related chromium compound $(\eta^5\text{-}C_5H_5)Cr(NO)_2(\eta^1\text{-}C_5H_5)$ are credited as being the first recognized examples of

[11] $(C_5H_5)_2ReH$ was first made by Fischer, but due to the lack of NMR instrumentation in München could not be adequately characterized (see [71]).

[12] For the η (eta) or the originally used h (hapto) nomenclature.

fluxional organometallic molecules, and in this respect were followed by many others. In a review published in 1975, Cotton described the main developments in this field, also including the unusual nonrigidity of metal carbonyls and their derivatives [78].

5.7 From Sandwich Complexes to Organometallic Dendrimers

The chemistry of mixed cyclopentadienyl(arene) transition metal complexes also became a topic of general interest in the late 1950s, and remains so to date. The first compound of this family was diamagnetic $(C_5H_4Me)Mn(C_6H_6)$, which was prepared by treatment of C_6H_5MgBr with in situ generated (not well characterized) methylcyclopentadienyl manganese chloride in tetrahydrofuran by Tom Coffield's group [79]. Shortly thereafter, Fischer and Kögler reported the synthesis of the paramagnetic complex $(C_5H_5)Cr(C_6H_6)$, providing the link between $Cr(C_5H_5)_2$ and $Cr(C_6H_6)_2$ [80]. Salts of cationic compounds of the general composition $[(C_5H_5)Fe(arene)]^+$ were prepared by Coffield [79] and Wilkinson [81] in the late 1950s, and their chemistry was thoroughly investigated by Aleksandr Nesmeyanov and his school in Moscow in the following two decades [82, 83]. In the course of these studies it was shown that the 18-electron cations $[C_5H_5Fe(arene)]^+$ could be reduced electrochemically to the 19-electron molecules $(C_5H_5)Fe(arene)$, which for benzene and sterically unprotected arenes are thermally unstable and disproportionate to give ferrocene and the respective arene. If, however, the aromatic ligand is peralkylated or sterically protected, the paramagnetic complexes $(C_5R_5)Fe(arene)$ ($R{=}H, CH_3$), prepared by reduction of the corresponding cations with sodium amalgam, proved to be quite stable and, as Didier Astruc (Fig. 5.6) has demonstrated in a series of elegant studies, can be used as electron-reservoirs. These are by definition compounds "which store or transfer electrons stoichiometrically or catalytically without decomposition" [84, 85]. One of several interesting observations was that by using $(C_5H_5)Fe(C_6Me_6)$ as the starting material, O_2 can be reduced to the superoxide radical anion O_2^-, the potassium salt of which was discovered by Gay-Lussac as early as in 1805. The oxidized product of the reaction of $(C_5H_5)Fe(C_6Me_6)$ with O_2 is the cyclohexadienyl iron(II) complex $(C_5H_5)Fe(C_6Me_5CH_2)$, a relative to ferrocene, containing an exocyclic carbon–carbon double bond that is not involved in the coordination to the metal [84, 85].

Another fascinating result of Astruc's work was that all the six methyl groups of the cation $[(C_5H_5)Fe(C_6Me_6)]^+$ (**6**) could be deprotonated stepwise by *t*BuOK to generate highly reactive intermediates which couple with electrophiles RX to give new ring-substituted products [86]. The potential of this methodology is shown by the synthesis of the sandwich complex **7** (see Scheme 5.2), which in the presence of hexamethylbenzene undergoes a photochemical ring-ligand exchange to regenerate **6** and affords the otherwise difficultly accessible free hexabutenylbenzene **8**. Similar to 1,3,5-$C_6H_3[C(CH_2CH{=}CH_2)_3]_3$,

Fig. 5.6 Didier Astruc (born 1946 in Versailles) studied chemistry at the University of Rennes, where he received his Ph.D. with Professor Rene Dabard in 1975. He then moved to MIT as a NATO Postdoctoral Fellow, where he worked with the 2005 Nobel laureate Richard R. Schrock. After being a Lecturer and Master Lecturer at the University Institute for Technology of Saint-Nazaire, he worked for the CNRS at Rennes where he became *Maitre de Recherche* in 1982. Since 1983 he is Professor of Chemistry at the University of Bordeaux I and has been promoted to the exceptional class of university professors in 1996. His research interests comprise preparative and mechanistic organometallic chemistry, catalysis, and electron transfer processes. More recently, he has developed the synthesis and supramolecular electronics of organometallic dendrimers. He is the author of "*Electron Transfer and Radical Processes in Transition-Metal Chemistry*" and of the standard textbook "*Organometallic Chemistry and Catalysis*". A recipient of several major research awards, Didier is also a senior member of the Institut Universitaire de France, a member of the Academia Europeae, London, and the German Academy Leopoldina, and a Fellow of the Royal Society of Chemistry (photo by courtesy from D. A.)

the latter is an excellent precursor for preparing new types of stable redox active dendrimers, of which an example is shown in Fig. 5.7 [87]. This giant dendrimer contains 243 ferrocenyl units at the periphery and displays a single wave in cyclic-voltammetric studies [88]. At present it is probably one of the largest

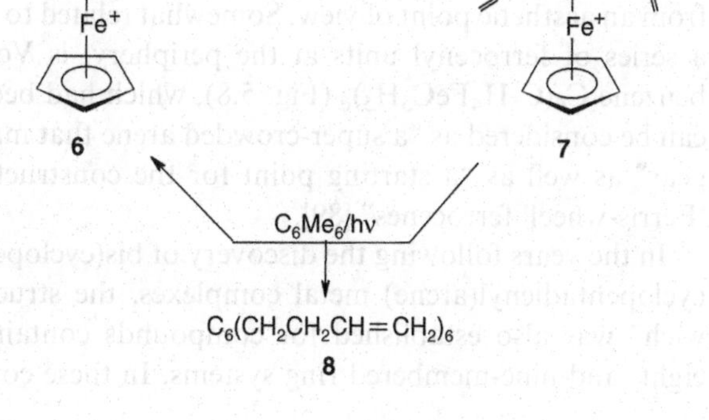

1) *t*BuOK
2) C_3H_5Br
Fe^+
6
Fe^+
7
$C_6Me_6/h\nu$
$C_6(CH_2CH_2CH{=}CH_2)_6$
8

Scheme 5.2 A simple synthetic route to hexabutenylbenzene **8**, being a useful starting material for the preparation of redox active dendrimers

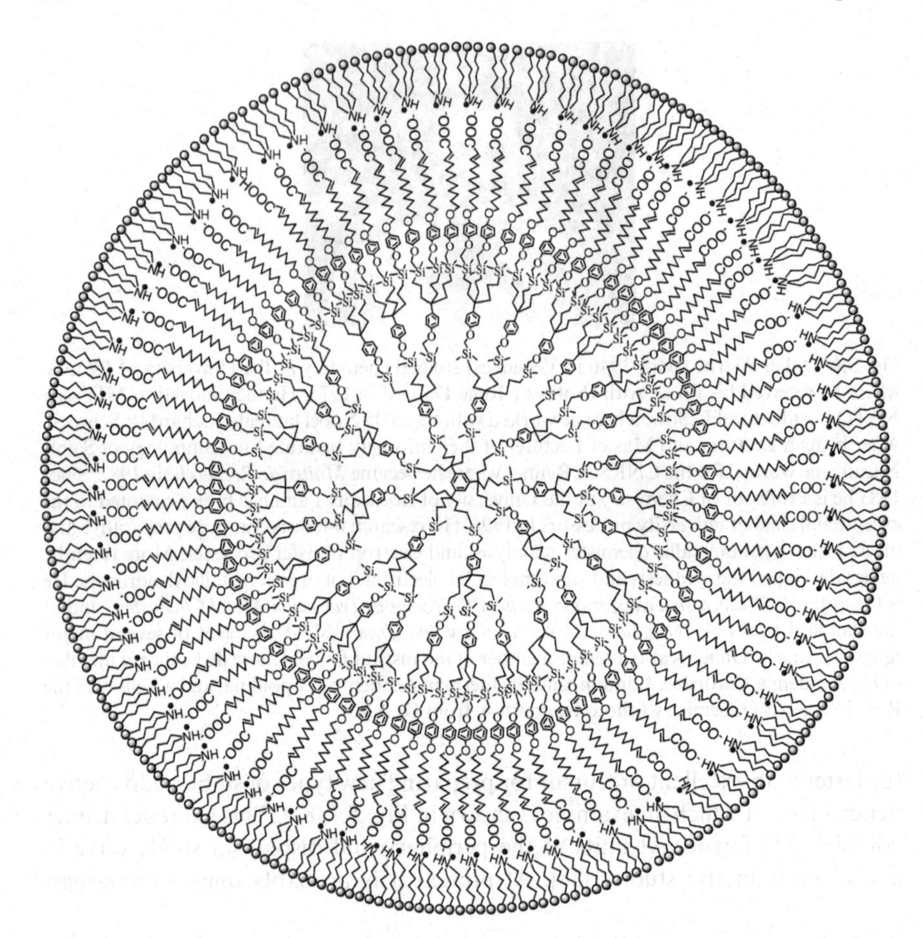

Fig. 5.7 A 243-ferrocenyl dendrimer prepared from a dendritic precursor, having 81 CO_2H groups at the periphery, and the tertiary amine $N[(CH_2)_4C_5H_4Fe(C_5H_5)]_3$ in THF at room temperature. The tiny balls at the periphery of the giant dendrimer represent $(C_5H_4)Fe(C_5H_5)$ units (from ref. 87; reproduced with permission of Wiley-VCH)

organometallic molecules and impressive not only from a scientific but also from an aesthetic point of view. Somewhat related to Astruc's compounds, with a series of ferrocenyl units at the periphery, is Vollhardt's hexa(ferrocenyl) benzene $C_6(C_5H_4FeC_5H_5)_6$ (Fig. 5.8), which had been sought for decades and can be considered as "a super-crowded arene that may function as a molecular gear" as well as "a starting point for the construction of cyclic hexa-decker 'Ferris-wheel' ferrocenes" [89].

In the years following the discovery of bis(cyclopentadienyl), bis(arene) and cyclopentadienyl(arene) metal complexes, the structural feature of a "sandwich" was also established for compounds containing three-, four-, seven-, eight- and nine-membered ring systems. In these compounds the metal centre

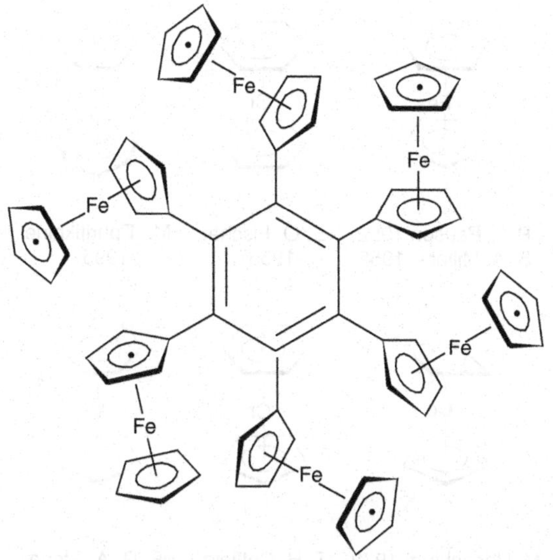

Fig. 5.8 Hexa(ferrocenyl)benzene, prepared in Peter Vollhardt's group by sixfold Negishi-type ferrocenylation of hexaiodobenzene and characterized by X-ray crystallography

is surrounded either by two equal or two different planar C_nH_n or C_nR_n ligands, as is illustrated by the two series of complexes shown in Fig. 5.9. The chromium derivative $(C_5H_5)Cr(C_7H_7)$ deserves particular attention as it is an isomer of bis(benzene)chromium and, analogous to this, easily oxidized to the paramagnetic $[(C_5H_5)Cr(C_7H_7)]^+$ cation [90]. This cation is isoelectronic with neutral $(C_5H_5)V(C_7H_7)$, which was prepared by King and Stone as early as in 1959 and has been the first representative of the $(C_5H_5)M(C_7H_7)$ family [91].

The only sandwich compound containing two C_nH_n or C_nR_n ring ligands with $n=3$ or 4 is, to the best of my knowledge, the nickel complex $Ni[C_4(C_6H_5)_4]_2$, occasionally described as the "smaller ring brother of ferrocene and bis(benzene)-chromium". Its synthesis, however, was preceded by the serendipitous generation of $[(C_4Me_4)NiCl_2]_2$, which has a unique place in the history of organometallic compounds, being the first transition metal complex with a π-bonded ring ligand that was unknown in its free, non-coordinated form (for the method of synthesis see Chap. 1). Almost at the same time as Criegee's report about the preparation and characterization of $[(C_4Me_4)NiCl_2]_2$ appeared [92], Walter Hübel (like E. O. Fischer, a former Ph.D. student of Hieber) communicated the isolation of several new iron carbonyl compounds, which were formed by treatment of $Fe(CO)_5$ or $Fe_3(CO)_{12}$ with diphenylacetylene [93, 94]. One of these compounds was an unusually stable yellow solid, for which the elemental analysis indicated the composition $(PhC_2Ph)_2Fe(CO)_3$. Two possible structures were taken into consideration: One in which the $Fe(CO)_3$ group is part of a delocalized organoiron heterocycle, and one in which the iron tricarbonyl group is coordinated to a planar

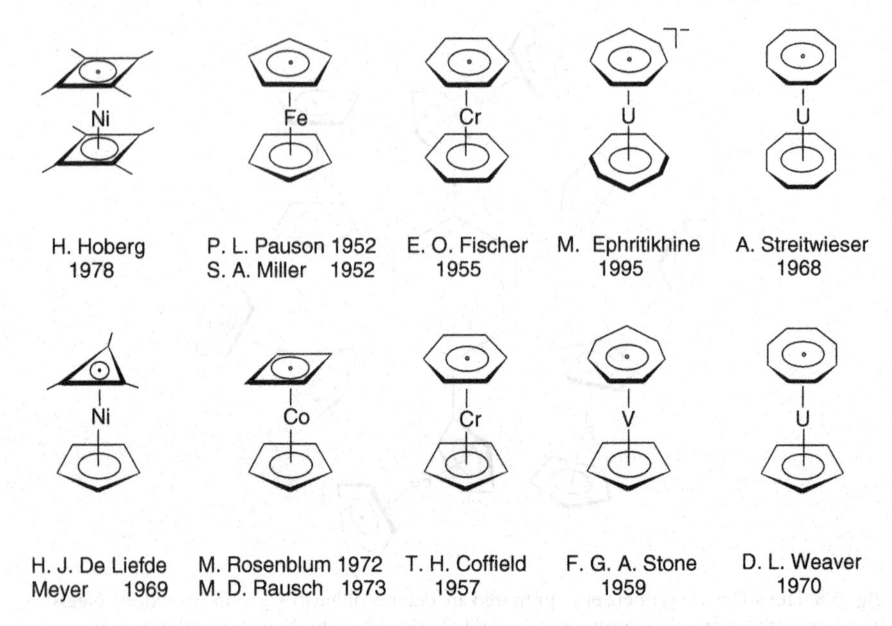

Fig. 5.9 Two series of sandwich-type compounds containing either two equal or two different planar carbocyclic ring systems (with the names of the principal investigators and the date of discovery)

tetraphenylcyclobutadiene ring. Richard Dodge and Verner Schomaker carried out an X-ray crystal structure analysis and confirmed the latter proposal [95].

5.8 The Taming of Cyclobutadiene: A Case of Theory before Experiment

Criegee's and Hübel's work found world-wide attention because it revealed that unknown, long-sought organic molecules can be trapped by coordination to transition metals. Even more importantly, it represented a rare "case of theory before experiment", as Dietmar Seyferth has pointed out [96], referring to the fact that a few years previously Christopher Longuet-Higgins and Leslie Orgel had published an article entitled "The Possible Existence of Transition Metal Complexes of cycloButadiene" [97]. Based on MO theory, Longuet-Higgins and Orgel suggested that formally zerovalent metals should form 18-electron compounds such as $(C_4H_4)Ni(CO)_2$ or $(C_4H_4)Fe(CO)_3$, and they predicted that cyclobutadiene metal complexes could be more stable than their cyclopentadienyl analogues. This seemed to be true for $[(C_4Me_4)NiCl_2]_2$ and $(C_4Ph_4)Fe(CO)_3$ as well as for $[(C_4Ph_4)PdX_2]_2$ and $(C_4Me_4)Co(C_5H_5)$, the latter two compounds being highlights of Peter Maitlis' work in the early 1960s [98, 99]. However, the question remained whether similar species with the *unsubstituted* cyclobutadiene as ligand would be stable as well.

At this time, in the early 1960s, Rowland Pettit entered the stage and remained the leading figure in the field of cyclobutadiene transition metal complexes until his untimely death. Before he began his work, he had already been interested in the chemistry of acyclic diolefin iron tricarbonyls (see Chap. 7), which he used, inter alia, for the preparation of novel π-allyl, π-pentadienyl and π-hexadienyl iron compounds [100]. Based on this experience, and being aware of Criegee's synthesis of $[(C_4Me_4)NiCl_2]_2$ from a dichlorocyclobutene precursor, Pettit suggested to his Ph.D. student George Emerson to carry out the reaction of $Fe_2(CO)_9$ with *cis*-3,4-dichlorocyclobutene. This was done in pentane at 30°C and, after removal of the solvent and vacuum distillation, a yellow liquid was obtained that crystallized on cooling from pentane as pale yellow prisms with a melting point of 26°C [101]. It was cyclobutadiene iron tricarbonyl. An alternate route, with the salt $Na_2[Fe(CO)_4]$ instead of $Fe_2(CO)_9$ as the starting material, proved to be less convenient for $(C_4H_4)Fe(CO)_3$ but worked well for $(C_4H_4)Ru(CO)_3$ and $(C_4H_4)M(CO)_4$ (M=Mo, W), with $Na_2[Ru(CO)_4]$ and $Na_2[M(CO)_5]$ as the precursors [102, 103]. Subsequently to Pettit's work, Myron Rosenblum and Marvin Rausch showed, that α-pyrone could also be used as a starting material to generate cyclobutadiene, and that complexes such as $(C_4H_4)Co(C_5H_5)$, $(C_4H_4)Rh(C_5H_5)$ and (C_4H_4) $V(C_5H_5)(CO)_2$ were accessible via this route [104–106].

Prior to the exact determination of the molecular structure of (C_4H_4)-$Fe(CO)_3$ [96], Pettit and his group had begun to develop the chemistry of this unique half-sandwich type complex in great breadth. Already in their first communication on the reactivity of $(C_4H_4)Fe(CO)_3$, it was stated that this compound "is aromatic in the sense that it undergoes electrophilic substitution reactions to yield a series of new cyclobutadiene complexes" [107]. In less than a decade, it was confirmed that almost all types of electrophilic substitution reactions, known for benzene and ferrocene, could be carried out with the cyclobutadiene iron complex. Once a key functional group had been introduced into the four-membered ring of $(C_4H_4)Fe(CO)_3$, standard functional group conversions were possible, as in the case of ferrocene [108]. Even more interesting were those reactions of $(C_4H_4)Fe(CO)_3$ (**9**, see Scheme 5.3) and its ring-substituted derivatives, in which free, transient cyclobutadienes were generated by oxidation, in particular with ceric ammonium nitrate. Several trapping reagents were used and cubane and homocubyl derivatives such as **10** and **11** could be prepared apart from bicyclohexenes and bicyclohexadienes [101, 108]. Further studies of Pettit's group were devoted to an understanding of the nature of the ground state of free cyclobutadiene, which very probably possesses a rectangular rather than a square structure and is therefore a singlet diene and not a triplet diradical. More recently, the chemistry of cyclobutadiene transition metal complexes was further highlighted when Akira Sekiguchi's group at the University of Tsukuba in Japan reported the isolation and structural characterization of the tetrasilacyclobutadiene iron tricarbonyl $(Si_4R_4)Fe(CO)_3$

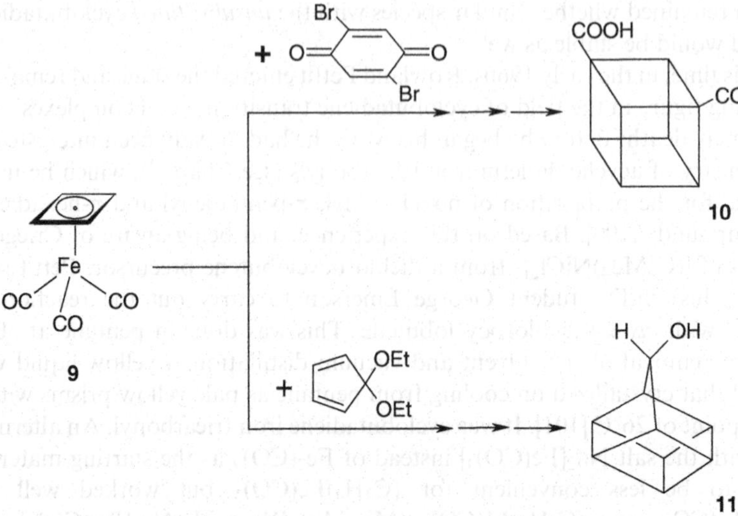

Scheme 5.3 Two examples illustrating the utility of $(C_4H_4)Fe(CO)_3$ as starting material for the preparation of otherwise hardly accessible organic molecules

(R=SiMe*t*Bu$_2$), which they emphatically described as "the silicon cousin of Pettit's $(C_4H_4)Fe(CO)_3$" [109]. The sila-analogue was obtained by treatment of $Na_2[Fe(CO)_4]$ with the 3,4-dibromo-cyclotetrasilene $R_4Si_4Br_2$ in THF and thus by a similar route as its "cousin" and various other metal complexes with C_4H_4, C_4Me_4 and C_4Ph_4 as ligands.

5.9 The Smaller and Larger Ring Brothers of Ferrocene

As already mentioned, the only metal sandwich with two cyclobutadiene ligands is the nickel complex $Ni(C_4Ph_4)_2$ (**13**) which was prepared by Heinz Hoberg and coworkers at the Max Planck Institute in Mülheim in 1978 [110]. Heinz Hoberg, one of Karl Ziegler's former Ph.D. students, was familiar with the handling of highly air- and moisture-sensitive organoaluminium compounds such as **12** and, without this experience, might not have been able to successfully perform the reaction shown in Scheme 5.4. After skilful work-up, the sandwich complex **13** was isolated in only 5% yield. The air-stable crystalline solid, which is inert towards CO at 110°C, has a melting point of 404°C and thus belongs to the thermally most stable organometallic compounds presently known.

While $Ni(C_4Ph_4)_2$ can be considered as the "smaller ring brother of ferrocene and bis(benzene)chromium", the "larger ring brother" is the uranium compound $U(C_8H_8)_2$. Its synthesis was reported by Ulrich Müller-Westerhoff and Andrew Streitwieser from the University of California at Berkeley in 1968, and it was the

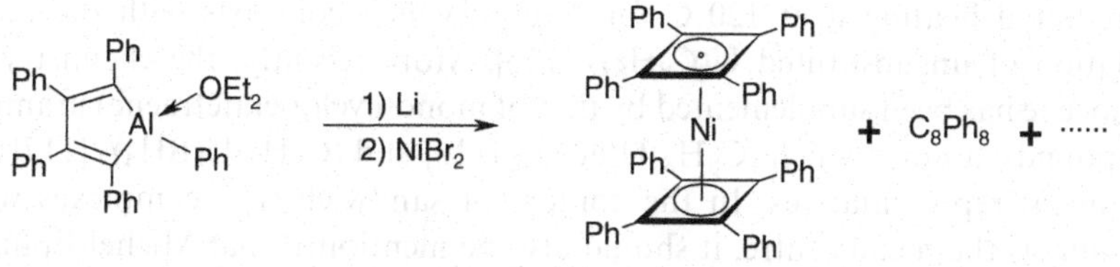

Ph, Ph, OEt$_2$, Al, Ph, Ph, Ph — 1) Li, 2) NiBr$_2$ → Ph, Ph, Ph, Ph, Ni, Ph, Ph, Ph, Ph + C_8Ph_8 + ...

12 13

Scheme 5.4 The one-pot synthesis of bis(tetraphenylcyclobutadiene)nickel **13**, being the only metal sandwich with two cyclobutadiene ligands

first sandwich complex of the general composition $M(C_8H_8)_2$ [111]. The key to success for obtaining $U(C_8H_8)_2$ was the utilization of the dipotassium salt of the 10-π-electron system $C_8H_8^{2-}$, which had been prepared and spectroscopically characterized by Tom Katz in 1960 [112]. Ulrich (Ulli) Müller-Westerhoff (see Fig. 2.31), who had already used the dipotassium salt for his fruitless attempts to generate octalene, treated UCl_4 with $K_2C_8H_8$ in THF and obtained beautiful green crystals, identified as $U(C_8H_8)_2$ by mass spectrometry. As he recalled 35 years later, Andrew Streitwieser and he celebrated this result with a bottle of champagne and christened "the new compound uranocene, to highlight its similarity to ferrocene" [113]. The isolation and characterization of $U(C_8H_8)_2$ was, without any doubt, an experimental masterpiece, since the compound is extremely air-sensitive, going up in flames when exposed to air and being rapidly decomposed by aqueous bases and acids. Despite this sensitivity, Ken Raymond and Allen Zalkin were able to determine its crystal structure and proved the D_{8h} molecular symmetry with two planar octagonal rings coordinated to uranium at the centre. They concluded from this result that there "is no question that this is a pseudo-aromatic ring and that di-π-cyclooctatetraeneuranium or 'uranocene' is an authentic π sandwich complex of the 5f transition series" [114]. One should keep in mind, however, that apart from the structural analogy there is also an important difference between $Fe(C_5H_5)_2$ and $U(C_8H_8)_2$ as the latter has two unpaired electrons and therefore is paramagnetic.

Following the preparation and characterization of uranocene, extensive synthetic studies were carried out to further develop this new area of actinide organometallic chemistry. Besides Streitwieser and his group, other laboratories in the US and Europe were involved and succeeded within a relatively short period of time (mainly between 1969 and 1974) in isolating thorium, protactinium, neptunium and plutonium compounds of the general composition $M(C_8H_8)_2$ [113]. While the cyclooctatetraene dianion route was used in most cases, $Th(C_8H_8)_2$, $Pu(C_8H_8)_2$ and $U(C_8H_8)_2$ as well were also accessible by treatment of finely divided metals with cyclooctatetraene. Both thorocene and plutonocene are isomorphous with the uranium analogue. Among the numerous substituted uranocenes, 1,1',3,3',5,5',7,7'-octaphenyluranocene deserves particular attention as it sublimes unchanged at 400°C, and its oxidation by air required heating it at 120°C for 4 days, which contrasts with the facile oxidation of unsubstituted $U(C_8H_8)_2$ [115]. More recently, the chemistry of uranocene has been supplemented by that of mono(cyclooctatetraene) uranium compounds, among which $(C_8H_8)U(acac)_2$ [116] and $(C_8H_8)U(BH_4)_2$ [117] are prominent representatives. In the context of sandwich-type complexes with uranium as the metal centre, it should also be mentioned that Michel Ephritikhine and coworkers at the *Laboratoire de Chimie de l'Uranium* in Saclay near Paris reported the preparation of a stable salt of the anion $[U(C_7H_7)_2]^-$ in 1995. This is the first (and as far as I know) still the only structurally characterized sandwich compound with two planar C_7H_7 ring ligands [118].

5.10 Sandwiches with P_5 and Heterocycles as Ring Ligands

In addition to transition metal complexes of the general composition $(C_nR_n)M(C_mR'_m)$ (R and R'═H, Me, Ph, etc.), the sandwich family has also been extended to compounds in which instead of two carbocyclic ligands, one or two *heterocyclic* ring systems are linked to the metal centre. Some relatives of ferrocene, in which up to five CH units of $Fe(C_5H_5)_2$ are replaced by nitrogen, phosphorus or arsenic, are depicted in Fig. 5.10. The pioneers in this field were Peter Pauson, Francois Mathey, John Nixon and Otto Scherer [119–124], the latter two of whom demonstrated that phosphaalkynes and white phosphorus can be used as precursors for the preparation of phosphacyclopentadienyl complexes. The challenging quest remains, of course, the synthesis of decaphosphaferrocene $Fe(P_5)_2$, considered to be the "holy grail" in this area. Despite some unsuccessful attempts, it should be a promising goal since DFT calculations predict that the bond dissociation energy yielding the iron atom and two cyclo-P_5 ligands is nearly the same as for ferrocene [125]. A step nearer to reaching this goal was accomplished in 2002 by John Ellis (see Fig. 4.5), who prepared stable salts of the diamagnetic $[Ti(P_5)_2]^{2-}$ dianion, the first carbon-free sandwich complex [126]. Regarding analogues of bis(benzene)chromium, a real landmark was the preparation of bis(pyridine)chromium, carried out by Christoph Elschenbroich in 1988 (see Scheme 5.5). The decisive synthetic step was the oxidative displacement of the trimethylsilyl substituents from the two six-membered rings of **14** in the presence of water, thus generating the $[Cr(C_5H_5N)_2]^+$ cation as an intermediate, which was subsequently reduced by sodium dithionite to **15** [127]. A decade before, Joseph Lagowski from the University of Texas at Austin had reported the synthesis of a relative of compound **14** with methyl instead of trimethylsilyl substituents at the two rings [128]. The Austin team had also applied the metal atom/ligand vapour cocondensation technique and had used the same apparatus as Peter Timms for the preparation of $Cr(PF_3)_6$, $Fe(PF_3)_5$, $Ni(PF_3)_4$ and other metal complexes with PF_3 and PH_3 as ligands [129].

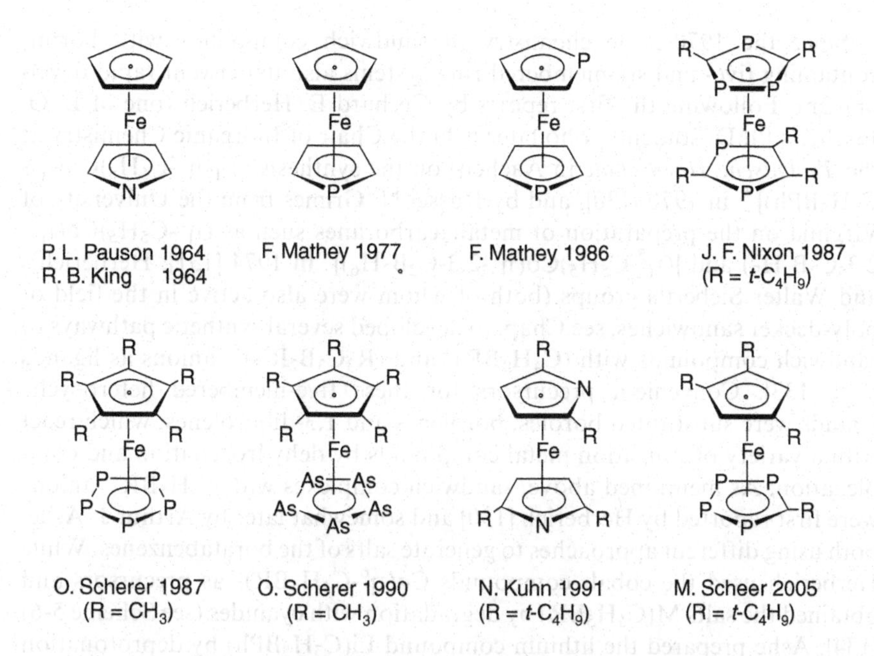

Fig. 5.10 Iron sandwich complexes, which are structurally related to ferrocene but contain either one or two planar heterocyclic five-membered rings (with the names of the principal investigators and the date of discovery)

Scheme 5.5 Preparation of bis(pyridine)chromium **15** via the ring-substituted derivative **14**, which is obtained by a metal atom/ligand vapour cocondensation reaction

Since the 1970s, the chemistry of sandwich compounds with boron-containing five- and six-membered ring systems also underwent rapid development. Following the first reports by Gerhard E. Herberich (one of E. O. Fischer's Ph.D. students, who later held the Chair of Inorganic Chemistry at the *Technische Hochschule* in Aachen) on the synthesis of $[(\eta^5\text{-}C_5H_5)Co(\eta^6\text{-}C_5H_5BPh)]^+$ in 1970 [130], and by Russel N. Grimes from the University of Virginia on the preparation of metallacarboranes such as $(\eta^5\text{-}C_5H_5)Co(\eta^5\text{-}2,3\text{-}C_2B_3H_7)$ and $[(\eta^5\text{-}C_5H_5)Co(\eta^5\text{-}2,3\text{-}C_2B_3H_6)]^-$ in 1974 [131], Herberich's and Walter Siebert's groups (both of whom were also active in the field of poly-decker sandwiches, see Chap. 6) developed several synthetic pathways to sandwich compounds with $(C_4H_4BR)^-$ and $(R_3C_3B_2R'_2)^{2-}$ anions as ligands [132, 133]. Convenient precursors for these five-membered heterocyclic ligands were substituted boroles, borolenes and 1,3-diborolenes, which react with a variety of transition metal compounds by dehydrogenation and complexation. As mentioned above, sandwich complexes with $C_5H_5BR^-$ anions were first reported by Herberich [130] and somewhat later by Arthur J. Ashe, both using different approaches to generate salts of the boratabenzenes. While Herberich used the cobalt compounds $Co(\eta^6\text{-}C_5H_5BR)_2$ as precursors and obtained the salts $M(C_5H_5BR)$ by degradation with cyanides (see Scheme 5-6) [134], Ashe prepared the lithium compound $Li(C_5H_5BPh)$ by deprotonation of 1-phenylbora-2,5-cyclohexadiene, the latter being obtained from the corresponding stannacyclohexadiene derivative [135]. The alkali metal boratabenzenes react cleanly with transition metal halides or other starting materials to yield sandwich as well as half-sandwich type compounds with $[C_5H_5BPh]^-$ and $[C_5H_5BMe]^-$ as ligands [134]. Ashe's group also prepared the unsubstituted bis(boratabenzene)iron complex $Fe(\eta^6\text{-}C_5H_5BH)_2$ – a close relative of ferrocene–via $[1,3\text{-}C_5H_6B(OMe)]Fe(CO)_3$ and $Fe[C_5H_5B(OMe)]_2$ as precursors [136]. With regard to sandwich compounds with boron-containing six-membered rings, another interesting facet is that, while attempts to obtain borazine analogues of bis(benzene)chromium of composition $Cr(R_3B_3N_3R'_3)_2$ failed, related half-sandwich complexes $(R_3B_3N_3R'_3)Cr(CO)_3$ (R and R' = alkyl) could be prepared in our laboratory as early as in 1967 [137, 138]. They are significantly less stable, both thermodynamically and kinetically, than their

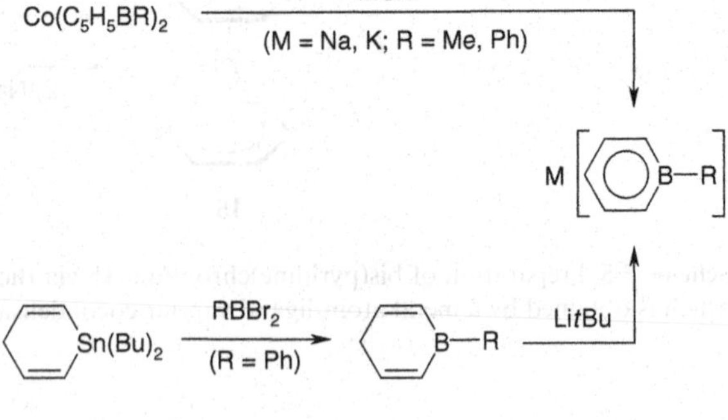

Scheme 5.6 Two routes to generate alkali salts of substituted boratabenzenes, used as precursors for the synthesis of corresponding sandwich complexes

benzene chromium tricarbonyl counterparts. Based on thermochemical studies by Joseph Connor (who was a postdoc in Fischer's group in the 1960s and later became Professor of Inorganic Chemistry at the University of Kent in Canterbury, UK), the dissociation enthalpy D(Cr-ring) of $(B_3N_3Me_6)Cr(CO)_3$ is estimated at 105 kJ/mol compared with 206 kJ/mol for $(C_6Me_6)Cr(CO)_3$ [139].

5.11 Two Highlights from the 21st Century

At the beginning of the twenty-first century, more than 50 years after the discovery of ferrocene, the chemistry of sandwich complexes is still alive and as fascinating as ever. This can be illustrated by two papers published in 2004 and 2006. The first stems from Ernesto Carmona's laboratory in Sevilla, which has become one of the top addresses for organometallic chemistry in Europe in recent years. Since the late 1990s, Carmona (Fig. 5.11) and his group have been interested in the chemistry of beryllocene, which had been prepared by Fischer and Hofmann in 1959 and which occupied a unique place among the main group metallocenes because its molecular structure was a matter of debate for several decades [140]. Given that interest, and taking into consideration that the permethylated zincocene $Zn(C_5Me_5)_2$ has the same unsymmetrical structure with one η^1- and one η^5-bonded ring as $Be(C_5H_5)_2$, Carmona's group attempted to prepare the ethyl derivative $(C_5Me_5)Zn(C_2H_5)$ by comproportionation of $Zn(C_5Me_5)_2$ and $Zn(C_2H_5)_2$. Apart from the required product, they isolated the dinuclear sandwich complex $Zn_2(\eta^5\text{-}C_5Me_5)_2$, which contains zinc in the oxidation state +1 and represents the first isolated compound with a direct, relatively short Zn–Zn bond (see Fig. 5.12) [141]. Another elegant route (reaction of $ZnCl_2$ with KC_5Me_5 and KH) allowed the preparation of $Zn_2(\eta^5\text{-}C_5Me_5)_2$ in gram quantities and opened up the possibility of studying the chemistry of this unique molecule [142]. The almost linear arrangement ring centre–metal–metal– ring centre with a metal–metal bond unsupported by bridging ligands has no precedent, neither among the transition metal nor the main group metallocene families.

The second paper, which received equal attention, reported the synthesis of an unprecedented type of a mononuclear metallocene with an f-element [143]. Up to 2006, the general knowledge was that metallocene derivatives of the lanthanides and actinides are found exclusively in a bent-sandwich configuration, whatever the 4f or 5f metal, the oxidation state, the electronic charge, and the nature and number of auxiliary ligands may be [144]. This statement had to be revised after Michel Ephritikhine (who, as mentioned above, also contributed significantly to the development of organouranium chemistry with C_7H_7 and C_8H_8 as π-bonded ligands) isolated salts of the bis(pentamethylcyclopentadienyl) complex $[(\eta^5\text{-}C_5Me_5)_2U(NCMe)_5]^{2+}$. The exceptionally stable tetraphenyloborate was obtained in good to excellent yield either by treatment of

Fig. 5.11 Ernesto Carmona (born in 1948) obtained his Ph.D. in 1974 from the University of Sevilla under Professor Francisco González. His interest in organometallic chemistry originated from his stay as a postdoctoral fellow in Geoffrey Wilkinson's laboratory at Imperial College London. In 1977, after 3 years in London, he returned to his alma mater at Sevilla, where he is now Professor of Inorganic Chemistry. His research interests comprise C–H and C–C bond activation, the coordination chemistry of dinitrogen and carbon dioxide, transition metal olefin, alkyl, allyl and carbene complexes, and metallocene chemistry of the transition metal and main group elements. In recent years, he delivered, among others, the Seaborg Lecture, the Pacific Northwest Inorganic Chemistry Lecture, the Goldschmidt-Hermanos Elhúyar Lecture, the Catalán-Sabatier Lecture and the Sir Geoffrey Wilkinson Lecture. In 2007, he also received the Luigi Sacconi Medal from the Italian Chemical Society (photo by courtesy from E. C.)

$(C_5Me_5)_2UI_2$ with $TlBPh_4$ or of $(C_5Me_5)_2U(CH_3)_2$ with $(HNEt_3)BPh_4$ in acetonitrile [143]. As shown by X-ray diffraction analysis, the dication has two pentamethylcyclopentadienyl rings which are parallel and, equally important, equidistant from and parallel to the plane defined by the metal centre and the nitrogen atoms of the five acetonitrile ligands. These ligands form an girdle

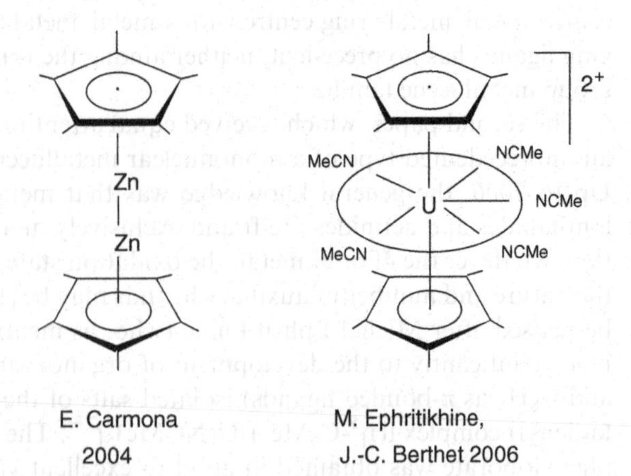

Fig. 5.12 The structure of two novel sandwich complexes prepared in the first decade of the twenty-first century

around the $U(C_5Me_5)_2$ fragment (Fig. 5.12), the linear arrangement of which was formerly unknown for a bis(cyclopentadienyl) compound of uranium or any other f-element. The novel structural feature raises the question about the nature of the metal–cyclopentadienyl bonding which could be different to that of the numerous bent 4f and 5f metallocene derivatives.

5.12 Brintzinger's Sandwich-Type Catalysts

Despite the fact that from almost all transition metals, including the f-elements, cyclopentadienyl complexes are known, only a few have been applied in organic synthesis and industrial processes. The most important of those are the bent bis(cyclopentadienyl) titanium and zirconium dihalides, of which the dichloride $(C_5H_5)_2TiCl_2$ together with Et_2AlCl as activator has been used as catalyst for the polymerization of ethene as early as in the mid 1950s. At that time, David Breslow at the Hercules Research Centre in Wilmington and Giulio Natta at the *Istituto di Chimica Industriale del Politecnico* in Milano found that titanocene derivatives, which are soluble in aromatic hydrocarbons, could be used instead of $TiCl_4$, in combination with aluminium alkyls, to give Ziegler-type catalysts [145–147]. However, although these catalysts as well as analogous zirconocene systems worked under homogeneous conditions, the activity was low. Thus, they were only scarcely used in industry during the next two decades. This changed after it was observed that water, previously considered to be a catalyst poison for Ziegler–Natta systems, led to a remarkable rise in activity [148]. Subsequent investigations on the influence of water, and on the course of the reactions of water with organoaluminium species, by Walter Kaminsky and Hansjörg Sinn resulted in the isolation of methylalumoxanes (MAO), which are probably cyclic and in part acyclic oligomers $(MeAlO)_n$ containing five to 20 aluminium atoms [149, 150]. With the defined $(C_5H_5)_2TiMe_2$/MAO system, the productivity for polyethylene could be raised significantly and, in contrast to the catalyst derived from $(C_5H_5)_2TiCl_2$ and Et_2AlCl, also α-olefins could be polymerized. Unfortunately, these polymers were atactic and initially found only limited use.

The breakthrough in the worldwide efforts to develop catalytic systems, able to direct the polymerization of α-olefins to isotactic and syndiotactic products, came with the elegant work of Hans Brintzinger (Fig. 5.13) and his students at around 1980 [151–154]. They found that if instead of metallocenes of the general composition $(C_5H_5)_2MX_2$ (M=Ti, Zr; X=Cl, Me, etc.) analogues with conformationally constrained ethylene-bridged ligands such as ethylenebis(indenyl) or ethylenebis(tetrahydroindenyl) were used, the complexes would be chiral, and they assumed that the chirality could be retained even under catalytic conditions. Soon this was confirmed, and it was shown that these so-called *ansa*-metallocenes, when activated with MAO, give highly isotactic polymers from propene and other α-olefins [155–161]. Following the early reports,

Fig. 5.13 Hans Herbert Brintzinger (born in 1935 at Jena, the place of my alma mater) studied Chemistry at Tübingen and Basel, and received his Ph.D. degree with Hans Erlenmeyer and Silvio Fallab from the University of Basel in Switzerland in 1960. After he finished his *Habilitation* at Basel in 1964, he became *Privatdozent* (Lecturer) and continued to carry out research in bioinorganic chemistry. In 1965, he moved to the University of Michigan at Ann Arbor, became Associate Professor that same year, and was promoted to Full Professor in 1970. At Ann Arbor, his research group became interested in the fixation of dinitrogen by transition metal complexes as well as in the structure of the so-called "titanocene". It turned out that this complex has not the proposed structure $Ti(C_5H_5)_2$ but is a dimer with a bridging fulvalenediyl ligand. In 1972, he accepted an appointment as Professor of Chemistry at the University of Konstanz, where he has been ever since. At Konstanz, his research interests focussed on preparative and mechanistic organometallic chemistry, involving the synthesis of $(C_5H_5)_2W(CO)_2$ (the first compound containing one η^5- and one η^3-bonded cyclopentadienyl ligand) and the first chiral *ansa*-metallocenes of titanium, zirconium and hafnium. This work was recognized worldwide and in a short period of time found application in the polymer industry. Hans Brintzinger received several major awards including the Alwin Mittasch Medal for Research in Catalysis, the ACS Award in Organometallic Chemistry, and the prestigious Karl Ziegler Prize from the German Chemical Society (photo by courtesy from H. H. B.)

hundreds of chiral *ansa*-metallocenes were prepared and created a "revolution in catalysis" [162]. The unique feature of the corresponding catalysts is that all their active sites are the same, which allows "polymer chemists to do something that has never before been possible using traditional, heterogeneous, Ziegler-Natta catalysts. That is, to predict accurately the properties of the resulting polymers by knowing the structure of the catalyst used during their manufacture" [162]. Scheme 5.7 illustrates how, by simply changing the ligand sphere of *ansa*-zirconocenes, isotactic, syndiotactic and atactic polypropylene is formed in the presence of MAO as the cocatalyst. It is now generally known that with the tailor-made Brintzinger-type metallocenes in combination with MAO not

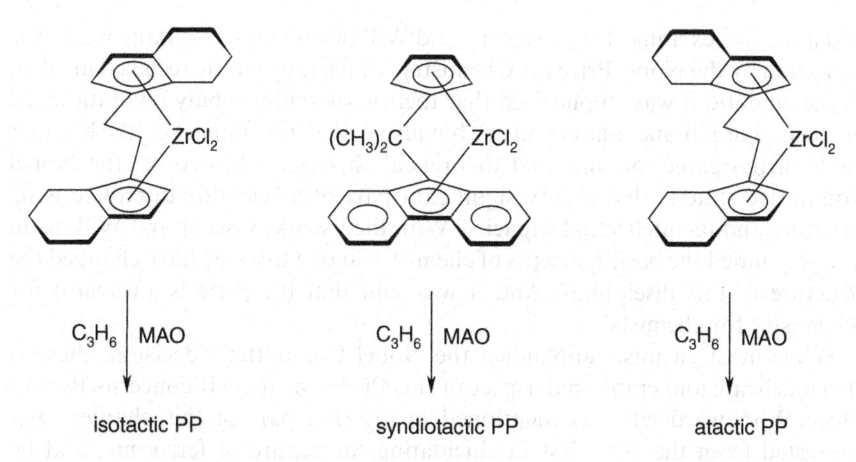

Scheme 5.7 Three examples of *ansa*-zirconocene dichlorides directing the polymerization of propene, in the presence of MAO, to isotactic, syndiotactic and atactic polypropylene (PP)

only the tacticity but also the whole architecture of olefin polymers can be controlled, and that with these catalysts novel olefin block copolymers can be obtained [163]. Given some recent data from industry [164], it appears that metallocene catalysts are in many ways a more attractive alternative to the traditional Ziegler–Natta systems.

5.13 Woodward and the Nobel Prize

At the end of this chapter, it is appropriate to recall that before the two papers on ferrocene by Pauson and Miller appeared, the treatment of organometallic chemistry in textbooks was mainly limited to the synthesis of compounds of main group metals, and the use of Grignard reagents in synthetic organic chemistry. This changed quickly in the 1950s after a whole series of cyclopentadienyl and arene metal complexes had been prepared and the unique structural and bonding characteristics of these compounds discovered. As a result, organometallic chemistry, in particular that of the transition metals, underwent an explosive growth that finally led to numerous new types of complexes containing metal–carbon bonds. This development was primarily spurred by the unusual coordination capabilities of the ubiquitous cyclopentadienyl ring, and the estimate, made in 1991, that "more than 80% of all known organometallic complexes of transition metals contain a substituted or unsubstituted cyclopentadienyl ligand", is probably still true [165]. The human driving forces from the beginning in 1952 until the mid 1960s were undoubtedly Ernst Otto Fischer and Geoffrey Wilkinson and without their firm competition, the progress in this field would not have been achieved. Although the competition was not always friendly, "to put it mildly" [166], it was according to Al Cotton "science at its best and most exciting" [71]. Fischer's and Wilkinson's epoch-making work was honoured by the Nobel Prize for Chemistry, which they jointly received in 1973. In the laudatio it was emphasized that their work had not only revolutionized the field of organometallic chemistry but also had a great impact on the broader fields of inorganic, organic and theoretical chemistry. Moreover, the Nobel Committee stated "that a very significant part of a scientific discipline is its structure and its intellectual capacity. With their work, Fischer and Wilkinson have expanded the basic concepts of chemistry and in this way have changed the structure of this discipline". And it was said that the prize is an award for "chemistry for chemists".

While most chemists applauded the Nobel Committee's decision, there is also a delicate and emotional aspect of this Prize's history. It concerns Robert (Bob) Woodward who, as mentioned in the first part of this chapter, was interested from the very first in elucidating the nature of ferrocene, and he was the first to suggest that it could behave like a typical aromatic molecule. He was on sabbatical leave in England when the Nobel Committee announced the winners of the 1973 Nobel Prize in chemistry. In an unpublished letter to the Chairman of the Nobel Committee for Chemistry, discovered about 20 years after his death in 1979, he reacted on October 26th, 1973 to the press release as follows: "The notice in the The Times of London (October 24, p. 5) of the award of this year's Nobel Prize in Chemistry leaves me no choice but to let you know, most respectfully, that you have – inadvertently, I am sure – committed a grave injustice" [13]. And by quoting several newspaper articles, describing Fischer's and Wilkinson's work without mentioning his contributions to the ferrocene story, he continued: "The problem is that there were two seminal ideas in this field – first the proposal of the unusual and hitherto unknown sandwich structure, and second, the prediction that such structures would display unusual, 'aromatic' characteristics. Both of these concepts were simply, completely, and entirely mine, and mine alone. Indeed, when I, as a gesture to a friend and junior colleague interested in organometallic chemistry, invited Professor Wilkinson to join me and my colleagues in the simple experiments which verified my structure proposal, his initial reaction to my views was close to derision... But in the event, he had second thoughts about his initial scoffing view of my structural proposal and its consequences, and all together we published the initial seminal communication that was written by me. The decision to place my name last in the roster of the authors was made, by me alone, again as a courtesy to a junior staff colleague of independent status" [13].

Geoffrey Wilkinson provided a vastly different view and although he agreed that both, Woodward and he, felt from the beginning that the linear arrangement for the rings and the metal centre in ferrocene as proposed by Kealy and Pauson was wrong, he emphasized that it was a collaborative effort in which both parties contributed to the fundamental ideas [6]. In the closing part of his letter to the Nobel Committee, Woodward mentioned that he had not been able "to find a complete account of the ancillary material released to the press", and he continued "that quite possibly this material may well make a clear acknowledgment – ignored by the press – of my definite contributions in those respects. Should this be true ... the problem is much minimized". However, neither the award citation nor the ancillary material released to the press mentioned Woodward by name. Arne Fredga, then Chairman of the Nobel Committee for Chemistry, replied to Woodward's letter by saying that "the committee does not make available to the press information about a newly elected Nobel Laureate ... and it is customary not to mention co-workers and co-authors who are not sharing the prize, and this rule has been followed also in the present case". While Woodward, nevertheless, further objected to the fact that he was passed over for that prize [13], the majority of the organometallic community agrees that "the subject matter was one of the best choices the Nobel Committee ever made" [71]. The fact remains that Fischer and Wilkinson received the Nobel Prize for their extensive investigations on the chemistry of organometallic sandwich compounds, not the initial discovery alone. And thus, as Thomas Zydowsky summarized in his account [13], "a story is left, albeit an incomplete and unresolvable one, which speaks to the emotions of the people behind the scientific advances". In my opinion, it illustrates the human factor in scientific research.

Biographies

Biographies of Ernst Otto Fischer and Franz Hein see Chap. 2.

Peter Pauson (Fig. 5.14) is considered not only by his friends, but also by his colleagues and students, to be a charming and admired person. He was born on the 30th of July 1925 in Bamberg, which is in the northern part of Bavaria and one of the most beautiful mid-sized towns in Germany. After visiting primary and secondary schools at Bamberg, he moved with the rest of his family to the UK in 1939 as a refugee from the Nazi persecution of Jewish citizens. In 1942, he entered Glasgow University to study chemistry. His inclination "to the organic side" [5] was greatly enhanced by the inspiring teachings of Thomas S. Stevens, who is well known as the discoverer of the Stevens rearrangement and the Bamford–Stevens reaction. In 1946, Pauson received a research studentship from Sheffield University, where he did his doctoral work in the group of Robert D. Haworth. In his thesis he elucidated the structure of "purpurogallin", which he confirmed to be a disubstituted benzotropolone, and prepared 5-methyltropolone, the first synthetic and fully characterized simple tropolone.

Pauson obtained his Ph.D. from the University of Sheffield in 1949 and then moved to the Duquesne University in Pittsburgh, where he was offered the post of a temporary Assistant Professor without interview. Following his interests in non-benzenoid systems, he attempted to prepare fulvalene (see Scheme 5.1) but, as mentioned above, isolated ferrocene instead. After he finished his 2-years term at Duquesne, he spend the academic year 1951–1952 at the University of Chicago as a postdoctoral fellow working on peroxide chemistry with Morris S. Kharasch, in those days the "pope" in this field. In 1953, he was appointed to a DuPont

Fig. 5.14 Peter Pauson at the time when he held the Freeland Chair of Chemistry at the University of Strathclyde (photo by courtesy from P. P.)

Fellowship at Harvard, where he had the chance to do independent research. While his application for the fellowship was based on a proposal to synthesize the alkaloid colchicine (a tropolone), he soon became involved in the work on metallocenes which, as he recalled, in those days was "in full swing in both Woodward's and Wilkinson's laboratories" [5]. At the end of the year at Harvard, he returned to the UK to take up a lectureship in organic chemistry at the University of Sheffield. He made the final move to Glasgow in 1959, when he was appointed to the Freeland Chair of Chemistry at the Royal College of Science and Technology which, in 1964, became the University of Strathclyde. In retrospect it is interesting to note that the Freeland Chair was established in 1830 and first occupied by Thomas Graham, one of the early scientific giants in the UK.

Within the chemical community, Pauson is first and foremost remembered as the "father of ferrocene" (Fig. 5.15). However, his contributions to the development of organo-transition metal and synthetic organic chemistry are widespread. His early work on metallocenes and cyclopentadienyl metal carbonyls was later extended by studies on butadiene and cyclohexadiene iron tricarbonyl, on cycloheptatriene and tropylium complexes of the group VI metals, on π-pyrrolyl and π-indolyl metal compounds, as well as by detailed investigations of the properties of these species. Since the 1970s, his name is also associated with the Pauson–Khand reaction [167, 168], which is a widely applicable synthetic route to cyclopentenone derivatives in a single step from an alkyne, an olefin and a CO molecule present as a ligand in the intermediate alkyne dicobalt hexacarbonyl complex. It is characteristic for Peter that, when I introduced him once at an international conference and mentioned the importance of the Pauson–Khand reaction, he corrected me by saying that it is the Khand reaction since it was not discovered by him but by his long-time coworker Ihsan Khand.

Fig. 5.15 Speakers and friends of Peter Pauson's gathering at the Annual Congress of the Royal Society of Chemistry in Glasgow in 1976, to celebrate the 25th anniversary of the discovery of ferrocene (from the left: Ernst Otto Fischer, Stephen Davies, Francois Mathey, Wolfgang Herrmann, Michael Lappert, Jack Lewis, William Watts, Peter Pauson, myself, Myron Rosenblum and Malcolm Green)

Pauson has received numerous awards beginning with the Tilden Lectureship of the Chemical Society in 1959 and including, more recently, the Makdougall–Brisbane Prize of the Royal Society of Edinburgh. Those who know him personally will testify to "his inspiring leadership, unfailing courtesy and good humour", as well as to his "practical concern for the well-being and career progression of his more junior associates" [169]. After he retired in 1995, he became a Leverhulme Emeritus Fellow at the University of Strathclyde.

Geoffrey Wilkinson (usually called Geoff by his students and colleagues) was born on the 14th of July 1921 in the village of Springside on the outskirts of Todmorden in West Yorkshire. The family had moved to this place around 1880 and Geoffrey was always very proud of the fact that he was a Yorkshireman. He often returned there and always enjoyed walking across the Moors near Todmorden and in the Lake District with family and friends.

Wilkinson's interest in chemistry was aroused early in life. Already at the age of 6 or 7 he was fascinated to see his father, a house painter and decorator, mixing his materials. One of his uncles ran a factory making inter alia Glauber's salt and he loved to go on Saturday mornings to tinker in the small laboratory at the factory. Having won a West Riding County Minor Scholarship in 1931, he went to Todmorden Secondary School and made most of his education there. This small school has had an unusual record of scholarly achievement, including two Nobel laureates (Wilkinson and John Cockcroft) within 25 years. After finishing school at Todmorden, he obtained a Royal Scholarship to the

Imperial College of Science and Technology in London in 1939. His main subject was chemistry, but he also studied geology as an ancillary discipline, and did so well that he won the prestigious Murchison Prize in Geology. While throughout his life he made a point of not caring about prizes or awards, he proudly showed visitors this impressive bronze medal.

In 1941, he graduated not only top of the Imperial College chemistry list, but also top of the consolidated list of all the University of London colleges. In the same year, he started the work for his doctoral thesis under the supervision of Henry V. A. Briscoe, who at that time was the only Professor of Inorganic Chemistry in the country. The topic, presumably related to wartime research, was "Some physico-chemical observations on hydrolysis in the homogeneous vapour phase". This rather obscure title conceals the fact that the main substrate studied was phosgene and thus Wilkinson later remarked that Briscoe "directed this Ph.D. research from a safe distance" [170]. Before he received his Ph.D., he was selected by Friedrich A. Paneth to join the British contingent for the Atomic Energy Project in Canada, where he worked from late 1942 until mid 1944 at the University of Montreal and then at Chalk River. His role was to identify the elements generated by the fission processes, and thus he had to become familiar in considerable detail with the descriptive chemistry of almost every element from the Periodic Table. It was this experience which provided an invaluable foundation for the subsequent ease in developing new chemistry of the transition metals and the lanthanides.

After the war, Wilkinson returned briefly to Britain to take his Ph.D. viva and then went to the Lawrence Livermore Laboratory at the University of California at Berkeley, where he was associated with Glenn Seaborg for nearly 4 years. He worked on the production of neutron-deficient isotopes of the transition elements and the lanthanides, and it was stated by Seaborg that Wilkinson personally produced more new isotopes than anyone else. One of his nuclear transmutations was that of platinum to gold, which caught the public imagination after a report in the San Francisco Chronicle entitled "*Scientist discovers gold mine in the cyclotron*" [170]. It was due to this, that he jokingly described himself as the first successful alchemist.

On the occasion of a visit to England to participate in the first post-war conference on nuclear and radiation chemistry, he was offered a research fellowship by Paneth which he eventually declined. Briscoe advised him that there might not be much future in this kind of nuclear chemistry since he would always be dependent on the physicists running the cyclotron. Wilkinson therefore accepted an offer from MIT to work as a research associate in 1950–1951 thereby returning to coordination chemistry. He prepared some unusual zerovalent nickel complexes, such as $Ni(PCl_3)_4$ and $Ni(PF_3)_4$, and also published a note on an intriguing haemoglobin–PF_3 adduct. In September 1951, he was appointed Assistant Professor of Chemistry at Harvard, where he started his seminal work on sandwich complexes that laid the cornerstone of the Nobel Prize.

After a sabbatical break of 9 months in Jannik Bjerrum's laboratory in Copenhagen, Wilkinson was appointed in 1955 to Briscoe's chair at Imperial College, which was still the only established chair of inorganic chemistry in Britain at the time. At age 34, he was one of the youngest professors ever to be appointed at that institution. In the following years, his interests rapidly diversified in several directions and, analogous to the development in Fischer's laboratory in München, he continued work on cyclopentadienyl metal compounds for less than a decade. Outstanding in the 1960s and 1970s was the discovery of the catalytic activity of $RhCl(PPh_3)_3$ (now called Wilkinson's catalyst) and the work on binary alkyls and aryls of the transition elements (see Chap. 9). In the final part of his research career, he also made important contributions to the chemistry of amido and alkoxide complexes of metals in high oxidation states and illustrated the analogies between oxo and imido compounds of the transition elements.

Geoffrey Wilkinson (Fig. 5.16) held the Chair at Imperial College for 32 years with great distinction. As described above, he was awarded the Nobel Prize for chemistry in 1973 together with Ernst Otto Fischer. Since he had a keen appreciation of history and highly admired Frankland, he initiated the renaming of his chair as the *Sir Edward Frankland Chair of Inorganic Chemistry* in 1978. He was the Head of Department from 1976 (the year he was knighted) to 1988 and was an effective and well-liked administrator. The success of his leadership lay in his humour and approachability, and his ability to delegate duties. The spirit in his research group "was more like that of an urgent gold rush in the West than the scholarly and disciplined calm expected in academia" [171]. His enthusiasm was always contageous, and it is no mere accident that a large number of his coworkers subsequently entered academic life. Wilkinson's research contributions (the list of publications comprises 557 entries) clearly illustrate one of his most

Fig. 5.16 Geoffrey Wilkinson at the time when he received the Nobel Prize (photo reproduced with permission of The Royal Society)

dearly held beliefs that "innovative and imaginative exploratory synthesis can be one of the most creative approaches for the discovery of new chemistry" [170]. He was proud of the fact that all of his research was "blue sky" and curiosity-driven. To present day students his name is best known for the excellent textbook *Advanced Inorganic Chemistry*, which he coauthored with his former student Al Cotton and which influenced enormously the content and teaching of inorganic chemistry courses throughout the world.

Geoffrey Wilkinson formally retired in 1988 but continued doing innovative research with a dedicated team. As part of the Centenary celebrations of Imperial College London, two Royal Society of Chemistry National Chemical Landmark plaques were unveiled in May 2007, commemorating the two Nobel Laureates Derek Barton and Geoffrey Wilkinson. Both had been students at the college, both had left and returned some 15 years later to take up professorial chairs there. Wilkinson died on the 26th of September 1996.

Rowland Pettit (or Rolly, as he was called by his students and many friends) was an outstanding creative scientist, who made significant contributions not only to organometallic but also to organic chemistry during the course of his all-too-short career. He was born on the 6th of February 1927 in Port Lincoln, a small town in Southern Australia, and carried out his chemistry studies at the University of Adelaide. After receiving his first Ph.D. in 1953, he was awarded an *Exhibition of 1851 Overseas Fellowship* to work for a second doctorate with Michael Dewar (see Fig. 7.6), who at that time was an Assistant Professor of Chemistry at Queen Mary College in London. During this period, Pettit's wide-ranging interests led to synthetic and theoretical studies on non-benzenoid aromatic compounds, including the elusive tropylium ion for which he provided the first rational synthesis. He obtained his second Ph.D. from the University of London in 1957, and in the same year accepted a faculty position at the University of Texas in Austin.

While at the beginning, he continued research in basic organic chemistry, he soon became interested in organometallic chemistry and, in particular, in the chemistry of organoiron compounds. His initial contribution to this topic, reported in 1959, was the synthesis and characterization of norbornadiene iron tricarbonyl, which was the first transition metal complex of a non-conjugated diolefin (see Chap. 7). This work was extended by studying the protonation of diene iron tricarbonyls to give novel cationic organoiron complexes, the $Fe(CO)_5$-catalysed rearrangement of 1,3-dienes, and – last but not least – the preparation of cyclobutadiene iron tricarbonyl. The ready release of cyclobutadiene from its iron complex led to important insights into the electronic and structural nature of the free ligand and provided a unique way to combine organometallic and synthetic organic chemistry. He was also interested in problems of practical interest, which emerged from his work on metal-catalysed carbon monoxide reduction and homologation. These studies significantly improved the understanding of the homogeneous water gas shift reaction and the mechanism of the heterogeneous Fischer–Tropsch process.

Fig. 5.17 Rowland Pettit exercising his passion after giving his Plenary Lecture at the X. Sheffield–Leeds Symposium on Organometallic Chemistry in July 1979 (photo by courtesy from Professor Peter Maitlis)

Rowland Pettit (Fig. 5.17) was not only a great scientist but also an inspiring mentor to his many graduate students and postdoctoral fellows. Moreover, he enjoyed life and had a fine sense of humor. I remember quite well that during the breaks at conferences or when lecture sessions were finished, he looked for people who liked to play pool and it was very hard to beat him at this game. Unfortunately, at those and other occasions he never seemed to cease smoking, which probably contributed to his early death on the 10th of December 1981 at the age of 54. His spirit, warmth and considerate personality will be remembered by all who had the privilege to meet and to talk with him.

References

1. R. S. Nyholm, "The Renaissance of Inorganic Chemistry", Inaugural Lecture delivered at University College London, 1 March 1956, published for the College by H. K. Lewis & Co. Ltd., London.
2. L. M. Venanzi, Coordination Chemistry in Europe since Alfred Werner, *Chimia* **48**, 16–22 (1994).
3. T. J. Kealy, and P. L. Pauson, A New Type of Organo-iron Compound, *Nature* **168**, 1039–1040 (1951).
4. S. A. Miller, J. A. Tebboth, and J. F. Tremaine, Dicyclopentadienyliron, *J. Chem. Soc.* **1952**, 632–635.
5. P. L. Pauson, Ferrocene – How it all began, *J. Organomet. Chem.* **637–639**, 3–6 (2001).
6. G. Wilkinson, The Iron Sandwich. A Recollection of the First Four Months, *J. Organomet. Chem.* **100**, 273–278 (1975).
7. E. O. Fischer, and R. Jira, How Metallocene Chemistry and Research Began in Munich, *J. Organomet. Chem.* **637–639**, 7–12 (2001).

8. E. O. Fischer, and W. Pfab, Cyclopentadien-Metallkomplexe, ein neuer Typ metallorganischer Verbindungen, *Z. Naturforsch., Part B*, **7**, 377–379 (1952).
9. G. Wilkinson, M. Rosenblum, M. C. Whiting, and R. B. Woodward, The Structure of Iron Bis-cyclopentadienyl, *J. Am. Chem. Soc.* **74**, 2125–2126 (1952).
10. P. Laszlo, and R. Hoffmann, Ferrocene: Ironclad History or Rashomon Tale?, *Angew. Chem. Int. Ed.* **39**, 123–124 (2000).
11. M. Rosenblum, The Early Ferrocene Days – a Personal Recollection, *J. Organomet. Chem.* **637–639**, 13–15 (2001).
12. R. B. Woodward, M. Rosenblum, and M. C. Whiting, A New Aromatic System, *J. Am. Chem. Soc.* **74**, 3458–3459 (1952).
13. T. M. Zydowsky, Of Sandwiches and Nobel Prizes: Robert Burns Woodward, *The Chemical Intelligencer* **2000**, 29–34.
14. P. F. Eiland, and R. Pepinsky, X-ray Examination of Iron Biscyclopentadienyl, *J. Am. Chem. Soc.* **74**, 4971 (1952).
15. J. D. Dunitz, and L. E. Orgel, Bis-cyclopentadienyl Iron: a Molecular Sandwich, *Nature* **171**, 121–122 (1953).
16. G. Wilkinson, The Preparation and Some Properties of the Cobalticinium Salts, *J. Am. Chem. Soc.* **74**, 6148–6149 (1952).
17. G. Wilkinson, P. L. Pauson, J. M. Birmingham, and F. A. Cotton, Bis-cyclolopentadienyl Derivatives of Some Transition Elements, *J. Am. Chem. Soc.* **75**, 1011–1012 (1953).
18. E. O. Fischer and R. Jira, Über den Di-cyclopentadienyl-Komplex des Kobalts, *Z. Naturforsch., Part B*, **8**, 1–2 (1953).
19. E. O. Fischer and R. Jira, Di-cyclopentadienyl-nickel, *Z. Naturforsch., Part B*, **8**, 217–219 (1953).
20. G. Wilkinson, F. A. Cotton, and J. M. Birmingham, On Manganese Cyclopentadienide and Some Chemical Reactions of Neutral Bis-cyclopentadienyl Metal Compounds, *J. Inorg. Nucl. Chem.* **2**, 95–113 (1956).
21. G. Wilkinson, Ferrocene, *Org. Synth.* **36**, 31–34 (1956).
22. J. Thiele, Über Abkömmlinge des Cyclopentadiens, *Ber. dtsch. chem. Ges.* **34**, 68–71 (1901).
23. G. Wilkinson, and F. A. Cotton, Cyclopentadienyl and Arene Metal Compounds, *Progr. Inorg. Chem.* **1**, 1–124 (1959).
24. E. O. Fischer, and H. P. Fritz, Compounds of Aromatic Ring Systems and Metals, *Adv. Inorg. Chem. Radiochem.* **1**, 55–115 (1959).
25. E. O. Fischer, and W. Hafner, Di-benzol-chrom, *Z. Naturforsch., Part B*, **10**, 665–668 (1955).
26. D. Seyferth, Editor's Page, *Organometallics* **21**, 1519 (2002).
27. W. Hafner, Aromatenkomplexe des Chroms und Vanadins, Ph.D. Thesis, Technische Hochschule München (1955).
28. E. O. Fischer, W. Hafner, and K. Öfele, Eine Synthese von Chromhexacarbonyl, *Chem. Ber.* **92**, 3050–3052 (1959).
29. E. Weiss, and E. O. Fischer, Zur Kristallstruktur und Molekelgestalt des Di-benzol-chrom(0), *Z. anorg. allg. Chem.* **286**, 142–145 (1956).
30. F. Hein, Notiz über Chromorganoverbindungen, *Ber. dtsch. chem. Ges.* **52**, 195–196 (1919).
31. F. Hein, Chromorganische Verbindungen, I. Mitteilung: Pentaphenylchromhydroxyd, *Ber. dtsch. chem. Ges.* **54**, 1905–1938 (1921).
32. D. Seyferth, Bis(benzene)chromium. 1. Franz Hein at the University of Leipzig and Harold Zeiss and Minoru Tsutsui at Yale, *Organometallics* **21**, 1520–1530 (2002).
33. F. Hein, Chromorganische Verbindungen, II. Mitteilung: Die abnorme Salzbildung des Pentaphenylchromhydroxyds; Tetraphenylchromsalze (Abstoßung einer Phenylgruppe), *Ber. dtsch. chem. Ges.* **54**, 2708–2727 (1921).
34. F. Hein, and W. Eißner, Über das Tetraphenylchrom $(C_6H_5)_4Cr$, *Ber. dtsch. chem. Ges.* **59**, 362–366 (1926).

35. F. Hein, and E. Markert, Über das Triphenylchrom, sowie zur Kenntnis des Tetraphenylchroms und Diäthylthalliums, *Ber. dtsch. chem. Ges.* **61**, 2255–2267 (1928).
36. F. Hein, Über die Organochromverbindungen und ihre Beziehung zur Komplexchemie des Chroms, *J. Prakt. Chem.* **132**, 59–71 (1932).
37. W. Klemm, and A. Neuber, Das magnetische Verhalten der Chromphenylverbindungen, *Z. anorg. allg. Chem.* **227**, 261–271 (1936).
38. F. Hein, Bemerkung zu der vorstehenden Abhandlung von W. Klemm und A. Neuber, *Z. anorg. allg. Chem.* **227**, 272 (1936).
39. F. A. Cotton, Alkyls and Aryls of Transition Metals, *Chem. Rev.* **55**, 551–594 (1955).
40. H. Zeiss, Arene Complexes of the Transition Metals, in: *Organometallic Chemistry*, ACS Monograph Series No. 147 (Ed. H. Zeiss, Reinhold Publishing Corp., New York, 1960, pp. 380–425).
41. H. Zeiss, and M. Tsutsui, π-Complexes of the Transition Metals. I. Hein's Polyaromatic Chromium Compounds, *J. Am. Chem. Soc.* **79**, 3062–3066 (1957).
42. H. Zeiss, and M. Tsutsui, Tetraphenylchromjodid, *Angew. Chem.* **67**, 282 (1955).
43. H. Zeiss, *Yale Sci. Mag.* **29**, 14 (1955).
44. E. O. Fischer, and D. Seus, Zur Frage der Struktur der Chrom-phenyl-Verbindungen, *Chem. Ber.* **89**, 1809–1815 (1956).
45. F. Hein, and K. Eisfeld, Darstellung von Benzol-diphenyl-chrom(I)-jodid mittels der reduzierenden Friedel-Crafts-Reaktion, *Z. anorg. allg. Chem.* **292**, 162–166 (1957).
46. W. Herwig, and H. H. Zeiss, Substituted Aromatic-chromium Complexes, *J. Am. Chem. Soc.* **78**, 5959 (1956).
47. F. Hein, Zur Frage der Struktur der Chrom-phenyl-Verbindungen. Bemerkungen zur Abhandlung E. O. Fischer and D. Seus, *Chem. Ber.* **89**, 1816–1821 (1956).
48. W. Herwig, and H. Zeiss, Triphenylchromium, *J. Am. Chem. Soc.* **79**, 6561 (1957).
49. F. Hein, and R. Weiss, Zur Existenz echter Organochromverbindungen, *Z. anorg. allg. Chem.* **295**, 145–152 (1958).
50. E. Uhlig, Seventy-five Years of π-Complexes of Chromium(I), *Organometallics* **12**, 4751–4756 (1993).
51. E. O. Fischer, and H. P. Fritz, π-Komplexe benzoider Systeme mit Übergangsmetallen, *Angew. Chem.* **73**, 353–364 (1961).
52. E. O. Fischer, and H. Brunner, Metallierung von Di-benzol-chrom, *Chem. Ber.* **95**, 1999–2004 (1962).
53. E. O. Fischer, and H. Brunner, Neue Di-benzol-chrom-Derivate, *Chem. Ber.* **98**, 175–185 (1965).
54. 53.C. Elschenbroich, Metalation of Dibenzenechromium by *N,N,N',N'*-Tetramethylethylenediamine Complexes of *n*-Butyllithium and Phenyllithium, *J. Organomet. Chem.* **14**, 157–163 (1968).
55. C. Elschenbroich, Bis(trimethylsilyl)benzol]chrom(0): Synthese, Eigenschaften und Elektronenaustausch mit seinem Radikalkation, *J. Organomet. Chem.* **22**, 677–683 (1970).
56. C. Elschenbroich, E. Bilger, J. Heck, F. Stohler, and J. Heinzer, Intramolekularer Ligand-Ligand-Elektronentransfer in paramagnetischen Bis(η^6-aren)chrom-Komplexanionen, *Chem. Ber.* **117**, 23–41 (1984).
57. C. Elschenbroich, J. Hurley, B. Metz, and G. Baum, Tetraphenylsilane as a Chelating Ligand: Synthesis, Structural Characterization, and Reactivity of the Tilted Bis(arene) Metal Complexes $[(C_6H_5)_2Si(\eta^6\text{-}C_6H_5)_2]M$ (M=V, Cr), *Organometallics* **9**, 889–897 (1990).
58. C. Elschenbroich, E. Schmidt, R. Gondrum, B. Metz, O. Burghaus, W. Massa, and S. Wocadlo, Metal π Complexes of Benzene Derivatives. Germanium in the Periphery of Bis(benzene)vanadium and Bis(benzene)chromium. Synthesis and Structure of New Heterametallocyclophanes, *Organometallics* **16**, 4589–4596 (1997).
59. C. Elschenbroich, and F. Stohler, Bis(diphenylphosphino-h^6-benzol)-chrom(0), *J. Organomet. Chem.* **67**, C51–C54 (1974).

60. P. S. Skell, and M. J. McGlinchey, Reactions of Transition Metal Atoms with Organic Compounds, *Angew. Chem. Int. Ed. Engl.* **14**, 195–199 (1975).
61. P. L. Timms, and T. W. Turney, Metal Atom Synthesis of Organometallic Compounds, *Adv. Organomet. Chem.* **15**, 53–112 (1977).
62. R. Middleton, J. R. Hull, S. R. Simpson, C. H. Tomlinson, and P. L. Timms, Formation of Complexes with Arenes, Trifluorophosphine, and Nitric Oxide, *J. Chem. Soc., Dalton Trans.* **1973**, 120–124.
63. M. L. H. Green, The Use of Atoms of the Group IV, V and VI Transition Metals for the Synthesis of Zerovalent Arene Compounds and Related Studies, *J. Organomet. Chem.* **200**, 119–132 (1980).
64. E. O. Fischer, and K. Öfele, Benzol-chrom-tricarbonyl, *Chem. Ber.* **90**, 2532–2535 (1957).
65. G. Natta, R. Ercoli, and F. Calderazzo, Synthesis of Chromotricarbonylarenes, *Chim. Ind. (Milano)* **40**, 287–289 (1958).
66. B. Nicholls, and M. C. Whiting, General Method for Preparing Tricarbonylchromium Derivatives of Aromatic Compounds, *Proc. Chem. Soc.* **1958**, 152.
67. E. O. Fischer, K. Öfele, H. Essler, W. Fröhlich, J. P. Mortensen, and W. Semmlinger, Über eine neue Synthese für Aromaten-metall-carbonyle, *Z. Naturforsch., Part B*, **13**, 458 (1958).
68. M. F. Semmelhack, Nucleophilic Addition to Arene-Metal Complexes, in: *Comprehensive Organic Synthesis* (Ed. B. Trost, Pergamon, Oxford, U.K., 1991, Vol. 4, pp. 517–549).
69. A. R. Pape, K.P. Kaliappan, and E. P. Kündig, Transition-Metal-Mediated Dearomatization Reactions, *Chem. Rev.* **100**, 2917–2940 (2000).
70. H. G. Schmalz, and F. Dehmel, Chromium-Arene Complexes, in: *Transition Metals for Organic Synthesis, Second Ed.* (Eds. M. Beller and C. Bolm, Wiley-VCH, Weinheim, 2006, Vol. 1, Chap. 3.12).
71. F. A. Cotton, Cyclopentadienyl-metal Chemistry in the Wilkinson Group, Harvard, 1952–1955, *J. Organomet. Chem.* **637–639**, 18–26 (2001).
72. G. Wilkinson, and J. M. Birmingham, Biscyclopentadienylrhenium Hydride – A New Type of Hydride, *J. Am. Chem. Soc.* **77**, 3421–3422 (1955).
73. L. T. Reynolds, and G. Wilkinson, π-Cyclopentadienyl Compounds of Uranium-IV and Thorium-IV, *J. Inorg. Nucl. Chem.* **2**, 246–253 (1956).
74. T. S. Piper, and G. Wilkinson, Alkyl and Aryl Derivatives of π-Cyclopentadienyl Compounds of Chromium, Molydenum, Tungsten, and Iron, *J. Inorg. Nucl. Chem.* **3**, 104–124 (1956).
75. B. F. Hallam, and P. L. Pauson, Dicyclopentadienyliron Dicarbonyl, *Chem. & Ind.* **1955**, 53.
76. F. A. Cotton, Proposed Nomenclature for Olefin-Metal and Other Organometallic Complexes, *J. Am. Chem. Soc.* **90**, 6230–6232 (1968).
77. M. J. Bennett, jr., F. A. Cotton, A. Davison, J. W. Faller, S. J. Lippard, and S. M. Morehouse, Stereochemically Nonrigid Organometallic Compounds. I. π-Cyclopentadienyliron Dicarbonyl σ-Cyclopentadiene, *J. Am. Chem. Soc.* **88**, 4371–4376 (1966).
78. F. A. Cotton, Fluxionality in Organometallics and Metal Carbonyls, *J. Organomet. Chem.* **100**, 29–41 (1975).
79. T. H. Coffield, V. Sandel, and R. D. Closson, Cyclopentadienyl-aromatic Sandwich Complexes of Manganese and Iron, *J. Am. Chem. Soc.* **79**, 5826 (1957).
80. E. O. Fischer, and H. P. Kögler, Cyclopentadienyl-chrom-benzol, *Z. Naturforsch., Part B*, **13**, 197–198 (1958).
81. M. L. H. Green, L. Pratt, and G. Wilkinson, Spectroscopic Studies of Some Organoiron Complexes, *J. Chem. Soc.* **1960**, 989–997.
82. A. N. Nesmeyanov, My Way in Organometallic Chemistry, *Adv. Organomet. Chem.* **10**, 1–78 (1972).
83. A. N. Nesmeyanov, N. A. Vol'kenau, and L. S. Kotova, Preparation and Study of Electroneutral Biphenyl Cyclopentadienyliron, *Koord. Khim.* **4**, 1699–1704 (1978).
84. D. Astruc, Organoiron Electron-Reservoir Complexes, *Acc. Chem. Res.* **19**, 377–383 (1986).
85. D. Astruc, Nineteen-electron Complexes and Their Role in Organometallic Mechanisms, *Chem. Rev.* **88**, 1189–1216 (1988).

86. D. Astruc, The Use of π-Organoiron Sandwiches in Aromatic Chemistry, *Topics Current Chemistry* **160**, 47–95 (1991).
87. S. Nlate, J. Ruiz, V. Sartor, R. Navarro, J.-C. Blais, and D. Astruc, Molecular Batteries: Ferrocenylsilylation of Dendrons, Dendritic Cores, and Dendrimers: New Convergent and Divergent Routes to Ferrocenyl Dendrimers with Stable Redox Activity, *Chem. Eur. J.* **6**, 2544–2553 (2000).
88. C. Ornelas, D. Méry, J.-C. Blais, E. Cloutet, J. R. Aranzaes, and D. Astruc, Efficient Mono- and Bifunctionalization of Polyolefin Dendrimers by Olefin Metathesis, *Angew. Chem. Int. Ed.* **44**, 7399–7404 (2005).
89. Y. Yu, A. D. Bond, P. W. Leonard, U. J. Lorenz, T. V. Timofeeva, K. P. C. Vollhardt, G. D. Whitener, and A. A. Yakovenko, Hexaferrocenylbenzene, *Chem. Comm.* **2006**, 2572–2574.
90. E. O. Fischer, and S. Breitschaft, Cyclopentadienyl-cycloheptatrienyl-chromium(0), *Angew. Chem. Int. Ed. Engl.* **2**, 100 (1963).
91. R. B. King, and F. G. A. Stone, π-Cyclopentadienyl-π-cycloheptatrienyl Vanadium, *J. Am. Chem. Soc.* **81**, 5263–5264 (1959).
92. R. Criegee, and G. Schröder, Ein Nickel-Komplex des Tetramethyl-Cyclobutadiens, *Angew. Chem.* **71**, 70–71 (1959).
93. W. Hübel, E. H. Braye, A. Clauss, E. Weiss, U. Krüerke, D. A. Brown, G. S. D. King, and C. Hoogzand, The Reactions of Metal Carbonyls with Acetylenic Compounds, *J. Inorg. Nucl. Chem.* **9**, 204–210 (1959).
94. W. Hübel, Organometallic Derivatives from Metal Carbonyls and Acetylenic Compounds, in: *Organic Synthesis via Metal Carbonyls* (Eds. I. Wender and P. Pino, Wiley Interscience, New York, 1968, Vol. I, pp. 273–342).
95. R. P. Dodge, and V. Schomaker, Crystal Structure of Tetraphenylcyclobutadiene Iron Tricarbonyl, *Nature* **186**, 798–799 (1960).
96. D. Seyferth, (Cyclobutadiene)iron Tricarbonyl – A Case of Theory before Experiment, *Organometallics* **22**, 2–20 (2003).
97. H. C. Longuet-Higgins, and L. E. Orgel, The Possible Existence of Transition-metal Complexes of *cyclo*Butadiene, *J. Chem. Soc.* **1956**, 1969–1972.
98. P. M. Maitlis, Cyclobutadiene-Metal Complexes, *Adv. Organomet. Chem.* **4**, 95–143 (1966).
99. P. M. Maitlis, Acetylenes, Cyclobutadienes and Palladium: A Personal View, *J. Organomet. Chem.* **200**, 161–176 (1980).
100. R. Pettit, and G. F. Emerson, Diene-Iron Carbonyl Complexes and Related Species, *Adv. Organomet. Chem.* **1**, 1–46 (1964).
101. G. F. Emerson, L. Watts, and R. Pettit, Cyclobutadiene- and Benzocyclobutadiene-Iron Tricarbonyl Complexes, *J. Am. Chem. Soc.* **87**, 131–133 (1965);
102. R. Pettit, Cyclobutadiene and Its Metal Complexes, *Pure Appl. Chem.* **17**, 253–272 (1968).
103. R. Pettit, The Role of Cyclobutadieneiron Tricarbonyl in the "Cyclobutadiene Problem", *J. Organomet. Chem.* **100**, 205–217 (1975).
104. M. Rosenblum, B. North, D. Wells, and W. P. Giering, Synthesis and Chemistry of h^4-Cyclobutadiene(h^5-cyclopentadienyl)cobalt, *J. Am. Chem. Soc.* **94**, 1239–1246 (1972).
105. S. A. Gardner, and M. D. Rausch, The Formation and Acetylation of π-Cyclopentadienyl-π-cyclobutadienerhodium, *J. Organomet. Chem.* **56**, 365–368 (1973).
106. M. D. Rausch, and A. V. Grossi, (η^4-Cyclobutadiene)(η^5-cyclopentadienyl)dicarbonylvanadium, *J. Chem. Soc., Chem. Comm.* **1978**, 401–402.
107. J. D. Fitzpatrick, L. Watts, G. F. Emerson, and R. Pettit, Cyclobutadieneiron Tricarbonyl. A New Aromatic System, *J. Am. Chem. Soc.* **87**, 3254–3255 (1965).
108. A. Efraty, Cyclobutadienemetal Complexes, *Chem. Rev.* **77**, 691–744 (1977).
109. K. Takanashi, V. Ya. Lee, M. Ichinohe, and A. Sekiguchi, A (Tetrasilacyclobutadiene)-tricarbonyliron Complex [{η^4-(tBu$_2$MeSi)$_4$Si$_4$}Fe(CO)$_3$]: The Silicon Cousin of Pettit's (Cyclobutadiene)tricarbonyliron Complex [{η^4-H$_4$C$_4$}Fe(CO)$_3$], *Angew. Chem. Int. Ed.* **45**, 3269–3272 (2006).

110. H. Hoberg, R. Krause-Göing, and R. Mynott, Bis(tetraphenylcyclobutadiene)nickel, *Angew. Chem. Int. Ed. Engl.* **17**, 123–124 (1978).
111. A. Streitwieser jr., and U. Müller-Westerhoff, Bis(cyclooctatetraenyl)uranium (Uranocene). A New Class of Sandwich Complexes That Utilize Atomic f Orbitals, *J. Am. Chem. Soc.* **90**, 7364 (1968).
112. T. J. Katz, The Cyclooctatetraenyl Dianion, *J. Am. Chem. Soc.* **82**, 3784–3785 (1960).
113. D. Seyferth, Uranocene. The First Member of a New Class of Organometallic Derivatives of the f Elements, *Organometallics* **23**, 3562–3583 (2004).
114. A. Zalkin, and K. N. Raymond, The Structure of Di-π-cyclooctatetraeneuranium (Uranocene), *J. Am. Chem. Soc.* **91**, 5667–5668 (1969).
115. A. Streitwieser jr., and R. Walker, Bis-π-(1,3,5,7-tetraphenylcyclooctatetraene)uranium, an Air Stable Uranocene, *J. Organomet. Chem.* **97**, C41–C42 (1975).
116. T. R. Boussie, R. M. Moore, A. Streitwieser jr., A. Zalkin, J. Brennan, and K. A. Smith, The Uranocene Half-Sandwich: ([8]Annulene)uranium(IV) Dichloride and Some Derivatives, *Organometallics* **9**, 2010–2016 (1990).
117. D. Baudry, E. Bulot, M. Ephritikhine, M. Nierlich, M. Lance, and J. Vigner, Monocyclooctatetraenyluranium(IV) Borohydrides. Crystal Structure of (η-C_8H_8)U (BH_4)($OPPh_3$), *J. Organomet. Chem.* **388**, 279–287 (1990).
118. T. Arliguie, M. Lance, M. Nierlich, J. Vigner, and M. Ephritikhine, Synthesis and Crystal Structure of [K($C_{12}H_{24}O_6$)][U(η-C_7H_7)$_2$], the First Cycloheptatrienyl Sandwich Compound, *J. Chem. Soc., Chem. Comm.* **1995**, 183–184.
119. O. Scherer, Complexes with Substituent-free Acyclic and Cyclic Phosphorus, Arsenic, Antimony, and Bismuth Ligands, *Angew. Chem. Int. Ed. Engl.* **29**, 1104–1122 (1990).
120. F. Mathey, The Chemistry of Phospha- and Polyphosphacyclopentadienide Anions, *Coord. Chem. Rev.* **137**, 1–52 (1994).
121. A. J. Ashe III, and S. Al-Ahmad, Diheteroferrocenes and Related Derivatives of the Group 15 Elements: Arsenic, Antimony, and Bismuth, *Adv. Organomet. Chem.* **39**, 325–353 (1996).
122. K. B. Dillon, F. Mathey, and J. F. Nixon, *Phosphorus: The Carbon Copy* (Wiley, New York, 1998).
123. O. J. Scherer, P_n and As_n Ligands: A Novel Chapter in the Chemistry of Phosphorus and Arsenic, *Acc. Chem. Res.* **32**, 751–762 (1999).
124. F. Mathey, Phosphaorganic Chemistry: Panorama and Perspectives, *Angew. Chem. Int. Ed.* **42**, 1578–1604 (2003).
125. J. Frunzke, M. Lein, and G. Frenking, Structures, Metal-Ligand Bond Strength, and Bonding Analysis of Ferrocene Derivatives with Group-15 Heteroligands Fe(η^5-E_5)$_2$ and FeCp(η^5-E_5) (E=N, P, As, Sb). A Theoretical Study, *Organometallics* **21**, 3351–3359 (2002).
126. E. Urnezius, W. W. Brennessel, C. J. Cramer, J. E. Ellis, and P. von R. Schleyer, A Carbon-Free Sandwich Complex [$(P_5)_2Ti]^{2-}$, *Science* **295**, 832–834 (2002).
127. C. Elschenbroich, J. Koch, J. Kröker, M. Wünsch, W. Massa, G. Baum, and G. Stork, η^6-Koordination in unsubstituiertem Pyridin: (η^6-Benzol)(η^6-pyridin)chrom und Bis(η^6-pyridin)chrom, *Chem. Ber.* **121**, 1983–1988 (1988).
128. L. H. Simons, P. E. Riley, R. E. Davis, and J. J. Lagowski, Bis(2,6-dimethylpyridine)-chromium. A π-Heterocyclic Complex, *J. Am. Chem. Soc.* **98**, 1044–1045 (1976).
129. P. L. Timms, Chemistry of Transition-metal Vapours. Part I. Reactions with Trifluorophosphine and Related Compounds, *J. Chem. Soc. A*, **1970**, 2526–2528.
130. G. E. Herberich, G. Greis, and H. F. Heil, Novel Aromatic Boron Heterocycle as Ligand in a Transition Metal π-Complex, *Angew. Chem. Int. Ed. Engl.* **9**, 805–806 (1970).
131. R. N. Grimes, D. C. Beer, L. G. Sneddon, V. R. Miller, and R. Weiss, Small Cobalt and Nickel Metallocarboranes from 2,3-$C_2B_4H_8$ and 1,6-$C_2B_4H_6$. Sandwich Complexes of the Cyclic $C_2B_3H_7^{2-}$ and $C_2B_3H_5^{2-}$ Ligands, *Inorg. Chem.* **13**, 1138–1146 (1974).

132. W. Siebert, Boron Heterocycles as Ligands in Transition-Metal Chemistry, *Adv. Organomet. Chem.* **18**, 301–340 (1980).
133. G. E. Herberich, Boron Rings Ligated to Metals, in: *Comprehensive Organometallic Chemistry II* (Eds. E. W. Abel, F. G. A. Stone, and G. Wilkinson, Pergamon Press, New York, 1995, Vol. 1, p. 197–216).
134. G. E. Herberich, and H. Ohst, Borabenzene Metal Complexes, *Adv. Organomet. Chem.* **25**, 199–236 (1986).
135. A. J. Ashe III, and P. Shu, The 1-Phenylborabenzene Anion, *J. Am. Chem. Soc.* **93**, 1804–1805 (1971).
136. A. J. Ashe III, W. Butler, and H. F. Sandford, Bis(borabenzene)iron. Nucleophilic and Electrophilic Substitution in Borabenzene-Iron Complexes, *J. Am. Chem. Soc.* **101**, 7066–7067 (1979).
137. R. Prinz, and H. Werner, Hexamethylborazene-chromium-tricarbonyl, *Angew. Chem. Int. Ed. Engl.* **6**, 91–92 (1967).
138. H. Werner, R. Prinz, and E. Deckelmann, Synthese und Eigenschaften von Hexaalkylborazolchromtricarbonylen, *Chem. Ber.* **102**, 95–103 (1969).
139. M. Scotti, H. Werner, D. L. S. Brown, S. Cavell, J. A. Connor, and H. A. Skinner, The Stability of the Borazole-to-Metal Bond in $R_3B_3N_3R'_3Cr(CO)_3$. Kinetic and Thermochemical Studies, *Inorg. Chim. Acta* **25**, 261–267 (1977).
140. R. Fernández, and E. Carmona, Recent Developments in the Chemistry of Beryllocenes, *Eur. J. Inorg. Chem.* **2005**, 3197–3206.
141. I. Resa, E. Carmona, E. Gutierrez-Puebla, and A. Monge, Decamethyldizincocene, a Stable Compound of Zn(I) with a Zn–Zn Bond, *Science* **305**, 1136–1138 (2004).
142. D. del Rio, A. Galindo, I. Resa, and E. Carmona, Theoretical and Synthetic Studies on $[Zn_2(\eta^5\text{-}C_5Me_5)_2]$: Analysis of the Zn–Zn Bonding Interaction, *Angew. Chem. Int. Ed.* **44**, 1244–1247 (2005).
143. J. Maynadié, J.-C. Berthet, P. Thuéry, and M. Ephritikhine, An Unpredented Type of Linear Metallocene with an f-Element, *J. Am. Chem. Soc.* **128**, 1082–1083 (2006).
144. H. Schumann, J. A. Meese-Marktscheffel, and L. Esser, Synthesis, Structure, and Reactivity of Organometallic π-Complexes of the Rare Earths in the Oxidation State Ln3 + with Aromatic Ligands, *Chem. Rev.* **95**, 865–986 (1995).
145. D. S. Breslow, US Pat. Appl. 537039 (1955).
146. D. S. Breslow, and N. R. Newburg, Bis(cyclopentadienyl)titanium Dichloride–Alkylaluminium Complexes as Catalysts for the Polymerization of Ethylene, *J. Am. Chem. Soc.* **79**, 5072–5073 (1957).
147. G. Natta, P. Pino, G. Mazzanti, and U. Giannini, A Crystalline Organometallic Complex Containing Titanium and Aluminium, *J. Am. Chem. Soc.* **79**, 2975–2976 (1957).
148. K. H. Reichert, and K. R. Meyer, Kinetics of Low Pressure Polymerization of Ethylene by Soluble Ziegler Catalysts, *Makromol. Chem.* **169**, 163–176 (1973).
149. A. F. Andresen, H.-G. Cordes, J. Herwig, W. Kaminsky, A. Merck, R. Mottweiler, J. Pein, H. Sinn, and H. J. Vollmer, Halogen-free Soluble Ziegler Catalysts for Ethylene Polymerization. Control of Molecular Weight by the Choice of the Reaction Temperature, *Angew. Chem. Int. Ed. Engl.* **15**, 630–632 (1976).
150. H. Sinn, and W. Kaminsky, Ziegler-Natta Catalysis, *Adv. Organomet. Chem.* **18**, 99–149 (1980).
151. H. Schnutenhaus, and H. H. Brintzinger, 1,1'-Trimethylene-bis(η^5-3-*tert*-butylcyclopentadienyl)titanium(IV) Dichloride, a Chiral *ansa*-Titanocene Derivative, *Angew. Chem. Int. Ed. Engl.* **18**, 777–778 (1979).
152. J. A. Smith, J. von Seyerl, G. Huttner, and H. H. Brintzinger, Molecular Structure and Proton Magnetic Resonance Spectra of Methylene- and Ethylene-Bridged Dicyclopentadienyltitanium Compounds, *J. Organomet. Chem.* **173**, 175–185 (1979).

153. F. R. W. P. Wild, L. Zsolnai, G. Huttner, and H. H. Brintzinger, Synthesis and Molecular Structures of Chiral *ansa*-Titanocene Derivatives with Bridged Tetrahydroindenyl Ligands, *J. Organomet. Chem.* **232**, 233–247 (1982).
154. H. H.Brintzinger, D. Fischer, R. Mülhaupt, B. Rieger, and R. M. Waymouth, Stereospecific Olefin Polymerization with Chiral Metallocene Catalysts, *Angew. Chem. Int. Ed. Engl.* **34**, 1143–1170 (1995).
155. K. Mashima, Y. Nakayama, and A. Nakamura, Recent Trends in the Polymerization of α-Olefins Catalyzed by Organometallic Complexes of Early Transition Metals, *Adv. Polym. Sci.* **133**, 1–51 (1997).
156. H. G. Alt, A. Köppl, Effect of the Nature of Metallocene Complexes of Group IV Metals on Their Performance in Catalytic Ethylene and Propylene Polymerization, *Chem. Rev.* **100**, 1205–1221 (2000).
157. G. W. Coates, Precise Control of Polyolefin Stereochemistry Using Single-Site Metal Catalysts, *Chem. Rev.* **100**, 1223–1252 (2000).
158. A. H. Hoveyda, Chiral Zirconium Catalysts for Enantioselective Synthesis, in: *Titanium and Zirconium in Organic Synthesis* (Ed. I. Marek, Wiley-VCH, Weinheim, 2002, pp. 180–229).
159. P. Corradini, G. Guerra, and L. Cavallo, Do New Century Catalysts Unravel the Mechanism of Stereocontrol of Old Ziegler-Natta Catalysts ?, Acc. Chem. Res. 37, 231–241 (2004).
160. J. A. Ewen, Mechanisms of Stereochemical Control in Propylene Polymerizations with Soluble Group 4B Metallocene/Methylalumoxane Catalysts, *J. Am. Chem. Soc.* **106**, 6355–6364 (1984).
161. W. Kaminsky, K. Külper, H. H. Brintzinger, and F. R. W. P. Wild, Polymerization of Propene and Butene with a Chiral Zirconocene and Methylaluminoxane as Cocatalyst, *Angew. Chem. Int. Ed. Engl.* **24**, 507–508 (1985).
162. A Revolution in Catalysis, *Chemistry in Britain* **1998** (February), 45–47.
163. M. Zintl, and B. Rieger, Novel Olefin Block Copolymers through Chain-Shuttling Polymerization, *Angew. Chem. Int. Ed.* **46**, 333–335 (2007).
164. M. Röper, Homogene Katalyse in der Chemischen Industrie, *Chemie in unserer Zeit* **40**, 126–135 (2006).
165. C. Janiak, and H. Schumann, Bulky or Supracyclopentadienyl Derivatives in Organometallic Chemistry, *Adv. Organomet. Chem.* **33**, 291–393 (1991).
166. L. Venanzi, *The* “Renaissance of Inorganic Chemistry”: *Recollections of a Participant*, Main lecture at the 33rd ICCC (Florence, 1998).
167. I. U. Khand, G. R. Knox, P. L. Pauson, W. E. Watts, and M. I. Foreman, Reaction of Acetylenehexacarbonyldicobalt Complexes, $(R^1C_2R^2)Co_2(CO)_6$, with Norbornene and its Derivatives, *J. Chem. Soc., Perkin Trans. 1*, **1973**, 977–981.
168. S. E. Gibson (née Thomas), and A. Stevenazzi, Pauson-Khand-Reaction: The Beginning of a Catalytic Era, *Angew. Chem. Int. Ed.* **42**, 1800–1810 (2003).
169. W. E. Watts, *J. Organomet. Chem.* **413**, xi–xii (1991).
170. M. A. Bennett, A. A. Danopoulos, W. P. Griffith, and M. L. H. Green, The Contributions to Original Research in Chemistry by Professor Sir Geoffrey Wilkinson FRS 1921–1996, *J. Chem. Soc., Dalton Trans.* **1997**, 3049–3060.
171. M. L. H. Green, and W. P. Griffith, Geoffrey Wilkinson and Platinum Metals Chemistry, *Platinum Metals Rev.* **42**, 168–173 (1998).

Chapter 6
One Deck More: The Chemical "Big Mac"

Experiments that yield light are more worthwhile than experiments that yield fruit.

Francis Bacon, British Philosopher (1561–1626)

Following the reports on the preparation of ferrocene and bis(benzene)chromium, a huge number of papers appeared during the next decade, which dealt not only with the synthesis of other sandwich-type complexes of the general composition $M(C_5H_5)_2$ and $M(C_6H_6)_2$ but also with the reactivity of these novel organometallic compounds. The majority of these studies were concerned to the so-called aromatic nature of ferrocene and its congeners which in many aspects resembled that of benzene and other arenes. Myron Rosenblum, one of the pioneers in this field, illustrated very convincingly in his monograph *Chemistry of the Iron Group Metallocenes* [1], that ferrocene in particular is extremely susceptible to ring substitution by various electrophiles. He also mentioned that, with regard to the mechanism of these reactions, there was the general belief that the attack of the electrophile EX takes place at the metal atom of the starting material **1**, thus generating the short-lived intermediate **A** which after rearrangement to **B** and formation of a C–E bond affords the substituted product **2** by elimination of HX (Scheme 6.1). There was no hint in any of those reports that the labile 16-electron intermediate **B** could loose the cyclopentadiene ligand and the so-formed $[Fe(C_5H_5)]^+$ cation could react with **1** to give the dinuclear complex $[Fe_2(C_5H_5)_3]^+$.

6.1 The Breakthrough: $[Ni_2(C_5H_5)_3]^+$

The possibility that a cationic species of the general composition $[M_2(C_5H_5)_3]^+$ could exist, was first mentioned in a paper published in 1964 [2]. Ernst Schumacher and Richard Taubenest, at that time at the University of Zürich, reported that the ions $[Fe_2(C_5H_5)_3]^+$ and $[Ni_2(C_5H_5)_3]^+$ can be generated from ferrocene and nickelocene in a mass spectrometer under appropriate conditions. The authors concluded that the precursor molecules, after being transformed to the

H. Werner, *Landmarks in Organo-Transition Metal Chemistry*,
Profiles in Inorganic Chemistry, DOI 10.1007/978-0-387-09848-7_6,

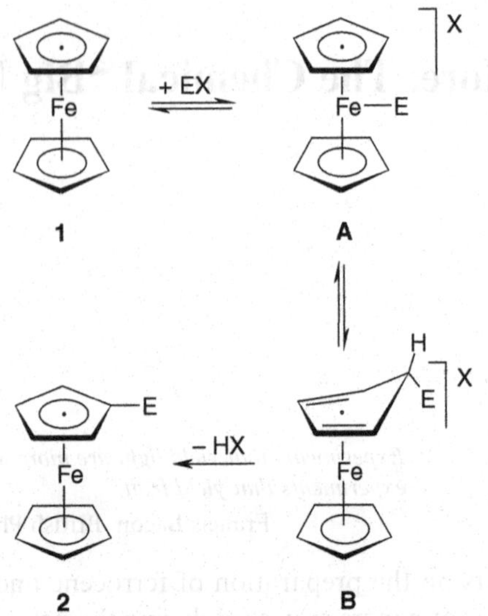

Scheme 6.1 Proposed mechanism for the electrophilic substitution of ferrocene **1** (EX = electrophilic reagent)

cations $[M(C_5H_5)_2]^+$, yield the monocyclopentadienyl metal fragments $[M(C_5H_5)]^+$ which undergo an ion–molecule reaction with the metallocenes to give the dinuclear complexes $[M_2(C_5H_5)_3]^+$. A more detailed mass spectrometric study with $Fe(C_5H_5)_2$ as the starting material supported this hypothesis, confirming that it is unlikely that the cation $[Fe_2(C_5H_5)_3]^+$ originates from a neutral ferrocene dimer since no such species had ever been detected [3].

In retrospect, it is ironic to it that when I met Ernst Schumacher in 1969 (he was then Professor at the University of Bern in Switzerland) we did not talk about the experiments he did at Zürich in the same building where I was at that time. Instead, his interest focussed on our work on borazine transition metal compounds and we discussed in some detail whether it would be possible to incorporate metal atoms like chromium or molybdenum between the layers of hexagonal boron nitride $(BN)_x$ in a similar way as it can be done with graphite. In the course of these discussions I did not mention that, after I had moved to Zürich, we had begun to investigate the reactivity of nickelocene towards both nucleophilic and electrophilic substrates. The reason was that we were still at the beginning, and while we had been able to prepare a series of monocyclopentadienyl nickel complexes from $Ni(C_5H_5)_2$ and Lewis bases, our attempts to obtain alkyl- or acyl-substituted nickelocenes by the Friedel–Crafts reaction failed.

It was in the context of this work that I suggested to Albrecht Salzer (see Fig. 2.20), who joined me in autumn of 1970, that he should study in the first part of his thesis the behaviour of nickelocene towards trityl chloride Ph_3CCl. The idea was that, provided the mechanistic scheme shown in Scheme 6.1 was correct, a cationic species analogous to **B** with nickel instead of iron should be

Scheme 6.2 Proposed mechanism for the reaction of nickelocene **3** with trityl chloride

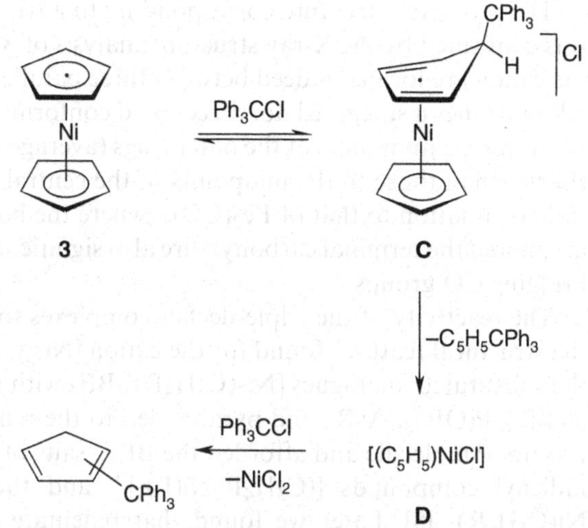

stable since it would obey the 18-electron rule. Therefore I asked Albrecht to treat $Ni(C_5H_5)_2$ (**3**) with Ph_3CCl in nitromethane, taking into consideration that in this solvent trityl chloride partially dissociates in Ph_3C^+ and Cl^-. When Albrecht did the experiment and used the starting materials in the molar ratio of 1:2, he obtained $NiCl_2$ and an isomeric mixture of 1- and 2-triphenylmethylcyclopentadiene [4]. To explain this result we assumed (Scheme 6.2) that the required cation $[(C_5H_5)Ni(C_5H_5CPh_3)]^+$ (**C**) was formed as an intermediate which reacts with nucleophilic chloride by ligand substitution to afford the half-sandwich type compound **D**, apart from the isolated $C_5H_5CPh_3$. Treatment of this labile 16-electron species with another equivalent of trityl chloride could then give $NiCl_2$ and a second molecule of $C_5H_5CPh_3$.

To suppress the proposed displacement of the diene in intermediate **C** with chloride and at the same time to trap the complex cation with a bulky anion, Albrecht performed the reaction of nickelocene with $[Ph_3C]PF_6$ next. The result was surprising indeed. Instead of the expected complex $[(C_5H_5)Ni(C_5H_5CPh_3)]PF_6$, Albrecht isolated a dark brown, remarkably stable solid which analysed as $[Ni_2C_{15}H_{15}]PF_6$. The yield was quantitative if a molar ratio of $Ni(C_5H_5)_2$ to $[Ph_3C]PF_6$ of 2:1 was employed [5]. An analogous product $[Ni_2C_{15}H_{15}]BF_4$ (**4**) could be prepared from **3** and either $[Ph_3C]BF_4$, $[Ph_2CH]BF_4$, $[C_7H_7]BF_4$, $[Me_3O]BF_4$ or HBF_4 as the electrophilic reagent. Moreover, from $Ni(C_5H_4R)_2$ (R = Me, *t*Bu) and HBF_4 in propionic anhydride the ring-substituted complexes $[Ni_2C_{15}H_{12}R_3]BF_4$ were obtained [6]. Based on the ^1H and ^{13}C NMR spectra of $[Ni_2C_{15}H_{15}]^+$, each of which displayed two sharp singlets with an intensity ratio of 2:1, we assumed that the dinuclear cation contained three symmetrically π-bonded cyclopentadienyl ligands of which two were equivalent.

The proposed structure corresponding to a triple-decker sandwich complex was confirmed by the X-ray structure analysis of **4** [7]. The two nickel atoms of the cation are located indeed between three parallel five-membered rings which adopt neither a staggered nor an eclipsed conformation. The distances from the metals to the midpoints of the outer rings (average value 1.728 Å) are ca. 0.06 Å shorter than those to the midpoints of the central ring ligand. It is therefore a similar situation to that of $Fe_2(CO)_9$, where the bond lengths between the iron atoms and the terminal carbonyls are also significantly shorter than those to the bridging CO groups.

The reactivity of the triple-decker complexes towards Lewis bases reflected the structural features found for the cation $[Ni_2(C_5H_5)_3]^+$. Treatment of **4** and the substituted analogues $[Ni_2(C_5H_4R)_3]BF_4$ with monodentate ligands L such as PR_3, $P(OR)_3$, AsR_3 and pyridine led to the removal of one "storey" of the precursor molecule and afforded the BF_4^- salts of the cationic monocyclopentadienyl compounds $[(C_5H_4R)Ni(L)_2]^+$ and the mononuclear sandwiches $Ni(C_5H_4R)_2$ [8]. Later we found that bidentate ligands L–L such as dppe, dipy, norbornadiene and 1,5-cyclooctadiene reacted similarly and gave the corresponding chelate complexes $[(C_5H_4R)Ni(L\text{–}L)]BF_4$. However, the most surprising result regarding the reactivity of the cation $[Ni_2(C_5H_5)_3]^+$ was obtained by Albrecht Salzer and his group in the 1990s [9]. They showed that the reaction of the triple-decker **4** with the "half-open" sandwich **5** yields the metallabenzene derivative **6** which is formed via the incorporation of the $[Ni(C_5H_5)]^+$ fragment between the terminal carbon atoms of the substituted pentadienyl ligand of the ruthenium-containing substrate (Scheme 6.3).

While our attempts to prepare analogues of **4** of the general composition $[(C_5H_5)Ni(\mu\text{-}C_5H_5)M(C_5H_5)]BF_4$ with M = Fe, Ru and Pd failed, we found that the synthetic procedure for **4** initially employed could be supplemented by others (Scheme 6.4). It was truly remarkable that whatever we used as the electrophilic reagent, most of them reacted with nickelocene to form the triple-decker cation $[Ni_2(C_5H_5)_3]^+$. Even with the in situ generated highly reactive 12-electron species $[Rh(C_8H_{12})]^+$ and $[Pd(C_3H_5)]^+$ we were unable to add a

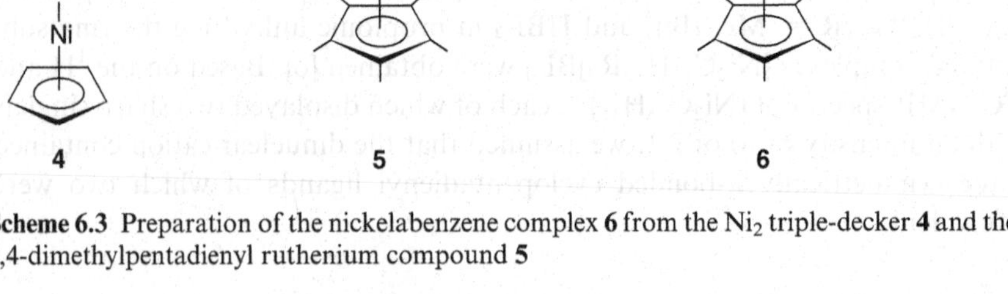

Scheme 6.3 Preparation of the nickelabenzene complex **6** from the Ni_2 triple-decker **4** and the 2,4-dimethylpentadienyl ruthenium compound **5**

Scheme 6.4 Preparative routes leading to the triple-decker sandwich complex **4**

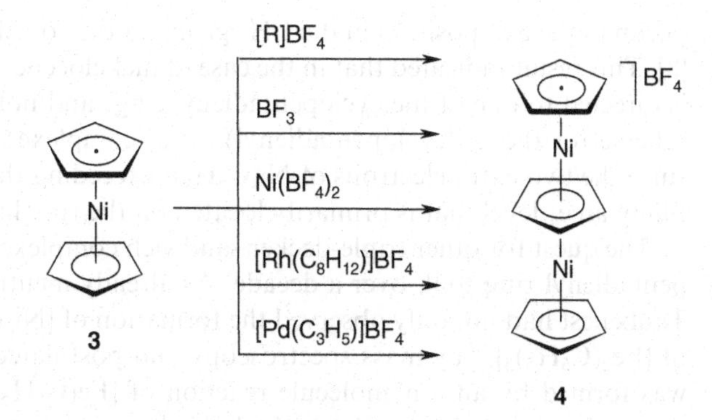

new "storey" onto $Ni(C_5H_5)_2$ and to prepare the dinuclear cations $[(C_5H_5)Ni(\mu\text{-}C_5H_5)Rh(C_8H_{12})]^+$ and $[(C_5H_5)Ni(\mu\text{-}C_5H_5)Pd(C_3H_5)]^+$ with a 30-electron configuration [8]. It seemed that one of the cyclopentadienyl rings of nickelocene could easily be abstracted by interaction with a Lewis acid to form the cation $[Ni(C_5H_5)]^+$ which readily reacts with $Ni(C_5H_5)_2$ to give $[Ni_2(C_5H_5)_3]^+$.

In attempting to prove the mechanistic scheme leading to the Ni_2 triple-decker sandwich, we were lucky in that Trevor Court joined us as a postdoctoral fellow in early 1972. Based on the proposal outlined in Scheme 6.1, we had already tried to prepare or, at least, to detect the 18-electron species $[(C_5H_5)Ni(C_5H_6)]^+$ by using instead of Ph_3CCl different acids such as HBF_4, HBF_3OH, HPF_6 or H_2SO_4 and solvents of different polarity but failed. Even at low temperatures we obtained either the dinuclear cation $[Ni_2(C_5H_5)_3]^+$ or an undefined mixture of products. While thinking about how to explain these results, Trevor asked why we did not use HF both as an acid and a solvent. At first I was somewhat amused, but when he told us that he as well as some other students working with Michael Dove at the University of Nottingham had prepared several otherwise not accessible transition metal compounds in liquid HF, I asked him to find out how to do this experiment. Supported by some NMR experts from the ETH Zürich, Trevor finally dissolved a sample of nickelocene in HF in a "spaghetti tube" (an NMR tube made from Teflon) and confirmed the formation of the mononuclear cation $[(C_5H_5)Ni(C_5H_6)]^+$ by ^1H NMR spectroscopy [10]. Since this diene complex was stable under these conditions only for a short period of time, Trevor attempted to precipitate the cation from the solution in HF as a stable BF_4^- or SbF_6^- salt by addition of a Lewis acid such as BF_3 or SbF_5. This procedure, however, did not lead to the isolation of the expected compounds $[(C_5H_5)Ni(C_5H_6)]X$ but gave the coordinatively unsaturated, highly air-sensitive complexes $[Ni(C_5H_5)]BF_4$ and $[Ni(C_5H_5)]SbF_6$ instead [10]. These complexes reacted instantaneously with nickelocene in nitromethane to afford the corresponding Ni_2 triple-decker sandwich in quantitative yield. In subsequent experiments, the labelled cation $[(C_5H_5)Ni(C_5H_5D)]^+$ was generated from $Ni(C_5H_5)_2$ and CF_3CO_2D at low temperature and it was shown by NMR spectroscopy that the deuterium occupies the *exo* position in the CH_2 group of the coordinated cyclopentadiene [8]. This result indicated that in the case of nickelocene the attack of the proton is directed to one of the cyclopentadienyl rings and not to the metal. The MO scheme for the bis(cyclopentadienyl) metal complexes supports this proposal since the two extra electrons of $Ni(C_5H_5)_2$ exceeding the 18-electron count are filling an e_1 level that is primarily located on the ring ligands.

The quest for other triple-decker sandwich complexes with a bridging cyclopentadienyl ring took over a decade. As already mentioned, Schumacher and Taubenest had not only observed the formation of $[Ni_2(C_5H_5)_3]^+$ but also that of $[Fe_2(C_5H_5)_3]^+$ by mass spectroscopy and postulated that the latter species was formed by an ion–molecule reaction of $[Fe(C_5H_5)]^+$ and $Fe(C_5H_5)_2$ [2]. Later, the cation $[Fe_2(C_5H_5)_3]^+$ had also been generated from the iron cluster $[(C_5H_5)Fe(\mu\text{-}CO)]_4$ by electron impact [11]. Based on MO calculations by Roald Hoffmann, it was predicted that the Fe_2 triple-decker sandwich should be thermodynamically more stable than the nickel counterpart since the two high energy, only slightly bonding e_1 sets occupied in $[Ni_2(C_5H_5)_3]^+$ would be unoccupied in $[Fe_2(C_5H_5)_3]^+$ [12]. Nevertheless, various attempts undertaken in our laboratory to isolate salts of the cation $[Fe_2(C_5H_5)_3]^+$ failed. We came to the conclusion that the formation of this species was mainly handicapped for kinetic reasons since the exceptionally stable ferrocene with its 18-electron configuration appeared to be reluctant to react with precursors of the cation $[Fe(C_5H_5)]^+$ forming the required triple-decker sandwich complex.

6.2 The Iron and Ruthenium Counterparts

The breakthrough in this field came in the late 1980s and early 1990s by the work of Margarita Rybinskaya at the Nesmeyanov Institute in Moscow and Gerhard Herberich at the *Technische Hochschule* in Aachen. In their pioneering studies, Rybinskaya's group showed that in contrast to the metallocenes $M(C_5H_5)_2$ the more electron-rich decamethylmetallocenes $M(C_5Me_5)_2$ (M = Fe, Ru, Os) undergo a stacking reaction with the 12-electron fragments $[Fe(C_5H_5)]^+$ and $[Ru(C_5R_5)]^+$, being generated from $[(C_5H_5)Fe(C_6H_6)]PF_6$ or $[(C_5R_5)Ru(MeCN)_3]PF_6$ (R = H, Me), affording the homo- and hetero-metallic triple-decker complexes **7**–**12** in good to excellent yield (Scheme 6.5) [13]. The X-ray structure analyses of **10**, **11** and **12** revealed, that the structure of these 30-valence electron compounds is similar to that of $[Ni_2(C_5H_5)_3]^+$, with the planar C_5Me_5 ring bridging the $M(C_5R_5)$ and $M'(C_5Me_5)$ moieties. An interesting facet of the preparative work carried out by Rybinskaya was that in complexes such as $[Ru_2(C_5Me_5)_3]^+$ the acidity of the CH_3 protons of the terminal C_5Me_5 rings was enhanced by the positive charge of the triple-decker cation, thus enabling the synthesis of substituted derivatives $[(C_5Me_4CH_2E)Ru(\mu\text{-}C_5Me_5)Ru(C_5Me_5)]^+$ (E = Me, $SiMe_3$, SEt) by stepwise deprotonation and electrophilic addition [14].

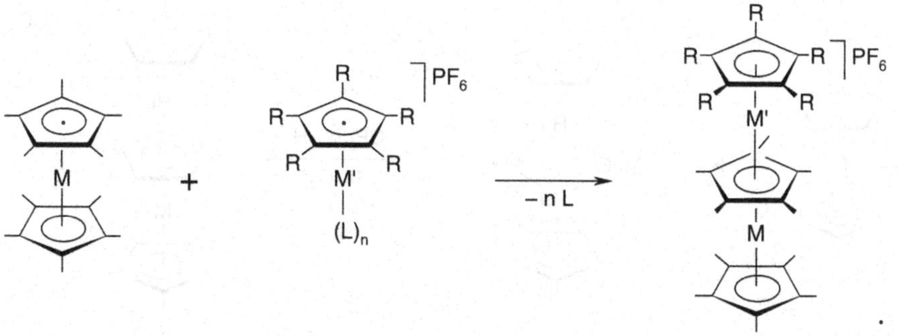

	M	M'	R
7	Fe	Fe	H
8	Ru	Fe	H
9	Os	Fe	H

	M	M'	R
10	Ru	Ru	H
11	Ru	Ru	Me
12	Os	Ru	H

7 – 12

Scheme 6.5 Preparation of homo- and hetero-metallic triple-decker sandwich complexes **7–12** containing a bridging pentamethylcyclopentadienyl ligand

The possibility of preparing analogues of the triple-decker complexes **7–12** containing an *unsubstituted* C_5H_5 ligand instead of C_5Me_5 was first confirmed by Herberich's group in 1993 [15]. The synthesis was somewhat tricky insofar as the reaction of $(C_5H_5)M(C_5Me_5)$ (M = Fe, Ru) with $[(C_5Me_5)Ru(NCMe)_3]PF_6$ in acetone led to an equilibrium with only 30% of the cation $[(C_5Me_5)M(\mu\text{-}C_5H_5)Ru(C_5Me_5)]^+$ present. Addition of benzene furnished the degradation of the dinuclear cation and led to the formation of $[(C_5Me_5)Ru(C_6H_6)]^+$. The equilibrium could be shifted, however, towards the triple-decker complex if the stacking reaction was performed in solvents with weak ligand properties such as diethyl ether or nitromethane, and if nucleophilic counter ions were avoided. Herberich showed that the combination $[(C_5Me_5)Ru(OMe)]_2/CF_3SO_3H$ in diethyl ether served as an efficient source for the $[Ru(C_5Me_5)]^+$ fragment which in turn gave the dinuclear complexes **13** and **14** upon addition of $(C_5H_5)M(C_5Me_5)$ (Scheme 6.6). The stacking reaction leading to **13** and **14** proceeded with remarkable regioselectivity since the bridging ligand was exclusively the C_5H_5 and not the C_5Me_5 unit.

The recent progress in the field of triple-decker sandwich complexes containing a bridging five-membered carbocyclic ligand has been highlighted by a paper of Rybinskaya and coworkers, who isolated a stable salt of the long-sought elusive cation $[Fe_2(C_5H_5)_3]^+$, 27 years after we reported the synthesis of $[Ni_2(C_5H_5)_3]PF_6$ [16]. The key to success was to generate the required fragment by visible light irradiation of $[(C_5H_5)Fe(C_6H_6)]PF_6$ in CH_2Cl_2 at 0 °C and then add ferrocene to give $[Fe_2(C_5H_5)_3]PF_6$. The permethylated analogue $[Fe_2(C_5Me_5)_3]PF_6$ was obtained by a similar route; in this case, however, at room temperature due to its greater stability. Rybinskaya's group also succeeded

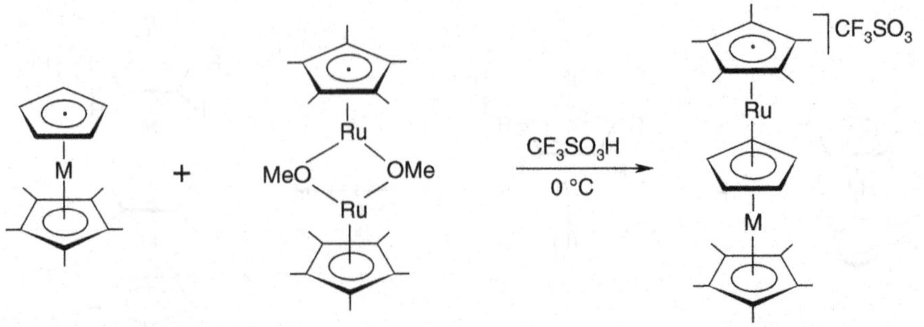

M + MeO Ru OMe Ru $\xrightarrow[0\,°C]{CF_3SO_3H}$ Ru M CF_3SO_3

13: M = Fe
14: M = Ru

Scheme 6.6 Preparation of the FeRu and Ru_2 triple-decker sandwiches **13** and **14** containing a bridging unsubstituted cyclopentadienyl ligand

in generating the heterometallic triple-decker $[(C_5H_5)Fe(\mu\text{-}C_5H_5)Co(C_4H_4)]^+$ from the in situ prepared ion $[Fe(C_5H_5)]^+$ and $(C_5H_5)Co(C_4H_4)$. The dinuclear FeCo cation is quite labile and loses ferrocene at room temperature, leaving a $[Co(C_4H_4)]^+$ fragment that reacts with $(C_5H_5)Co(C_4H_4)$ to afford the symmetrical complex $[(C_4H_4)Co(\mu\text{-}C_5H_5)Co(C_4H_4)]^+$. The stacking methodology has also been applied for the preparation of the first unsymmetrical 34-electron triple-decker $[(C_5H_5)Ni(\mu\text{-}C_5H_5)Co(C_6Me_6)]^+$ containing a central cyclopentadienyl ligand [17]. In this case, the in situ generated $[Co(C_6Me_6)]^+$ cation, being isolobal to $[Ni(C_5H_5)]^+$, reacted with nickelocene to give the desired product.

That cationic triple-decker compounds of the main group elements with a bridging C_5R_5 ligand also exist, was first shown by Alan Cowley at the beginning of the twenty-first century. The Sn_2 cation **15** (Scheme 6.7) was generated upon treatment of the electron-rich stannocene $Sn(C_5Me_5)_2$ with electrophilic $Ga(C_6F_5)_3$, while the cationic indium complex **16** resulted from the reaction of

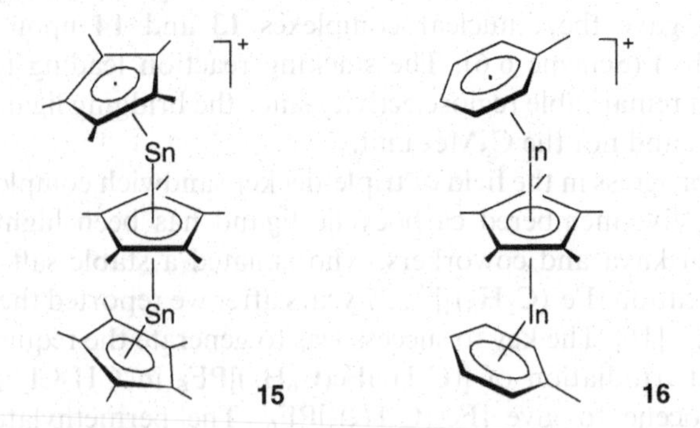

Scheme 6.7 Main group triple-decker cations **15** and **16** with a bent structure and a bridging pentamethylcyclopentadienyl ring

$[In(C_5Me_5)]_6$ with an equimolar mixture of $B(C_6F_5)_3$ and $H[B(C_6F_5)_3(OH)]$ in toluene [18]. In the latter case the solvent acts as the source of the terminal ligands. X-ray diffraction studies revealed that both **15** and **16** are bent triple-decker sandwiches with a linear arrangement between the two metal atoms and the centroid of the bridging five-membered ring. Typical for **15** and **16** is the *cisoid*-type geometry which is in contrast to the structure of the triple-decker anions $[Tl_2(C_5H_5)_3]^-$ and $[Cs_2(C_5H_5)_3]^-$, where the terminal cyclopentadienyl rings are *trans* disposed.

6.3 Arene-bridged Triple-Decker Sandwiches

The first examples of triple-decker sandwiches containing an arene-bridge were reported by Klaus Jonas from the Max Planck Institute in Mülheim in 1983 [19]. The paramagnetic complex **17** (Scheme 6.8) with a central benzene ligand was prepared in about 50% yield from the bis(allyl)vanadium(III) compound $(C_5H_5)V(C_3H_5)_2$ and an excess of 1,3-cyclohexadiene at elevated temperatures. As a by-product, the analogous $(C_5H_5)V(\mu\text{-}C_6H_5\text{-}n\text{Pr})V(C_5H_5)$ derivative was obtained. In the presence of toluene and mesitylene, complex **17** readily underwent an arene exchange which, remarkably, proceeded with retention of the triple-decker sandwich structure. Treatment of **17** (which has a 26-electron configuration) with allyl chloride or iodine quantitatively liberated the arene ligand. The X-ray structure analyses of **17** and the analogue $(C_5H_5)V(\mu\text{-}1,3,5\text{-}C_6H_3Me_3)V(C_5H_5)$ confirmed that the five- and six-membered rings are completely planar and essentially parallel to each other. The same structural feature was found for the dinuclear chromium complexes $Cr_2(C_6H_3R_3)_3$ (**18**; R = Me, *t*Bu), which were prepared by metal atom/ligand vapour synthesis [20, 21]. Similar to $[Fe_2(C_5H_5)_3]^+$, these neutral Cr_2 triple-decker sandwiches are 30-electron compounds while the recently reported benzene-bridged dication $[(C_5H_5)Ni(\mu\text{-}C_6H_6)Ni(C_5H_5)]^{2+}$ (**19**) has a 34-electron count. The latter is a structural counterpart of $[Ni_2(C_5H_5)_3]^+$. The synthesis of **19**, carried out by Malcolm Green's group at Oxford, was an exceptional piece of work and made possible

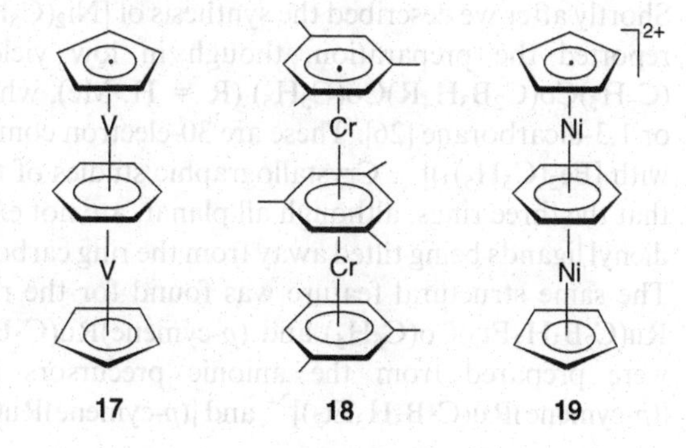

Scheme 6.8 Arene-bridged triple-decker sandwich complexes **17**–**19** with 26 (V_2), 30 (Cr_2) and 34 (Ni_2) valence electrons

by using the novel weakly coordinating counter ion $[\{B_3(\mu\text{-}O)_3\}(C_6F_5)_4]^-$ [22]. In addition to **19**, several other dinuclear complexes of the general composition $(C_5R_5)M(\mu\text{-}C_6R'_6)M(C_5R_5)$ (M = Fe, Co) have been described in the literature but they either contain a non-planar six-membered ring or the bridging ligand is slipped from a symmetrical $\mu\text{-}\eta^6{:}\eta^6$– to a $\mu\text{-}\eta^4{:}\eta^4$-coordination mode, resulting in a non-linear arrangement between the two metal atoms and the centroid of the arene [23]. These compounds showed an increased metal–arene lability, which allows an easy replacement of the central ring by other arenes or oligoolefins with retention of the *trans* stereochemistry.

By going from C_5H_5 to C_6H_6 and further to C_7H_7 and C_8H_8 there is a significant increase in ring size and thus small metals cannot coordinate in a η^7 or η^8 bonding mode as easily as they can in η^5 or η^6. Owing to this, it is not surprising that up to now true triple-decker complexes of the type $(\eta^n\text{-}C_nR_n)M(\mu\text{-}\eta^7{:}\eta^7\text{-}C_7H_7)M(\eta^n\text{-}C_nR_n)$ and $(\eta^n\text{-}C_nR_n)M(\mu\text{-}\eta^8{:}\eta^8\text{-}C_8H_8)M(\eta^n\text{-}C_nR_n)$ with a *trans* geometry at the planar middle deck are unknown for the d-block transition metals. A vast number of neutral or cationic compounds containing a bridging C_8H_8 ligand have been described but in all of those the eight-membered ring is non-planar and either $\eta^4{:}\eta^4$- or $\eta^5{:}\eta^5$-bonded to the metal centres [23]. True triple-decker sandwiches with a *trans*-disposed $M(\mu\text{-}\eta^8{:}\eta^8\text{-}C_8H_8)M$ unit were only prepared for barium and the lanthanides, that means for elements with large ionic radii. The europium and ytterbium complexes $(C_5Me_5)M(\mu\text{-}C_8H_8)M(C_5Me_5)$ [24] as well as the barium compound $(C_5H\textit{i}Pr_4)Ba(\mu\text{-}C_8H_8)Ba(C_5H\textit{i}Pr_4)$ [25] possess a *transoid* bent structure, the bending, however, being less than for the cationic species **15** and **16** with a bridging C_5Me_5 ligand.

6.4 "Big Macs" with Bridging P_5, P_6 and Heterocycles as Ligands

Analogues of the prototypical "Big Macs" such as **4** or **18**, which have heterocycles instead of carbocycles as bridging ligands, are manifold and nowadays known particularly with boron- and phosphorus-containing ring systems. Shortly after we described the synthesis of $[Ni_2(C_5H_5)_3)]PF_6$ [5, 6], Russel Grimes reported the preparation, though in low yield, of the cobalt complexes $(C_5H_5)Co(C_2B_3H_4R)Co(C_5H_5)$ (R = H, Me), where the middle ring was a 1,2- or 1,3-dicarborane [26]. These are 30-electron compounds and thus isoelectronic with $[Fe_2(C_5H_5)_3)]^+$. Crystallographic studies of the methyl derivative revealed that the three rings, although all planar, are not exactly parallel, the cyclopentadienyl ligands being tilted away from the ring carbons of the central ring by ca. 5°. The same structural feature was found for the related complexes (*p*-cymene)-$Ru(C_2B_3H_3Et_2)Co(C_5H_5)$ and (*p*-cymene)$Ru(C_2B_3H_3Et_2)Ru$(*p*-cymene), which were prepared from the anionic precursors $[(C_5H_5)Co(C_2B_3H_3Et_2)]^{2-}$ or $[(p\text{-cymene})Ru(C_2B_3H_3Et_2)]^{2-}$ and $[(p\text{-cymene})RuCl_2]_2$ by stacking reactions [27].

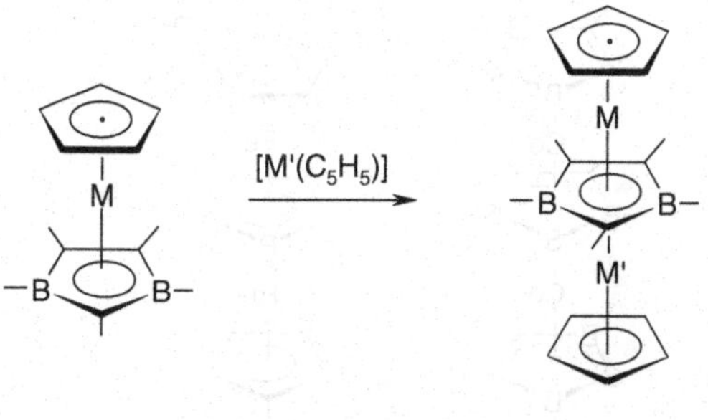

$[M'(C_5H_5)]$

20 – 24

	M	M'
20	Ni	Ni
21	Ni	Co
22	Co	Co
23	Co	Fe
24	Fe	Fe

Scheme 6.9 Preparative route to homo- and hetero-metallic triple-decker complexes **20–24** with cyclic carboranes as bridging ligands (the substituents at the boron and carbon atoms of the middle deck are alkyl groups)

A more general synthetic route to triple-decker sandwiches with a similar structure as Grimes' compounds was developed by Walter Siebert at the University of Heidelberg in the early 1980s [28, 29]. Siebert's group made use of the mononuclear sandwich compounds $(C_5H_5)M(R_3C_3B_2R'_2)$ or $(C_5H_5)M(R_3HC_3B_2R'_2)$ as the precursors which reacted with $M'(C_5H_5)$ fragments, generated from cyclopentadienyl metal carbonyls or related ethene metal derivatives, by a stacking process to give the homo- and hetero-metallic complexes **20–24** (Scheme 6.9) in moderate to good yield. While **23** has a 30-electron configuration, the other triple-decker sandwiches of this series possess 33 (**20**), 32 (**21**), 31 (**22**) and 29 (**24**) valence electrons and are thus paramagnetic, in agreement with the MO scheme [12]. In tetrahydrofuran, compound **20** can be reduced with potassium to the diamagnetic anion $[\mathbf{20}]^-$, which is the first anionic triple-decker sandwich compound of a transition metal and isoelectronic to cationic $[Ni_2(C_5H_5)_3)]^+$ [28].

Triple-decker sandwiches **25** [30], **26** [31], **27** [32], **28** [33], **29** [34] and **30** [35] with bridging five-membered C_4B, C_2B_2S, C_4P and C_2P_3 rings as well as with six-membered C_5B and C_4B_2 heterocycles have also been prepared (Scheme 6.10), in most cases by stacking reactions. They are diamagnetic and all obey the 30-electron rule. The "homoleptic" compounds **25** and **26**, in which not only the bridging but also the terminal ligands are heterocyclic boron-containing rings, are of particular interest. The Rh_2 compound **25** is a fascinating starting material as it offers, inter alia, the opportunity to prepare water-soluble salts of the sandwich-type anion $[Rh(C_4H_4BPh)_2]^-$ [36]. The related uncharged cobalt complex $(C_5H_5)Co(C_4H_4BPh)$ reacts with $[(C_5Me_5)Ir(acetone)_3]^{2+}$ by coordination of the $[Ir(C_5Me_5)]^{2+}$ unit to the borole ring to give the hetero-metallic dication $[(C_5H_5)Co(\mu\text{-}C_4H_4BPh)Ir(C_5Me_5)]^{2+}$, which has been the first stable triple-decker sandwich with iridium as one of the metal centres [37]. Recently, Aleksandr Kudinov at the Nesmeyanov Institute of Organoelement Compounds in Moscow applied the same methodology to obtain 30-electron

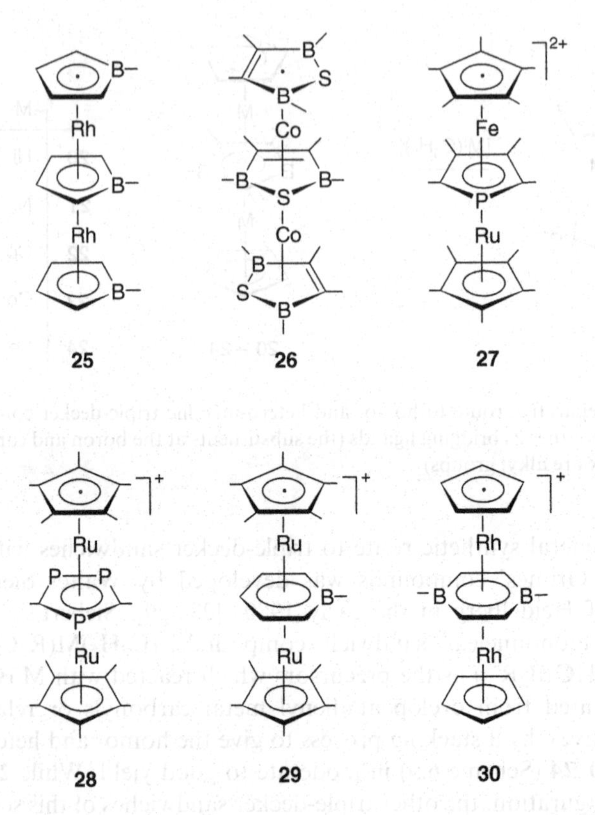

Scheme 6.10 Triple-decker sandwich complexes **25**–**30** with bridging five- and six-membered heterocyclic ligands (the substituents at the ring atoms are alkyl or aryl groups)

triple-decker complexes of the general composition $[(C_5H_5)Rh(\mu\text{-}C_4H_4BPh)M(ring)]^{2+}$, where the M(ring) unit is $[Co(C_5Me_5)]$, $[Ir(C_5Me_5)]$ or Ru(arene), by using the sandwich compound $(C_5H_5)Rh(C_4H_4BPh)$ as the precursor [38]. Detailed crystallographic studies revealed that the coordination of the $[M(ring)]^{2+}$ fragments to the borole ring leads to an elongation of the bonds within the heterocycle, which had not been observed (within experimental error) for complexes such as **4**, **10**, **14** or **17** with a bridging carbocyclic ligand.

The closest analogues of the triple-decker sandwiches with C_5H_5 and C_6H_6 ligands as the middle deck are undoubtedly provided by the dinuclear complexes **31** and **32**–**35** containing a symmetrical P_5 or P_6 ring in the bridge. It was Otto Scherer at the University of Kaiserslautern in Germany who illustrated in the 1980s that the slogan *Phosphorus: The Carbon Copy* [39] also holds in the field of the chemical "Big Macs". The appropriate synthetic route to obtain compounds such as **32**–**35** (Scheme 6.11) was to treat mono- or dinuclear cyclopentadienyl metal carbonyls with white phosphorus at high temperatures which gave the triple-decker sandwiches in low to moderate yield [40]. If $[(C_5Me_5)Fe(CO)_2]_2$ was used as starting material, the reaction with P_4

Scheme 6.11 Triple-decker sandwich complexes **31–35** with bridging P_5 and P_6 ring ligands (the substituents at the cyclopentadienyl rings are alkyl groups)

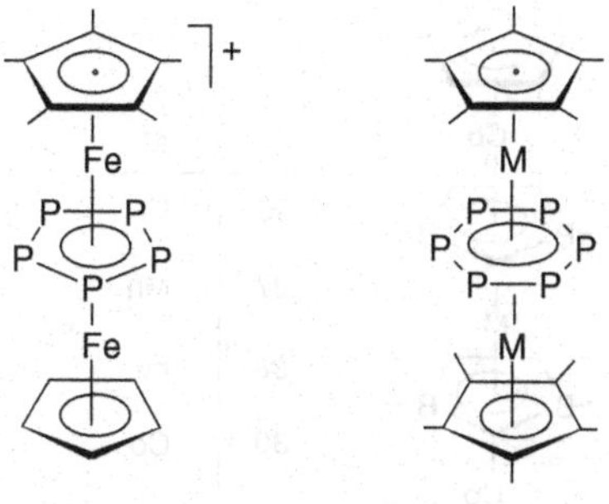

31 **32 – 35**

	M
32	V
33	Nb
34	Mo
35	W

afforded the ferrocene-analogue $(C_5Me_5)Fe(P_5)$, which then reacted with $[(C_5H_5)Fe(C_6H_6)]PF_6$ under UV-irradiation to afford **31**. The Fe(μ-As_5)Fe counterpart of **31** [41] as well as an uncharged analogue with a central Mo (μ-Sb_5)Mo unit also exist [42]. While in **31**, having a 30-electron count, and in the P_6-bridged triple-deckers **34** and **35**, having an 28-electron count, the middle deck is symmetric and completely planar, the P_6 ligand in the 26-electron complexes **32** and **33**, although planar, is somewhat distorted towards a μ-η^3:η^3-bonding mode. With regard to the formation of the previously unknown P_5 unit in the reactions of $[(C_5Me_5)Fe(CO)_2]_2$ and other cyclopentadienyl metal carbonyls with P_4, it is interesting to note that at the same time, when Scherer reported the synthesis of $(C_5Me_5)Fe(P_5)$, the first stable alkali-metal salts of the P_5^- anion were prepared from Na or $LiPH_2$ and P_4 [43]. Both crystallographic and theoretical studies confirmed that the P–P bond lengths and the P–P–P bond angles in **31** and NaP_5 are nearly identical [40].

6.5 Tetra-, Penta- and Hexa-Decker Sandwich Complexes

The molecular architecture found in the triple-decker complexes had also been extended to sandwich-type complexes of higher nuclearity, in which two, three or four five-membered heterocycles act as bridging ligands [29]. The methodology to prepare trinuclear compounds such as **36–42** (Scheme 6.12), developed by Walter Siebert, used the in situ generated anion $[(C_5H_5)Co\ (R_3C_3B_2R'_2)]^-$ as a coordinating unit which reacted with salts MX_2 to give the heterometallic tetra-decker sandwich complexes in high yield. With the exception of **42**, they are paramagnetic with high-spin configurations in compounds **36–39**. Homometallic complexes such as $(C_5H_5)Ni(\mu\text{-}R_3C_3B_2R'_2)$ Ni $(\mu\text{-}R_3C_3B_2R'_2)Ni(C_5H_5)$ were obtained by a similar route. Regarding the molecular structures of **36–42**, the striking feature is that the central metal M is slipped away from the midpoints of the bridging heterocycles in the direction of the single carbon atom, thus indicating a η^3- rather than a η^5-coordination of the C_3B_2 rings [29].

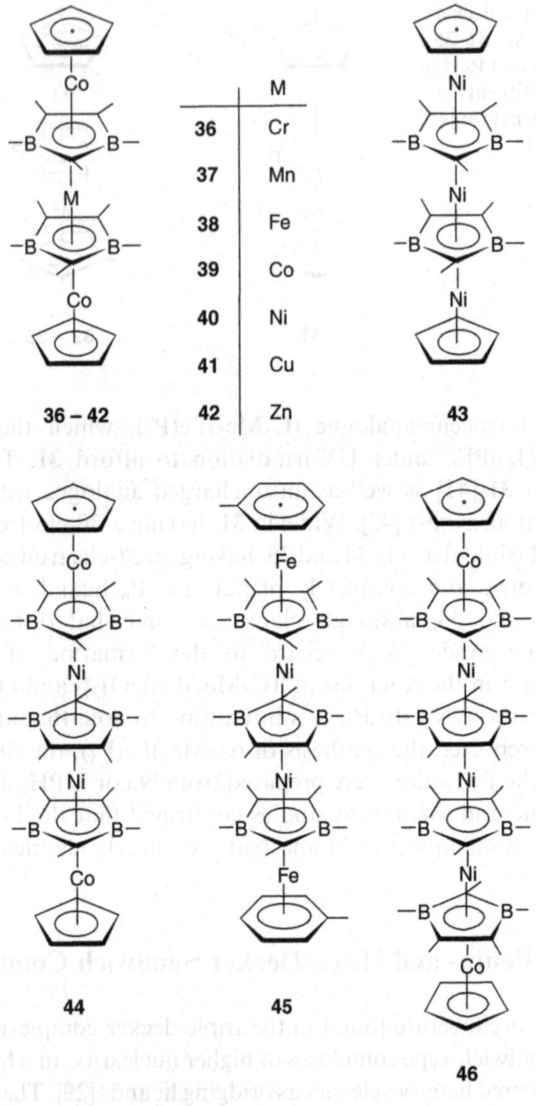

	M
36	Cr
37	Mn
38	Fe
39	Co
40	Ni
41	Cu
42	Zn

Scheme 6.12 Tetra-, penta- and hexa-decker sandwich complexes 36–46 with cyclic carboranes as bridging ligands (the substituents at the boron and carbon atoms of the heterocycles are alkyl groups)

The preparation of the tetra-decker complex **43** was achieved via reduction of $(C_5H_5)Ni(Et_4C_3B_2Me)$ with potassium to afford the corresponding anion which on treatment with $NiCl_2$ gives **43** [28]. Further stacking to penta- and hexa-decker complexes **44**, **45** and **46** was possible by treatment of the di- and trinuclear π-allyl nickel derivatives $(C_5H_5)Co(\mu\text{-}R_3C_3B_2R'_2)Ni(C_3H_5)$ and $(C_3H_5)Ni(\mu\text{-}R_3C_3B_2R'_2)Ni(\mu\text{-}R_3C_3B_2R'_2)Ni(C_3H_5)$ with either 2,

3-dihydro-1,3-diboroles or appropriate sandwich compounds, again illustrating the pronounced capability of five-membered C_3B_2 heterocycles to bridge two metal centres [29]. With $(C_3H_5)Ni(\mu\text{-}R_3C_3B_2R'_2)Ni(C_3H_5)_2$ as the precursor, Siebert's group also succeeded to obtain amorphous polymers of the approximate composition $[Ni(R_3C_3B_2R'_2)]_n$, which were paramagnetic and semiconducting [44]. Since EXAFS measurements revealed different Ni–Ni and Ni–C,B distances, it remains uncertain whether these polymers have the expected polydecker structure or not.

Could "Super Big Macs", in other words, tetra-, penta- or hexa-decker sandwich complexes exclusively with five- or six-membered *carbocyclic* ligands also be accessible? Theoretical studies have been carried out and, by using the fragment molecular orbital approach, in one of these Jemmis and Reddy concluded [45], that for tetra-decker complexes $(C_5H_5)M(\mu\text{-}C_5H_5)M'\text{-}(\mu\text{-}C_5H_5)M(C_5H_5)$ a symmetric structure with coaxial metal atoms should be stable if M is a late and M' an early transition metal. Previous attempts in our laboratory to prepare, for example, the compound $(C_5H_5)Co(\mu\text{-}C_5H_5)Cr\text{-}(\mu\text{-}C_5H_5)Co(C_5H_5)$ with a 42-electron count failed. Nevertheless, given my imperturbable optimism, as well as Roald Hoffmann's previous statement "that experimentalists have a good chemical intuition anyway, and that theory doesn't help very much" into account, [1] I am convinced that the chapter on polydecker sandwich complexes of the general composition $M_x(C_nR_n)_{x+1}$, with the carbocyclic ligands C_5R_5 and C_6R_6 in particular, is not closed as yet [46]. [2]

Notes

1. R. Hoffmann, letter to the author, dated June 7, 1976.
2. After this manuscript was finished, Frank Edelmann's group at the University of Magdeburg in Germany reported the synthesis and molecular structure of the ytterbium tetra-decker complex $(C_5Me_5)Yb(\mu\text{-}\eta^8,\eta^8\text{-}C_8H_5R_3)Yb(\mu\text{-}\eta^8,\eta^8\text{-}C_8H_5R_3)Yb(C_5Me_5)$ (R = $SiMe_3$) with an almost linear tetra-decker arrangement.

References

1. M. Rosenblum, *Chemistry of Iron Group Metallocenes, Part 1* (Wiley, New York, 1965, Chap. 4).
2. E. Schumacher, and R. Taubenest, "Tripeldecker-Sandwiches" aus Ferrocen und Nickelocen, *Helv. Chim. Acta* **47**, 1525–1529 (1964).
3. S. M. Schildcrout, High-Pressure Mass Spectra and Gaseous Ion Chemistry of Ferrocene, *J. Am. Chem. Soc.* **95**, 3846–3849 (1973).
4. H. Werner, G. Mattmann, A. Salzer, and T. Winkler, Elektronentransfer-Reaktionen von Dicyclopentadienylnickel und –cobalt mit Triphenylmethyl-chlorid, *J. Organomet. Chem.* **25**, 461–474 (1970).
5. H. Werner, and A. Salzer, Die Synthese eines ersten Doppel-Sandwich-Komplexes: Das Dinickeltricyclopentadienyl-Kation, *Synth. Inorg. Met.-Org. Chem.* **2**, 239–248 (1972).
6. A. Salzer, and H. Werner, A New Route to Triple-Decker Sandwich Compounds, *Angew. Chem. Int. Ed. Engl.* **11**, 930–932 (1972).

7. E. Dubler, M. Textor, H. R. Oswald, and A. Salzer, X-Ray Structure Analysis of the Triple-Decker Sandwich Complex Tris(η-cyclopentadienyl)dinickel Tetrafluoroborate, *Angew. Chem. Int. Ed. Engl.* **13**, 135 (1974).
8. H. Werner, New Varieties of Sandwich Complexes, *Angew. Chem. Int. Ed. Engl.* **16**, 1–9 (1977).
9. U. Bertling, U. Englert, and A. Salzer, From the Tripeldecker to a Metallabenzene: A New Generation of Sandwich Complexes, *Angew. Chem. Int. Ed. Engl.* **33**, 1003–1005 (1994).
10. T. L. Court, and H. Werner, Concerning the Mechanism of Formation of the Cationic Triple Decker Sandwich Complex $[Ni_2(C_5H_5)_3]^+$, and the Isolation of $[NiC_5H_5]BF_4$, *J. Organomet. Chem.* **65**, 245–251 (1974).
11. R. B. King, Electron Impact Studies on Cyclopentadienyliron Carbonyl Tetramer: A Novel Route to a "Triple-decker" Sandwich Ion, *Chem. Comm.* **1969**, 436–437.
12. J. W. Lauher, M. Elian, R. H. Summerville, and R. Hoffmann, Triple-Decker Sandwiches, *J. Am. Chem. Soc.* **98**, 3219–3224 (1976).
13. A. R. Kudinov, M. I. Rybinskaya, Yu. T. Struchkov, A. I. Yanovskii, and P. V. Petrovskii, Synthesis of the First 30-Electron Triple-Decker Complexes of the Iron Group Metals with Cyclopentadienyl Ligands. X-Ray Structure of $[(\eta\text{-}C_5H_5)Ru(\mu,\eta\text{-}C_5Me_5)Ru(\eta\text{-}C_5Me_5)]PF_6$, *J. Organomet. Chem.* **336**, 187–197 (1987).
14. A. R. Kudinov, A. A. Filchikov, and M. I. Rybinskaya, Deprotonation and Subsequent Functionalization of Methyl Groups in Cationic Ruthenium Triple-Decker Complexes, *Mendeleev Comm.* **1992**, 64–65.
15. G. E. Herberich, U. Englert, F. Marken, and P. Hofmann, Triple-Decker Complexes with Bridging Cyclopentadienyl Ligands and Novel Cyclopentadienyl Transfer Reactions, *Organometallics* **12**, 4039–4045 (1993).
16. A. R. Kudinov, A. A. Filchikov, P. V. Petrovskii, and M. I. Rybinskaya, 30-Electron Cationic Iron- and Cobalt-containing Triple-Decker Complexes with a Central Cyclopentadienyl Ligand. The First Synthesis of the Parent Triple-Decker Iron Complex with Cyclopentadienyl Ligands, $[(\eta^5\text{-}C_5H_5)Fe(\mu\text{-}\eta{:}\eta\text{-}C_5H_5)Fe(\eta^5\text{-}C_5H_5)]PF_6$, *Russ. Chem. Bull.* **48**, 1352–1355 (1999).
17. A. R. Kudinov, and M. I. Rybinskaya, New Triple-Decker Complexes Prepared by the Stacking Reactions of Cationic Metallofragments with Sandwich Compounds, *Russ. Chem. Bull.* **48**, 1615–1621 (1999).
18. A. H. Cowley, C. L. B. Macdonald, J. S. Silverman, J. D. Gorden, and A. Voigt, Triple-Decker Main Group Cations, *Chem. Comm.* **2001**, 175–176.
19. A. W. Duff, K. Jonas, R. Goddard, H.-J. Kraus, and C. Krüger, The First Triple-Decker Sandwich with a Bridging Benzene Ring, *J. Am. Chem. Soc.* **105**, 5479–5480 (1983).
20. W. M. Lamanna, Metal Vapor Synthesis of a Novel Triple-Decker Sandwich Complex: $(\eta^6\text{-Mesitylene})_2(\mu\text{-}\eta^6{:}\eta^6\text{-mesitylene})Cr_2$, *J. Am. Chem. Soc.* **108**, 2096–2097 (1986).
21. F. N. Cloke, K. A. E. Courtney, A. A. Sameh, and A. C. Swain, Bis(η-arene) Complexes of the Early Transition Metals Derived From the 1,3,5-Tri-*t*-butylbenzene Ligand, *Polyhedron* **8**, 1641–1648 (1989).
22. J. L. Priego, L. H. Doerrer, L. H. Rees, and M. L. H. Green, Weakly Coordinating Anions Stabilise the Unprecedented Monovalent and Divalent η-Benzene Nickel Cations $[(\eta^5\text{-}C_5H_5)Ni(\eta^6\text{-}C_6H_6)Ni(\eta^5\text{-}C_5H_5)]^{2+}$ and $[Ni(\eta^6\text{-}C_6H_6)_2]^{2+}$, *Chem. Comm.* **2000**, 779–780.
23. V. Beck, and D. O'Hare, Triple-Decker Transition Metal Complexes Bridged by a Single Carbocyclic Ring, *J. Organomet. Chem.* **689**, 3920–3938 (2004).
24. W. J. Evans, M. A. Johnston, M. A. Greci, and J. W. Ziller, Synthesis, Structure, and Reactivity of Unsolvated Triple-Decked Bent Metallocenes of Divalent Europium and Ytterbium, *Organometallics* **18**, 1460–1464 (1999).
25. H. Sitzmann, M. D. Walter, and G. Wolmershäuser, A Triple-Decker Sandwich Complex of Barium, *Angew. Chem. Int. Ed.* **41**, 2421–2422 (2002).

26. D. C. Beer, V. R. Miller, L. G. Sneddon, R. N. Grimes, M. Mathew, and G. J. Palenik, Triple-Decked Sandwich Compounds. Planar $C_2B_3H_5^{4-}$ Cyclocarborane Ligands Analogous to $C_5H_5^-$, *J. Am. Chem. Soc.* **95**, 3046–3048 (1973).
27. R. N. Grimes, Small Carboranes as Building Blocks in Designed Organometallic Synthesis, *Pure Appl. Chem.* **63**, 369–372 (1991).
28. W. Siebert, Boron Heterocycles as Ligands in Transition-Metal Chemistry, *Adv. Organomet. Chem.* **18**, 301–340 (1980).
29. W. Siebert, 2,3-Dihydro-1,3-diborole-metal Complexes with Activated Carbon-Hydrogen Bonds. Building Blocks for Multilayer Sandwich Compounds, *Angew. Chem. Int. Ed. Engl.* **24**, 943–958 (1985).
30. G. E. Herberich, B. Hessner, W. Boveleth, H. Lüthe, R. Saive, and L. Zelenka, A Novel and General Route to (η^5-Borole)metal Complexes: Compounds of Manganese, Ruthenium, and Rhodium, *Angew. Chem. Int. Ed. Engl.* **22**, 996 (1983).
31. W. Siebert, T. Renk, K. Kirnberger, M. Bochmann, and C. Krüger, μ-Thiadiborolenebis (η-cyclopentadienyliron) – Preparation and Structure of a New Triple-Decker Sandwich, *Angew. Chem. Int. Ed. Engl.* **15**, 779–780 (1976).
32. G. E. Herberich, and B. Ganter, Synthesis of the First Triple-Decker Complexes with a Bridging Phospholyl Ligand. Structures of the Triple-Decker Complex [(μ-C_4Me_4P)-{Fe($C_5Me_4CH_2C_6H_{11}$)}(RuCp*)]CF_3SO_3 and the Related Sandwich Complex Fe(C_4Me_4P)($C_5Me_4CH_2C_6H_{11}$), *Organometallics* **16**, 522–524 (1997).
33. P. B. Hitchcock, J. A. Johnson, and J. F. Nixon, Syntheses and Structures of Diruthenium Triple-Decker Complexes with Bridging 1,2,4-Triphospholyl and 1-Arsa-3,4-Diphospholyl Anions, *Organometallics* **14**, 4382–4389 (1995).
34. G. E. Herberich, U. Englert, and D. Pubanz, Der Boratabenzol-Ring als Brückenligand in Tripeldeckerkomplexen. Synthese und Struktur von [(μ,η^6,η^6-C_5H_5BMe)(RuCp*)$_2$] CF_3SO_3·0.5CH_2Cl_2 und Übertragung eines Boratabenzolrings von einem RuCp*-Fragment auf ein RhCp*-Fragment, *J. Organomet. Chem.* **459**, 1–9 (1993).
35. G. E. Herberich, B. Hessner, G. Huttner, and L. Zsolnai, A Triple-Decker Sandwich Complex of Rhodium with Diborabenzene as the Bridging Ligand, *Angew. Chem. Int. Ed. Engl.* **20**, 472–473 (1981).
36. G. E. Herberich, U. Büschges, B. Hessner, and H. Lüthe, (η^5-Borol)rhodium-Komplexe und Nucleophiler Abbau von μ-(η^5-1-Phenylborol)-bis[(η^5-1-phenylborol)rhodium], *J. Organomet. Chem.* **312**, 13–25 (1986).
37. G. E. Herberich, B. A. Dunne, and B. Hessner, Triple-Decker Complexes as Intermediates in Ring Ligand Transfers, *Angew. Chem. Int. Ed. Engl.* **28**, 737–738 (1989).
38. D. A. Loginov, D. V. Muratov, P. V. Petrovskii, Z. A. Starikova, M. Corsini, F. Laschi, F. de Biani Fabrizi, P. Zanello, and A. R. Kudinov, Reactions of the *B*-Phenylborole Complex [CpRh(η^5-C_4H_4BPh)] with Metalloelectrophiles [(ring)M]$^{2+}$, *Eur. J. Inorg. Chem.* **2005**, 1737–1746.
39. K. B. Dillon, F. Mathey, and J. F. Nixon, *Phosphorus: The Carbon Copy*, (Wiley, New York, 1998).
40. O. J. Scherer, Complexes with Substituent-free Acyclic and Cyclic Phosphorus, Arsenic, Antimony, and Bismuth Ligands, *Angew. Chem. Int. Ed. Engl.* **29**, 1104–1122 (1990).
41. O. J. Scherer, P_n and As_n Ligands: A Novel Chapter in the Chemistry of Phosphorus and Arsenic, *Acc. Chem. Res.* **32**, 751–762 (1999).
42. H. J. Breunig, N. Burford, and R. Rösler, Stabilization of a Pentastibacyclopentadienyl Ligand in the Triple-Decker Sandwich Complexes [{(η^5-1,2,4-*t*$Bu_3C_5H_2$)Mo}$_2$(μ,η^5-Sb_5)] and [(η^5-1,2,4-*t*$Bu_3C_5H_2$)Mo(μ,η^5-Sb_5)Mo(η^5-1,4-*t*Bu_2-2-MeC_5H_2)], *Angew. Chem. Int. Ed.* **39**, 4148–4150 (2000).
43. M. Baudler, D. Düster, and D. Ouzounis, Existenz und Charakterisierung des Pentaphosphacyclopentadienid-Anions P_5^-, des Tetraphosphacyclopentadienid-Ions P_4CH^-, und des Triphosphacyclobutadienid-Ions $P_3CH_2^-$, *Z. anorg. allg. Chem.* **544**, 87–94 (1987).

44. T. Kuhlmann, S. Roth, J. Roziere, and W. Siebert, Polymeric (η^5,μ-2,3-Dihydro-1, 3-diborolyl)nickel, the First Multilayer Sandwich Compound, *Angew. Chem. Int. Ed. Engl.* **25**, 105–106 (1986).
45. E. D. Jemmis, and A. C. Reddy, Electronic Structure and Bonding in Tetradecker Sandwich Complexes, *J. Am. Chem. Soc.* **112**, 722–727 (1990).
46. A. Edelmann, S. Blaurock, V. Lorenz, L. Hilfert, and F. T. Edelmann, *Angew. Chem. Int. Ed.* **46**, 6732–6734 (2007).

Chapter 7
The Binding of Ethene and Its Congeners: Prototypical Metal π-Complexes

Erkennen heisst, das äusserliche Wahrgenommene mit den inneren Ideen zusammen zu bringen und ihre Übereinstimmung zu beurteilen.[1]

Johannes Kepler, German Astronomer (1571–1630).

7.1 From 1827 to the 1930s: In the Footsteps of Zeise

Ever since the preparation of Zeise's salt $K[(C_2H_4)PtCl_3]\cdot H_2O$ and the verification of its exact composition (for details see Chap. 3), the nature of the bonding of ethene to a transition metal centre had been hotly debated. Therefore, it is not surprising that until the middle of the twentieth century metal complexes containing ethene or other olefinic ligands occupied a unique place in coordination chemistry. The reason, why ethene was regarded as completely different in its bonding capability compared with conventional ligands such as ammonia, water or chloride, was, that it did not contain a lone pair of electrons. In 1927, the eminent coordination chemist Paul Pfeiffer, a former coworker of Alfred Werner and later the director of the *Chemisches Institut* at the University of Bonn, described Zeise's salt as $K[C_2H_4\cdots PtCl_3]$, thereby implying that the olefinic carbon atoms had some kind of *secondary valence* [1]. Pfeiffer suggested that this type of bonding should also be applied to the dimer $[(C_2H_4)PtCl_2]_2$, for which a chloride-bridged structure **I** was proposed (see Fig. 7.1). In contrast to this, Morris Kharasch and Thomas Ashford considered the dimer as an analogue of "organic molecular compounds" and assumed a ring structure **II** with ethylene bridges and platinum–carbon σ-bonds [2].

If the Kharasch–Ashford proposal had been correct, the existence of stereoisomers would be expected in the case of unsymmetrically substituted olefins [2, 3]. Moreover, if a monosubstituted olefin were bound to platinum via a Pt—C σ-bond, the carbon atom containing the substituent would become a

[1] In English: "Perception means bringing together observations and the inner ideas, and reconciling their agreement".

H. Werner, *Landmarks in Organo-Transition Metal Chemistry*,
Profiles in Inorganic Chemistry, DOI 10.1007/978-0-387-09848-7_7,

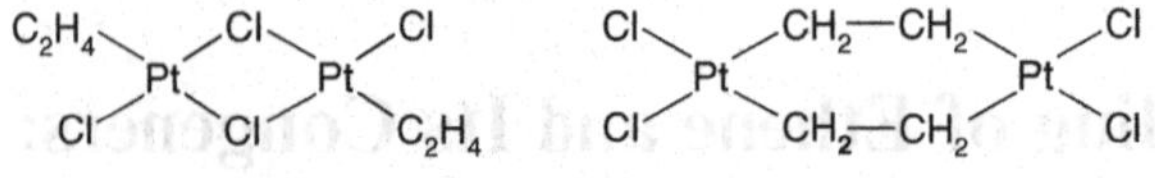

Fig. 7.1 Structural motifs for the dinuclear complex $[(C_2H_4)PtCl_2]_2$, proposed by Pfeiffer in 1927 (I) and Kharasch and Ashford in 1936 (II)

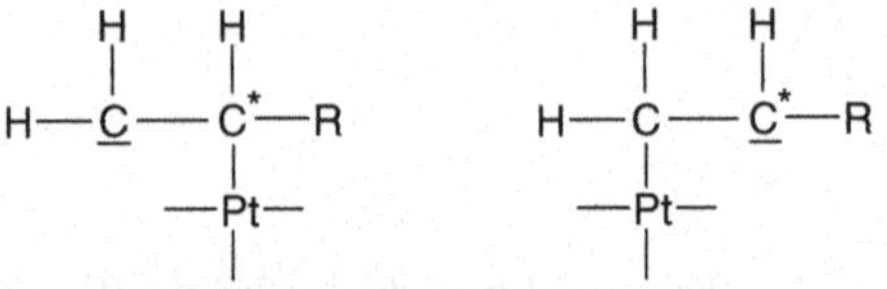

Fig. 7.2 Expected optically active olefin platinum(II) complexes, provided that the monosubstituted olefin were bonded to the metal via a Pt—C σ-bond

chiral centre and thus the corresponding complex should form enantiomers (Fig. 7.2). However, several attempts to separate metal compounds with substituted olefins as ligands into diastereomers or enantiomers failed [3].

The fact, that transition metal complexes exist not only with ethene and its congeners but also with functionalized monoolefins as ligands, dates back to Zeise, who obtained a yellow solid named "acechlorplatin" by treatment of $PtCl_4$ with acetone [4]. This compound analysed as $(C_6H_{10}O)PtCl_2$ and was assumed to contain mesityl oxide (4-methylpent-3-en-2-one) as the olefinic ligand. This seemed to be confirmed after Wilhelm Prandtl and Karl Hofmann reported in 1900, that "acechlorplatin" could be prepared directly from $H_2[PtCl_6]$ and excess mesityl oxide (Fig. 7.3) [5]. However, a more recent spectroscopic investigation by Robert Gillard's group at Cardiff has shown, that instead of mesityl oxide the isomeric isomesityl oxide (4-methylpent-4-en-2-one) is coordinated to platinum and that compound **1**, obtained from $Na_2[PtCl_4]$ and mesityl oxide, is identical to that formed with isomesityl oxide [6]. The reaction of **1** with triphenylphosphine gave pure isomesityl oxide by ligand exchange in almost quantitative yield.

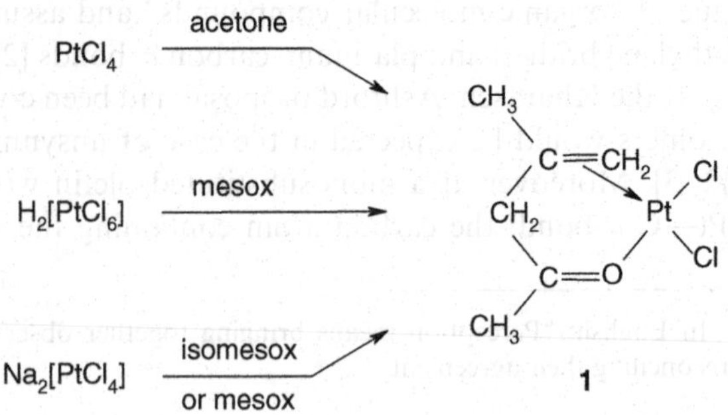

Fig. 7.3 Preparative routes to "acechlorplatin" (**1**), in which isomesityl oxide (isomesox) and not mesityl oxide (mesox) is bonded to $PtCl_2$ in a chelating fashion

Beginning in the 1880s, a series of studies had also been made to identify the nature of platinum complexes with allylamines as ligands, first prepared by Carl Liebermann and C. Paal from the corresponding allylammonium hexachloroplatinate in boiling water [7]. These compounds were initially described as "$CH_2{=}CHCH_2NRR' \cdot HCl \cdot PtCl_2$" and the crucial question was, whether the allylamine preferred to coordinate via the C=C double bond or the amino group. Even after elaborate investigations up to the 1950s, chemists from the famous Russian school, such as Anna Hel'man and A. Rubinstein, had different views about the type of bonding of allylamines, and it was not before the mid 1960s, that this problem was solved [8]. Based on IR and UV spectroscopic studies as well as on chemical evidence, Robert Denning and Luigi Venanzi showed that in the allylamine platinum halides it is indeed the C=C double bond and not the amine group which coordinates to the metal centre [9].

Besides allylamines, other functionalized monoolefins were also used as ligands in platinum(II) complexes in the early days. In 1900, Einar Biilmann prepared the allylalcohol analogue of Zeise's salt $K[(CH_2{=}CHCH_2OH)PtCl_3]$ from $K_2[PtCl_4]$ and showed that by the same route similar compounds with unsaturated acids as ligands were accessible as well [10]. Subsequent studies by Pfeiffer confirmed that a variety of unsaturated alcohols, aldehydes and esters equally reacted with $K_2[PtCl_4]$ to give anions of the general composition $[(\text{olefin})PtCl_3]^-$, which could be easily precipitated with bulky cations such as *trans*-$[CoCl_2(en)_2]^+$ or $[Co(C_2O_4)(en)_2]^+$ to give stable crystalline salts [11]. Allylbromide and allylethers behaved similarly as the above mentioned functionalized olefins.

The first transition metal complex with a *diolefin* as the ligand also contained platinum as the metal centre. In 1908, Hofmann reported that the reaction of $K_2[PtCl_4]$ with dicyclopentadiene in propanol afforded "*clusters of needles*" which analysed as $C_{10}H_{12}PtCl_2$ [12]. Based on the bonding theories of that time, an addition of a chlorine atom and a PtCl radical to one of the double bonds was postulated, in analogy to the behaviour of mercury salts HgX_2 towards olefinic systems. It is quite remarkable that this proposal was accepted for almost 50 years until John Doyle and Hans Jonassen showed, on the basis of spectroscopic studies and ligand exchange reactions, that $(C_{10}H_{12})PtCl_2$ is a four-coordinate platinum(II) complex with the diolefin as a chelating ligand [13]. Later this structure was confirmed crystallographically [14].

In the late 1930s and the 1940s, the reactivity of butadiene towards platinum chlorides was also thoroughly investigated, particularly by Anna Hel'man in Moscow [15, 16]. She prepared several salts of the dianion $[(C_4H_6)Pt_2Cl_6]^{2-}$ and postulated, that in the anionic species the butadiene behaves as a bridge between the two $PtCl_3^-$ units. Mononuclear compounds $(C_4H_6)PtCl_2(NH_3)$ and $(C_4H_6)PtCl_2(py)$ with the butadiene as a monodentate ligand were obtained by addition of ammonia or pyridine to solutions of the free acid $H[(C_4H_6)PtCl_3]$, which was generated by treatment of $[(C_4H_6)PtCl_2]_2$ with hydrochloric acid [15, 16]. Hel'man had also prepared the neutral complex $[(C_4H_6)PtCl_2]_2$, which she considered to contain the diolefinic ligands in bridging positions. In the 1950s, Slade and Jonassen described a less stable isomer of this molecule, which they obtained from $[(C_2H_4)PtCl_2]_2$ and butadiene by ligand exchange [17]. Since the IR spectrum displayed a typical vibrational band for a free C=C double bond at 1608 cm^{-1}, they suggested that in this dimer the diolefin was coordinated as a terminal ligand in a monodentate fashion.

In retrospect, it is rather surprising that in contrast to the mono- and diolefin platinum(II) complexes, which had been extensively studied at least since the turn of the twentieth century, the chemistry of related palladium(II) compounds was barely investigated. Although early predictions were found in the literature about the existence of ethene palladium(II) derivatives similar to Zeise's platinum(II) complexes [8], it was not before 1938 when Kharasch and coworkers prepared a series of dinuclear chloro-bridged palladium(II) compounds of the type $[(olefin)PdCl_2]_2$ from $(PhCN)_2PdCl_2$ by displacement of the benzonitrile ligands [18]. These compounds were also accessible in good yield by addition of the olefin to a suspension of $PdCl_2$ in aprotic solvents. As more recent studies have shown, olefin palladium(II) complexes are in general less stable but more reactive than their platinum(II) counterparts. The enhanced reactivity is an important aspect for the palladium-catalysed oxidation of ethene to acetaldehyde in the Wacker process, in which the initial step consists in the replacement of a chloride ion in $[PdCl_4]^{2-}$ by ethene [19, 20].

Apart from platinum in particular and palladium, prior to the 1950s only a few olefin complexes of other transition metals have been described. Both Marcellin Berthelot [21] and Wilhelm Manchot [22] studied the absorption of ethene by CuCl in acidic, neutral and basic solutions in the first decade of the twentieth century and found that under normal conditions the ethene content could not be raised above 0.17 mole percent. However, under a pressure of 2–3 bar the ratio of $CuCl:C_2H_4$ could be increased to 1:1 [22]. In the 1930s, solid "$CuCl \cdot C_2H_4$" and analogous compounds with propene and isobutene were isolated, which proved to be quite labile and readily dissociated unless they were kept under a high-external pressure of the corresponding olefin [23, 24]. More stable 1:1 adducts were formed with olefins containing functional groups, the latter being eventually coordinated to copper(I) in addition to the C=C double bond.

Also in the 1930s, detailed studies about the thermodynamic stability of adducts of silver(I) with olefins were carried out by Howard Lucas and coworkers, who determined the equilibrium constants between the hydrated Ag^+ ion and the corresponding cationic olefin silver(I) complex in dilute aqueous solutions of silver nitrate [25]. In the context of this work, Saul Winstein and Lucas made an initial attempt to describe the interaction between Ag^+ and an olefin by quantum mechanics [26]. Assisted by Linus Pauling, they explained the existence of olefin silver(I) compounds in terms of resonance stabilization between the mesomeric forms shown in Fig. 7.4. Following this idea, Kenneth Pitzer proposed a "side-on coordination" of Ag^+ to the olefin in 1945 and explained the stability of the corresponding 1:1 adducts as due to an "argentated double bond", in analogy to his concept of the "protonated double bond" [27]. He postulated that the unoccupied s-orbital of silver(I) allowed the formation of a bond with the olefin, similar to the s-orbital of the proton.

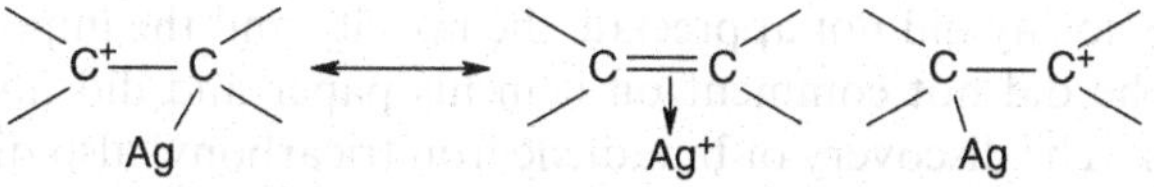

Fig. 7.4 Mesomeric forms of mononuclear olefin silver(I) compounds, proposed by Winstein and Lucas in 1938

7.2 Reihlen's Strange Butadiene Iron Tricarbonyl

The difficulties encountered in providing a satisfactory explanation of the bonding of unsaturated organic molecules to transition metal atoms were probably also the main reason why, despite the impressive work by Mond, Manchot and Hieber (see Chap. 4) in the first half of the twentieth century, studies on the reactivity of metal carbonyls towards olefins had only occasionally been performed. In the late 1920s, Hans Reihlen, at that time an Associate Professor of Inorganic Chemistry at the University of Tübingen, Germany, attempted to resolve the problem whether the CO groups in $Fe(CO)_5$ were attached to the metal through the carbon or the oxygen atom by investigating the chemical properties of iron pentacarbonyl. Since he anticipated that the iron carried a positive charge of +2, he assumed that it would donate two electrons to two of the five CO ligands to form CO^- units, which he considered as pseudohalogen entities. If this view had been correct, the substitution of the CO groups in $Fe(CO)_5$ should have stopped at the $L_2Fe(CO)_3$ stage, since "otherwise the iron would be mono- or zerovalent" [28].

To prove his hypothesis, Reihlen treated iron pentacarbonyl with cyclohexene, isobutene and styrene in a sealed tube at elevated temperatures but observed no reaction. When he used butadiene instead of monoolefins, he isolated a light yellow liquid of unpleasant odour, which was soluble in organic solvents but insoluble in water. The elemental analysis was in agreement with the formula $(C_4H_6)Fe(CO)_3$ and the molecular weight determination showed that the compound was monomeric. Reihlen concluded that the great similarity of $(C_4H_6)Fe(CO)_3$ with $Fe(CO)_5$ indicated that at least the two displaced CO ligands were bonded to iron as Fe—CO and suggested two cyclic structures **III** and **IV** for the butadiene complex (Fig. 7.5). He favoured structure **IV**, firstly in view of Hein's so-called "polyphenylchromium compounds" (see Chap. 5), and secondly because he considered the formation of a five-membered ring involving the butadiene and the iron atom as a stabilizing factor for the molecule [28].

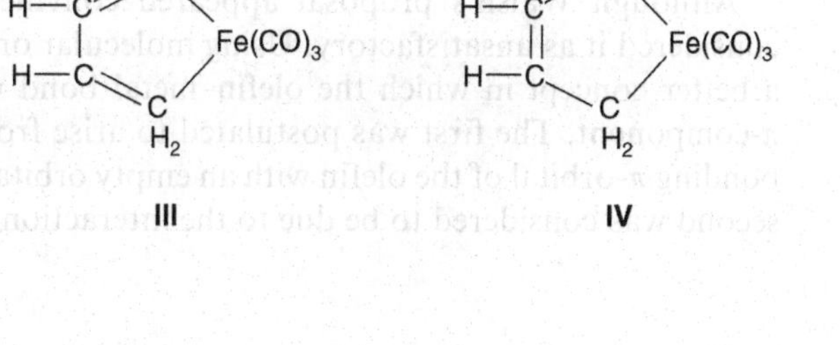

Fig. 7.5 Two possible cyclic structures **III** and **IV** of butadiene iron tricarbonyl, proposed by Reihlen in 1930

Reihlen obviously did not appreciate the novelty and the importance of his results, since he did not comment on it in his paper and did not pursue the matter further. The discovery of butadiene iron tricarbonyl also did not stimulate the interest of Reihlen's contemporaries, presumably because the type of bonding both in olefin metal complexes and metal carbonyls was not well understood. Moreover, the new compound was of limited stability and began to decompose within a few days at room temperature, even when light and oxygen were excluded [28]. It was possibly for this reason that in Krause and von Grosse's extensive monograph on organometallic compounds, published in 1937, as well as in the metal carbonyl chapter of the second edition of Emeléus and Anderson's textbook, published in 1952, the preparation of $(C_4H_6)Fe(CO)_3$ was not mentioned at all [29, 30]. Similar to Schützenberger's studies on carbonyl platinum complexes in the nineteenth century (see Chap. 3), Reihlen's investigations apparently made no impact on organometallic chemistry at the time and, despite the importance of Reppe's work on the interaction of iron carbonyls with acetylenes, (see Chap. 4) no further papers on $(C_4H_6)Fe(CO)_3$ were published during the next 28 years.

7.3 Michael Dewar's "Landmark Contribution"

The breakthrough towards an understanding of the nature of the bonding between olefinic ligands and transition metals occurred in 1950. At a Colloque International, held in Montpellier, France, Michael Dewar (Fig. 7.6) presented a paper entitled "*A Review of the π-Complex Theory*", which was published in the *Bulletin de la Société de Chimie de France* in 1951 [31]. A small section of Dewar's review and a comment stimulated by a question from Saul Winstein, were crucial for what is now being referred to as the Dewar–Chatt–Duncanson model in organometallic chemistry [32]. The radical idea that olefins might be able to donate their π-electrons to form dative bonds, a property usually associated with lone-pair donors, had been suggested by Dewar already in 1945, when he explained the *trans* addition of an electrophilic reagent such as Br_2 to olefins via a bromonium–olefin cation as an intermediate [33, 34]. Two years later, similar π-complex structures had been proposed by Arthur Walsh for ethene oxide, cyclopropane and also for the 1:1 adducts of olefins with silver cations [35]. Walsh thought that the bond between the olefin and the metal can be compared precisely with that of a ligand such as ammonia, as the π-electrons of, for example, ethene occupy an orbital with an ionization potential of 10.45 eV, which is nearly identical to that of the free electron pair of the NH_3 molecule (10.8 eV).

Although Walsh's proposal appeared convincing at first sight, Dewar considered it as unsatisfactory. Using molecular orbital theory, he developed a better concept in which the olefin–metal bond was divided in a σ- and a π-component. The first was postulated to arise from an overlap of the filled bonding π-orbital of the olefin with an empty orbital of the metal, whereas the second was considered to be due to the interaction of a filled d-orbital of the

Fig. 7.6 Michael James Stewart Dewar (1918–1997) received his doctorate from Oxford University with Sir Robert Robinson in 1942. After staying at Oxford as a postdoctoral fellow until 1945, he accepted a position as a research director in industry at Maidenhead near London, where he wrote his influential book "*The Electronic Theory of Organic Chemistry*". This appeared in 1949 and presented a landmark as it was the first general account of organic chemistry in terms of molecular orbital theory. In 1951, at the age of 33, he was appointed to a Chair at Queen Mary College, London, where he continued his theoretical work. Since he did not enjoy the administrative duties of chairing a big department, he accepted a professorship from the University of Chicago in 1959. There he stayed until 1963, when he was appointed to the first Robert A. Welch Chair in Chemistry at the University of Texas at Austin. At that institution he focussed his energy totally on the development of increasingly sophisticated semi-empirical methods for quantitative molecular orbital computations which were tested in a wide variety of applications. In 1989, he moved to a half-time appointment at the University of Florida at Gainesville, from which he retired in 1994. Elected a Fellow of the Royal Society and of the American National Academy of Science, the number of his awards serves as a nearly complete list of those available to organic chemists worldwide. It characterizes him best when he said in his monograph "*A Semiempirical Life*", published in 1992: "Chemistry is, or should be, fun" (photo reproduced with permission of Professor John Murrell)

metal with the vacant antibonding π-orbital of the olefin. Dewar drew this schematically as depicted in Fig. 7.7 and emphasized that "the phases of the lobes of the orbitals being indicated to show the symmetry properties. The s-orbital of Ag^+ has the wrong symmetry for interaction with the antibonding π-MO, and likewise the d-orbital has the wrong symmetry for interaction with the bonding π-MO. The two molecular orbitals are therefore distinct". And he continued: "The combination of these two oppositely-directed dative molecular bonds should leave the olefin much less charged than it would be in a normal π-complex; this would account for the low reactivity of the π-complexes from olefins with metals, where the binding energy of the d-electrons is low, and also for the differences in reactivity of different metals since the stabilities of the two bonds will be affected differently by changes in overall structure" [31].

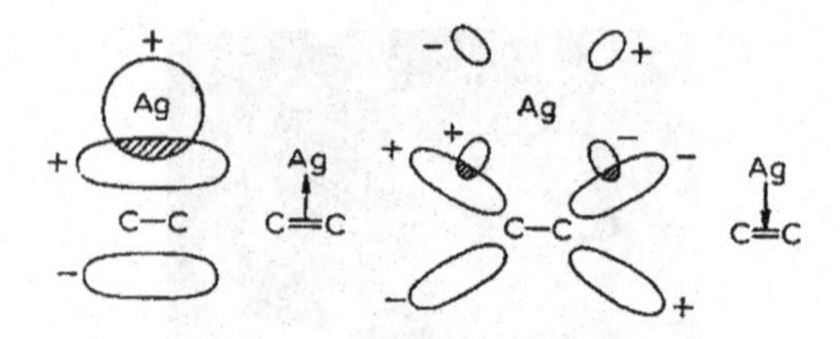

Fig. 7.7 Bonding model proposed by Dewar for olefin silver(I) compounds, showing the orbitals used in the combination of the olefin with the silver(I) cation

Dewar's "landmark contribution" [32] did not receive much attention at the time it was published, possibly because the author did not seek to establish the experimental evidence for his model in subsequent publications. He seemed not to be very interested in the field of transition metal chemistry and was probably not aware that his description of the bonding in olefin silver complexes was supported by Raman studies reported a decade previously. In 1941, Harvey Taufen and coworkers had found that the olefin remained largely unchanged in its coordination to Ag^+ and that the C=C bond was weakened only slightly by the formation of the olefin silver complex [36]. In contrast to Dewar, Joseph Chatt knew this paper and mentioned the results in a review on the mercuration of olefins, which like Dewar's article was also published in 1951 [37]. In his paper, Chatt made a clear distinction between the olefin silver and olefin platinum complexes and argued that, in contrast to the ionized olefin silver(I) salts, in the olefin platinum(II) compounds the metal is present in a covalent state and not as an ion. He also believed that for Ag^+ the d-shell was core like and not available in the manner necessary to stabilize the olefin–platinum bond [37][2].

7.4 The Dewar–Chatt–Duncanson Model

The partial confusion arising after Dewar's and Chatt's reviews were published, was resolved after Chatt and Duncanson reported in 1953 in the *Journal of the Chemical Society* the results of infrared spectroscopic studies on a range of olefin platinum(II) complexes [38]. In this highly cited paper they proposed, with particular reference to Dewar's model, that in the olefin platinum(II) complexes the σ-type bond would be formed by overlap of the filled π-orbital of the olefin with a vacant $5d6s6p^2$ hybrid orbital of the platinum atom, and the π-type bond by overlap of a filled 5d6p hybrid orbital of the metal with the empty antibonding π-orbital of the olefin (Fig. 7.8). In addition, Chatt and Duncanson illustrated how the model could be used to interpret not only the physical properties of the olefin platinum compounds, such as the spectroscopic data and dipole moments, but also their reactivity and their greater stability compared to the olefin silver salts.

[2] Interestingly, more recent MO calculations on silver(I) and copper(I) complexes have indicated that there is much less back donation between these metals and the olefin than in the related platinum compounds, and that the interaction might be primarily electrostatic (see [32], p. 4).

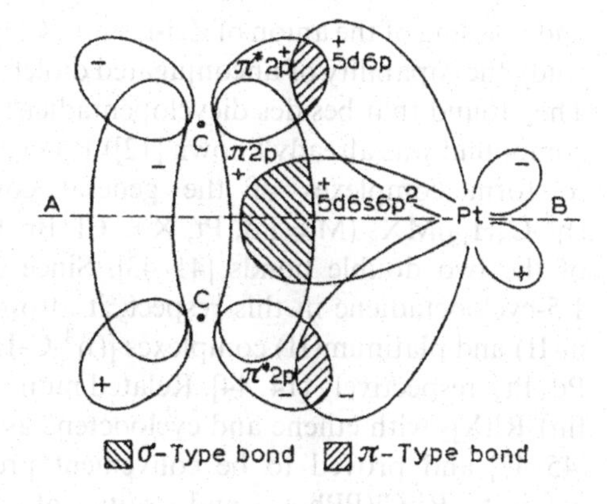

Fig. 7.8 Bonding model proposed by Chatt and Duncanson for Zeise's salt, showing the orbitals used in the combination of ethene with platinum(II) (from ref. 38; reproduced with permission of The Royal Society of Chemistry)

The Chatt–Duncanson paper received immediate attention, and quite soon in papers as well as in textbooks the interpretation of the bonding in olefin metal complexes was referred to the Dewar–Chatt–Duncanson model. While most of the inorganic chemical community accepted this terminology, some others were irritated, as was in particular Dewar. Later he wrote: "Indeed, the orbital diagram given by Chatt and Duncanson is identical with the prototype (meaning his model presented in [31]), apart from the shapes of the stylized curves used to represent orbitals and the addition of specific quantum numbers. Since this application represents only a special case of a more general theory, it would seem more appropriate to refer to it as the 'π-complex theory of metal–olefin complexes' rather than to use any specific designation. However, if the latter is preferred, it seems inappropriate to include the names of Chatt and Duncanson in it" [39].

The association of specific names with a theoretical model or a special compound is a complex phenomenon indeed and usually not controlled by the individuals involved. In his leading paper to celebrate the 50th anniversary of the publication of Dewar's seminal review, Michael Mingos expressed his view that "Chatt's association with the model certainly cannot be justified on the grounds that he made an important conceptual contribution to our theoretical understanding, but he did illustrate how the model could provide a valuable insight into a range of chemical properties associated with this exciting and expanding area of chemistry" [32]. Nevertheless, for many of his contemporaries it was important that Chatt incorporated the Dewar bonding model into a more general picture which covered all aspects of synergistic bonds, such as those between transition metal atoms and CO or PF_3. Chatt also applied the model to explain the *trans*-effect and to other concepts arising from ligand field theory as well as organometallic and coordination chemistry. Thus, despite the criticism by several theoreticians the description as the Dewar–Chatt–Duncanson model survived and was recently also used by Gregory Kubas to explain the bonding in metal dihydrogen complexes [40].

The Chatt–Duncanson paper, however, deserves merit not only on its own but also because it led to wide-ranging studies on the chemistry of mono-, di- and oligoolefin metal complexes. Given their interest in the structure and bonding of the anion of Zeise's salt, Chatt's group began in the mid 1950s to study the capability of unconjugated diolefins to coordinate to d^8 metal centres. They found that besides dicyclopentadiene (of which the dichloroplatinum(II) compound was already known [12]) 1,5-cyclooctadiene is a very suitable ligand to form complexes of the general composition $[(\eta^4\text{-}C_8H_{12})RhX]_2$ and $(\eta^4\text{-}C_8H_{12})MX_2$ (M = Pd, Pt; X = Cl, Br, I) due to the favourable disposition of the two double bonds [41–43]. Since norbornadiene is comparable with 1,5-cyclooctadiene in this respect, it affords analogous rhodium(I), palladium(II) and platinum(II) complexes $[(\eta^4\text{-}C_7H_8)RhX]_2$ and $(\eta^4\text{-}C_7H_8)MX_2$ (M = Pd, Pt), respectively [14, 44]. Related monoolefin rhodium(I) compounds $[(\text{olefin})_2RhX]_2$ with ethene and cyclooctene as ligands were prepared in the 1960s [45–47] and proved to be convenient precursors not only for Wilkinson's catalyst $RhCl(PPh_3)_3$ and salts of the catalytically active cations $[(\text{olefin})_2Rh(S)_2]^+$ (S = solvent = THF, CH_3CN, acetone) [48, 49] but also for numerous other rhodium(I) derivatives [48–50].

7.5 An Exciting Branch: Mono- and Oligoolefin Metal Carbonyls

The 1950s also witnessed the rebirth of the chemistry of olefin metal carbonyls. The leader was Peter Pauson (see Fig. 5.14) who, after he returned from the United States to England, started a research project on the reactivity of iron carbonyls towards conjugated diolefins. Having previously prepared ferrocene and $(\eta^5\text{-}C_5H_5)Fe(CO)_2(\eta^1\text{-}C_5H_5)$ (see Chap. 5), he first attempted to obtain $(\eta^4\text{-}C_5H_6)Fe(CO)_3$ from $Fe(CO)_5$ and cyclopentadiene but failed. In continuation of this work, he repeated the preparation of butadiene iron tricarbonyl and, contrary to Reihlen, found that the compound could be isolated as pale yellow crystals melting at 19°C, which were indefinitely stable when stored in a refrigerator. The UV and IR spectra suggested that it contained an intact, unaltered butadiene ligand, which was not in agreement with Reihlen's preferred structure **IV** (see Fig. 7.5). Since the compound was inert towards maleic anhydride and hydrogen, even in the presence of hydrogenation catalysts, and since it could be recovered unchanged from its solutions in pyridine and strong acids, Pauson suggested the structure shown in Fig. 7.9 [51]. He assumed that the molecule contains a planar (or almost planar) butadiene ligand in a *cisoid* configuration with the iron atom below the C_4H_6 plane and equidistant from the four olefinic carbon atoms. Moreover, he believed that such a structure "will lead to a metal–carbon bond of a type more closely related to that in ferrocene" [51]. This proposal was substantiated by an X-ray diffraction analysis by Owen Mills, who found that the butadiene ligand was indeed planar with nearly identical C—C distances of 1.45–1.46 Å and C—C—C bond angles of 118°. Mills concluded that these results were in agreement with Hallam and Pauson's idea of a "complete delocalisation of the π-electrons from the formal diene structure" [52].

Pauson's work on the structure and bonding of butadiene iron tricarbonyl launched a series of studies making the chemistry of olefin metal carbonyls an

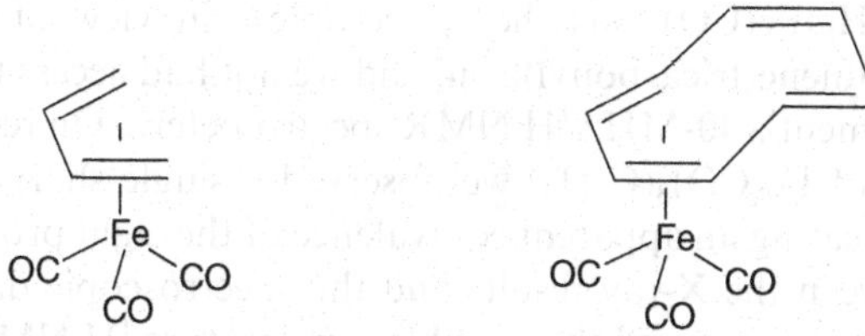

Fe OC CO CO Fe OC CO CO

Fig. 7.9 The molecular structures of butadiene and cyclooctatetraene iron tricarbonyl determined by X-ray structure analysis (carbon and hydrogen atoms of the olefinic ligands are omitted for clarity)

attractive field of research worldwide. In less than 4 years, 1,3-cyclohexadiene iron tricarbonyl (the first iron carbonyl with a cyclic conjugated diolefin) [51], norbornadiene iron tricarbonyl and norbornadiene molybdenum tetracarbonyl (the first metal carbonyl derivatives with a non-conjugated diolefin) [53, 54], mono- and dinuclear cyclooctatetraene iron carbonyls [55–57], the 1,3,5-cycloheptatriene metal tricarbonyls [58–60] and the 1,5-cyclooctadiene metal tetracarbonyls [61–63] of chromium, molybdenum and tungsten were prepared and characterized by IR and NMR spectroscopy. In the same period of time, the structures of $(\eta^4\text{-}C_8H_8)Fe(CO)_3$, $(\mu\text{–}\eta^4\text{:}\eta^4\text{-}C_8H_8)Fe_2(CO)_6$ [64] and $(\eta^6\text{-}C_7H_8)Mo(CO)_3$ [65] were determined by X-ray diffraction. A noteworthy detail is that the cyclooctatetraene iron carbonyls as well as the 1,5-cyclooctadiene chromium, molybdenum and tungsten tetracarbonyls were independently prepared by three different research groups, reflecting the competition and the activity in this area at that time.

To get an idea of the problems associated with a correct structural proposal of a novel organometallic complex in those days, it is interesting to read the papers reporting the preparation and characterization of cyclooctatetraene iron tricarbonyl. This compound is a red, sublimable solid, which is readily soluble in organic solvents and thermally stable up to about 160°C. It is also not air-sensitive and can be recovered in nearly quantitative yield after UV irradiation in an oxygen atmosphere for 20 h [66]. The IR spectrum did not indicate the presence of uncoordinated double bonds and the ^{1}H NMR spectrum displayed a single resonance at room temperature. Based on these results, various structures were proposed, most of which assumed a planar or a tub-shaped eight-membered ring containing a completely delocalized π-electron system [67]. However, none of these proposals (the authors of which were subdivided by Lipscomb into "those who prepared the compounds, those who could not await our promised X-ray diffraction results, and those who did not publish") [64] was correct. In contrast to the predictions of a theoretical study [68], the olefinic ligand in $(\eta^4\text{-}C_8H_8)Fe(CO)_3$ turned out to have a configuration which was previously unknown and is best described as a bent eight-membered ring (see Fig. 7.9). The $Fe(CO)_3$ group is linked to only one half of the ring, and the bonding is obviously analogous to that in butadiene iron tricarbonyl. Gordon Stone recalled in his autobiography "*Leaving No Stone Unturned*", that "ironically we would almost certainly have assigned correctly it (the structure of $(C_8H_8)Fe(CO)_3$) as the η^4 complex, in view of the suggested structure for butadiene(tricarbonyl)iron, had we not had access to the Harvard chemistry department's 40-MHz ^{1}H NMR spectrometer... On recording the ^{1}H NMR spectrum of $Fe(CO)_3(C_8H_8)$ we observed a single sharp peak at room temperature, indicating an apparent equivalence of the eight protons" [69]. The discrepancy between the X-ray results and the spectroscopic data was solved after Cotton's group measured the variable-temperature ^{1}H NMR spectrum of $(\eta^4\text{-}C_8H_8)Ru(CO)_3$ some years later [70]. They found that a limiting spectrum was reached at about −145°C and showed, with the aid of computer-simulated spectra, that the fluxional behaviour of both the iron and the ruthenium compounds is due to 1,2-shifts of the metal around the eight-membered ring.

Apart from the di- and oligoolefin iron tricarbonyl complexes, which nowadays are frequently used in organic synthesis [71, 72], the chemistry of the readily accessible cycloheptatriene chromium and molybdenum tricarbonyls **2** and **3** was the focus of intense research efforts as well. Only a few months after the synthesis of **2** and **3** was published [58, 59], both Hyp Dauben and Peter Pauson reported that these compounds react with triphenylmethyl tetrafluoroborate in methylene chloride to give the tropylium complexes **4** and **5** in excellent yield (Scheme 7.1) [73, 74]. Later this method of hydride abstraction was also used for the preparation of the tropylium cation itself and subsequently led to the generation of several cationic π-complexes of iron, manganese and cobalt [71, 72]. The reactions of the cations of **4** and **5** with nucleophilic

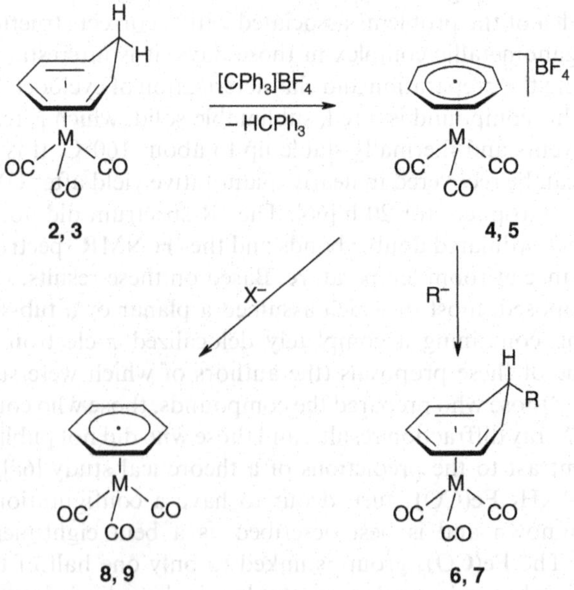

Scheme 7.1 Preparation of cationic tropylium metal tricarbonyls and their reactions with anionic nucleophiles (**2, 4, 6, 8**: M = Cr; **3, 5, 7, 9**: M = Mo; for R^- and X^- see text)

reagents such as hydride, methoxide, hydrogen sulfide or diethyl methylmalonate, which were studied in particular by Pauson [75], proceeded probably via an attack of the nucleophile at the metal atom and afforded the substituted cycloheptatriene complexes **6** and **7** with the substituent R in the *endo* position. In contrast, the reactions of **4** and **5** with sodium cyclopentadienide or sodium diethyl malonate unexpectedly led to a ring contraction and gave exclusively the benzene metal tricarbonyls **8** and **9** [72]. The coordinated six-membered ring derives entirely from the tropylium ligand as was confirmed by labelling and cross-over experiments [75].

The mechanism of the unusual ring contraction leading to **8** and **9** remains incompletely understood. The initial step probably gives the expected products $(\eta^6\text{-}C_7H_7X)M(CO)_3$, which may then react with the basic substrate X^- by proton abstraction to yield the short-lived anionic intermediate $[(C_7H_6X)\text{-}M(CO)_3]^-$, that eliminates the CX^- fragment. The role of the $M(CO)_3$ unit, although not clear as yet, must be significant since treatment of free cyclohepatrienes C_7H_7X with NaC_5H_5 or $NaCH(CO_2Et)_2$ does not generate benzene. Regarding the reactivity of **3**, it is interesting to note that in contrast to $[CPh_3]^+$, which interacts by *hydride* abstraction from the seven-membered ring, strong acids such as HBF_4 react with the neutral complex by *protonation* of the ring and formation of $[(\eta^5\text{-}C_7H_9)Mo(CO)_3]BF_4$, the corresponding cation having not an 18- but a 16-electron configuration [76].

Following the synthesis of di- and oligoolefin metal carbonyls, the preparation of monoolefin metal carbonyls of the general composition $(olefin)Fe(CO)_4$ and $(olefin)M(CO)_5$ (M = Cr, Mo) was also attempted. From Reihlen's work it was already known that iron pentacarbonyl did not react with cyclohexene, isobutene and styrene, even upon heating the reaction mixture in a sealed tube at elevated temperatures [28]. Under these conditions, $Fe(CO)_5$ proved also to be inert in the presence of ethene. However, if $Fe_2(CO)_9$ was used as the precursor in the reaction with ethene, a mixture of $Fe(CO)_5$ and $(C_2H_4)Fe(CO)_4$ was obtained which could only be separated with considerable loss of material [77]. The ethene complex is an orange-yellow oil, which decomposes at room temperature, but can be stored for prolonged periods of time at –80°C. Analogous compounds $(olefin)Fe(CO)_4$ with propene and styrene as ligands were prepared by irradiation of a solution of $Fe(CO)_5$ and the corresponding olefin [78], using a methodology that was also applied by Fischer's group for the preparation of $(C_5H_5)Mn(CO)_2(C_2H_4)$ and $(mesitylene)Cr(CO)_2(C_2H_4)$, respectively [79–81]. More recently, Martyn Poliakoff at the University of Nottingham succeeded in isolating $(C_2H_4)Cr(CO)_5$ for the first time by the UV photolysis of $Cr(CO)_6$ in supercritical ethene (scC_2H_4) in a flow reactor at room temperature. The decisive fact was that the product precipitated rapidly from scC_2H_4 under a pressure of ethene and, once in the solid state, proved to be much more stable than in solution. In Martyn's opinion the ethene complex "promises to be a useful source of '$Cr(CO)_5$'...with the added advantage that C_2H_4 (is easy) to remove from solution" [82].

The photochemical route provided access not only to olefin metal carbonyls with coordinated ethene, propene, cyclooctene or styrene but also to those with functionalized olefins as ligands. The general observation is that electron-withdrawing groups such as F, CN, CHO or CO_2R stabilize the metal–olefin bond while electron-donating groups such as alkyl, OR or NR_2 weaken it. Apart from mono(olefin) compounds such as (olefin)$Fe(CO)_4$ and (olefin)$Cr(CO)_5$, some (in most cases rather labile) bis(olefin) complexes of the general composition $(olefin)_2M(CO)_4$ (M = Mo, W) have also been described [83–85]. Related tris(olefin) compounds $(olefin)_3M(CO)_3$ of the group VI elements are still unknown. In the early 1960s, Fischer's group succeeded in preparing the bis(butadiene) and bis(1,3-cyclohexadiene) metal dicarbonyls $(\eta^4\text{-}C_4H_6)_2Mo(CO)_2$ and $(\eta^4\text{-}C_6H_8)_2M(CO)_2$ (M = Cr, Mo), using the mesitylene complexes $(\eta^6\text{-}1,3,5\text{-}C_6H_3Me_3)M(CO)_3$ instead of $M(CO)_6$ as starting materials [86, 87].

7.6 Schrauzer's Early Studies on Homoleptic Olefin Nickel(0) Complexes

At the same time when the crucial steps in the development of the chemistry of olefin metal carbonyls were done, the synthesis of the first transition metal complexes containing exclusively mono-, di- or oligoolefinic ligands was reported. Gerhard Schrauzer, in those days working for his *Habilitation* at the University in München, attempted to explore the catalytic activity of nickel(0) compounds derived from nickel tetracarbonyl, and for this purpose treated $Ni(CO)_4$ with various olefins under reflux conditions. With acrylonitrile as the substrate, he isolated a red, highly air-sensitive product analysing as $Ni(CH_2{=}CHCN)_2$, for which he postulated a π-complex structure not involving the coordination of the nitrile group to the metal [88, 89]. The electron-deficient nature of $Ni(CH_2{=}CHCN)_2$ was shown via the reaction with triphenylphosphine, which instantly gave the 1:1 and 1:2 adducts $Ni(CH_2{=}CHCN)_2(PPh_3)$ and $Ni(CH_2{=}CHCN)_2(PPh_3)_2$, respectively. Fumaronitrile, cinnamonitrile and acrolein behaved similar and reacted with $Ni(CO)_4$ to give the corresponding bis(olefin) nickel(0) complexes. In contrast, the reaction of $Ni(CO)_4$ with olefins such as ethene, propene or styrene proceeded differently and did not afford compounds of the general type $Ni(olefin)_n$ ($n = 2, 3$ or 4). With regard to the catalytic activity of the new nickel(0) compounds, the acrylonitrile complex $Ni(CH_2{=}CHCN)_2$ was found to be an active catalyst for the cooligomerization of acetylene with acrylonitrile to heptatrienenitrile and the tetramerization of acetylene to cyclooctatetraene, similar to Reppe's nickel carbonyl derivative $Ni(CO)_3(PPh_3)$ [90].

In subsequent work, Schrauzer's group also prepared a series of quinone nickel compounds from $Ni(CO)_4$ among which bis(duroquinone) nickel(0) (**10**, see Scheme 7.2) was thoroughly investigated [91, 92]. In contrast to the bis(olefin) complexes $Ni(CH_2{=}CHX)_2$ (X = CN, CHO), the $Ni(DQ)_2$ analogue is not air-sensitive and thermally stable until 205°C. The spectroscopic data of **10** indicate a significant charge transfer from the metal to a low-lying, vacant molecular orbital of the quinone which is in agreement with the stability of the

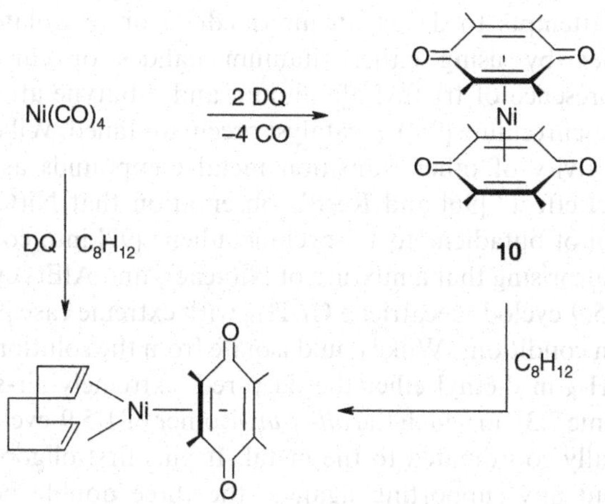

Scheme 7.2 Preparation of bis(duroquinone) nickel(0) (**10**) and the related 1,5-cyclooctadiene(duroquinone) complex **11** (the methyl groups of the quinone ligands are omitted for clarity)

complex. The reaction of **10** with di- and oligoolefins such as norbornadiene, 1,5-cyclooctadiene, cyclooctatetraene or dicyclopentadiene in high-boiling solvents led to the formation of the mixed-ligand compounds (DQ)Ni(olefin), which were also obtained directly from $Ni(CO)_4$, duroquinone and excess di- or oligoolefin [91, 92]. In these compounds, of which the 1,5-cyclooctadiene derivative **11** was characterized crystallographically [93], the olefins behave as bidentate ligands and thus the metal attains a closed-shell configuration.

7.7 Wilke's Masterpieces and the "Naked Nickel"

The breakthrough in the field of olefin nickel(0) complexes containing olefins without electron-withdrawing substituents was achieved by Günther Wilke from the *Max-Planck-Institut für Kohlenforschung* at Mülheim in Germany. Given his previous cooperation with Karl Ziegler and his interest in the mechanism of olefin polymerization reactions, Wilke reported in 1957 that, while a catalyst prepared from $TiCl_4$ and diethylaluminium chloride rapidly converted ethene into polyethylene, it reacted – totally unexpectedly – under the same conditions with butadiene to yield exclusively *trans,trans,cis*-1,5,9-cyclododecatriene [94]. To explain the mechanism of this novel and, from the point of organic synthesis, particularly interesting cyclotrimerization, Wilke assumed that initially a butadiene metal complex was formed as an intermediate, which rearranged via C—C coupling of the coordinated diolefins to the 12-membered carbon ring, that is still linked to the metal. In the last step the ring could be displaced by excess butadiene to give the free cyclotrimer and regenerate the catalyst.

After early attempts to detect an intermediate or to isolate a butadiene metal π-complex by using either titanium halides or chromyl chloride (which in the presence of triethyl aluminium and 2-butyne afforded bis(hexamethylbenzene)chromium [95]) as catalyst precursor failed, Wilke investigated the catalytic activity of other transition metal compounds as well. Taking Ziegler's "nickel effect" [96] and Reed's observation that $Ni(CO)_4$ catalysed the dimerization of butadiene to 1,5-cyclooctadiene [97] into consideration, it was not really surprising that a mixture of $Ni(acac)_2$ and $AlEt_3$ cyclotrimerized butadiene to 1,5,9-cyclododecatriene $C_{12}H_{18}$ with extreme ease [98, 99]. Under carefully chosen conditions, Wilke could isolate from the solution of $Ni(acac)_2$, $AlEt_3$ and $C_{12}H_{18}$ in diethyl ether the deep-red, extremely air-sensitive compound **12** (Scheme 7.3), in which the *all-trans*-isomer of 1,5,9-cyclododecatriene was symmetrically coordinated to the metal. In this first oligoolefin nickel(0) complex without any supporting ligands, the three double bonds adopt a propeller-like arrangement around the nickel atom in which the propeller can have a left-hand or a right-hand screw [100]. The prediction, that this structure should be less stable than that in which the six carbon atoms of the double bonds lie in the same plane as the zerovalent nickel [101], was confirmed by the reaction of **12** with pure *all-cis*-1,5,9-cyclododecatriene, which gave the thermally stable, almost colourless isomer **13** [102]. Under the same conditions, **12** also reacted with CO, isonitriles and tertiary phosphines to afford the 1:1-adducts (*all-trans*-$C_{12}H_{18}$)Ni(L) (L═CO, CNR, PR_3) in nearly quantitative yields. The conversion of the 16-electron starting material to the 18-electron compounds (*all-trans*-$C_{12}H_{18}$)Ni(L) was accompanied by a displacement of the nickel atom out of the plane of the triene resulting in an almost tetrahedral coordination sphere around the metal centre [103]. With the chiral phosphine

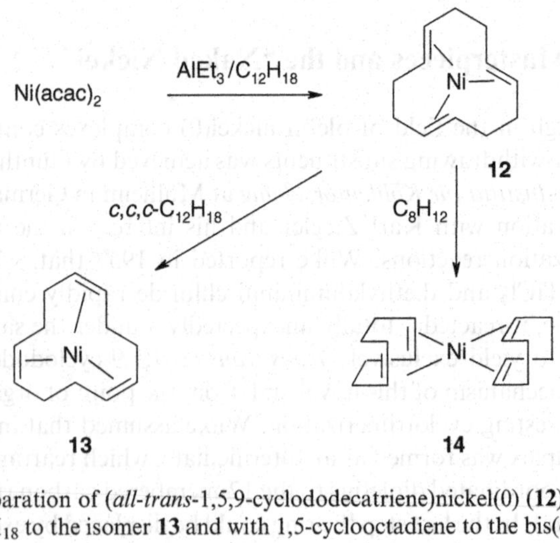

Ni(acac)$_2$ —— $AlEt_3/C_{12}H_{18}$ ——→ **12**; **12** —— c,c,c-$C_{12}H_{18}$ ——→ **13**; **12** —— C_8H_{12} ——→ **14**

Scheme 7.3 Preparation of (*all-trans*-1,5,9-cyclododecatriene)nickel(0) (**12**) and its reactions with *all-cis*-$C_{12}H_{18}$ to the isomer **13** and with 1,5-cyclooctadiene to the bis(diolefin) nickel(0) complex **14**

PMe_2(Ment), a mixture of the corresponding diastereoisomers (*all-trans*-$C_{12}H_{18}$)Ni[PMe_2(Ment)] was obtained, which could be separated by fractional crystallization. Abstraction of the chiral phosphine by the π-allyl nickel(II) dimer [(η^3-C_3H_5)NiBr]$_2$ led to the two enantiomers of **12**, which do not racemize at room temperature [103]. They are probably the simplest optically active transition metal complexes known.

While carbon monoxide, isonitriles and phosphines reacted with **12** to yield the 1:1 adducts (*all-trans*-$C_{12}H_{18}$)Ni(L), treatment of **12** with 1,5-cyclooctadiene gave the yellow, sublimable bis(diolefin) complex **14** by complete displacement of the triene ligand. This closed-shell compound was also prepared from Ni(acac)$_2$ and $AlEt_3$ or $AlEt_2$(OEt) in the presence of the diene [98, 99]. Moreover, it could be obtained by vapour co-condensation of nickel atoms and 1,5-cyclooctadiene, as Skell reported some years later [104]. The X-ray diffraction analysis of **14** revealed, that the nickel atom is coordinated by the four C═C double bonds of the two diene molecules in a distorted tetrahedral geometry with Ni—C distances that are slightly longer than in the 16-electron compound **12** [105]. The reaction of **12** with cyclooctatetraene at room temperature gave a sparely soluble black product of composition [Ni(C_8H_8)]$_n$, which, contrary to the original proposal [98, 99], is not a polymer but the dimer **15**. The two nickel atoms are sandwiched between two essentially planar eight-membered rings, which are coordinated in a bis(allylic) η^3,η^3 fashion (Fig. 7.10) [106]. Treatment of **12** with cyclooctatetraene at −78°C afforded a golden-yellow compound instead of a black material, which analysed as Ni(C_8H_8)$_2$ and is assumed to be structurally analogous to **14**. At room temperature it looses one molecule of cyclooctatetraene to generate **15**.

Although the bis(diolefin) complex **14** has a closed-shell configuration, it proved to be rather labile. Under mild conditions it did not only react with CO and triphenylphosphine to give Ni(CO)$_4$ and (η^4-C_8H_{12})Ni(PPh$_3$)$_2$, but also proved to be an excellent catalyst particularly for C—C coupling reactions. It is now used worldwide for this purpose [103]. Already in the first stages of their work, Wilke's group found that **14** catalysed the trimerization of butadiene very efficiently, forming a mixture of the 1,5,9-cyclododecatriene isomers. The 16-electron compound **12** behaved similarly and upon treatment with butadiene at room temperature gave the cyclic trimer $C_{12}H_{18}$. However, if this reaction was carried out at −40°C, compound **12** took up exactly three molecules of

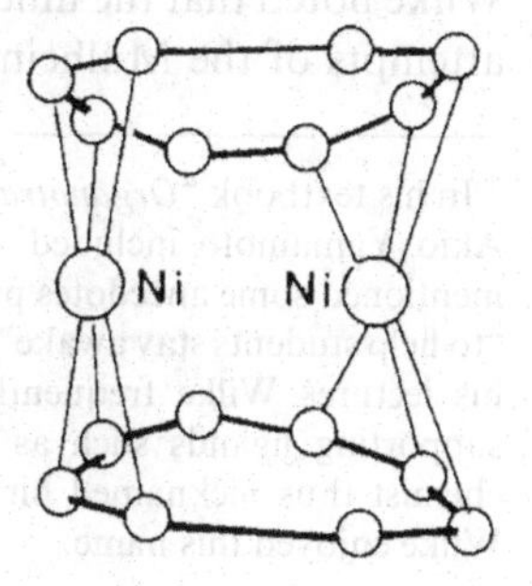

Fig. 7.10 Schematic representation of the molecular structure of the dinuclear nickel(0) complex $Ni_2(C_8H_8)_2$ (**15**); the *smaller balls* represent the CH units of the eight-membered rings

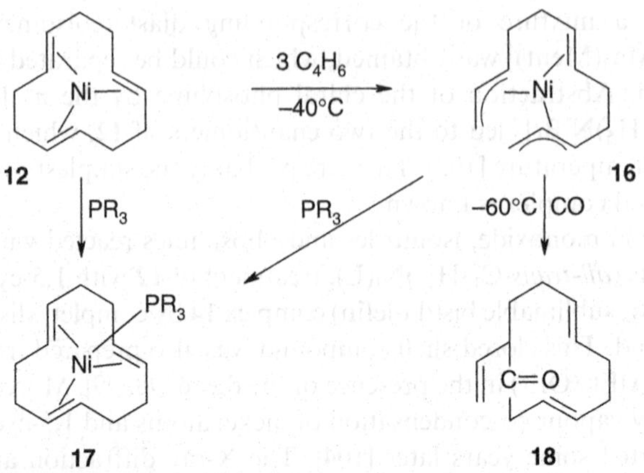

Scheme 7.4 Preparation of the non-cyclic nickel(II) complex **16** and its reactions with tertiary phosphines to give the nickel(0) compound **17** and with carbon monoxide to yield the cyclic unsaturated ketone **18**

butadiene and furnished a red–brown solid **16** with the same empirical composition as the starting material (Scheme 7.4) [107, 108]. Based on the NMR spectroscopic data and the result of the reaction with H_2 at room temperature, which led to metallic nickel and quantitative amounts of *n*-dodecane, Wilke assumed that in the course of the ring ligand displacement the three entering butadiene units couple together to yield a C_{12} chain, that is still coordinated to nickel. This chain contains a central, *trans*-configured C=C double bond and two terminal π-allyl groups. In the presence of excess butadiene, **16** underwent ring closure and gave *all-trans*-1,5,9-cyclododecatriene. The transient or "naked" nickel(0)[3] formed in this step reacts immediately with C_4H_6 by oxidative addition to regenerate **16**, which, similar to **14**, is also an excellent catalyst for the cyclotrimerization of butadiene. Tertiary phosphines reacted with **16** to afford the monophosphine complexes **17**, which were identical to those prepared from **12** and PR_3. Treatment of **16** with carbon monoxide at room temperature gave $Ni(CO)_4$ and 1,5,9-$C_{12}H_{18}$, while the low temperature reaction led to the dienone **18** via insertion of CO into the carbon skeleton [107, 108]. In his seminal review "*Contributions to Organo-Nickel Chemistry*" [103], Wilke noted that the unique structure and high-reactivity of **16** also initiated the attempts of the Mülheim group to prepare bis(η^3-allyl) nickel, which was the

[3] In his textbook "*Organotransition Metal Chemistry*" (Wiley Interscience, New York, 1986) Akio Yamamoto included several short paragraphs entitled "Intermezzos", in which he mentioned some anecdotes presented during his classes at the Tokyo Institute of Technology "to help students stay awake". In one of those intermezzos, Akio reported that in the course of his lectures Wilke frequently called the catalytically active nickel complexes having no supporting ligands such as CO, PR_3 or cyclopentadienyl "naked nickel". An American chemist thus nicknamed him "a man who brought sex into chemistry" and it seems that Wilke enjoyed this name.

key to opening up the rich and extensive field of the synthetically and industrially important π-allyl transition metal complexes [109].

Apart from the synthesis of **14**, the unfavourable spacial arrangement of the three double bonds in **12** also allowed the preparation of $Ni(C_2H_4)_3$, which was the first transition metal compound of the general composition $M(C_2H_4)_n$. It has already been mentioned, that earlier attempts by Schrauzer to obtain $Ni(C_2H_4)_n$ ($n = 3$ or 4) from $Ni(CO)_4$ failed [90].[4] The reaction of **12** with ethene was carried out in diethyl ether at 0°C and, after the solution was cooled to −78°C, gave colourless needles that were extremely air-sensitive and decomposed above 0°C [110]. Norbornene formed a thermally more stable product of composition $Ni(C_7H_{10})_3$, which was characterized crystallographically [96]. In agreement with a theoretical study [101], $Ni(C_7H_{10})_3$ possesses a trigonal-planar geometry with the nickel atom coordinated in *exo* position to the C═C double bonds.

The olefin ligands of $Ni(C_2H_4)_3$ could easily be displaced by tertiary phosphines and, if these were used in excess, the 18-electron compounds $Ni(PR_3)_4$ were formed. However, if $Ni(C_2H_4)_3$ was treated with equimolar amounts of tricyclohexylphosphine, the monosubstituted derivative $Ni(C_2H_4)_2(PCy_3)$ was obtained. A perfect planar arrangement of the three ligands around the metal centre was confirmed by an X-ray diffraction analysis [111]. The structurally related monoolefin compound $Ni(C_2H_4)(PPh_3)_2$ was prepared from $Ni(acac)_2$ and $AlEt_2(OEt)$ in the presence of triphenylphosphine and used as starting material for the synthesis of analogous nickel(0) complexes of the general composition $Ni(L)(PPh_3)_2$ ($L{=}C_2F_4$, styrene, (*E*)-stilbene, 2-butyne, diphenylacetylene, etc.). By passing O_2 through a solution of $Ni(C_2H_4)(PPh_3)_2$ in toluene or diethyl ether at −78°C, even the dioxygen nickel derivative $Ni(O_2)(PPh_3)_2$ could be generated [112]. The catalytic activity of $Ni(C_2H_4)_3$ was illustrated by Wilke's group inter alia via the preparation of the industrially important triethylaluminium from $Al(nBu)_3$ and ethene, which in the presence of catalytic amounts of $Ni(C_2H_4)_3$ gave the two products $AlEt_3$ and butene in quantitative yield [110]. Owing to this result, Wilke deduced that the formation of $Ni(C_2H_4)_3$ might be the key step in the dimerization of ethene into butene using the "nickel effect" [96].

---

[4] In one of his essays [113], Wolfgang Herrmann (see Fig. 9.4) referred the reader to a paper by Paul Sabatier (Nobel laureate for Chemistry 1912), in which he mentioned that he attempted to coordinate molecules such as N_2O, NO, NO_2, ethene, acetylene, etc. to nickel in a similar way as it was done by Mond for the preparation of nickel tetracarbonyl. Sabatier observed that "when ethylene was passed through a column containing nickel, no reaction occurs at ordinary temperature....but above 300°C the column starts to glow and the gas begins to decompose; not only carbon and hydrogen are produced, but also larger amounts of ethane. The latter compound must have come from hydrogenation of ethylene resulting from some special property of nickel, which appeared to be a hydrogenation catalyst". Though no olefin nickel complex was formed (possibly as a short-lived intermediate), this observation initiated a series of further investigations which finally led to the Nobel prize.

7.8 Stone and the Family of Olefin Palladium(0) and Platinum(0) Compounds

The next important steps in the field of olefin complexes of zerovalent group X metals were taken by Gordon Stone at the University of Bristol. Given his earlier interest in the chemistry of phosphine palladium(0) and platinum(0) compounds such as $M(PPh_3)_4$, $M(PPh_2Me)_4$ (M = Pd, Pt), $Pt(PEt_3)_3$ and Pt[(*E*)-stilbene]$(PMe_3)_2$ [114], and being aware of the impressive potential of $Ni(C_2H_4)_3$ and $Ni(\eta^4\text{-}C_8H_{12})_2$ as catalysts, Stone was interested in developing useful synthetic routes to "ligand free" olefin palladium(0) and platinum(0) complexes [115].[5] Since he found that the preparative method for $Pt(\eta^4\text{-}C_8H_{12})_2$ (**19**), reported by Jörn Müller and Peter Göser in 1967 [116] (see below), was difficult to repeat, he suggested to his coworker John Spencer to reduce the well known precursor $(\eta^4\text{-}C_8H_{12})PtCl_2$ to **19** with the uncommon reducing agent $Li_2C_8H_8$ (Scheme 7.5). As he recalled [115], he anticipated that $Li_2C_8H_8$ should have the advantage that the cyclooctatetraene, released after electron transfer from the dianion, did not coordinate to platinum(0) in the presence of an excess of 1,5-cyclooctadiene.

This expectation was fulfilled and John Spencer could prepare **19** via the new route in about 70% yield [117]. The platinum(0) compound was found to be oxidatively and thermally more stable than its nickel counterpart **14**, and the dry solid could even be handled in air. As shown crystallographically, the metal atom in **19** is, analogous to **14**, in a distorted tetrahedral environment as a consequence of the disposition of the four C=C double bonds. The bite angle

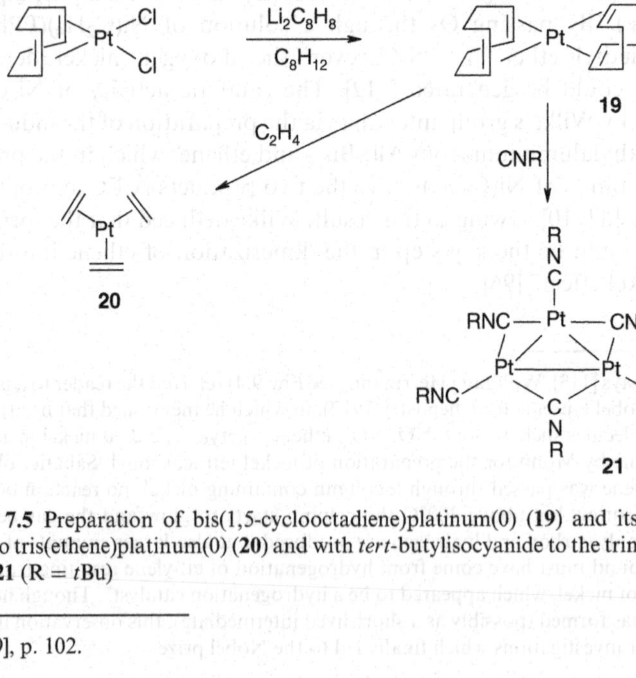

Pt Cl Cl $Li_2C_8H_8$ C_8H_{12} Pt **19** C_2H_4 CNR Pt **20** R N C RNC — Pt — CNR Pt Pt RNC CNR C N R **21**

Scheme 7.5 Preparation of bis(1,5-cyclooctadiene)platinum(0) (**19**) and its reactions with ethene to tris(ethene)platinum(0) (**20**) and with *tert*-butylisocyanide to the trinuclear platinum cluster **21** (R = *t*Bu)

[5] see [69], p. 102.

between the platinum and the chelating diolefins is, however, only 86° rather than 110° as in an undistorted $Pt(L)_4$ molecule, and this may in part account for the easy displacement of the cyclooctadiene ligands by ethene to give $Pt(C_2H_4)_3$ (**20**). Norbornene and *trans*-cyclooctene reacted similarly with **19** and afforded the 16-electron complexes $Pt(C_7H_{10})_3$ and $Pt(C_8H_{14})_3$, respectively. Unexpectedly, the reaction of **19** with *tert*-butylisocyanide led to the formation of the trinuclear platinum cluster **21**, in which three of the six isocyanide ligands are in bridging positions [118]. From **19** and sterically demanding phosphines, such as PCy_3 or $PMe\textit{t}Bu_2$, the coordinatively unsaturated 14-electron complexes $Pt(PR_3)_2$ were obtained, the size of the phosphine probably preventing the aggregation to tri- or oligonuclear species [115].

The tris(ethene) complex **20**, which forms white crystals that can be sublimed at 20°C in the presence of ethene and kept for weeks at –20°C [117], proved to have an extensive chemistry too. Similar to Wilke's olefin nickel(0) compounds, it provided a convenient source for "naked" platinum and was used inter alia for the preparation of a variety of alkyne platinum(0) complexes. Among these, the binary compounds **22** and **23** deserve special attention (Fig. 7.11). In the former, the coordination sphere around the metal atom corresponds, as in **19**, to a distorted tetrahedron with an angle of 82° between the two planes containing the alkyne carbon atoms [119]. In this compound, the alkyne behaves not as a 2- but as a 4-π-electron donor ligand. In the dinuclear complex, the platinum atoms and the ligating carbon atoms of the bridging alkyne adopt a dimetallatetrahedrane framework with, however, a Pt-to-Pt separation that suggests little or no direct metal–metal bonding [115]. With regard to bridging units between two platinum(0) centres, a serendipitous, though very significant result from Stone's laboratory was obtained from the reaction of **19** with the fluorinated olefin $CF_3CF{=}CF_2$. The dinuclear carbene (or alkylidene) complex **24**, which arises from a fluorine migration to generate the bridging bis(trifluoromethyl)-methylene ligand, was isolated and characterized [120].

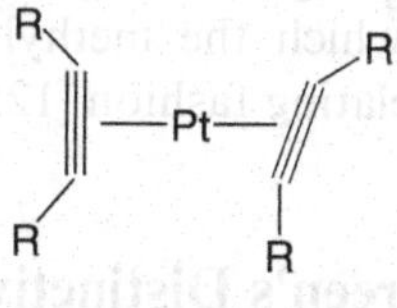

R R Pt R R
22 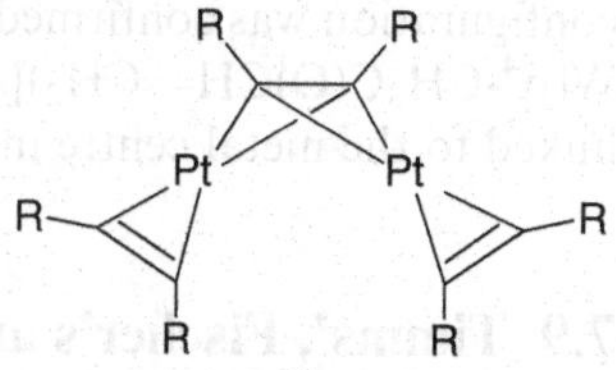
R R Pt Pt R R R R
23 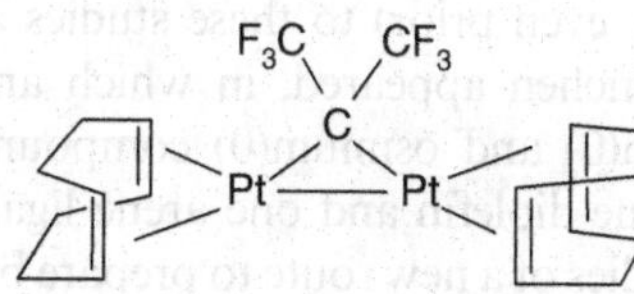
F_3C CF_3 C Pt Pt
24

Fig. 7.11 The molecular structures of three noteworthy platinum(0) complexes prepared from $Pt(C_2H_4)_3$ and alkynes (**22**, **23**), and from $Pt(C_8H_{12})_2$ and the olefin $CF_3CF{=}CF_2$ (**24**)

In addition to **19**, Stone's group also prepared the corresponding palladium compound $Pd(\eta^4-C_8H_{12})_2$ from $(\eta^4-C_8H_{12})PdCl_2$ and $Li_2C_8H_8$, using pure 1,5-cyclooctadiene as solvent [117]. The isolation of the compound was an experimental masterpiece since the work-up of the reaction mixture had to be done under an atmosphere of propene and during the whole procedure, including the recrystallization of the crude product from liquid propene, the temperature had to be kept at –25°C or below. Similar to its platinum analogue **19**, $Pd(\eta^4-C_8H_{12})_2$ readily reacts with norbornene, *trans*-cyclooctene and ethene to give the monoolefin complexes $Pd(C_7H_{10})_3$, $Pd(C_8H_{14})_3$ and $Pd(C_2H_4)_3$ by displacement of the diene ligands [117]. Independently, Geoffrey Ozin from the University of Toronto in Canada developed a direct route to $Pd(C_2H_4)_3$ via the cocondensation of palladium atoms with C_2H_4 in a C_2H_4/inert gas matrix at 15 K [121]. Moreover, he found that under slightly different conditions the coordinatively less saturated species $Pd(C_2H_4)_2$ and $Pd(C_2H_4)$ could be generated; they were characterized by IR and UV–vis spectroscopy.

Parallel to Stone's work, Peter Timms (who was also on the staff at the University of Bristol in those days) applied the metal/ligand cocondensation technique to prepare $Pd(\eta^4-C_8H_{12})_2$ from palladium vapour and 1,5-cyclooctadiene on a gram scale [122] and also succeeded in obtaining $Fe(\eta^4-C_8H_{12})_2$ via this route [123]. This 16-electron compound is thermally very unstable and rapidly decomposes to metallic iron above –20°C. Treatment of $Fe(\eta^4-C_8H_{12})_2$ with trifluorophosphine gave the stable complex $(\eta^4-C_8H_{12})Fe(PF_3)_3$, regarded as the analogue of Stone's $(\eta^4-C_8H_{12})Fe(CO)_3$ [123]. Also in the mid 1970s, Philip Skell reported the preparation of $Mo(\eta^4-C_4H_6)_3$ and $W(\eta^4-C_4H_6)_3$ from the corresponding metal vapours and described these tris(butadiene) complexes emphatically as "unique in the literature of organometallic compounds" [124]. $Mo(\eta^4-C_4H_6)_3$ and $W(\eta^4-C_4H_6)_3$ can be handled in air and are thermally stable up to 130–135°C. Although both compounds were carefully characterized by spectroscopic means, it is not yet clear, whether they have an octahedral or a trigonal prismatic geometry. The latter (rather unusual) configuration was confirmed crystallographically for the tungsten(0) derivative $W[\eta^4-CH_3C(O)CH{=}CH_2)]_3$, in which the methyl vinyl ketone ligands are linked to the metal centre in a chelating fashion [125].

7.9 Timms', Fischer's and Green's Distinctive Shares

However, in the 1960s and early 1970s the chemistry of di- and oligoolefin metal(0) complexes was not limited to Wilke's, Stone's, Timms' and Skell's work, because simultaneously (or even prior) to these studies a series of papers from Fischer's laboratory in München appeared, in which an unusual method for preparing iron(0), ruthenium(0) and osmium(0) compounds with either one diolefin and one triolefin, or one diolefin and one arene ligand was described. In the context of mechanistic studies of a new route to prepare bis(arene) metal complexes [126], it was found that the reaction of $FeCl_3$, $RuCl_3$ and $OsCl_3$ with isopropylmagnesium bromide in diethyl ether generated highly reactive intermediates of the supposed composition $M(iPr)_3(OEt_2)_n$, which in the presence of an excess of 1,3-cyclohexadiene under UV irradiation gave the closed shell complexes $(\eta^4\text{-}C_6H_8)M(\eta^6\text{-}C_6H_6)$ [127, 128]. The isolated yield of 35% obtained for the iron compound drops to 0.2% for the osmium analogue, possibly due to the different stabilities of the alkyl metal intermediates. The thermal stability of the complexes rises significantly from Fe to Os, and the same is true for the stability towards oxygen.

The photochemical reactions of $FeCl_3$ and $RuCl_3$ with *i*PrMgBr and mixtures of 1,5-cyclooctadiene and cycloheptatriene, or 1,5-cyclooctadiene and 1,3,5-cyclooctatriene, led to the formation of iron(0) and ruthenium(0) complexes of the general composition $(\eta^4\text{-}C_8H_{12})M(\eta^6\text{-}C_7H_8)$ (**25**) and $(\eta^4\text{-}C_8H_{12})M(\eta^6\text{-}C_8H_{10})$ (**26, 27**) (Scheme 7.6) [127, 128]. In addition, a ruthenium compound analysing as $Ru(C_7H_8)_2$, which contains one norbornadiene and one 1,3,5-cycloheptatriene ligand, was obtained. The isopropyl Grignard method, mainly developed by Jörn Müller (a Ph.D. student of E. O. Fischer who later held the Chair of Inorganic and Analytical Chemistry at the *Technische Universität* in Berlin), could also be used to prepare paramagnetic $V(\eta^6\text{-}C_7H_8)_2$, the first bis(triolefin) metal(0) compound [129], and diamagnetic $Pt(\eta^4\text{-}C_8H_{12})_2$ (**19**), as mentioned above. In the course of preparing **19**, Müller and Göser could isolate the dialkyl platinum(II) complex $(\eta^4\text{-}C_8H_{12})PtiPr_2$ as the initially formed product from $(\eta^4\text{-}C_8H_{12})PtCl_2$ and *i*PrMgBr at low temperature in the dark, which supported the idea that in the reaction of metal halides with the isopropyl Grignard reagent alkyl-transition metal species were involved as intermediates. Similar to the 18-electron cycloheptatriene metal tricarbonyls $(\eta^6\text{-}C_7H_8)M(CO)_3$ (M=Cr, Mo), the 17-electron compound $V(\eta^6\text{-}C_7H_8)_2$ reacted with $(CPh_3)BF_4$ to give the mono- and bis(tropylium) complexes $[(\eta^6\text{-}C_7H_7)V(\eta^6\text{-}C_7H_8)]BF_4$ and $[V(\eta^6\text{-}C_7H_7)_2](BF_4)_2$ by hydride abstraction [129].

Some of the latest, though significant, results in the emerging field of zerovalent olefin metal complexes were obtained by Malcolm Green (Fig. 7.12) almost at the end of the twentieth century. Following in the footsteps of Skell and Timms, Malcolm was the main driving force in the late 1970s and 1980s to apply the metal/ligand vapour cocondensation technique to high-melting transition metals,

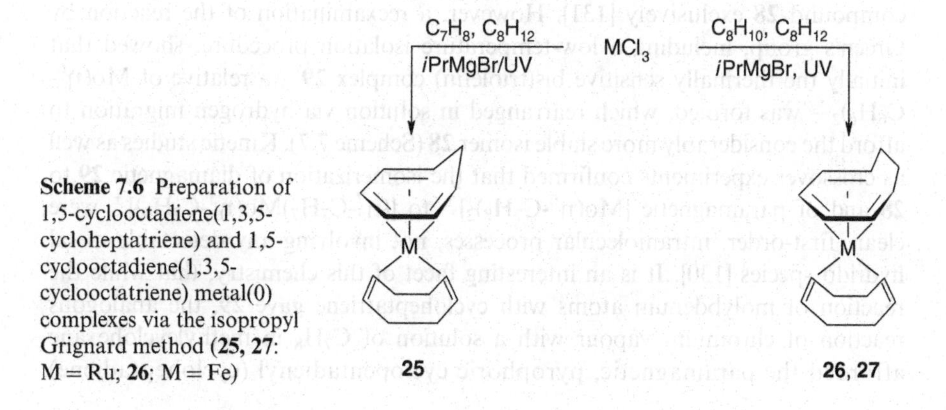

Scheme 7.6 Preparation of 1,5-cyclooctadiene(1,3,5-cycloheptatriene) and 1,5-cyclooctadiene(1,3,5-cyclooctatriene) metal(0) complexes via the isopropyl Grignard method (**25, 27**: M = Ru; **26**: M = Fe)

Fig. 7.12 Malcolm L.H. Green (born in 1936) obtained his Ph.D. in 1959 from Imperial College London with Geoffrey Wilkinson. After one year as a postdoctoral fellow at Imperial College, he was Assistant Lecturer at Cambridge University from 1960 to 1963, and then moved to Oxford where he was appointed Septcentenary Fellow of Inorganic Chemistry at Balliol College. In 1985, he was elected Fellow of the Royal Society and became Professor and Head of the Inorganic Chemistry Laboratory at Oxford University in 1989. His research interests include inter alia the chemistry of metallocenes, C—H bond activation, agostic carbon–hydrogen–transition metal interactions, multigram-scale metal atom vapour synthesis techniques, and more recently, the organometallic chemistry of fullerenes and the chemistry of metal complexes in carbon nanotubes. He has published more than 600 papers and patents and is without any doubt one of the most creative inorganic chemist of his generation. Malcolm's work has been widely recognized by numerous distinguished lectureships and awards including the ACS Award in Inorganic Chemistry, the Sir Edward Frankland Prize of the Royal Society of Chemistry, and the Karl Ziegler Prize of the *Gesellschaft Deutscher Chemiker*. More recently, he became Professor Emeritus at Oxford University, but is still running a relatively large research group (photo by courtesy of M. L. H. G.)

particularly to those of the group IV, V and VI elements. During this work, he treated inter alia molybdenum atoms with cycloheptatriene, using an electron-gun furnace [130]. This reaction of vapourized molybdenum with C_7H_8 had been originally reported by Timms to give the η^7-cycloheptatrienyl(η^5-cycloheptadienyl) compound **28** exclusively [131]. However, a reexamination of the reaction by Green's group, including a low-temperature isolation procedure, showed that initially the thermally sensitive bis(triolefin) complex **29** – a relative of Mo(η^6-$C_6H_6)_2$ – was formed, which rearranged in solution via hydrogen migration to afford the considerably more stable isomer **28** (Scheme 7.7). Kinetic studies as well as crossover experiments confirmed that the isomerization of diamagnetic **29** to **28** and of paramagnetic $[Mo(\eta^6\text{-}C_7H_8)_2]^+$ to $[(\eta^7\text{-}C_7H_7)Mo(\eta^5\text{-}C_7H_9)]^+$ were clean, first-order, intramolecular processes, not involving any detectable metal hydrido species [130]. It is an interesting facet of this chemistry, that while the reaction of molybdenum atoms with cycloheptatriene gave **29**, the analogous reaction of chromium vapour with a solution of C_7H_8 in methylcyclohexane afforded the paramagnetic, pyrophoric cyclopentadienyl (cycloheptadiene)

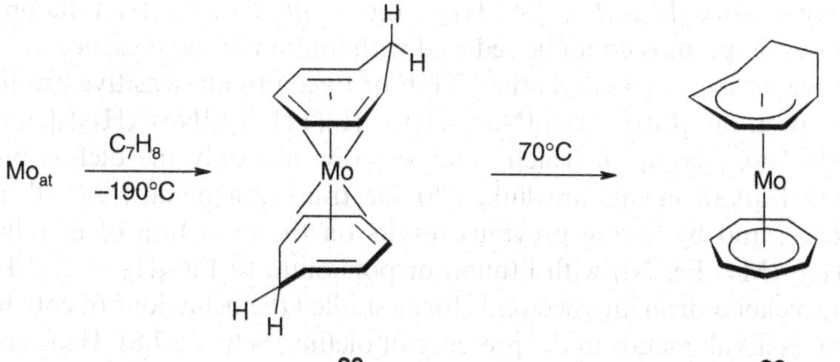

Scheme 7.7 Preparation of bis(η^6-cycloheptatriene)molybdenum(0) **29** via the metal/ligand vapour cocondensation technique and its thermal rearrangement to the η^7-cycloheptatrienyl(η^5-cycloheptadienyl) isomer **28**

complex (η^7-C_7H_7)Cr(η^4-C_7H_{10}) instead of the expected diamagnetic compounds Cr(η^6-C_7H_8)$_2$ or (η^7-C_7H_7)Cr(η^5-C_7H_9) [131].

Whereas Mo(η^6-C_7H_8)$_2$ (**29**) could be prepared via the metal/ligand vapour cocondensation technique [130], the corresponding bis(triolefin) complex Zr(η^6-C_7H_8)$_2$ could not be obtained by this route. The isomer (η^7-C_7H_7)Zr(η^5-C_7H_9) was formed instead, and there was no evidence for the generation of Zr(η^6-C_7H_8)$_2$ from zirconium atoms and cycloheptatriene as an intermediate [132]. Green's group showed, however, that the desired 16-electron bis(triolefin) derivative Zr(η^6-C_7H_8)$_2$ could be prepared by reduction of $ZrCl_4$ with sodium amalgam in the presence of cycloheptatriene [133]. The compound rearranged thermally to (η^7-C_7H_7)Zr(η^5-C_7H_9) and reacted with trimethylphosphine to afford the 18-electron complex (η^7-C_7H_7)Zr(η^5-C_7H_9)(PMe_3). The X-ray diffraction analysis of Zr(η^6-C_7H_8)$_2$ revealed a non-parallel arrangement of the two triolefin ligands with a dihedral angle of 25.6° between the best planes of the C_6 parts of the two seven-membered rings [133]. This bent structure is rather surprising when compared with the normal parallel arrangement of the rings in Mo(η^6-C_7H_8)$_2$ and in bis(arene) metal(0) complexes as well.

7.10 A Recent Milestone: Jonas' Olefin Analogues of Hieber's Metal Carbonylates

Finally, at the end of this chapter on the chemistry of olefin metal π-complexes it is worth mentioning that, in analogy to Hieber's metal carbonylate anions $[Fe(CO)_4]^{2-}$ and $[Co(CO)_4]^-$, the corresponding ethene derivatives $[Fe(C_2H_4)_4]^{2-}$ and $[Co(C_2H_4)_4]^-$ have also been prepared. This skilful work was carried out in the late 1970s and the 1980s at the *Max-Planck-Institut für Kohlenforschung* in Mülheim by Klaus Jonas, who was a former Ph.D. student of Wilke's and, prior to his own work, strongly involved in the research on olefin nickel(0) complexes [96, 102, 110]. In the course of his studies on the reactivity of $Ni(C_2H_4)_3$, $Ni(\eta^4\text{-}C_8H_{12})_2$ and Ni(*all,trans*-$C_{12}H_{18}$), he observed that these compounds could be reduced with lithium in the presence of coordinating agents such as diethyl ether, THF or tmeda to air-sensitive products of the composition $[Li(tmeda)]_2[Ni(C_2H_4)_3]$, $[Li(THF)_2]_2[Ni(C_8H_{12})_2]$ and $[Li(tmeda)]_2[Ni(C_{12}H_{18})]$, in which, unexpectedly, not only the olefins but also the ligated alkali metals are linked to the transition metal [134]. With this experience, and by taking previous results on the reduction of metallocenes $M(C_5H_5)_2$ (M = Fe, Ni) with lithium or potassium to LiC_5H_5 or KC_5H_5 and metallic nickel or iron into account, Jonas studied the behaviour of cobaltocene **30** towards alkali metals in the presence of olefins (Scheme 7.8). He found that the reaction of **30** with excess lithium in THF at –30°C generated the salt-like intermediate **31**, which upon treatment with ethene and 1,5-cyclooctadiene gave the half-sandwich-type compounds **32** and **33** in high yield [135]. Subsequent degradation of **32** and **33** with alkali metals in the presence of THF or tmeda afforded the corresponding salts of the anionic olefin cobalt complexes **34**–**36**, of which **36** is isoelectronic (and probably isostructural) to $Ni(\eta^4\text{-}C_8H_{12})_2$. The cyclopentadienyl ring of **33** could also be displaced by lithium in nitrogen atmosphere to yield the bis(dinitrogen) complex **37**. As Jonas described, all these degradation processes "proceed practically without side reactions and almost quantitatively" and "the alkali metal cyclopentadienides can be

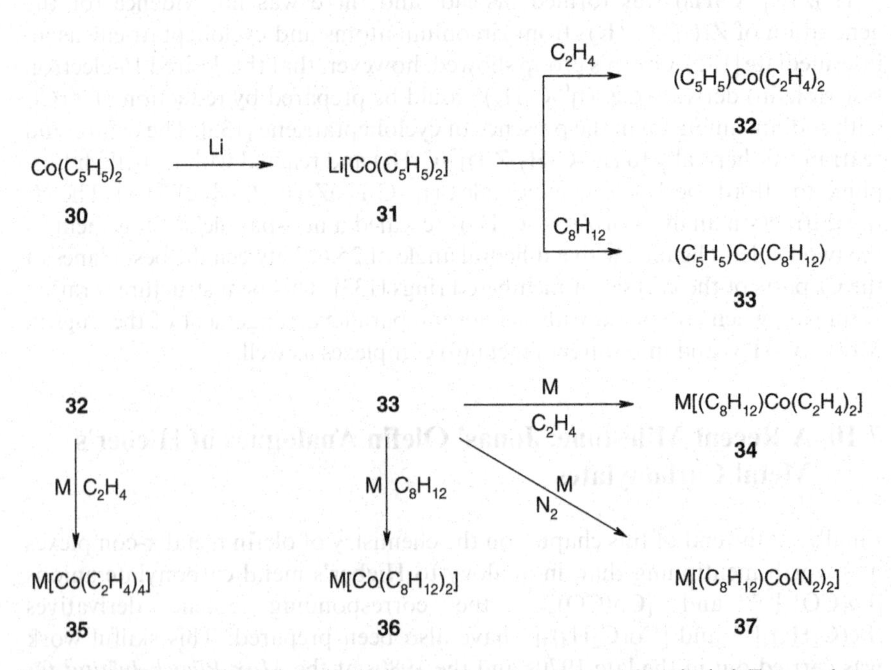

Scheme 7.8 Preparation of the half-sandwich type cyclopentadienylcobalt(I) complexes **32** and **33** from cobaltocene and their degradation with alkali metals M (Li or K) in the presence of THF or tmeda to give the anionic olefin cobalt(–I) complexes **34**–**37** (the coordination of THF and tmeda to the cations M^+ is omitted for clarity)

separated from the cobalt complexes... without difficulty" [135]. Despite the extreme sensitivity of $[Li(THF)_2][Co(\eta^4\text{-}C_8H_{12})_2]$, its molecular structure could be determined crystallographically [136].

In further exploring the potential of the reductive displacement reaction of the bis(cyclopentadienyl) complexes with alkali metals, Jonas discovered that not only cobaltocene (a 19-electron compound) but also ferrocene (a closed-shell compound) could be reduced with lithium in the presence of tmeda and ethene [137, 138]. With two equivalents of Li, the monoanion $[(\eta^5\text{-}C_5H_5)Fe(C_2H_4)_2]^-$, isoelectronic to $(\eta^5\text{-}C_5H_5)Co(C_2H_4)_2$, was formed, while with four equivalents of the alkali metal the dianion $[Fe(C_2H_4)_4]^{2-}$ (of which the isoelectronic molecule $Ni(C_2H_4)_4$ is unknown) was obtained. The X-ray crystal structure analysis of $[Li(tmeda)]_2[Fe(C_2H_4)_4]$ revealed a distorted octahedral environment around the iron atom with the four olefins occupying the equatorial and the two lithium atoms the apical sites [137, 138]. The reaction of $[Li(tmeda)]_2[Fe(C_2H_4)_4]$ with 1,5-C_8H_{12} afforded the corresponding cyclooctadiene complex $[Li(tmeda)]_2[Fe(\eta^4\text{-}C_8H_{12})_2]$, completing the series of isoelectronic species $Ni(\eta^4\text{-}C_8H_{12})_2$, $[Co(\eta^4\text{-}C_8H_{12})_2]^-$ and $[Fe(\eta^4\text{-}C_8H_{12})_2]^{2-}$, respectively [139, 140].

Parallel to Jonas' research, anionic ethene nickel complexes were also prepared by Wilke's group. With $Ni(C_2H_4)_3$ as the starting material, it was found that this compound did not only react with lithium in the presence of tmeda to give $[Li(tmeda)]_2[Ni(C_2H_4)_3]$ (see above), but upon treatment with trialkyl (hydrido)-aluminates and -gallates $MHM'R_3$ ($M = Li, Na$; $M' = Al, Ga$; $R = Me, Et$, etc.) also afforded dinuclear compounds such as $[M(tmeda)_2]$ $[HNi_2(C_2H_4)_4]$ via thermally labile anionic species $[R_3M'—H—Ni(C_2H_4)_2]^-$ as intermediates [141]. In the dinuclear anion $[HNi_2(C_2H_4)_4]^-$, the Ni–H–Ni unit is not linear and, due to the relatively short Ni–Ni distance, can be regarded as a 3-centre-4-electron system. Despite the different coordination numbers of the nickel atoms, the bonding pattern of $[HNi_2(C_2H_4)_4]^-$ is thus similar to that of the carbonyl complex $[HNi_2(CO)_6]^-$. Surprisingly, the dinuclear anion $[HNi_2(C_2H_4)_4]^-$ does not add ethene and is therefore not involved as an intermediate in the polymerization of ethene via the "nickel effect" [103, 110].

Biographies

Biography of Ernst Otto Fischer – see Chap. 2, for that of Wilhelm Christoph Zeise see Chap. 3, and for the biographies of Peter Pauson and Geoffrey Wilkinson see Chap. 5.

For almost four decades, **Joseph Chatt** was one of the most important contributors to the development of modern inorganic and organometallic chemistry. He was born on the 6th of November 1914 in Horden, County Durham, England into a family that had farmed there for many generations. Neither at home nor at his small village school was he exposed to science, but he stayed regularly with an uncle who was the Chief Scientist at a steelworks near Newcastle-upon-Tyne during the summer. Although he later recognized that it was during those vacations that his interest in science was first aroused, he attributed his specific dedication to chemistry to the fact that the family's farm was in the vicinity of Caldbeck Fells, an area famous for its wide range of minerals. With the support of a teacher at his secondary school, he learned how to carry out analyses to identify the minerals and reveal which elements were present. This aroused his fascination for inorganic chemistry in particular, which he retained throughout his life.

At the age of 14, soon after he entered The Nelson School at Wigton as a fee-paying pupil, Chatt's father died, and it seemed that he had to leave the school. However, his potential had already been recognized, and a grant from the County Education Committee enabled him to stay on. He matriculated in just over 2 years and his outstanding performance brought him various local scholarships. This meant that he could consider going to Cambridge, where he finally found an enlightened admissions tutor at Emmanuel College who agreed to accept him. After he graduated with First Class Honours in 1937, he decided to carry out research in inorganic chemistry with Frederick. G. Mann. Initially, he worked on halogen-bridged phosphine palladium complexes, and then moved on to a more general study of arsine and phosphine compounds, mainly of the transition elements.

He received his Ph.D. in summer 1940 and firmly wished to stay in fundamental research. He was offered a post at St. Andrews University which he could not take up because of World War II. Instead he was assigned to a "war problem" at Cambridge involving the synthesis of the then unknown 1,3,5,7-tetranitronaphthalene, which Sir Robert Robinson anticipated to be an exceptionally effective explosive. In 8 months he managed to make 100 g of the compound but was then dispatched to the Research Department of Woolwich Arsenal, at that time evacuated to the University College at Swansea. There he spent most of his time in the library where he eventually found R. N. Keller's review on olefin metal complexes [3]. He remembered that Mann had mentioned in his lectures that olefins were the only ligating species that did not possess a lone pair of electrons, which prompted him to work on their complexes as soon as he was free to do so.

Following a short time in industry, he took up an ICI fellowship at Imperial College London but was frustrated by the inadequate facilities in the labs, largely a legacy of the war. At a reception of the Royal Society, he was lucky to meet R. M. Winter, the Controller of Research in ICI, who invited him to join the company. He accepted and was given a place at the Butterwick Research Laboratories, a new fundamental research establishment, later always referred to as "the Frythe". In an interview given about 45 years later, he recalled that "it was there that I did most of the work for which I am known" [142].

At the Frythe, Chatt began the systematic studies on olefin complexes and, based on Dewar's seminal ideas, established the currently accepted model for the binding of olefins to transition metals. At the initial stage, much of the work was carried out by himself but from the early 1950s onwards he attracted some very able young people to work with him. As a result of his achievements and his success in persuading ICI to allocate the necessary resources, the group had a total staff of about 40 by the time it was disbanded. Apart from the investigations on olefin (and also alkyne) metal compounds, Chatt's group was among the first to use tertiary phosphines to stabilize complexes of transition metals in low-oxidation states. Moreover, they showed that, contrary to the general belief, alkyl and aryls of the late transition elements could be obtained. Another important achievement that arose from the work at the Frythe, was the classification of metal centres in complexes in terms of their Lewis acidities, which laid the foundation for the widely used concept of hard and soft Lewis acids and bases, so successfully developed by Ralph Pearson.

When ICI decided to reorganize its research activities in 1962, that included closure of the Frythe, Chatt was expected to move to the Heavy Organic Chemicals Division at Runcorn. He suspected, however, that he would have much less freedom there in his choice of projects and thus decided to leave. While he was considering offers from some major universities from the United States, the British Agricultural Research Council decided to establish an interdisciplinary research unit to study the mechanism of biological fixation of nitrogen. Due to his reputation as one of the leading inorganic chemists in the country, Chatt (Fig. 7.13) was invited to be the director of "the Unit" (as it was generally called), being provisionally housed at Queen Mary College in London. Somewhat later, the Unit moved to the newly founded University of Sussex near Brighton, where it was initially housed in the laboratories of the School of Molecular Sciences and later in a specially designed building attached to that school. In less than a decade, the Unit not only became the leading centre in the world for research in a very important and highly competitive field but also a model for organization of

Fig. 7.13 Joseph Chatt at the time when he was the director of "the Unit" at the University of Sussex (photo from ref. 121; reproduced with permission of The Royal Society)

interdisciplinary research, with inorganic chemists, physical chemists, biochemists, microbiologists and geneticists all focused on a central problem.

Much of the Unit's success stemmed from Chatt's style of leadership and his firm dedication to fundamental research. His moral was that "if you want really good new science, choose a dedicated scientist and let him get on with it" [143]. Given this credit, the chemists in the Unit generated a large body of novel transition metal chemistry, particularly an extensive series of dinitrogen complexes, and succeeded in identifying the sequence of reactions through which coordinated N_2 was converted to ammonia. The work of the biological group, led by John Postgate, was equally successful and opened up a new area of investigations in biological nitrogen fixation. When Chatt retired from the directorship in 1980, the Unit remained largely unchanged until it was moved to Norwich and became the Nitrogen Fixation Laboratory.

The importance of Chatt's work was recognized by many awards including the Tilden, Liversidge and Nyholm Lectureships of the Royal Society of Chemistry, the American Chemical Society Award for Distinguished Service in the Advancement of Inorganic Chemistry, and the Wolf Prize for Chemistry, the first to be awarded to someone resident outside North America. He became a Fellow of the Royal Society in 1961, received its Davy Metal in 1979, and was appointed Commander of the British Empire in 1978. After his retirement from the Unit, he continued to participate in research in the School of Molecular Science and went to the university regularly. He died on the 19th of May 1994.

Günther Wilke is described by many of his contemporaries as the "master of homogeneous catalysis". Following the footsteps of Franz Fischer and Karl Ziegler, he was the Director of the prestigious *Max-Planck-Institut für Kohlenforschung* in Mülheim for more than two decades and was most influential in making this institution the second main centre of organometallic chemistry in Germany, after the renowned *Technische Hochschule* (later *Technische Universität*) in München.

Wilke was born on the 23th of February 1925 in Heidelberg, Germany, and carried out his undergraduate and postgraduate studies at the *Technische Hochschule* in Karlsruhe and the university of his hometown. At Heidelberg he worked for his doctoral thesis with the eminent organic chemist Karl Freudenberg and received a Ph.D. in 1951. Following a recommendation from his supervisor, he accepted an invitation by Professor Karl Ziegler that same year to work as a research associate at the Max-Planck-Institut in Mülheim. In the course of his studies on the preparation and reactivity of organoaluminium compounds, he observed that with catalysts, formed in situ from aluminium alkyls and transition metal halides or derivatives thereof, butadiene can be converted into high-molecular weight 1,2-polybutadiene. Modification of these catalysts led directly to the discovery of the cyclotrimerization of butadiene yielding *trans,trans,cis*-1,5,9-cyclododecatriene. This result was of great interest for those divisions of the chemical industry manufacturing polymers since the cyclotrimer could easily be converted to the valuable synthetic fibre *Nylon 12*.

Given the importance of this work, Wilke's group grew to about ten coworkers within a short period of time. To accommodate these people, a provisional laboratory was built in less than 2 months in the courtyard of the institute. It was in this famous "hut" where the work on olefin nickel and π-allyl transition metal complexes began, which established the world-wide reputation of the "Wilke school". After finishing the *Habilitation* as an external candidate at the *Technische Hochschule* in Aachen in 1960, Wilke was promoted to *Wissenschaftliches Mitglied* of the Max Planck Society and became Professor of Chemistry at the newly founded University of Bochum in 1963. Three years later he was invited to a chair of organic chemistry at the ETH in Zürich. To counter this prestigious offer, he was appointed as Co-director at the Mülheim institute in 1967 and was promised to succeed Karl Ziegler upon his retirement. This promotion happened in 1969, and from then until spring 1993 Günther Wilke (Fig. 7.14) was the Director of the *Max-Planck-Institut* and chairman of the *Studiengesellschaft Kohle*.

Wilke's research activities in the field of olefin and π-allyl transition metal complexes continued until his retirement. In his own view, the most important result of his ground-breaking work on the nickel-catalysed cyclooligomerization of diolefins was that the direction of the catalytic reaction could be controlled by the presence of additional ligands which, due to their electronic and steric properties, modify the course of the process [144]. One typical example was that whereas "naked nickel" cyclotrimerized butadiene to isomers of 1,5,9-cyclododecatriene, the 1:1 complex of 1,5,9-cyclododecatrienenickel(0) and tris(2-phenyl-phenyl)phosphite catalysed the cyclodimerization of butadiene to 1,5-cyclooctadiene in nearly quantitative yield. Apart from cyclooligomerizations, Wilke applied the concept of "directing the course of a catalytic reaction" [103] also to other systems and developed catalysts for the dimerization of ethene and propene with turnover numbers that are comparable with those found for fast enzymatic processes. With phosphines having chiral substituents, high degrees of optical induction, for example, in the codimerization of norbornene and ethene forming *exo*-vinylbicycloheptane, could be achieved.

Almost from the beginning of his independent research in the mid 1950s, Wilke attracted a great number of graduate students and postdocs, several of

Fig. 7.14 Günther Wilke at the time when he retired as the Director of the *Max-Planck-Institut für Kohlenforschung* at Mülheim (photo by courtesy from G. W.)

whom later embarked upon an academic career. He regularly lectured at the University of Bochum, where he held an adjunct faculty position and was the Dean of the Faculty in 1968 and 1969. Moreover, he was the Vice President of the Max Planck Society from 1978 to 1990, the President of the German Chemical Society in 1980 and 1981, and the President of the *Nordrhein-Westfälische Akademie der Wissenschaften* from 1994 to 1998. Apart from his memberships in several European academies, he received numerous awards among which the Emil Fischer Medal and the Karl Ziegler Prize of the German Chemical Society, the Willard-Gibbs Medal of the American Chemical Society, and the Sir Edward Frankland Prize of the Royal Society of Chemistry deserve special attention. In 2000, he received the *Große Verdienstkreuz mit Stern des Verdienstordens der Bundesrepublik Deutschland*, the highest honour which can be granted to a German citizen from the President of the Federal Republic of Germany.

Together with Joseph Chatt, Peter Pauson and Geoffrey Wilkinson, **F. Gordon A. Stone** belongs to the pioneers of organometallic chemistry in the UK. In a congratulatory letter to him on the occasion of his 65th birthday, Henry Gilman, one of the pacemakers of inorganic chemistry in the United States, wrote: "You are a chemist's chemist, and as such will long serve as a classical reminder of innovative and truly profound research". And he added: "Your delightful sense of humour is penetrating and most friendly".[6] These remarks could apply equally as well to other facets of his personality, with "penetrating" being the suitable term.

Stone was born on the 19th of May 1925 in Exeter, a city in the southwest of England. His father was a government servant who helped in introducing him to the elementary principles of mathematics. After living with his parents for some years in East Kent, he returned to Exeter in 1940 where he stayed with his maternal grandparents and attended Exeter School. Although this school, like many in Britain in those days, operated under considerable difficulties, he developed an early enthusiasm for chemistry. After receiving the Higher School Certificate with exams in chemistry, physics, mathematics and English literature, he planned to study chemistry at either Imperial or King's College of London University. However, he was rejected by these institutions because he had failed the examination in his best subject, chemistry. So he spent a year at the Medway Technical College at Gillingham in Kent studying chemistry full-time. After he passed the final exam, he applied for admission to three of the Cambridge colleges and was accepted by all three. He decided in favour of Christ's, which he entered as an undergraduate at the end of 1945.

After Stone received his B.A. degree with First Class Honours in 1948, he did postgraduate work with Harry J. Emeléus, studying the chemistry of boron hydrides. To begin with he had to construct a Stock-type high-vacuum system from soda glass and to prepare the starting materials for the synthesis of diborane. He recalled that it took him "almost a year to generate a millimole

[6] [69], Editor's Note, p. xix.

of B_2H_6, probably the first sample of this compound ever to be made in the United Kingdom" [145]. Although he "accomplished relatively little" with his research project, he decided to continue with boron chemistry after receiving the Ph.D. in 1952 and joined Anton Burg's group at the University of Southern California as a Fulbright scholar. He stayed in Los Angeles for 2 years, but as a result of a conversation with Eugene G. Rochow moved to Harvard University in February 1954 joining Rochow's group as a postdoctoral fellow. Only a few months later, he was offered the position of Instructor in Inorganic Chemistry and was thus able to start some independent research. While his first graduate students, William Graham (for biography see Chap. 9) and Herbert Kaesz (now Emeritus Professor at UCLA) began to work on aspects of boron chemistry, Stone's interest soon moved into the area of organo-transition metal chemistry, partly because his laboratories were in the same building as those of Wilkinson's.

As a consequence of Wilkinson's departure to the UK, Stone was appointed to the vacant assistant professorship in inorganic chemistry at Harvard. He pioneered the development of fluorocarbon metal chemistry including the discovery of the first transition metal heterocycle with a FeC_4F_8 five-membered ring. In the following years his group expanded significantly and performed remarkable studies on the hydrometallation of C=C bonds, on pentafluorophenyl derivatives of metals and non-metals, on cyclooctatetraene iron complexes and on sandwich compounds, including the synthesis of the first neutral complex with a cycloheptatrienyl ligand. In September 1962, he returned to England to join the staff of the chemistry department at Queen Mary College, London but remained there for only a few months. Due to rapid expansion in the British university system in that period, he received an offer in 1963 from Bristol University, which he accepted to becoming the first occupant of a newly created chair in inorganic chemistry.

When he arrived at Bristol, there were very few members of staff with interests in inorganic chemistry. These included Edward Abel (later to become Professor at the University of Exeter, and subsequently President of the Royal Society of Chemistry) and Peter Woodward. With the support of these persons, and the appointment of Michael Green and Peter Timms, and later of Michael Bruce, Selby Knox, Judith Howard and John Spencer, in the following years Stone built up what became known as the "Bristol school of inorganic chemistry". It was, and still is, one of the most outstanding and productive centres of chemical research in the UK. Among the highlights of Stone's work at Bristol were the systematic design of metal clusters, subsequently rationalized in Roald Hoffmann's isolable analogy, the synthesis of the first complexes with the anti-aromatic 8π-electron system pentalene, the development of the chemistry of "naked" platinum, and last but not least the organometallic chemistry of *closo-* and *nido*-carboranes. As Anthony Hill emphasized [145], it is certainly true that "Stone has recently pioneered something of a renaissance in the chemistry of monocarbollide complexes".

Gordon Stone (Fig. 7.15) received numerous awards including the Tilden, Mond and Frankland Lectureships as well as the Longstaff Medal from the Royal Society of Chemistry, and the Davy Medal from the Royal Society. He was elected Fellow of the Royal Society in 1976 and appointed Commander of

Fig. 7.15 Gordon Stone at the time before he moved to Baylor University in the United States (photo by courtesy from F. G. A. S.)

the British Empire in 1990. Apart from his duties as Professor of Inorganic Chemistry and Head of the Bristol department, he served on several committees of the Royal Society of Chemistry and chaired in the late 1980s the University Grants Committee's review of the state of chemistry in British universities. The "Stone Report" came out with a number of recommendations, which sparked off a lively discussion in the whole academic community but showed little effect on the political level. After Stone retired from Bristol in 1990, he accepted a Welch Chair at Baylor University in Texas. There he is still continuing his research on fundamental organometallic chemistry and regularly extending his list of primary publications, which presently comprises more than 700 journal articles, excluding reviews. When he was asked in an interview by Peter Maitlis to what he attributed his success, he simply said: "Hard work and a large number of excellent coworkers from all over the world" [145].

References

1. P. Pfeiffer, *Organische Molekülverbindungen* (Verlag Ferdinand Enke, Stuttgart, 1927, p. 209).
2. M. S. Kharasch, and T. A. Ashford, Coordination Compounds of Platinous Halides with Unsaturated (Ethylene) Substances, *J. Am. Chem. Soc.* **58**, 1733–1738 (1936).
3. R. N. Keller, Coordination Compounds of Olefins with Metallic Salts, *Chem. Rev.* **28**, 229–267 (1941).
4. W. C. Zeise, Ueber Acechlorplatin, nebst Bemerkungen über einige andere Producte der Einwirkung zwischen Platinchlorid und Aceton, *J. Prakt. Chem.* **20**, 193–234 (1840).
5. W. Prandtl, and K. A. Hofmann, Ueber Platin-Kohlenstoff-Verbindungen, *Ber. dtsch. chem. Ges.* **33**, 2981–2983 (1900).

6. R. D. Gillard, B. T. Heaton, and M. F. Pilbrow, 4-Methylpent-4-en-2-one (Isomesityl Oxide) Complex of Platinum(II), *J. Chem. Soc. A* **1970**, 353–355.
7. C. Liebermann, and C. Paal, Ueber Derivate des Allylamins, *Ber. dtsch. chem. Ges.* **16**, 523–534 (1883).
8. M. Herberhold, *Metal π-Complexes, Vol. II: Complexes with Mono-Olefinic Ligands* (Elsevier, Amsterdam, 1972, Chapter 2).
9. R. G. Denning, and L. M. Venanzi, Platinum Complexes with Unsaturated Amines. Part I. Complexes with Allylamines, *J. Chem. Soc.*3241–3247 (1963).
10. E. Biilmann, Ueber die Einwirkung von Allylalkohol auf Kaliumplatochlorid, *Ber. dtsch. chem. Ges.* **33**, 2196–2201 (1900).
11. P. Pfeiffer, and H. Hoyer, Komplexverbindungen der Äthylenkörper mit Platinsalzen, *Z. anorg. allg. Chem.* **211**, 241–248 (1933).
12. K. A. Hofmann, and J. von Narbutt, Verbindungen von Platinchlorür mit Di-cyclopentadien, *Ber. dtsch. chem. Ges.* **41**, 1625–1628 (1908).
13. J. R. Doyle, and H. B. Jonassen, The Structure of Dicyclopentadienedichloroplatinum(II), *J. Am. Chem. Soc.* **78**, 3965–3967 (1956).
14. N. C. Baenziger, J. R. Doyle, G. F. Richards, and L. C. Carpenter, *Advances in the Chemistry of the Coordination Compounds* (McMillan, New York, 1961, p. 136).
15. A. D. Hel'man, Complex Compounds of Platinum and Butadiene (Bivinyl), *Dokl. Akad. Nauk SSSR* **23**, 532–536 (1939).
16. A. D. Hel'man, Reaction of Unsaturated Molecules with Sodium Chloroplatinate, *Dokl. Akad. Nauk SSSR* **31**, 761–764 (1941).
17. P. E. Slade Jr., and H. B. Jonassen, Coordination Compounds of Butadiene with Platinum(II), Palladium(II) and Copper(I) Halides, *J. Am. Chem. Soc.* **79**, 1277–1279 (1957).
18. M. S. Kharasch, R. C. Seyler, and F. R. Mayo, Coordination Compounds of Palladous Chloride, *J. Am. Chem. Soc.* **60**, 882–884 (1938).
19. J. Smidt, W. Hafner, R. Jira, J. Sedlmeier, R. Sieber, R. Rüttinger, and H. Kojer, Katalytische Umsetzungen von Olefinen an Platinmetall-Verbindungen, *Angew. Chem.* **71**, 176–182 (1959).
20. L. Hintermann, Wacker Type Oxidations, in. *Transition Metals for Organic Synthesis, Second Ed.* (Eds. M. Beller and C. Bolm, Wiley-VCH, Weinheim, 2004, Vol. 2, Chapter 2.8).
21. M. Berthelot, Observations relatives à l'action des sels cuivreux sur les carbures d'hydrogène et sur l'oxyde de carbone, *Ann. Chim. Phys.* **23**, 32–39 (1901).
22. W. Manchot, and W. Brandt, Ueber die Cuproverbindungen des Aethylens und des Kohlenoxyds, *Liebigs Ann. Chem.* **370**, 286–296 (1909).
23. H. Tropsch, and W. J. Mattox, Absorption of Ethylene by Solid Cuprous Chloride, *J. Am. Chem. Soc.* **57**, 1102–1103 (1935).
24. E. R. Gilliland, J. E. Seebold, J. R. FitzHugh and P. S. Morgan, Reaction of Olefins with Solid Cuprous Halides, *J. Am. Chem. Soc.* **61**, 1960–1962 (1939).
25. W. F. Eberz, H. J. Welge, D. M. Yost, and H. J. Lucas, The Rate of Hydration of Isobutene in the Presence of Silver Ion. The Nature of the Isobutene–Silver Complex, *J. Am. Chem. Soc.* **59**, 45–49 (1937).
26. S. Winstein, and H. J. Lucas, The Coordination of Silver Ion with Unsaturated Compounds, *J. Am. Chem. Soc.* **60**, 836–847 (1938).
27. K. S. Pitzer, Electron Deficient Molecules. I. The Principle of Hydroboron Structures, *J. Am. Chem. Soc.* **67**, 1126–1132 (1945).
28. H. Reihlen, A. Gruhl, G. von Hessling, and O. Pfrengle, Über Carbonyle und Nitrosyle, *Liebigs Ann. Chem.* **482**, 161–182 (1930).
29. E. Krause and A. von Grosse, *Die Chemie der metall-organischen Verbindungen* (Gebrüder Bornträger, Berlin, 1937).
30. H. J. Emeléus, and J. S. Anderson, *Modern Aspects of Inorganic Chemistry* (Routledge and Kegan Paul Ltd., London, 1952, Chapter 8).

31. M. J. S. Dewar, A Review of the π-Complex Theory, *Bull. Soc. Chim. Fr.* **1951**, C71–C79 (1951).
32. D. P. M. Mingos, A Historical Perspective on Dewar's Landmark Contribution to Organometallic Chemistry, *J. Organomet. Chem.* **635**, 1–8 (2001).
33. M. J. S. Dewar, Mechanism of the Benzidine and Related Rearrangements, *Nature* **156**, 784 (1945).
34. M. J. S. Dewar, The Mechanism of Benzidine-type Rearrangements, and the Role of π-Electrons in Organic Chemistry, *J. Chem. Soc.* **1946**, 406–408.
35. A. D. Walsh, Structures of Ethylene Oxide and Cyclopropane, *Nature* **159**, 165 (1947).
36. H. J. Taufen, M. J. Murray, and F. F. Cleveland, Effect of Silver Ion Coordination Upon the Raman Spectra of Some Unsaturated Compounds, *J. Am. Chem. Soc.* **63**, 3500–3503 (1941).
37. J. Chatt, The Addition Compounds of Olefins with Mercuric Salts, *Chem. Rev.* **48**, 7–43 (1951).
38. J. Chatt, and L. A. Duncanson, Infrared Spectra and Structure: Attempted Preparation of Acetylene Complexes, *J. Chem. Soc.* **1953**, 2939–2947.
39. M. J. S. Dewar, and G. P. Ford, Relationship Between Olefinic π Complexes and Three-Membered Rings, *J. Am. Chem. Soc.* **101**, 783–791 (1979).
40. G. J. Kubas, Metal–Dihydrogen and σ-Bond Coordination: The Consummate Extension of the Dewar-Chatt-Duncanson Model for Metal–Olefin π Bonding, *J. Organomet. Chem.* **635**, 37–68 (2001).
41. J. Chatt, L. M. Vallarino, and L. M. Venanzi, Diene Complexes of Platinum(II). The Structure of Hofmann and von Narbutt's [Dicyclopentadiene(RO)PtCl], *J. Chem. Soc.* **1957**, 2496–2505.
42. J. Chatt, L. M. Vallarino, and L. M. Venanzi, Some Diene Complexes of Palladium(II) and Their Alkoxy Derivatives, *J. Chem. Soc.* **1957**, 3413–3416.
43. J. Chatt, and L. M. Venanzi, Diene Complexes of Rhodium(I), *J. Chem. Soc.* **1957**, 4735–4741.
44. E. W. Abel, M. A. Bennett, and G. Wilkinson, Norbornadiene-Metal Complexes and Some Related Compounds, *J. Chem. Soc.* **1959**, 3178–3182.
45. R. Cramer, μ-Dichlorotetraethylenedirhodium(I), *Inorg. Chem.* **1**, 722–723 (1962).
46. L. Porri, A. Lionetti, G. Allegra, and A. Immirzi, Dibutadienerhodium(I) Chloride, *Chem. Comm.* **1965**, 336–337.
47. G. Winkhaus, and H. Singer, Rhodium(I)-Komplexe mit Mono- und Diolefinen, *Chem. Ber.* **99**, 3602–3609 (1966).
48. R. R. Schrock, and J. A. Osborn, Catalytic Hydrogenation Using Cationic Rhodium Complexes. I. Evolution of the Catalytic System and the Hydrogenation of Olefins, *J. Am. Chem. Soc.* **98**, 2134–2143 (1976).
49. H. Werner, M. E. Schneider, M. Bosch, J. Wolf, J. H. Teuben, A. Meetsma, and S. I. Troyanov, Cationic and Neutral Diphenyldiazomethanerhodium(I) Complexes as Catalytically Active Species in the C—C Coupling Reaction of Olefins with Diphenyldiazomethane, *Chem. Eur. J.* **6**, 3052–3059 (2000).
50. B. Rybtchinski, S. Oevers, M. Montag, A. Vigalok, H. Rozenberg, J. M. L. Martin, and D. Milstein, Comparison of Steric and Electronic Requirements for C—C and C—H Bond Activation. Chelating vs. Nonchelating Case, *J. Am. Chem. Soc.* **123**, 9064–9077 (2001).
51. B. F. Hallam, and P. L. Pauson, Butadiene- and *cyclo*Hexadiene-iron Tricarbonyls, *J. Chem. Soc.* **1958**, 642–645.
52. O. S. Mills, and G. Robinson, The Structure of Butadieneiron Tricarbonyl, *Proc. Chem. Soc.* **1960**, 421–422.
53. R. Burton, M. L. H. Green, E. W. Abel, and G. Wilkinson, Some New Organo-iron Complexes, *Chem. & Ind.* **1958**, 1592.

54. R. Pettit, Metal Complexes with 2,2,1-Bicycloheptadiene, *J. Am. Chem. Soc.* **81**, 1266 (1959).
55. T. A. Manuel, and F. G. A. Stone, Cyclooctatetraene-Iron Complexes, *Proc. Chem. Soc.* **1959**, 90.
56. A. Nakamura, and N. Hagihara, Cyclooctatetraeneiron Tricarbonyl, *Bull. Chem. Soc. Jpn.* **32**, 880–881 (1959).
57. M. D. Rausch, and G. N. Schrauzer, Cyclooctatetraeneiron Tricarbonyl and Cyclooctatetraenediiron Hexacarbonyl, *Chem. & Ind.* **1959**, 957–958.
58. E. W. Abel, M. A. Bennett, and G. Wilkinson, Cycloheptatriene Metal Complexes, *Proc. Chem. Soc.* **1958**, 152–153.
59. E. W. Abel, M. A. Bennett, R. Burton, and G. Wilkinson, The *cyclo*Heptatriene-Metal Complexes and Related Compounds, *J. Chem. Soc.* **1958**, 4559–4563.
60. M. A. Bennett, L. Pratt, and G. Wilkinson, Proton Resonance Spectra of Cycloheptatriene Complexes of Group VI Metals, *J. Chem. Soc.* **1961**, 2037–2044.
61. T. A. Manuel, and F. G. A. Stone, 1,5-Cyclooctadienetungstentetracarbonyl, *Chem. & Ind.* **1959**, 1349–1350.
62. M. A. Bennett, and G. Wilkinson, Some New Olefin Complexes of Mo(0) and Ru(II), *Chem. & Ind.* **1959**, 1516.
63. E. O. Fischer, and W. Fröhlich, Zur Komplexbildung des Cyclooctadiens-(1.5) mit Metallhexacarbonylen, *Chem. Ber.* **92**, 2995–2998 (1959).
64. B. Dickens, and W. N. Lipscomb, Molecular and Valence Structures of Complexes of Cyclooctatetraene with Iron Carbonyl, *J. Chem. Phys.* **37**, 2084–2093 (1962).
65. J. D. Dunitz, and P. Pauling, Struktur des Cycloheptatrien-molybdän-tricarbonyls, $C_7H_8Mo(CO)_3$, *Helv. Chim. Acta* **43**, 2188–2197 (1960).
66. T. A. Manuel, and F. G. A. Stone, Cyclooctatetraene Iron Tricarbonyl and Related Compounds, *J. Am. Chem. Soc.* **82**, 366–372 (1960).
67. E. O. Fischer, and H. Werner, *Metal π-Complexes*, (Elsevier, Amsterdam, 1966, Vol. I, p. 127).
68. F. A. Cotton, Bonding of Cyclooctatetraene to Metal Atoms: Simple Theoretical Considerations, *J. Chem. Soc.* **1960**, 400–406.
69. F. G. A. Stone, *Leaving No Stone Unturned* (American Chemical Society, Washington, DC, 1993, p. 36).
70. F. A. Cotton, Fluxional Organometallic Molecules, *Acc. Chem. Res.* **1**, 257–265 (1968).
71. H. Alper, Organic Syntheses with Iron Pentacarbonyl, in: *Organic Synthesis via Metal Carbonyls* (Eds. I. Wender and P. Pino, Wiley Interscience, New York, 1977, Vol. 2, pp. 545–593).
72. A. J. Pearson, *Metallo-organic Chemistry* (Wiley Interscience, New York, 1985, Chapters 7 and 8).
73. H. J. Dauben Jr., and L. R. Honnen, π-Tropenium-molybdenum-tricarbonyl Tetrafluoroborate, *J. Am. Chem. Soc.* **80**, 5570–5571 (1958).
74. J. D. Munro, and P. L. Pauson, Reactions of Tricarbonyltropyliumchromium Perchlorate with Anions: A Novel Rearrangement, *Proc. Chem. Soc.* **1959**, 267.
75. J. D. Munro, and P. L. Pauson, Cycloheptatriene- and Tropylium-Metal Complexes. Part I. The "Normal" Reaction of Tricarbonyltropyliumchromium Salts with Anions, *J. Chem. Soc.* **1961**, 3475–3486.
76. A. Salzer, and H. Werner, Elektrophile und nucleophile Additionsreaktionen an Siebenringliganden in Tricarbonylmolybdän-Komplexen: Synthese von $[C_7H_9Mo(CO)_3]BF_4$, $C_7H_9Mo(CO)_3Cl$ und $[C_7H_9P(C_6H_5)_3]BF_4$, *J. Organomet. Chem.* **87**, 101–108 (1975).
77. H. D. Murdoch, and E. Weiss, Eisencarbonylkomplexe des Äthylens und Hexatriens, *Helv. Chim. Acta* **46**, 1588–1594 (1963).
78. E. Koerner von Gustorf, M. C. Henry, and C. Di Pietro, Photochemische Umsetzung von $Fe(CO)_5$ mit Vinylchlorid, Styrol, Propylen und Vinyläthyläther, *Z. Naturforsch., Part B,* **21**, 42–45 (1966).

79. H. P. Kögler, and E. O. Fischer, Cyclopentadienyl-mangan-äthylen-dicarbonyl, *Z. Naturforsch., Part B,* **15**, 676 (1960).
80. E. O. Fischer, and M. Herberhold, Photochemische Substitutionsreaktionen an Cyclopentadienyl-mangan-tricarbonyl, in: *Essays in Coordination Chemistry, Exper. Suppl. IX* (Birkhäuser Verlag, Basel, 1964, p. 259–305).
81. E. O. Fischer, and P. Kuzel, Mesitylen-chrom(0)-äthylen-dicarbonyl, *Z. Naturforsch., Part B,* **16**, 475–476 (1961).
82. J. A. Banister, S. M. Howdle, and M. Poliakoff, Preparative-scale Organometallic Chemistry in Supercritical Fluids; Isolation of $[Cr(CO)_5(C_2H_4)]$ as a Stable Solid at Room Temperature, *J. Chem. Soc., Chem. Comm.* **1993**, 1814–1815.
83. I. W. Stolz, G. R. Dobson, and R. K. Sheline, Acetylene and Olefin Derivatives of Group VI Metal Carbonyls, *Inorg. Chem.* **2**, 1264–1267 (1963).
84. E. Koerner von Gustorf, and F.-W. Grevels, Photochemistry of Metal Carbonyls, Metallocenes, and Olefin Complexes, *Fortschr. Chem. Forsch.* **13**, 366–450 (1969).
85. F.-W. Grevels, J. Jacke, and S. Özkar, Photoreactions of Group 6 Metal Carbonyls with Ethene: Synthesis of *trans*-(η^2-Ethene)$_2$M(CO)$_4$ (M = Cr, Mo, W), *J. Am. Chem. Soc.* **109**, 7536–7537 (1987).
86. E. O. Fischer, H. P. Kögler, and P. Kuzel, Über neue π-Komplexe des Butadiens mit einfachen und substituierten Metallcarbonylresten, *Chem. Ber.* **93**, 3006–3013 (1960).
87. M. Herberhold, Ph.D. Thesis, Universität München, 1963.
88. G. N. Schrauzer, Bisacrylonitrile Nickel and Related Complexes from the Reaction of Nickel Tetracarbonyl with Compounds Containing Activated Double Bonds, *J. Am. Chem. Soc.* **81**, 5310–5312 (1959).
89. G. N. Schrauzer, Zur Kenntnis von Bis-acrylnitril-nickel(0), *Chem. Ber.* **94**, 642–650 (1961).
90. G. N. Schrauzer, Some Advances in the Organometallic Chemistry of Nickel, *Adv. Organomet. Chem.* **2**, 1–48 (1964).
91. G. N. Schrauzer, and H. Thyret, Zur Kenntnis von Bis-durochinon-nickel(0) and Cyclooctatetraene-durochinon-nickel(0), *Z. Naturforsch., Part B,* **16**, 353–356 (1961).
92. G. N. Schrauzer, and H. Thyret, Neuartige "Sandwich"-Verbindungen des Nickel(0). Zur Kenntnis von Durochinon-Nickel(0)-Komplexen mit cyclischen Dienen, *Z. Naturforsch., Part B,* **17**, 73–76 (1962).
93. M. D. Glick, and L. F. Dahl, Structure and Bonding in 1,5-Cyclooctadiene-duroquinone-nickel, *J. Organomet. Chem.* **3**, 200–221 (1965).
94. G. Wilke, Synthesen in der Cyclododekan-Reihe, *Angew. Chem.* **69**, 397–398 (1957).
95. G. Wilke, Cyclooligomerization of Butadiene and Transition Metal π-Complexes, *Angew. Chem. Int. Ed. Engl.* **2**, 105–115 (1963).
96. K. Fischer, K. Jonas, P. Misbach, R. Stabba, and G. Wilke, The "Nickel Effect", *Angew. Chem. Int. Ed. Engl.* **12**, 943–953 (1973).
97. H. W. B. Reed, The Catalytic Cyclic Polymerisation of Butadiene, *J. Chem. Soc.* **1954**, 1931–1941.
98. G. Wilke, Neues über cyclische Butadien-Oligomere, *Angew. Chem.* **72**, 581–582 (1960).
99. B. Bogdanovic, M. Kröner, and G. Wilke, Olefin-Komplexe des Nickel(0), *Liebigs Ann. Chem.* **699**, 1–23 (1966).
100. D. J. Brauer, and C. Krüger, The Three-dimensional Structure of *trans,trans,trans*-1,5,9-Cyclododecatrienenickel, *J. Organomet. Chem.* **44**, 397–402 (1972).
101. N. Rösch, and R. Hoffmann, Geometry of Transition Metal Complexes with Ethylene or Allyl Groups as the Only Ligands, *Inorg. Chem.* **13**, 2656–2666 (1974).
102. K. Jonas, P. Heimbach, and G. Wilke, 1,5,9-Cyclododecatriene Complexes of Nickel, *Angew. Chem. Int. Ed. Engl.* **7**, 949–950 (1968).
103. G. Wilke, Contributions to Organo-Nickel Chemistry, *Angew. Chem. Int. Ed. Engl.* **27**, 185–206 (1988).

104. P. S. Skell, J. J. Havel, D. L. Williams-Smith, and M. J. McGlinchey, Reactions of Nickel Atoms with Unsaturated Hydrocarbons, *J. Chem. Soc., Chem. Comm.* **1972**, 1098–1099.
105. P. W. Jolly, and G. Wilke, *The Organic Chemistry of Nickel*, Organonickel Complexes (Academic Press, New York, 1974, Vol. I, p. 258).
106. D. J. Brauer, and C. Krüger, The Stereochemistry of Transition Metal Cyclooctatetraenyl Complexes: Di-h^3,h$^{3'}$-cyclooctatetraenedinickel, a Sandwich Compound with Two Enveloped Nickel Atoms, *J. Organomet. Chem.* **122**, 265–273 (1976).
107. G. Wilke, M. Kröner, and B. Bogdanovic, Ein Zwischenprodukt der Synthese von Cyclododecatrien aus Butadien, *Angew. Chem.* **73**, 755–756 (1961).
108. B. Bogdanovic, P. Heimbach, M. Kröner, and G. Wilke, Zum Reaktionsablauf der Cyclotrimerisation von Butadien-(1.3), *Liebigs Ann. Chem.* **727**, 143–160 (1969).
109. G. Wilke, B. Bogdanovic, P. Hardt, P. Heimbach, W. Keim, M. Kröner, W. Oberkirch, K. Tanaka, E. Steinrücke, D. Walter, and H. Zimmermann, Allyl-Transition Metal Systems, *Angew. Chem. Int. Ed. Engl.* **5**, 151–164 (1966).
110. K. Fischer, K. Jonas, and G. Wilke, Tris(ethylene)nickel(0), *Angew. Chem. Int. Ed. Engl.* **12**, 565–566 (1973).
111. C. Krüger, and Y.-H. Tsay, The Molecular and Crystal Structure of Bis(ethylene) (tricyclohexylphosphine)nickel, *J. Organomet. Chem.* **34**, 387–395 (1972).
112. G. Wilke, and G. Herrmann, Ethylenebis(triphenylphosphine)nickel and Analogous Complexes, *Angew. Chem.* **74**, 693–694 (1962).
113. W. A. Herrmann, 100 Years of Metal Carbonyls: a Serentipitous Chemical Discovery of Major Scientific and Industrial Impact, *J. Organomet. Chem.* **383**, 21–44 (1990).
114. F. G. A. Stone, Synthetic Applications of d^{10} Metal Complexes, *J. Organomet. Chem.* **100**, 257–271 (1975).
115. F. G. A. Stone, "Ligand-Free" Platinum Compounds, *Acc. Chem. Res.* **14**, 318–325 (1981).
116. J. Müller, and P. Göser, Bis(1,5-cyclooctadiene)platinum(0), *Angew. Chem. Int. Ed. Engl.* **6**, 364–65 (1967).
117. M. Green, J. A. K. Howard, J. L. Spencer, and F. G. A. Stone, Synthesis of Ethylene, Cyclo-octa-1,5-diene, Bicyclo [2.2.1]heptene, and *trans*-Cyclo-octene Complexes of Palladium(0) and Platinum(0); Crystal and Molecular Structure of Tris(bicyclo2.2.1]heptene)platinum, *J. Chem. Soc., Dalton Trans.* **1977**, 271–277.
118. M. Green, J. A. K. Howard, M. Murray, J. L. Spencer, and F. G. A. Stone, Synthesis and Crystal and Molecular Structure of Tris-μ-(*t*-butylisocyanide)-tris-(*t*-butylisocyanide)-*triangulo*-triplatinum, *J. Chem. Soc., Dalton Trans.* **1977**, 1509–1514.
119. N. M. Boag, M. Green, D. M. Grove, J. A. K. Howard, J. L. Spencer, and F. G. A. Stone, The Synthesis and Crystal Structure of Bis(diphenylacetylene)platinum, and Studies on Related Compounds, *J. Chem. Soc., Dalton Trans.* **1980**, 2170–2181.
120. M. Green, A. Laguna, J. L. Spencer, and F. G. A. Stone, Bis(η-cyclo-octa-1,5-diene)-platinum and –palladium with Fluoroolefins, *J. Chem. Soc., Dalton Trans.* **1977**, 1010–1016.
121. H. Huber, G. A. Ozin, and W. J. Power, Synthesis and Characterization of Reactive Intermediates in the Palladium Atom-Ethylene System, $(C_2H_4)_nPd$ (where n = 1, 2, or 3), *Inorg. Chem.* **16**, 979–983 (1977)
122. R. M. Atkins, R. Mackenzie, P. L. Timms, and T. W. Turney, The Preparation of Palladium-Olefin Complexes from Palladium Vapour, *J. Chem. Soc., Chem. Comm.* **1975**, 764.
123. R. Mackenzie, and P. L. Timms, Reaction of Metal Atoms with Solutions; Preparation of Bis(cyclo-octa-1,5-diene)iron(0), *J. Chem. Soc., Chem. Comm.* **1974**, 650–651.
124. P. S. Skell, E. M. Van Dam, and M. P. Silvon, Reactions of Tungsten and Molybdenum Atoms with 1,3-Butadiene. Tris(butadiene)tungsten and —molybdenum, *J. Am. Chem. Soc.* **96**, 626–627 (1974).
125. R. E. Moriarty, R. D. Ernst, and R. Bau, Structure of Tris(methyl vinyl ketone) tungsten, *J. Chem. Soc., Chem. Comm.* **1972**, 1242–1243.

126. E. O. Fischer, J. Müller, and P. Kuzel, Eine neue Synthese von Dibenzolchrom, *Rev. Chim., Acad. Rep. Populaire Roumaine* 7, 827–834 (1962).
127. E. O. Fischer, and J. Müller, Metall-π-Komplexe des Rutheniums und Osmiums mit 6- und 8-gliedrigen cyclischen Oligoolefinen, *Chem. Ber.* **96**, 3217–3222 (1963).
128. J. Müller, and E. O. Fischer, Neue π-Komplexe des Eisen(0) und Ruthenium(0) mit cyclischen Olefinen, *J. Organomet. Chem.* **5**, 275–282 (1966) , and references cited therein.
129. J. Müller, and B. Mertschenk, Dicycloheptatrienvanadin(0), *J. Organomet. Chem.* **34**, C41–C42 (1972).
130. M. H. L. Green, P. A. Newman, and J. A. Bandy, A Comparison of the Rates of Intramolecular Hydrogen Migration in the Molecules $[Mo(\eta\text{-}C_7H_8)_2]^{n+}$ to give $[Mo(\eta\text{-}C_7H_7)(\eta\text{-}C_7H_9)]^{n+}$, n = 0 or 1: Crystal Structures of $[Mo(\eta\text{-}C_7H_8)_2]$, $[Mo(\eta\text{-}C_7H_7)(\eta\text{-}C_7H_9)]$, and $[Mo(\eta\text{-}C_7H_8)_2]BF_4$, *J. Chem. Soc., Dalton Trans.* **1989**, 331–343.
131. P. L. Timms, and T. W. Turney, Reactions of Transition-metal Vapours with Cycloheptatriene and Cyclooctatetraene, *J. Chem. Soc., Dalton Trans.* **1976**, 2021–2025.
132. F. G. N. Cloke, M. L. H. Green, and P. J. Lennon, Synthesis of d^2 η-Cycloheptatrienyl-η-cycloheptadienyl-zirconium and -hafnium Using the Metal Vapours, *J. Organomet. Chem.* **188**, C25–C26 (1980).
133. J. C. Green, M. L. H. Green, and N. M. Walker, Synthesis, Crystal Structure and Reactions of Zerovalent 16-Electron Bis(η-cycloheptatriene)zirconium, *J. Chem. Soc., Dalton Trans.* **1991**, 173–180.
134. K. Jonas, Dilithium-nickel-olefin Complexes. Novel Bimetallic Complexes From a Transition Metal and a Main Goup Metal, *Angew. Chem. Int. Ed. Engl.* **14**, 752–753 (1975).
135. K. Jonas, Reactive Organometallic Compounds Obtained from Metallocenes and Related Compounds and Their Synthetic Applications, *Angew. Chem. Int. Ed. Engl.* **24**, 295–311 (1985).
136. K. Jonas, R. Mynott, C. Krüger, J. C. Sekutowski, and Y.-H. Tsay, Bis(η-1,5-cyclooctadiene)cobalt Lithium, *Angew. Chem. Int. Ed. Engl.* **15**, 767–768 (1976).
137. K. Jonas, and L. Schieferstein, Simple Approach to Lithium- and Zinc-metalized η-Cyclopentadienyliron-olefin Complexes, *Angew. Chem. Int. Ed. Engl.* **18**, 549–550 (1979).
138. K. Jonas, L. Schieferstein, C. Krüger, and Y.-H. Tsay, Tetrakis(ethylene)irondilithium and Bis(μ^4-1,5-cyclooctadiene)irondilithium, *Angew. Chem. Int. Ed. Engl.* **18**, 550–551 (1979).
139. K. Jonas, and C. Krüger, Alkali Metal-Transition Metal π-Complexes, *Angew. Chem. Int. Ed. Engl.* **19**, 520–538 (1980).
140. K. Jonas, Alkali Metal-Transition Metal π-Complexes, *Adv. Organomet. Chem.* **19**, 97–122 (1981).
141. K. R. Pörschke, W. Kleimann, G. Wilke, K. H. Claus, and C. Krüger, Synthesis and Structure of $[Na(tmeda)_2]^+[HNi_2(C_2H_4)_4]^-$, *Angew. Chem. Int. Ed. Engl.* **22**, 991–992 (1983).
142. G. J. Leigh, A Celebration of Inorganic Lives: Interview of Joseph Chatt, *Coord. Chem. Rev.* **108**, 4–25 (1991).
143. C. Eaborn, and G. J. Leigh, Joseph Chatt, C.B.E., *Biogr. Mem. Fellows R. Soc.* **42**, 95–110 (1996).
144. G. Wilke, Contributions to Homogeneous Catalysis 1955–1980, *J. Organomet. Chem.* **200**, 349–364 (1980).
145. P. M. Maitlis, and A. F. Hill, An Interview with Peter Maitlis and A Selection of Higlights from Gordon Stone's Research by Anthony Hill, *Inorg. Chim. Acta* **358**, 1345–1357 (2005).

Chapter 8
Metal Carbenes and Carbynes: The Taming of "Non-existing" Molecules

Mir kommen die Wege, auf denen die Menschen zur Erkenntnis gelangen, fast ebenso bewunderungswürdig vor wie die Natur der Dinge selbst.[1]

Johannes Kepler, German Astronomer (1571–1630).

8.1 The Search for Divalent Carbon Compounds

The search for carbenes, which by definition are "divalent carbon compounds with two nonbonding electrons on one carbon atom" [1] began more than 150 years ago. At a time when the tetravalency of carbon was not an established fact, the French chemist Jean Baptiste André Dumas attempted to prepare methylene CH_2 by treatment of methanol with water-abstracting reagents such as P_2O_5 or concentrated sulfuric acid, assuming that methanol is a 1:1 adduct of CH_2 and H_2O [2, 3]. In the following decades, a series of similar studies was performed, aimed at obtaining CH_2 inter alia from CH_3Cl or CH_2I_2 by elimination of HCl or I_2, respectively [4]. Systematic investigations on the generation of methylene and related carbenes were initiated by Hermann Staudinger (Nobel laureate for chemistry, 1953) at the beginning of the twentieth century using, in particular, diazo compounds and ketenes as starting materials [5]. This work was continued by a number of eminent chemists such as Hans Meerwein, Jack Hine, Philip Skell, Wolfgang Kirmse, and William von E. Doering and is documented in a standard handbook [6] as well as in several monographs [7–10]. However, despite the explosive growth of knowledge about carbenes, which were in general described as "short-lived and highly reactive intermediates" [1], attempts to isolate compounds of the general composition CR_2 or CRR' (where R and R' are monovalent substituents), undertaken before 1990, remained unsuccessful.

The door to isolable carbenes was partially opened by Hans-Werner Wanzlick from the *Technische Universität* in Berlin. In 1970, he postulated that the

[1] In English: "*I think the ways by which people gain knowledge, are almost as wonderful as the nature of the things themselves.*"

H. Werner, *Landmarks in Organo-Transition Metal Chemistry*,
Profiles in Inorganic Chemistry, DOI 10.1007/978-0-387-09848-7_8,

imidazolium salt **1** was deprotonated by *tert*-butoxide to give the cyclic diaminocarbene **2**, which could not be isolated but trapped with oxygen and phenyl isothiocyanate to afford the ketone **3** and the zwitterion **4**, respectively [11]. Prior to this report, Wanzlick had already observed that the diphenyl-substituted imidazolium salt **5** reacted with *tert*-butoxide to generate the carbene **6**, which in solution dimerized to the substituted tetraaminoethene, and in the presence of mercury acetate gave the dicationic bis(carbene) complex **7** (see Scheme 8.1) [12]. Parallel to this work, Karl Öfele from the *Technische*

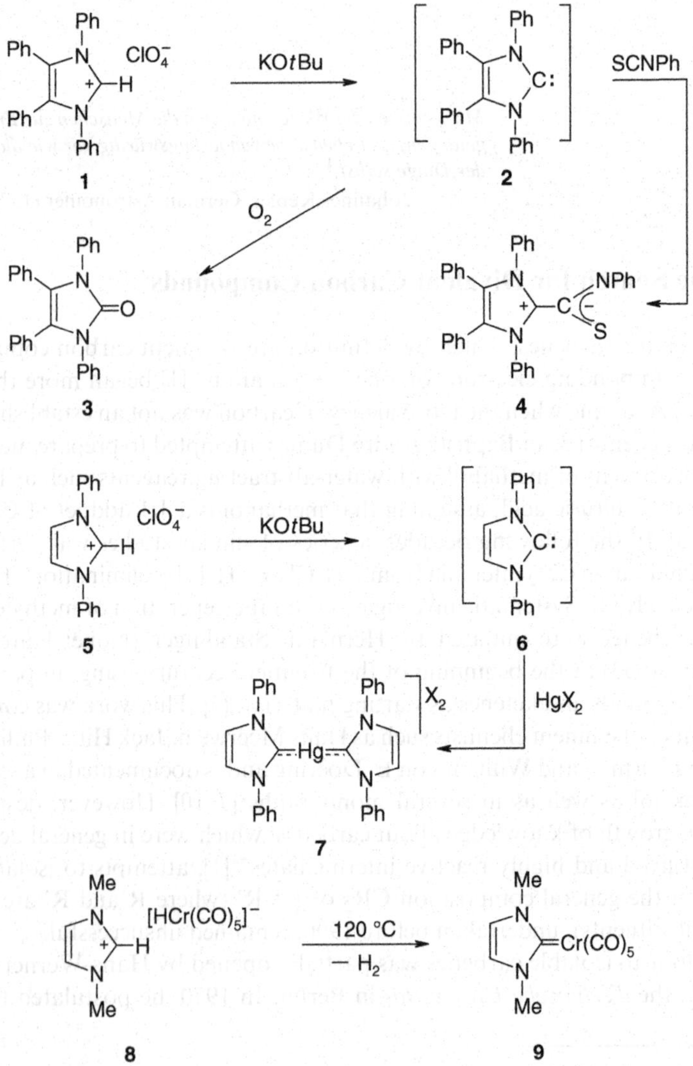

Scheme 8.1 Generation of Wanzlick's diaminocarbenes **2** and **6** and preparation of the carbene complexes **7** and **9** from the imidazolium salts **5** and **8**

Universität in München (for his studies on arene chromium tricarbonyls see Chap. 5) prepared the diaminocarbene chromium pentacarbonyl **9** from the imidazolium salt **8** by heating the starting material at 120°C [13]. Although Wanzlick's and Öfele's results received considerable attention, the opportunity to isolate stable carbenes was missed.

8.2 From Wanzlick's and Öfele's Work to Arduengo's Carbenes

The first paper on the synthesis of a thermally stable carbene, which could be kept in a bottle, appeared in 1991. It was entitled "*A Stable Crystalline Carbene*" and the main author was Anthony Arduengo (Fig. 8.1) from the Experimental Station of DuPont at Wilmington in the US [14]. He had set out to find a simple preparative route to industrially important imidazole-2-thiones and for this purpose treated imidazolium salts with strong bases and then added sulfur to form the required C═S bond [15, 16]. The ease with which these reactions succeeded, even in dry air, led Arduengo to believe that the respective carbenes, formed by deprotonation of the imidazolium ions, were less unstable than expected. Since the dimerization, as observed in the case of **6**, was thought to

Fig. 8.1 Anthony J. Arduengo, III (born in 1952) studied chemistry at the Georgia Institute of Technology where he received his B.S. in 1974 and his Ph.D. in 1976. After a year with DuPont, he moved to the University of Illinois–Urbana accepting a faculty position in the Department of Organic Chemistry. In 1984 he returned to DuPont, working as a member of the research staff and later advancing to Research Leader, Research Fellow, and eventually Research Supervisor. Since 1999 he holds the Saxon Chair in Organic Chemistry at the University of Alabama in Tuscaloosa as well as an adjunct professorship at the *Technische Universität* in Braunschweig, Germany. Arduengo received the gold medal for *Excellence in Main Group Chemistry* from the International Committee of Main Group Chemistry in 1996 and a Senior Alexander-von-Humboldt Award that same year (photo by courtesy from A. J. A.)

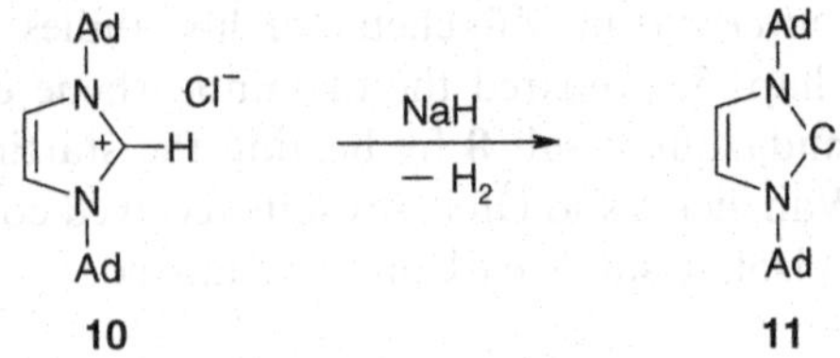

Ad N Cl⁻ +H N Ad **10** NaH − H₂ Ad N C: N Ad **11**

Scheme 8.2 Preparation of the first isolable diaminocarbene **11** from the imidazolium salt **10** as the precursor (Ad = adamantyl)

be retarded if more sterically demanding substituents than phenyl were bonded to the nitrogen atoms of the five-membered ring, salts of the respective imidazolium ion with two adamantyl groups were prepared. The deprotonation of the chloride **10** with NaH and catalytic amounts of dimethyl sulfoxide in THF proceeded smoothly and gave the carbene **11** as white crystals in 96% yield (Scheme 8.2) [14]. While the unusual melting point of 240–241°C (!) left some doubts, whether the product was what had been expected, an X-ray diffraction study proved that it was indeed the carbene.

Ardungo's work created an *incredible renaissance* [17] in carbene chemistry and released an avalanche of papers, including reviews and monographs, in an unusually short period of time. At present it is hard to have a complete overview of the field and to know how many stable, well characterized carbenes exist. Apart from *N*-heterocyclic carbenes such as **11**, related cyclic five-membered carbenes with one or three nitrogen atoms, with one nitrogen and one sulfur atom, with one nitrogen and two phosphorus atoms, with two nitrogen atoms in saturated five- and six-membered rings, and with methyl instead of aryl substituents at the carbon and nitrogen atoms have been prepared [15, 16, 18–22]. Even acyclic carbenes such as $C(N\textit{i}Pr_2)_2$ or $C(N\textit{i}Pr_2)(OMes)$ have been obtained and structurally characterized. The work in the late 1990s was highlighted by Arduengo's report about the isolation of the cyclic carbene **2**, being considered as "the realization of Wanzlick's dream" [23]. The tetraphenylimidazol-2-ylidene **2** is a white crystalline solid that is stable at room temperature in the absence of moisture and oxygen and melts at 199–202°C with decomposition.

8.3 The Breakthrough: Fischer's Metal Carbenes

Before we discuss the plethora of transition metal complexes containing *N*-heterocyclic carbenes as ligands, most of which were prepared in the 1990s, we have to go back to the late 1950s and the pioneering work of Ernst Otto Fischer in Munich. In the course of the *gold rush* [24] in the area of sandwich-type compounds, Fischer aimed at finding routes for the ring-metalation of cyclopentadienyl- and arene–transition metal complexes, and in this context asked his Ph.D. student Anton Wirzmüller around 1955 to study the reaction of $(C_5H_5)Mn(CO)_3$ with phenyl lithium, hoping that the lithiated derivative $(C_5H_4Li)Mn(CO)_3$ would be generated. However, instead of the expected compound, a labile red intermediate was formed, which could not be isolated, and upon addition of HCl decomposed. Although it was already known in those days that treatment of metal carbonyls with lithium alkyls or aryls gave metal acylates, Wirzmüller's attempts to characterize the lithium-containing product failed [24].

But Fischer did not forget what had been observed and, as Joseph A. Connor (see Chap. 5) expressed it, "he had a nose". Thus a few years after Wirzmüller finished his Ph.D. thesis, Fischer proposed to his graduate student Alfred Maasböl to do a "walk through a wild forest, not being sure where we would end" [24]. Though he was not convinced that a transition metal compound with a coordinated hydroxycarbene such as $C(OH)C_6H_5$ would be stable, he assumed that it might be possible to isolate a derivative with an alkoxy or aryloxy group instead of OH. As we know today, this assumption was correct. Maasböl reacted inter alia tungsten hexacarbonyl **12** with methyl and phenyl lithium and, after protonation of the in situ formed tungsten acylates **13a,b**, treated the corresponding hydroxycarbenes **14a,b** with diazomethane to afford the desired methoxy(methyl)- and methoxy(phenyl)carbene complexes **15a,b** (Scheme 8.3) [25]. A few years later, Rudolf Aumann (another Ph.D. student of Fischer, who in the 1970s became Professor of Organic Chemistry at the University of Münster, Germany) showed that the acylates **13a,b** can be directly converted to the tungsten carbenes **15a,b** via methylation with Meerwein's oxonium salt $[Me_3O]BF_4$ [26]. While the ^{1}H NMR spectra of **15a,b** could not unambiguously confirm that a C(OMe)R unit was coordinated to the metal, the X-ray crystal structure analysis of **15b**, carried out by Owen Mills at Manchester [27], vindicated Fischer's proposal. Based on the relatively short W—C and C—O distances, the data also revealed that the electron deficiency of the carbene carbon atom is at least in part compensated by a donating p(π)—p(π) bond of the methoxy oxygen and by π-back bonding from the metal to the empty p-orbital of the carbene carbon atom.

With the 1964 paper by Fischer and Maasböl [25], published 27 years earlier than the seminal report by Arduengo, the era of carbene–transition metal chemistry began. Between the middle of 1964 and the end of 1973 (when Fischer presented his Nobel Lecture) more than 50 papers were published from Fischer's

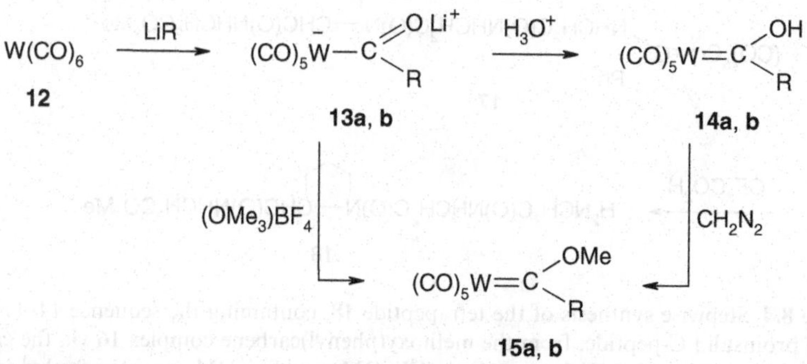

$W(CO)_6$ (**12**) $\xrightarrow{LiR}$ $(CO)_5\overset{-}{W}{-}C(=O)R\ Li^+$ (**13a, b**) $\xrightarrow{H_3O^+}$ $(CO)_5W{=}C(OH)R$ (**14a, b**)

13a, b $\xrightarrow{(OMe_3)BF_4}$ $(CO)_5W{=}C(OMe)R$ (**15a, b**) $\xleftarrow{CH_2N_2}$ **14a, b**

Scheme 8.3 Preparation of the first Fischer-type transition metal carbenes **15a,b** via the metal acylates **13a,b** and the labile metal hydroxycarbenes **14a,b** as intermediates (**a**: R = Me; **b**: R = Ph)

laboratory, reporting the preparation, spectroscopic data, structure and reactivity of carbene metal complexes [28–31]. The route by which a coordinated CO was converted to an acylate unit, followed by alkylation to generate the carbene ligand was applied to chromium, molybdenum, manganese, technetium, rhenium, iron, and nickel carbonyls. Instead of Meerwein's reagents, acyl halides, trimethylsilyl halides, $MeOSO_2F$, $n Bu_2BCl$, or $(C_5H_5)_2TiCl_2$ were used as electrophiles [28–32]. Nucleophilic attack at a carbonyl ligand of $Cr(CO)_6$ also occurred with $LiNEt_2$ or LiOEt to give the alkoxy(amino)- and bis(alkoxy) carbene complexes $Cr[C(OR)(NEt_2)](CO)_5$ and $Cr[C(OEt)_2](CO)_5$, after alkylation with $[R_3O]BF_4$ (R = Me, Et). Most of the Fischer carbenes prepared by this strategy are quite stable, sublimable, and readily soluble in common organic solvents. In compounds of the general composition $M[C(OMe)R](L)_n$, the methoxy group can be replaced by an NHR', NR'_2, SR' or SeR' unit according to an addition/elimination mechanism [33, 34] to give, respectively, the corresponding carbene complexes $M[C(NHR')R](L)_n$, $M[C(NR'_2)R](L)_n$, $M[C(SR')R](L)_n$, and $M[C(SeR')R](L)_n$. However, as Fischer emphasized in an account, the "synthesis of the last two series of complexes requires some experimental skill" [31].

The replacement of the methoxy by an amino group in the complexes $M[C(OMe)R](L)_n$ had also been extended to amino acid esters and esters of small peptides, indicating that the $M[C(R)](L)_n$ fragment can be used as a protecting group for the NH_2 moiety. The protecting group can be removed under mild conditions, e. g., with acetic acid at room temperature, leading to aldehydes and metal carbonyls as by-products. With the chromium carbene **16** as the starting material, not only common di- and tripeptides but also the tetrapeptide **18**, containing the sequence 14–17 of the human proinsulin C-peptide, was prepared in this way via complex **17** as an intermediate (see Scheme 8.4) [35, 36]. Compound **16**

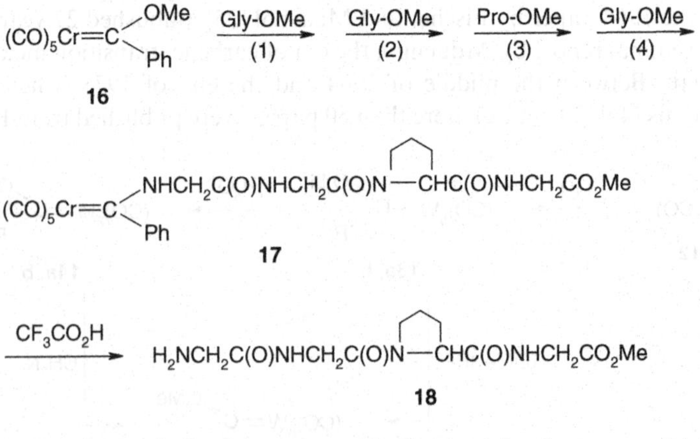

$(CO)_5Cr{=}C(OMe)Ph$ (**16**) $\xrightarrow[(1)]{\text{Gly-OMe}}$ $\xrightarrow[(2)]{\text{Gly-OMe}}$ $\xrightarrow[(3)]{\text{Pro-OMe}}$ $\xrightarrow[(4)]{\text{Gly-OMe}}$

$(CO)_5Cr{=}C(Ph)NHCH_2C(O)NHCH_2C(O)N{-}CHC(O)NHCH_2CO_2Me$ (**17**)

$\xrightarrow{CF_3CO_2H}$ $H_2NCH_2C(O)NHCH_2C(O)N{-}CHC(O)NHCH_2CO_2Me$ (**18**)

Scheme 8.4 Stepwise synthesis of the tetrapeptide **18**, containing the sequence 14–17 of the human proinsulin C-peptide, from the methoxy(phenyl)carbene complex **16** via the carbene chromium derivative **17** as the intermediate (Gly-OMe and Pro-OMe were the methyl esters of glycine and proline). The C—N coupling reactions in steps (1)–(4) were carried out using the DCCD/HOSU method, explained in [35]

as well as the molybdenum and tungsten analogues have also been used for the transfer of the C(OMe)Ph unit to olefins, leading to the formation of functionalized cyclopropanes. In elegant studies, Fischer's group showed that these transfer reactions do not proceed via a free methoxy(phenyl)carbene but that the metal atom is involved in the C—C coupling step [37, 38]. Subsequent studies by Charles Casey at the University of Wisconsin at Madison, using a different starting material, were in agreement with this result [39]. That complexes containing a C(OMe)R ligand can be used as valuable *carbene sources* was illustrated by the preparation of $Fe[C(OMe)Ph](CO)_4$ from $Fe(CO)_5$ and $(C_5H_5)Mo[C(OMe)Ph](CO)(NO)$ and by the reaction of **16** with Seyferth's reagent $PhHgCCl_3$ leading to the substituted styrene derivative $PhC(OMe){=}CCl_2$ in good yield [40, 41].

8.4 The Next Highlight: Fischer's Metal Carbynes

Attempts to replace the methoxy group of the carbene ligand in compound **16** by a halogen instead of an amino or thioalkyl unit led to the discovery of the first carbyne–transition metal complex [42]. The result was like finding a new *gold mine* and received coverage in most chemistry journals in 1974. To understand this, it is worth to remember that prior to the synthesis of **15a,b** the majority of organometallic chemists did not anticipate that carbene metal complexes existed, and the idea that a carbyne species CR (which was even unknown as a transient) could be linked to a transition metal was beyond people's imagination. Thus, when Fischer's Ph.D. student Gerhard Kreis carried out the reaction **16** with boron tribromide at low temperature, he expected to obtain the carbene complex $Cr[C(Br)C_6H_5](CO)_5$. From the reaction mixture he isolated a yellow, thermolabile, and air-sensitive solid, for which the C, H, Br, and Cr analyses were more or less in agreement with the expected composition. However, the oxygen content was 2.7% lower than calculated and this, in Fischer's words, "made us suspicious". In an essay, published in 1989 [24], he reasoned in his typical manner that "some people believe that it is very old-fashioned to do total analyses, but ... I wish that some people would do it more often. Spectroscopic and structural investigations alone are sometimes not sufficient and there were many examples where atoms were missed and formulae had to be reformulated. This can never occur if you have a total analysis." In the case of the product obtained from **16** and BBr_3, the total analysis revealed that the molecule contained only four oxygen atoms (certainly meaning four CO ligands instead of five) and, by taking into account that the IR spectrum displayed a band assigned to a Cr—Br stretching mode, the structure shown in Scheme 8.5 was proposed for compound **19.** However, due to the novelty of the structural type, Fischer first hesitated to publish the results and it was not before Gottfried Huttner (a former Ph.D. student of Fischer, who at that time did his *Habilitation* at the *Technische Universität* in München and later became Professor of Inorganic Chemistry first at the University of Konstanz and then at the University in Heidelberg) determined the structure of $Cr(CC_6H_5)(CO)_4I$ (prepared by Gerhard

Scheme 8.5 Preparation of the first Fischer-type transition metal carbyne **19** from the methoxy(phenyl)carbene complex **16** as the precursor

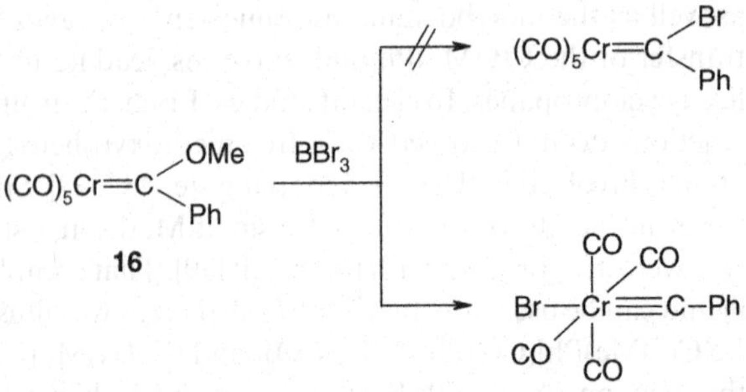

Kreis from **16** and BI_3) [43] that the seminal paper about the synthesis and characterization of the first carbyne metal complex was submitted to *Angewandte Chemie*. When Fischer went to Stockholm in December 1973, he had a ball-and-stick model of $Cr(CCH_3)(CO)_4I$ in his luggage, demonstrating to the audience of his Nobel Lecture that a transition metal compound with a metal–carbon triple bond is not a fantasy, but really exists.

A new field of research in organometallic chemistry was now open, and in the following decade more than 200 articles were published on "Fischer-type Carbyne Complexes". The results have been surveyed in several reviews [44–48] and in a monograph [49], which was presented to Ernst Otto Fischer on his 70th birthday. His short comment on the impressive development was: "We planted a sapling and it became a forest" [24].

8.5 Öfele's, Casey's and Chatt's Routes to Metal Carbenes

However, before we continue discussing some highlights of the chemistry of metal carbynes, we should return to the 1960s and early 1970s and summarize the progress made in the field of metal carbenes. In the introductory part of this chapter, Wanzlick's and Öfele's work on the preparation of the first complexes containing *N*-heterocyclic carbenes, published in 1968, has already been mentioned. By using metal carbonylate anions as precursors for novel organometallic compounds, Öfele reported also in 1968, that the dichlorocyclopropene derivative **20** reacted with $Na_2[Cr(CO)_5]$ in THF at low temperature to afford the pentacarbonyl chromium(0) complex **21** (Scheme 8.6) [50]. This compound, which is thermally significantly more stable than **16**, represented the first example of a metal carbene, in which the carbene carbon atom is not linked to a π-donating heteroatom such as oxygen, sulfur or nitrogen. Somewhat later, the synthesis of **21** was followed by Casey's report on the preparation of the tungsten carbene **23**, making use of the electrophilicity of the carbene carbon atom in Fischer's carbene compound **15b** as the precursor [51, 52]. The existence of intermediate **22**, initially supported by the spectroscopic characterization of

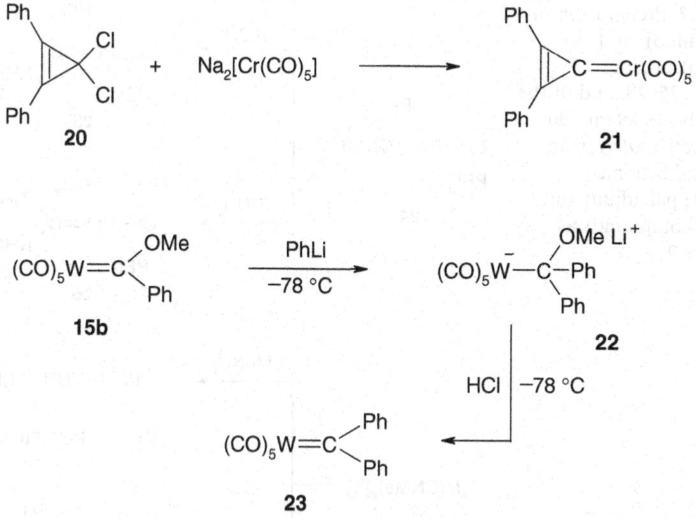

Scheme 8.6 Preparation of the metal carbene complexes **21** and **23** having no heteroatoms at the carbene carbon atom

ylide-type complexes such as M[C(OMe)(R)(L)]$(CO)_5$ (L = tertiary phosphines or sterically restricted tertiary amines) [28–31], was later confirmed by the isolation of this compound at low temperatures [53].

A completely different method of generating a carbene ligand was published by Joseph Chatt's group in 1969 [54]. They found that the reaction of isocyanide platinum(II) compounds such as **24** with primary alcohols or primary amines led to the formation of the carbene complexes **25** and **26** by addition of the alcohol O—H or the amine N—H bond to the C—N multiple bond of the isocyanide (Scheme 8.7). Via this route, Chatt as well as others prepared a whole series of neutral and cationic alkoxy(amino)- and bis(amino)carbene derivatives of almost all late transition metals including gold(I) and gold(III).[2] Even tetracarbene complexes such as **27** and **28** could be obtained in this way [55]. With respect to this methodology it is an irony of fate that as early as 1915 Tschugajeff and Skanawy–Grigorjewa studied the reaction of $[Pt(CNMe)_4]X_2$ with hydrazine and isolated a yellow solid, for which a peculiar structure was proposed [56]. It was only after Chatt's first paper appeared in 1969, that the correct structure of the product as the cyclic carbene complex **30** was established [57] and that of the palladium counterpart **29** confirmed by an X-ray diffraction analysis [58].

[2] See [32], pp. 8–11.

Scheme 8.7 Preparation of alkoxy(amino)- and bis(amino)carbene complexes **25–28**, and of the cyclic carbenes **29** and **30** (*Tschugajeff's salts*) from neutral and cationic isocyanide palladium and platinum compounds as precursors

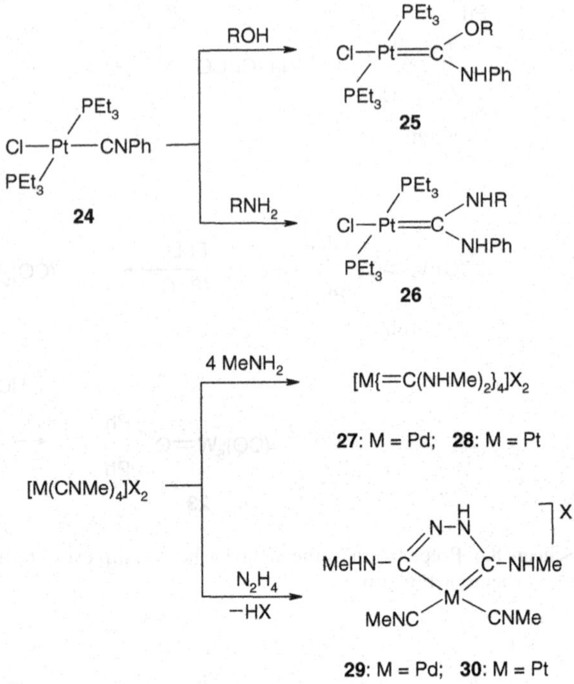

8.6 Lappert's Seminal Work on Bis(amino)carbene Complexes

A useful extension to the preparative procedures for metal carbenes, developed by Fischer and Chatt, was discovered by Michael Lappert in 1971 [59–64]. Following his interest in the organometallic chemistry of free radicals, Lappert observed that electron-rich olefins such as **31a,b** (**a**: R = alkyl; **b** = aryl) readily give rise to cations [**31a,b**]$^{+}$ or [**31a,b**]$^{2+}$, in which, according to MO calculations, the C—C double bond is significantly weakened. Although there was no evidence that the neutral compounds **31a,b** dissociated in solution generating a nucleophilic carbene, it was speculated that upon heating or in the presence of a suitable catalyst the C—C bond of **31a,b** was split and the bis(amino)carbene either trapped with an electrophile or fixed in the coordination sphere of a transition metal. While the tetrakis(amino) olefins proved to be inert in boiling xylene, the reaction of **31b** (R = C_6H_5) with the chloro-bridged platinum(II) compound **32**, known to undergo bridge-cleavage with amines or phosphines, gave the mononuclear carbene complex **33** (Scheme 8.8), which was characterized crystallographically [59].

In an extremely short period of time, Lappert's group prepared more than 100 bis(amino)carbene complexes of almost all transition metals of groups VI, VIII, IX, X, and XI [60–62]. With regard to the reaction conditions, the general observation was that the *N*-arylolefins **31b** were less reactive than the *N*-alkyl

Scheme 8.8 Preparation of the bis(amino)carbene platinum(II) complex **33** from the electron-rich olefin **31b** and the chloro-bridged platinum(II) compound **32** as the precursors

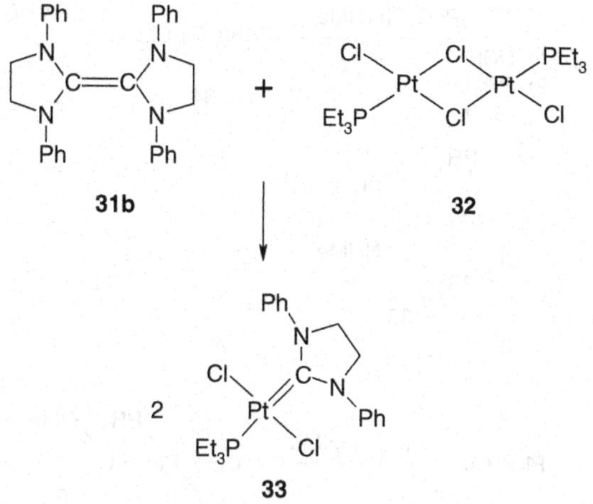

counterparts **31a**, and that the latter were the preferred starting materials for the formation of the respective carbene ligands. Very early in his research in this area, Lappert also indicated that bis(amino)carbenes and tertiary phosphines display similar behavior in their coordination ability, and he emphasized this similarity by noting that the two series of complexes of the general composition $M(\mathbf{31})_{n-m}(L)_m$ and $M(PR_3)_{n-m}(L)_m$ (where L represents a monodentate ligand, n the coordination number of M, and m the number of ligands L) were prepared via analogous routes. Moreover, he showed that in addition to **31a,b** open-chain olefins such as $C_2(NMe_2)_4$ and $C_2(SMe)_4$ can be used as precursors for the generation of the non-cyclic carbenes $C(NMe_2)_2$ and $C(SMe)_2$, which are extremely labile as free molecules but can be fixed and stabilized at a metal center. Under milder conditions than those required for C—C bond cleavage, **31a** ($R = CH_3$) as well as $C_2(NMe_2)_4$ may act as chelating ligands and form complexes such as *cis*-[W(CO)$_4$(*N,N'*-**31a**)] and *cis*-[Mo(CO)$_4${*N,N'*-$C_2(NMe_2)_4$}], respectively [62].

Lappert's group was also the first to use imidoyl chlorides (Vilsmeyer reagents) and electron-rich geminal dihalides as precursors for coordinated carbenes [65–68]. In the case of PhC(Cl)═NMe and the dimeric rhodium compound **34** as starting materials, they found that two different carbene rhodium(III) complexes **35** and **36** were obtained, depending on whether the hydrogen chloride was rigorously excluded or not (Scheme 8.9). In both processes, the decisive step probably consists of an oxidative addition of the imidoyl chloride to the rhodium(I) center, followed by attack of either HCl or a second molecule of the Vilsmeyer reagent on the five-coordinate intermediate. For the formation of the cationic carbene platinum(II) compound **38**, this stepwise procedure with **37** as intermediate was exemplified [69]. With dialkyl(chloromethylene) ammonium chlorides such as [Me_2N═CHCl]Cl and [Me_2N═CCl_2]Cl as precursors, two series of Cr(0), Mo(0), Fe(0), Mn(I), Rh(III), Ir(III), and Pt(IV) complexes with

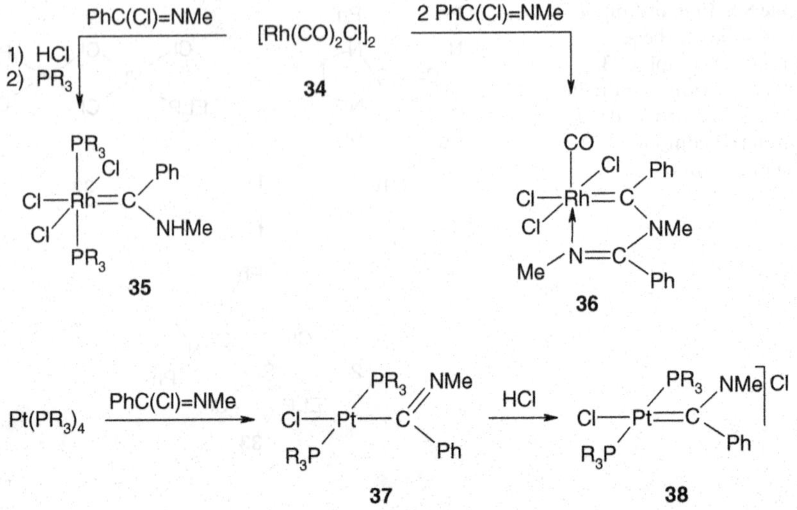

Scheme 8.9 Preparation of the octahedral aminocarbene rhodium(III) complexes **35** and **36** and the square-planar aminocarbene platinum(II) complex **38** from the imidoyl chloride PhC(Cl)=NMe as the precursor (for **35**: $PR_3 = PMe_2Ph$; for **37** and **38**: $PR_3 = PPh_3$)

either $CH(NMe_2)$ or $C(Cl)(NMe_2)$ as ligands were prepared and several of them structurally characterized by X-ray diffraction [60–64]. The preparation of $Cr[C(Cl)(NMe_2)](CO)_5$ from $Na_2[Cr(CO)_5]$ and $[Me_2N{=}CCl_2]Cl$ is reminiscent of Öfele's synthesis of **21** (see Scheme 8.6), which was published as early as in 1968 [50]. Referring to Lappert's work, it is also worth mentioning that in the same year that Casey described the synthesis of the tungsten carbene $W(CPh_2)(CO)_5$ (**23**) [51], Lappert reported the preparation of the related chromium and molybdenum pentacarbonyl compounds $M(SnR'_2)(CO)_5$ and $M(PbR'_2)(CO)_5$ (M = Cr, Mo) with two bulky $CH(SiMe_3)_2$ substituents at the tin(II) or lead(II) atom [70].

While Lappert was undoubtedly the first to describe transition metal complexes with secondary carbenes $CH(NR_2)$ (R = Me, Et, *i*Pr) as ligands [67, 68], in retrospect it is interesting to note that prior to the seminal paper by Dewar on the bonding of olefins to metal atoms (see Chap. 7), Chatt had proposed quite erroneously that in Zeise's salt ethene must be linked to platinum(II) unsymmetrically and coordinated to the metal by a π-dative bond to only one carbon atom. Therefore, he suggested the incorrect structure $[Pt({=}CHCH_3)Cl_3]^-$ for the anion of Zeise's salt and supported his proposal (1) by reference to the analogy of Pauling's model for the metal–carbonyl bond, and (2) by indicating that Zeise's salt hydrolyzed on heating to acetaldehyde [71]. Later he abandoned the carbene structure after some critical discussions with Christopher Ingold and after he studied the IR spectrum of Zeise's salt. In his autobiography [72], Gordon Stone referred to Chatt's idea about the binding of methylcarbene to a transition metal by pointing out that 35 years later "history has almost come full circle.... with the discovery that the μ-CHMe alkylidene fragment present in certain platinum-tungsten dimetal compounds readily rearranges to ligated ethylene".

8.7 A Big Step: Schrock's Metal Carbenes and Carbynes

An important stimulus for the chemistry of carbene and carbyne metal complexes was provided by Richard (Dick) Schrock's first communication about the preparation of a stable tantalum carbene in 1974 [73].[3] In those days, Schrock was a research associate at the Central Research and Development Department at DuPont in Wilmington in the group of George Parshall and had chosen to explore the organometallic chemistry of tantalum. Since similar to tungsten, its neighbor in group VI of the Periodic Table, this element is relatively stable in its highest possible oxidation state, the preparation of $Ta(CH_3)_5$ seemed to be feasible. Schrock was in particular encouraged in the pursuit of this goal after Wilkinson described in spring 1973 the synthesis of $W(CH_3)_6$ (see Chap. 9), which was reasonably stable and could be characterized spectroscopically.

Schrock's starting point for the preparation of $Ta(CH_3)_5$ (**40**) was a paper by Juvinall [74], who had reported the low-yield synthesis of $(CH_3)_3TaCl_2$ (**39**) from $TaCl_5$ and $Zn(CH_3)_2$. When Schrock repeated this work, he found that this compound could be prepared quantitatively in pentane on a large scale and stored for long periods at –40°C in the solid state. It reacted with two equivalents of methyl lithium to yield **40** (Scheme 8.10), which was isolated as a yellow, crystalline, low-melting solid [75]. Unfortunately, this compound was much less stable than $W(CH_3)_6$ and decomposed above 0°C to yield mainly methane, probably by a complex mechanism [76].

Based on his experience with **40**, Schrock decided to use neopentyl CH_2CMe_3 instead of CH_3 as alkyl substituent, taking into consideration that **40** contains 15 sterically unprotected C—H bonds, which could possibly interact with the metal center of a second $Ta(CH_3)_5$ molecule to initiate decomposition. In an attempt to prepare $Ta(CH_2CMe_3)_5$ via addition of two equivalents of $LiCH_2CMe_3$ to the starting material **41** (see Scheme 8.10), he obtained an orange, crystalline product that was thermally stable and melted at about 70°C [73]. Its mass spectrum

[3] At this point, the reader should be reminded that there are two systems of nomenclature for carbon–metal double and triple bonds. In contrast to Fischer, who named any unit of the general composition CR(R') a *carbene*, and analogously any CR unit a *carbyne*, independent whether or not R and R' contains a heteroatom such as oxygen, sulfur, nitrogen or chlorine [28–31, 43], Schrock later preferred the names *alkylidene* for CR(R') and *alkylidyne* for CR, provided that R and R' are either hydrogen, an alkyl, aryl or trimethylsilyl group. With regard to alkylidene he argued that a primary alkylidene CHR is derived from a primary alkyl ligand CH_2R and a secondary alkylidene CR_2 from a secondary alkyl ligand CHR_2 [84]. Nevertheless, up to now both terms carbenes/alkylidenes and carbynes/alkylidynes occur as synonyms in the literature. In this monograph, the terms carbene and carbyne have been preferentially used.

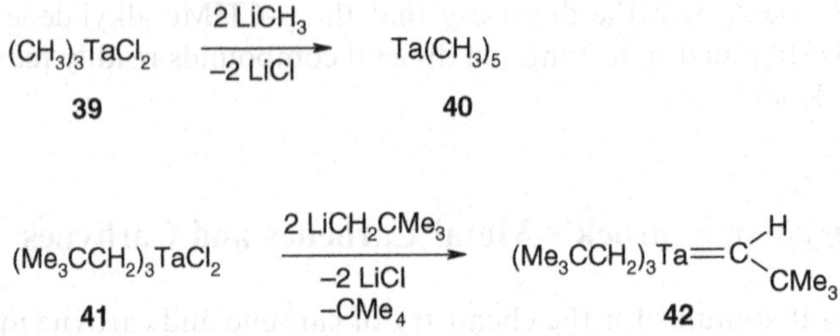

$$(CH_3)_3TaCl_2 \xrightarrow[-2\,LiCl]{2\,LiCH_3} Ta(CH_3)_5$$

39 **40**

$$(Me_3CCH_2)_3TaCl_2 \xrightarrow[-2\,LiCl,\ -CMe_4]{2\,LiCH_2CMe_3} (Me_3CCH_2)_3Ta{=}C(H)(CMe_3)$$

41 **42**

Scheme 8.10 Preparation of pentamethyltantalum(V) **40** and of the first Schrock-type transition metal carbene **42** from the tris(alkyl) tantalum(V) compounds **39** and **41** as the precursors

immediately revealed that it was not the expected penta(neopentyl)tantalum(V), but a compound the molecular weight of which was consistent with the formula $(Me_3CCH_2)_3Ta(CHCMe_3)$. It was sensitive to oxygen, water, and several functionalities, and it reacted with ketones and aldehydes in a Wittig-type fashion to yield polymeric $[(Me_3CCH_2)_3TaO]_x$ and the expected olefin [77]. This behavior appeared to be reminiscent to that of phosphorus ylides and revealed that the metal atom of the new compound **42** was highly electron-deficient. Referring to the unusual properties of **42**, Schrock stated in his Nobel Lecture that this first stable mononuclear alkyl metal carbene "appeared to behave as if the metal were in its highest possible oxidation state with the Ta = C bond being polarized so that the metal is relatively positive and the carbon relatively negative, opposite to what is found in Fischer-type carbene complexes" [78]. Although the exact mechanism of formation of **42** is still unknown, Schrock assumed that a C—H hydrogen in α-position of either intermediate $(Me_3CCH_2)_4TaCl$ or $Ta(CH_2CMe_3)_5$ [79] was activated by the metal and subsequently removed as a proton by a neopentyl group [78]. Support for this proposal was provided by the reaction of **42** with *n*-butyl lithium which gave the anionic carbyne species $[(Me_3CCH_2)_3Ta(CCMe_3)]Li$ by elimination of butane [80]. The attempted intramolecular α hydrogen abstraction from **42** by loss of another equivalent of neopentane to give $(Me_3CCH_2)_2Ta(CCMe_3)$ did not occur, probably because of coordinative unsaturation in the potential product molecule.

The next highlight in Schrock's work was the synthesis of the first isolable methylene transition metal complex **45** [81]. The initiating step was the preparation of the bis(cyclopentadienyl) tantalum(V) derivative **43** from the dichloro precursor **39** (already used for the preparation of **40**) and two equivalents of TlC_5H_5 (Scheme 8.11). With Tebbe's work on the catalytic activity of the 1:1 adduct of $[(C_5H_5)_2Ti(CH_2)]$ and $AlMe_2Cl$ in mind, Schrock treated **43** with excess trimethyl aluminium and isolated a product, which based on the spectroscopic data was formulated as $(C_5H_5)_2Ta(CH_2AlMe_3)(CH_3)$. Attempts to abstract the $AlMe_3$ unit from the CH_2AlMe_3 ligand led to rapid decomposition. To circumvent this problem, Schrock remembered the analogy between main group and transition metal

Scheme 8.11 Preparation of the first stable methylene transition metal complex **45** from **39** via the isolable tantalum(V) intermediates **43** and **44**

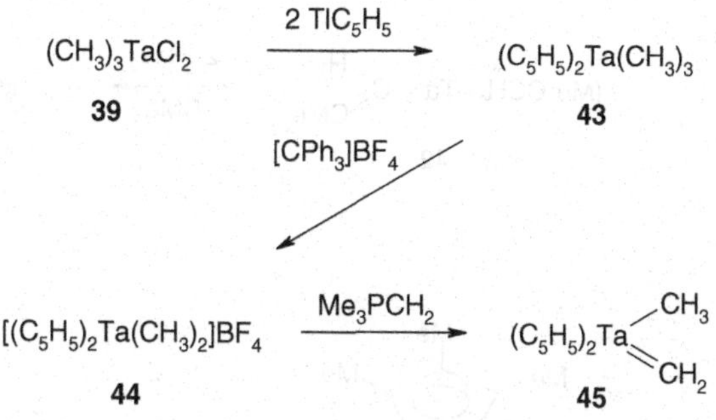

$(CH_3)_3TaCl_2$ (**39**) $\xrightarrow{2\ TlC_5H_5}$ $(C_5H_5)_2Ta(CH_3)_3$ (**43**) $\xrightarrow{[CPh_3]BF_4}$ $[(C_5H_5)_2Ta(CH_3)_2]BF_4$ (**44**) $\xrightarrow{Me_3PCH_2}$ $(C_5H_5)_2Ta(CH_3)(=CH_2)$ (**45**)

elements of group V, and thus in the next step prepared the *tantalonium* salt **44** from **43** and trityl tetrafluoroborate. The cation of **44** was readily deprotonated with Wittig's ylid Me_3PCH_2 leading to **45** in high yield [81, 82]. After recrystallization from toluene/pentane, the novel tantalum methylene compound was isolated as "shimmering greenish-white needles", which "decompose slowly in the solid state at 25°C" [83]. The related ethylidene complex $(C_5H_5)_2Ta(CHCH_3)(CH_3)$ was obtained from $(C_5H_5)_2Ta(PMe_3)(CH_3)$ and Et_3PCHCH_3 by displacement of the phosphine ligand, presumably via $(C_5H_5)_2Ta\{CHCH_3(PEt_3)\}(CH_3)$ as intermediate [84]. It could not be prepared directly from an ethyl tantalum precursor due to the preference of β- over α-hydrogen abstraction to give an ethene tantalum derivative. The X-ray structure analysis of **45** revealed a $Ta{=}CH_2$ bond length of 2.02 Å, which was 0.22 Å shorter than the $Ta{-}CH_3$ distance [82]. Later, almost the same difference in the Ta—C and Ta=C bond lengths was found in the phenylcarbene complex $(C_5H_5)_2Ta(CHPh)(CH_2Ph)$, prepared from $(PhCH_2)_3TaCl_2$ and two equivalents of TlC_5H_5 [85]. With regard to the isolation and characterization of **45**, it is worth mentioning that prior to Schrock's work Rowland Pettit, Malcolm Green, and others had attempted to prepare transition metal compounds with a $M{=}CH_2$ fragment, but could only observe those species as intermediates. Pettit generated the cation $[(C_5H_5)Fe(CH_2)(CO)_2]^+$ from $(C_5H_5)Fe(CO)_2(CH_2OCH_3)$ and HBF_4 and in the presence of cyclohexene trapped the methylene ligand to yield norcarane [86]. Moreover, at the same time that Schrock reported the synthesis of **45**, Wolfgang Herrmann (for biography see Chap. 9) attempted to prepare the manganese methylene complex $(C_5H_5)Mn(CH_2)(CO)_2$ from $(C_5H_5)Mn(CO)_2(THF)$ and diazomethane, but obtained the dinuclear derivative $[(C_5H_5)Mn(CO)_2]_2(\mu\text{-}CH_2)$. It was the first compound with a bridging CH_2 unit between two transition metal atoms [87, 88]. With Ph_2CN_2 and $PhC[C(O)Ph]N_2$ instead of CH_2N_2 as the precursors, Herrmann obtained the mononuclear complexes $(C_5H_5)Mn(CPh_2)(CO)_2$ and $(C_5H_5)Mn[C(Ph)C(O)Ph](CO)_2$, of which the latter was characterized crystallographically [89, 90].

In contrast to Fischer's compounds **15a,b** and **16** and their analogues with C(NHMe)R, $C(NMe_2)R$ or C(SMe)R as ligands, in which the carbene carbon atom is electrophilic [28–31], the carbene carbon atom is nucleophilic in

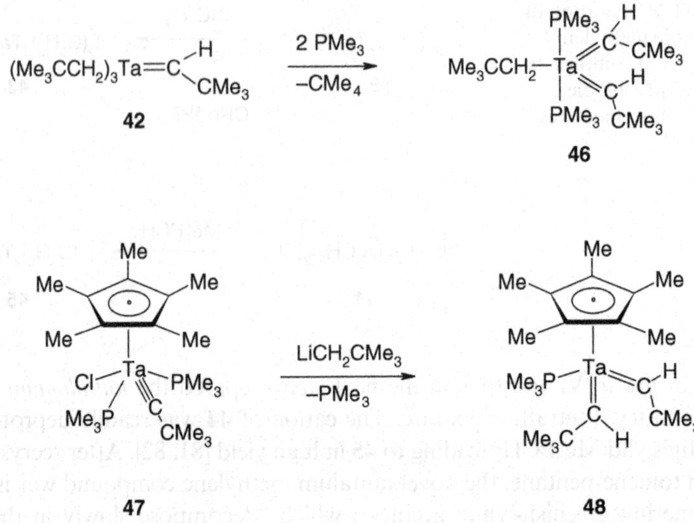

$(Me_3CCH_2)_3Ta{=}C(H)CMe_3$ **42** $\xrightarrow[-CMe_4]{2\ PMe_3}$ $Me_3CCH_2{-}Ta(PMe_3)_2({=}CHCMe_3)_2$ **46**

$(C_5Me_5)Ta(\equiv CCMe_3)(PMe_3)_2Cl$ **47** $\xrightarrow[-PMe_3]{LiCH_2CMe_3}$ $(C_5Me_5)Ta(PMe_3)({=}CHCMe_3)_2$ **48**

Scheme 8.12 Preparation of the first bis(carbene) tantalum complexes **46** and **48** from mono(carbene) or mono(carbyne) tantalum compounds as starting materials

Schrock's complexes **42** and **45** and related compounds of group V and VI elements with ligated methylene and secondary carbenes. This was first illustrated by the reactions of **45** with $AlMe_3$, Me_3SiBr, and CD_3I, which readily gave the isolable products $(C_5H_5)_2Ta(CH_2AlMe_3)(CH_3)$, $[(C_5H_5)_2Ta(CH_2SiMe_3)(CH_3)]Br$ and $(C_5H_5)_2Ta(CH_2CD_2)I$, respectively. The deuterated ethene complex is probably formed via the bis(alkyl) derivative $(C_5H_5)_2Ta(CH_2CD_3)(CH_3)I$, followed by elimination of CH_3D [84]. Addition of PMe_3 to **45** did not yield an ylide complex (as **16** does) but initiated a rearrangement, which finally produced $(C_5H_5)_2Ta(C_2H_4)(CH_3)$ and $(C_5H_5)_2Ta(PMe_3)(CH_3)$. Contrary to **45**, the tris(neopentyl) compound **42** reacted with PMe_3 by loss of neopentane and formation of **46**, which was the first bis(carbene) tantalum complex (Scheme 8.12) [91]. The niobium analogue $Nb(CHCMe_3)_2(PMe_3)_2Cl$ as well as a bis(neopentylidene) tantalum derivative $Ta(C_6H_2Me_3)(CHCMe_3)_2(PMe_3)_2Cl$ with a σ-bonded mesityl group were also prepared and the expected trigonal bipyramidal structure with the phosphine ligands in the axial positions confirmed by X-ray structure analysis [84].

The half-sandwich type bis(carbene) complex **48** became accessible by a different route and was obtained by treatment of the tantalum carbyne **47** with neopentyl lithium [91]. Already at that stage, the general observation was that the methylene and secondary carbene ligands in the high-valent niobium and tantalum complexes were generated by α-hydrogen abstraction from an alkyl ligand that did not contain a hydrogen in the β-position. Schrock concluded that apparently in the methyl, neopentyl, and benzyl compounds "an α-hydrogen atom is drawn toward a "semibridging" position from which it can be more readily removed as a 'proton' by another alkyl ligand" [84]. The proposed *semibridging* M···H—C interaction was later described by Maurice Brookhart and Malcolm Green as "agostic" and the small M—C—H angle as well as the relatively short M—H distance, mainly found in electron-deficient alkyl complexes of the elements of groups IV to VII, confirmed by a wealth of structural data [92, 93]. It was also substantiated that an *intramolecular* α-hydrogen abstraction appears to be favored for Ta, Mo, W, and Re, and that the neopentyl ligand CH_2CMe_3 is more readily converted to a secondary carbene than the CH_2SiMe_3 analogue. Another interesting result emerging from Schrock's work was that α-hydrogen abstraction could not only be induced by steric crowding but also by addition of a Lewis base, as was shown, for example, by the instantaneous reaction of $(Me_3CCH_2)_2TaCl_3$ (which is stable in pentane) with THF, which gave the octahedral tantalum(III) carbene complex $Ta(CHCMe_3)(THF)_2Cl_3$ in quantitative yield [94].

In 1978, when Schrock reported the synthesis of the first bis(carbene) complexes of niobium and tantalum [91], he also discovered a route to trisalkyl tungsten and molybdenum carbynes [95–97]. What could be viewed as an attempt to prepare either $W(CH_2CMe_3)_6$ or $(Me_3CCH_2)_4W(CHCMe_3)$ from WCl_6 and $LiCH_2CMe_3$ led actually to the formation of the four-coordinate tungsten complex **49** in low yield. Later, with $W(OMe)_3Cl_3$ and Me_3CCH_2MgCl as the starting materials, the yield could be improved to about 50% and the reaction carried out on a relatively large scale providing 20–30 g of the product (Scheme 8.13) [96, 97]. The analogous molybdenum carbyne **50** was obtained from MoO_2Cl_2 and six equivalents of $LiCH_2CMe_3$ in 35% yield [98]. The mechanism of formation of **49** and **50** is believed to involve the conversion of a CH_2CMe_3 to a $CHCMe_3$ ligand by an α-hydrogen abstraction at some stage, followed by conversion of the $CHCMe_3$ into a $CCMe_3$ ligand in a second α-hydrogen abstraction step [99, 100]. Both **49** and **50** are extremely air-sensitive solids, which are highly soluble in hydrocarbons and can be distilled in a good vacuum. Like the tantalum carbene **42**, the tungsten carbyne **49** melts at approximately 70°C. A close relative of **49** with $SiMe_3$ instead of CMe_3 groups at the alkyl and carbyne carbon atoms was prepared by Wilkinson from WCl_6 and six equivalents of $LiCH_2SiMe_3$ at low temperature [101].

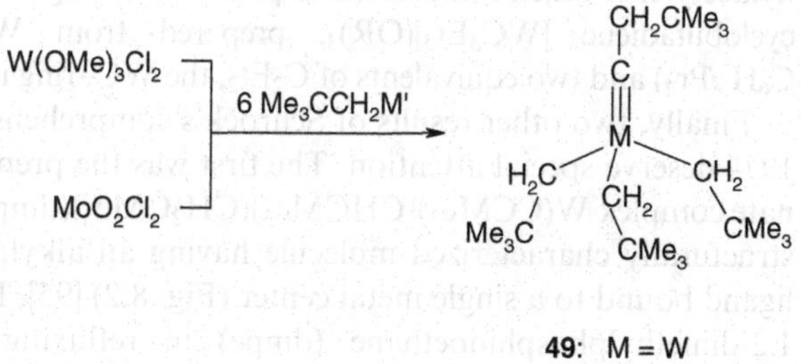

$W(OMe)_3Cl_3$ / MoO_2Cl_2 → 6 Me_3CCH_2M' → $(Me_3CCH_2)_3M{\equiv}CCMe_3$

49: M = W
50: M = Mo

Scheme 8.13 Preparation of the four-coordinate carbyne tungsten and molybdenum complexes **49** and **50** from tungsten(VI) and molybdenum(VI) compounds as starting materials (M' = Li or MgCl)

The chemistry of molybdenum and tungsten compounds with $M{\equiv}CCMe_3$ as the molecular building block began to develop rapidly after the reactions of **49** and **50** with HCl in the presence of 1,2-dimethoxyethane (dme) were carried out [96–98]. A series of $M(CCMe_3)X_3$ complexes, partly stabilized by dme, and with sterically demanding alkoxide ligands X such as $OCMe_3$, O-2,6-$C_6H_3i Pr_2$ or $OCMe(CF_3)_2$ were prepared and subsequently investigated as catalysts for alkyne metathesis. It was already known from Tom Katz's work in the 1970s [102], that Fischer-type carbenes such as **15a** and **16** (see Scheme 8.3) as well as Casey's diphenylcarbene **23** (see Scheme 8.6) catalyzed the polymerization of acetylenes, possibly via substituted metallacyclobutenes as intermediates. Schrock showed in the 1980s that the neopentylidyne molybdenum and tungsten complexes $M(CCMe_3)[OCMe(CF_3)_2]_3$ (M = Mo, W) were particularly effective in the metathesis of 3-heptyne and gave the symmetrical alkynes at a very high rate [96, 97]. He also found that the compounds $W(CCMe_3)(OR)_3$ underwent protonation by hydrogen halides, phenols or carboxylic acids to yield tungsten(IV) carbenes $W(CHCMe_3)(OCMe_3)_2X_2$ (X = Cl, Br, OPh, OC_6F_5, $MeCO_2$, $PhCO_2$) [96, 97] that were closely related to those reported by John Osborn with OCH_2CMe_3 as alkoxide ligands [103, 104].

The preparative routes to tungsten carbynes were further complemented by Schrock's discovery of "the remarkable reaction" [97] between $W_2(OCMe_3)_6$ (containing a W≡W triple bond) and internal alkynes C_2R_2, which gave the tris(alkoxide) compounds $W(CR)(OCMe_3)_3$ at room temperature [105, 106]. Even complexes that contained β-hydrogen atoms in the carbyne ligand could be prepared. Simple nitriles such as acetonitrile and benzonitrile reacted similarly, generating $W(CR)(OCMe_3)_3$ (R = Me, Ph) and polymeric $[W(N)(OCMe_3)_3]_x$ in equal ratio [105, 106]. With respect to the mechanism of the metathetical reaction of $W_2(OCMe_3)_6$ with alkynes, Schrock assumed that the crucial intermediate is a 1,3-dimetallacyclobutadiene $L_nM(\mu\text{-}CR)_2ML_n$, a type of species that was already known from Wilkinson's work with $L_n = (Me_3SiCH_2)_2$, $R = SiMe_3$ and M = Nb and Ta [107]. He also proposed that the initial precursor could be a cagelike $W_2(OR)_6(C_2R'_2)$ unit, analogous to those obtained by Malcolm Chisholm from $M_2(OCHMe_2)_6(py)_2$ (M = Mo, W) and acetylene [108]. While Schrock's attempts to isolate a 1,3-dimetallacyclobutadiene failed, he succeeded in preparing a stable complex of composition $(\eta^5\text{-}C_5H_5)W[C(Ph)C(CMe_3)C(Ph)]Cl_2$ from $(\eta^5\text{-}C_5H_5)W(CCMe_3)Cl_2$ and diphenylacetylene, which contains a non-planar WC_3 ring [109]. In the related tungstenacyclobutadiene $[WC_3Et_3](OR)_3$, prepared from $W(CCMe_3)(OR)_3$ (R = 2,6-$C_6H_3i Pr_2$) and two equivalents of C_2Et_2, the WC_3 ring is strictly planar [96, 97].

Finally, two other results of Schrock's comprehensive work carried out since 1974 deserve special attention. The first was the preparation of the five-coordinate complex $W(CCMe_3)(CHCMe_3)(CH_2CMe_3)$(dmpe) (**51**), which was the first structurally characterized molecule having an alkyl, a carbene and a carbyne ligand bound to a single metal center (Fig. 8.2) [95]. It was formed from **49** and 1,2-dimethylphosphinoethane (dmpe) in refluxing toluene by α-hydrogen abstraction. An X-ray diffraction study of **51** revealed W—C distances of 1.81,

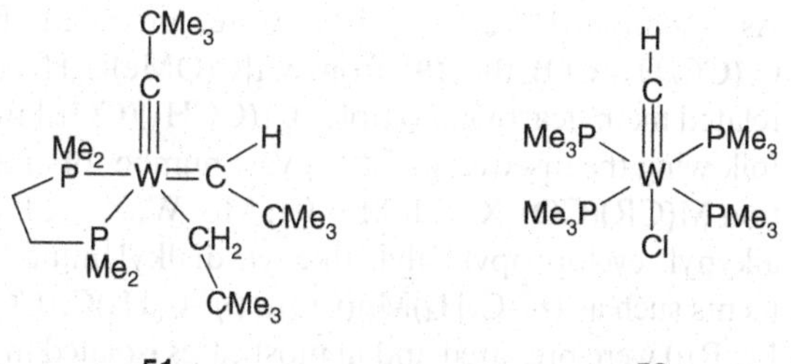

Fig. 8.2 Schematic representations of the molecular structures of **51**, the first transition metal complex having an alkyl, a carbene and a carbyne ligand, and of **52**, the first transition metal compound containing a terminal M≡CH unit

1.98, and 2.26 Å, illustrating quite nicely the presence of one triple, one double, and one single bond between tungsten and carbon [110]. The thermal reaction of **51** in neat dmpe gave, apart from the olefin $Me_3CCH{=}CHCMe_3$, the stable *trans*-configured carbyne(hydrido) compound $W(CCMe_3)(dmpe)_2H$ instead of the expected bis(carbyne) tungsten derivative $W(CCMe_3)_2(dmpe)_2$ [111].

The second remarkable result was the synthesis of the tungsten complex **52**, which was the first example of a stable transition metal compound with a terminal M≡CH unit. It was obtained upon treatment of $W(PMe_3)_4Cl_2$ with two equivalents of $AlMe_3$ in the presence of tmeda in moderate yield [112]. The proposed scheme for the formation of **52** includes the methyl tungsten(II) compound $W(CH_3)(PMe_3)_4Cl$ as an intermediate, which by α-hydride migration was thought to generate a $W(CH_2)(H)(Cl)$ species that in turn yielded the W(CH) complex **52** by elimination of H_2. The reaction of **52** with gaseous HCl at low temperature gave pentagonal-bipyramidal $W(CH)(H)(PMe_3)_3Cl_2$, whereas protonation of **52** with CF_3SO_3H afforded a grossly distorted methylene tungsten cation $[W(CH_2)(PMe_3)_4Cl]^+$; the latter reacted with $Na(BH_3CN)$ to give stable $W(CH)(H)(PMe_3)_3(Cl)(BH_3CN)$ [111]. To explain the results of the protonation reactions of **52** with HCl on one hand, and with CF_3SO_3H on the other, Schrock suggested that "there is not as much of a difference between a complex containing a grossly distorted alkylidene ligand and a true alkylidyne hydride complex as might first appear, and the factors that determine whether one or the other is formed are likely to be relatively subtle" [111]. In this respect it is interesting to note that in more recent work on the chemistry of molybdenum and tungsten complexes containing triamidoamine ligands, $[(R'NCH_2CH_2)_3N]^{3-}$ (R' = $SiMe_3$, C_6F_5), Schrock observed a quantitative conversion of $[(Me_3SiNCH_2CH_2)_3N]W(CH_3)$ to $[(Me_3SiNCH_2CH_2)_3N]W({\equiv}CH)$ by loss of H_2. The reaction was found to be irreversible under ambient conditions and occurred even in the solid state at 90°C under vacuum [113].

8.8 Fischer and His Followers

Despite Schrock's extensive work on tantalum, molybdenum, and tungsten carbyne complexes (e.g., **47**, **49**, **50**, **51**, and **52**) [96, 97, 99, 100], it must be recalled, however, that they were not the first compounds with a metal–carbon triple bond.

As mentioned earlier, they were preceded by Fischer's preparation of $Cr(CC_6H_5)(CO)_4Br$ (**19**) from $Cr[C(OMe)C_6H_5)](CO)_5$ (**16**) and BBr_3 and the related methylcarbyne complex $Cr(CCH_3)(CO)_4I$ in 1973 [28–31, 42]. In the decade following the discovery of **19**, a vast number of analogues of the general composition $M(CR)(CO)_4X$ with M = Cr, Mo, W; X = Cl, Br, I; and R = alkyl, aryl, vinyl, alkynyl, cyclopropyl, silyl, thienyl, dialkylamino and even organometallic fragments such as $(\eta^5\text{-}C_5H_4)Mn(CO)_3$, $(\eta^6\text{-}C_6H_5)Cr(CO)_3$ or $(\eta^5\text{-}C_5H_4)M(C_5H_5)$ (M = Fe, Ru) were prepared and in most cases isolated in high yield [114]. Apart from an alkoxy substituent (see Scheme 8.5), the leaving group in the carbene metal precursor could be alkyl- or dialkylamino, siloxy, acetoxy or mercapto, and instead of boron trihalides other Lewis acids such as aluminium or gallium trihalides could be used. The reactions of Fischer-type carbenes such as **15a,b** with BF_3 did not lead to the formation of $W(CR)(CO)_4F$ but instead gave *trans*-$W(CR)(CO)_4(BF_4)$ (**53**), from which the weakly bound BF_4^- ligand could easily be displaced by anionic (CN^-, SCN^-) and neutral ($AsPh_3$, CN*t*Bu, H_2O) nucleophiles to afford products of the type **54** and **55** (Scheme 8.14). Cationic chromium complexes similar to **55** with L = CO were obtained from $Cr[C(OEt)(NMe_2)](CO)_5$ and BCl_3 or BF_3, and the same route was used for the preparation of related nickel, iron, manganese, rhenium, and chromium compounds such as $[Ni(CN\textit{i}Pr_2)(CO)_2(PPh_3)]BCl_4$, $[Fe(CN\textit{i}Pr_2)(CO)_3(PPh_3)]BCl_4$, $[(\eta^5\text{-}C_5H_5)M(CR)(CO)_2]BX_4$ (M = Mn, Re; X = Cl, F), and $[(\eta^6\text{-}C_6H_6)Cr(CPh)(CO)_2]BCl_4$ [114]. Addition of anions such as CN^-, SCN^-, OCN^-, F^-, Cl^-, Br^-, I^-, OR^-, SPh^-, $SePh^-$, $TePh^-$ or $Co(CO)_4^-$ to the carbyne carbon atom of, for example, $[(\eta^5\text{-}C_5H_5)M(CR)(CO)_2]^+$ (M = Mn, Re) produced a large number of neutral carbene complexes $(\eta^5\text{-}C_5H_5)M[C(X)R](CO)_2]$, which were hardly accessible by a different route [115]. The reaction of $[Cr(CNMe_2)(CO)_5]BF_4$ with $LiNMe_2$ and $NHMe_2$ gave $Cr[C(NMe_2)_2](CO)_5$, which was the first transition metal carbene with an open-chain diaminocarbene ligand [116].

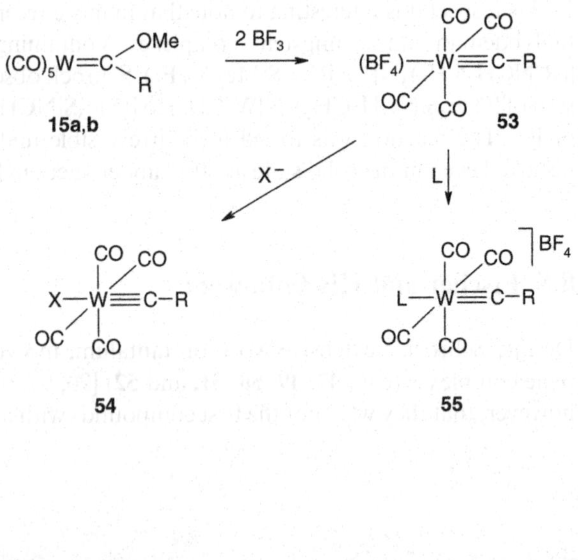

$(CO)_5W{=}C(OMe)R$ (**15a,b**) $\xrightarrow{2\,BF_3}$ $(BF_4)W(CO)_4{\equiv}C{-}R$ (**53**)

53 $\xrightarrow{X^-}$ $X{-}W(CO)_4{\equiv}C{-}R$ (**54**)

53 $\xrightarrow{L}$ $[L{-}W(CO)_4{\equiv}C{-}R]BF_4$ (**55**)

Scheme 8.14 Preparation of neutral and cationic tungsten carbynes **53–55** from the carbene complexes **15a,b** as the precursor (R = Me, Ph; for X^- and L see text)

An interesting observation was made by Fischer's group in 1976, when gaseous BCl_3 was passed into a solution of $Cr[C(OEt)(NEt_2)](CO)_5$ at –20°C [117]. Instead of the expected chromium carbyne $Cr(CNEt_2)(CO)_4Cl$ (which was formed in small amounts), the main product was the carbene complex $Cr[C(Cl)(NEt_2)](CO)_5$, an analogue of Lappert's $Cr[C(Cl)(NMe_2)](CO)_5$ mentioned above [62]. The carbene complex $Cr[C(Cl)(NEt_2)](CO)_5$ rearranged in solution at 30–50°C quantitatively by first-order kinetics to the chromium carbyne $Cr(CNEt_2)(CO)_4Cl$ [118]. The rate of this reaction proved to be independent of the presence of excess CO or of chloride ions, making an intramolecular migration of Cl from carbon to chromium most likely. Other chromium carbynes of the general composition $Cr(CNR_2)(CO)_4X$ with X = Cl, Br, I, SeR, TePh, $SnPh_3$ and $PbPh_3$ were prepared from the corresponding carbenes $Cr[C(X)(NR_2)](CO)_5$ via the cationic intermediates $[Cr(CNR_2)(CO)_5]^+$ in a similar way [114]. For the synthesis of structurally related thiocarbyne complexes $Cr(CSR)(CO)_4I$, Robert Angelici (who was, as mentioned in Chap. 2, a postdoc in Fischer's group in 1962/63 and subsequently became Professor of Inorganic Chemistry at Iowa State University in Ames, US) used the thiocarbonyl anion $[Cr(CS)(CO)_4I]^-$ as the precursor and treated it with $MeSO_3F$, CF_3CO_2Me or acetic acid anhydride as the electrophile to generate the thiocarbyne ligand [119]. The protonation of octahedral isocyanide molybdenum, tungsten and rhenium derivatives with HBF_4, H_2SO_4, HSO_3F, etc., carried out by Armando Pombeiro, and Raymond Richards and their coworkers, gave a series of neutral and cationic aminocarbyne complexes [120], among which *trans*-$[Re(CNH_2)(dppe)_2Cl]BF_4$ as the first compound with the unsubstituted CNH_2 ligand was the most prominent [121].

Following an earlier note from Fischer's laboratory [122], Andreas Mayr (in those days working at the State University of New York in Stony Brook) described an useful extension to the established methodology for the preparation of Fischer-type metal carbynes in the mid 1980s. He found that acyl complexes of the general composition $[M\{C(O)R\}(CO)_5]^-$, as obtained by Fischer from $M(CO)_6$ and LiR in the initial step of the metal carbene synthesis, could be converted directly into corresponding metal carbynes $M(CR)(CO)_4X$ (M = Cr, Mo, W; R = alkyl, aryl; X = Cl, Br, CF_3CO_2) by treating with carbon electrophiles such phosgene, diphosgene, oxalyl chloride as well as bromide, and trifluoroacetic anhydride [123, 124]. The oxide abstraction from the acyl moiety occurred between –70 and 0°C and gave the products in almost quantitative yield. The air-sensitive and in some cases thermolabile tetracarbonyls $M(CR)(CO)_4X$ reacted with excess of mono- or bidentate ligands L or L—L to afford the tricarbonyl $M(CR)(CO)_3(L)X$ as well as the dicarbonyl complexes $M(CR)(CO)_2(L)_2X$ and $M(CR)(CO)_2(L—L)X$, most of which were more stable than the unsubstituted precursors and could be handled in air at room temperature [123–125]. With ligands such as $P(OMe)_3$, dppe [123, 124] and isocyanides [126], non-CO containing compounds $M(CR)(L)_4X$ and $M(CR)(L—L)_2X$ were also obtained. The oxidation of $M(CR)(CO)_4Br$ (M = Mo, W; R = Me, Ph) with bromine in the presence of dme gave the high-valent metal carbynes M(CR)(dme)Br_3, thus closing the gap between Fischer-type and Schrock-type complexes with M≡CR moieties [47].

8.9 Using the Isolobal Analogy: Metal Complexes with Bridging Carbenes and Carbynes

The potential of metal carbynes to form di- and trinuclear transition metal complexes and even high-nuclearity metal clusters by addition reactions with coordinatively unsaturated metal species was discovered by Gordon Stone in the 1980s. Inspired by Roald Hoffmann's "isolobal analogy" [127], Stone used Fischer's molybdenum and tungsten carbynes **56a,b** and **57a,b** in particular to prepare a host of dimetal compounds of the general composition **58** with bridging CR' units (see Scheme 8.15) [128]. The possibility of changing the ligands in the fragments M'(L)$_n$, provided the fragments remain isolobal with CH_2, influenced the reactivity of the unsaturated three-membered ring and allowed the formation of complexes with bridging RC=CH_2, RC=S, RC=C=O or RCC(Ph)C(Ph) units. The reaction of **57a** (R = *p*-Tol) with dicobalt octacarbonyl proceeded similarly to that of diphenylacetylene with $Co_2(CO)_8$ and gave exclusively the trinuclear complex **59**. Related cluster compounds with three different metal atoms were obtained from some of the dimetal precursors **58** and revealed either a *trimetallatetrahedrane* or a *butterfly* structure [129, 130].

The most fascinating development in Stone's group, however, emerged from the isolation of the trinuclear PtW_2 compound **60**, which was prepared from **57a**

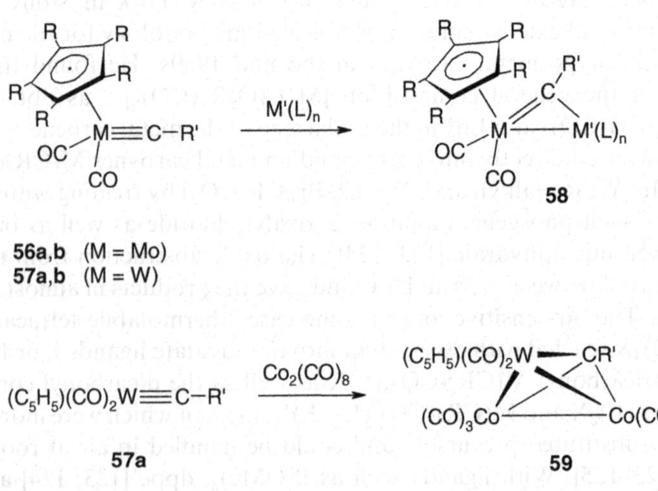

56a,b (M = Mo)
57a,b (M = W)

$(C_5H_5)(CO)_2W{\equiv}C{-}R'$ **57a** → ($Co_2(CO)_8$) **59**

Scheme 8.15 Stone's synthetic route to obtain dinuclear and trinuclear transition metal complexes such as **58** (with M'(L)$_n$ = $Pt(PR_3)_2$, $(C_5Me_5)Rh(CO)$, $(C_6Me_6)Cr(CO)_2$, $Fe(CO)_4$, etc.) and **59** using the isolobal analogy (**a**: R = H, **b**: R = Me; R' = *p*-Tol)

(R' = *p*-Tol) and $Pt(C_2H_4)_3$ in the ratio of 2:1 [131]. As the X-ray structure analysis confirmed, **60** had a nearly linear W—Pt—W spine, which was bridged by two carbyne units and semibridged by two CO ligands. The two three-membered rings were found to be orthogonal to each other, and the overall structure is strikingly similar to that of $Pt(C_2Ph_2)_2$ (see Chap. 7). Compound **60** as well as the PdW_2, NiW_2 and MMo_2 analogues (M = Ni, Pd, Pt) are unsaturated and could be used as precursors to obtain a series of oligonuclear clusters containing chains or rings of metal atoms in a stepwise manner. Scheme 8.16 shows two representative examples, **61** and **62**, which were prepared from **60** and $(C_5Me_5)Cu(THF)$, and from **60**, $Pt(C_8H_{12})_2$ and **57a**, respectively [131]. It is worth mentioning, that Stone's group did not only use Fischer-type carbynes but also Fischer-type carbenes such as **15a**, **16** and $(C_5H_5)Mn[C(OMe)Ph](CO)_2$ as precursors for the synthesis of several homo- and heterodinuclear compounds having bridging carbene ligands [128].

The first examples of carbene and carbyne complexes with an electron-rich (d^8) metal center were reported by Warren Roper (Fig. 8.3) in the early 1980s. His interest, however, in the chemistry of transition metal compounds with metal–carbon multiple bonds had arisen much earlier after listening to lectures by E. O. Fischer on metal carbenes [132]. Fascinated by Fischer's results, he set out to prepare ruthenium and osmium complexes with a terminal methylene ligand and, as the first step to reach this goal, reported in the mid 1970s the synthesis of the Ru(II) and Os(II) compounds $RuCl_2\{CH(NHR)\}(CO)(PPh_3)_2$ and $[OsCl\{CH(SMe)\}(CO)_2(PPh_3)_2]BF_4$ [133, 134]. The reaction of the thioformyl derivative $OsCl[CH({=}S)](CO)_2(PPh_3)_2$, which was an intermediate in the preparation of $[OsCl\{CH(SMe)\}(CO)_2(PPh_3)_2]BF_4$, with $NaBH_4$ gave the thioformaldehyde complex $Os(CH_2S)(CO)_2(PPh_3)_2$ [135], which, however, could not be converted to the desired methylene osmium(0) derivative

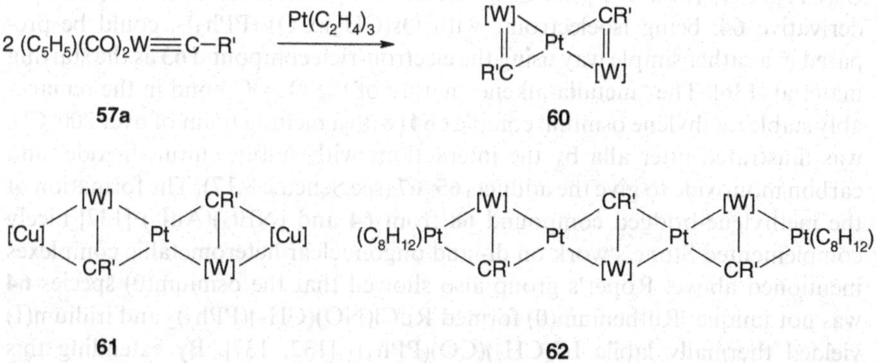

$$2\ (C_5H_5)(CO)_2W{\equiv}C{-}R' \xrightarrow{Pt(C_2H_4)_3} \mathbf{60}$$

57a **60**

61 **62**

Scheme 8.16 Preparation of the trinuclear PtW_2 complex **60**, and schematic representations of the molecular structures of the pentanuclear PtW_2Cu_2 and the heptanuclear Pt_4W_3 compounds **61** and **62**, both prepared from **60** as a precursor ([W] = $W(CO)_2(C_5H_5)$, [Cu]=$Cu(C_5Me_5)$, R' = *p*-Tol)

Fig. 8.3 Warren R. Roper (born in 1938) studied chemistry at the University of Canterbury in Christchurch, New Zealand, and completed his Ph.D. in 1963 under the supervision of Cuthbert J. Wilkins. He then undertook postdoctoral research with James P. Collman at the University of North Carolina at Chapel Hill in the US, and returned to New Zealand as Lecturer in Chemistry at the University of Auckland in 1966. In 1984, he was appointed Professor of Chemistry at the University of Auckland and became Research Professor of Chemistry at the same institution in 1999. His research interests are widespread with the emphasis on synthetic and structural inorganic and organometallic chemistry. Special topics have been low oxidation state platinum group metal complexes, oxidative addition reactions, migratory insertion reactions, metal–carbon multiple bonds, metallabenzenoids and more recently compounds with bonds between platinum group metals and the main group elements boron, silicon, and tin. His achievements were recognized by the Royal Society of Chemistry through the Organometallic Chemistry Award and the Centenary Lectureship. He was elected a Fellow of the Royal Society of New Zealand and of the Royal Society London, and was awarded the degree Doctor of Science (*honoris causa*) by the University of Canterbury in 1999 (photo by courtesy from W. R. R.)

$Os(CH_2)(CO)_2(PPh_3)_2$ by sulfur abstraction. In contrast, the chloro(nitrosyl) derivative **64**, being isoelectronic with $Os(CH_2)(CO)_2(PPh_3)_2$, could be prepared in a rather simple way using the electron-rich compound **63** as the starting material [136]. The "metalla-alkene" nature of the Os=C bond in the remarkably stable methylene osmium complex **64** (with a melting point of over 200°C!), was illustrated inter alia by the interactions with sulfur, sulfur dioxide, and carbon monoxide to give the adducts **65**–**67** (see Scheme 8.17). The formation of the methylene-bridged compound **68** from **64** and $(NEt_4)(AuI_2)$ [137] nicely complemented Stone's work on di- and oligonuclear heterometallic complexes mentioned above. Roper's group also showed that the osmium(0) species **64** was not unique: Ruthenium(0) formed $RuCl(NO)(CH_2)(PPh_3)_2$ and iridium(I) yielded thermally labile $IrI(CH_2)(CO)(PPh_3)_2$ [132, 137]. By extending this work, our group prepared in the 1990s a series of remarkably stable methylene ruthenium(0) and osmium(0) complexes of the general composition $MCl(NO)(CH_2)(PR_3)_2$ (M = Ru, Os; PR_3 = $P\textit{i}Pr_3$, $P\textit{i}Pr_2Ph$), using the four-coordinate compounds $MCl(NO)(PR_3)_2$ as the precursors [138, 139]. Although

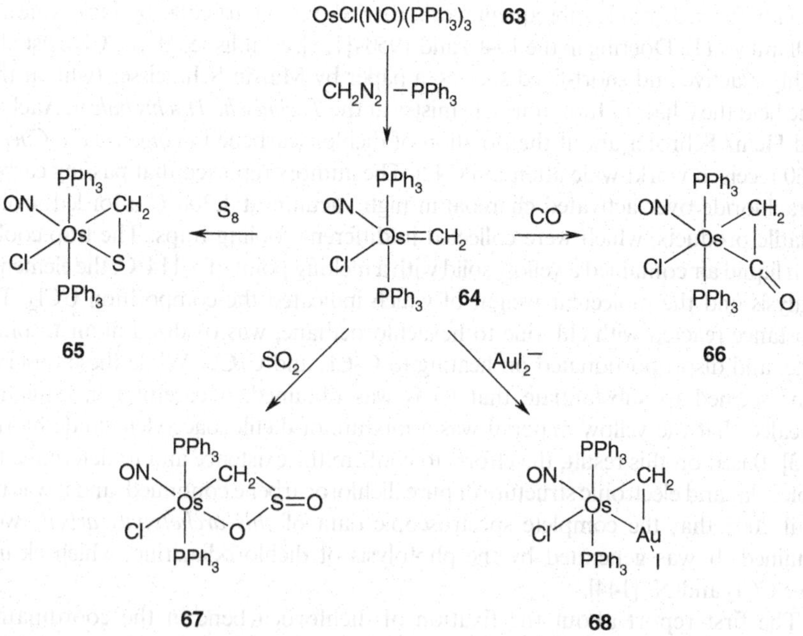

Scheme 8.17 Preparation of the methylene osmium(0) complex **64** and its reactions with S_8, CO, SO_2 and $[AuI_2]^-$ (as precursor of AuI) to form the 1:1 adducts **65**–**68**

the latter were coordinatively unsaturated and contained only 16 electrons in the valence shell, they could be isolated and their molecular structure determined by X-ray crystallography.

8.10 The Seemingly Existing CCl_2 and Its Generation at Transition Metal Centers

In addition to the series of $M(CH_2)$ derivatives, the elements of groups VIII and IX have also provided various well characterized dihalocarbene transition metal complexes [140]. Although this work was mainly carried out in the 1980s and 1990s, in the context of "taming" CCl_2 and its congeners, a view back into history is of some interest. In the introductory part of this chapter, it has already been mentioned that in the ninteenth century several chemists tried to obtain CH_2 and other carbenes from tetravalent carbon compounds as precursors. One of those chemists was Anton Geuther (a former assistant of Friedrich Wöhler at the University of Göttingen and, from 1863 until his death in 1889, the director of the *Chemisches Institut der Universität Jena*) who attempted to prepare CCl_2 from chloroform and potassium ethoxide by elimination of HCl [141]. Although the expected reaction took place (and in a variation of the procedure is still used in organic synthesis to generate CCl_2), Geuther as well as his contemporaries could not prove that dichlorocarbene was actually formed. While subsequent work, carried out in particular by Jack Hine and William von E. Doering in the 1940s and 1950s [7, 8], established that CCl_2 existed as highly reactive and short-lived species, a paper by Martin Schmeisser (who at that time held the Chair of Inorganic Chemistry at the *Technische Hochschule* in Aachen) and Heinz Schröter about the isolation of dichlorocarbene in *Angewandte Chemie* 1960 received world-wide attention [142]. The authors reported that passing carbon tetrachloride over activated charcoal in high vacuum at 1,300°C afforded several volatile products, which were collected in different cooling traps. The trap cooled with liquid air contained a yellow solid with a melting point of −114°C, the elemental analysis and the molecular weight of which indicated the composition CCl_2. The substance reacted with chlorine to hexachloroethane, was oxidized in air to phosgene, and disproportionated on heating to C_2Cl_6 and C_6Cl_6. While these observations seemed to substantiate that CCl_2 was obtained, a careful reinvestigation revealed that the yellow material was a mixture of dichloroacetylene and chlorine [143]. Based on this result, the efforts to confirm the existence and to determine the molecular and electronic structure of pure dichlorocarbene continued, and it was not until 2005 that the complete spectroscopic data of *this archetypal carbene* were obtained. It was generated by the photolysis of dichlorodiazirine, which cleanly gave CCl_2 and N_2 [144].

The first report about the fixation of dichlorocarbene in the coordination sphere of a transition metal appeared in 1977 [145]. In the course of investigating the chemistry of iron(II) tetraphenylporphyrin, Fe(TPP), Daniel Mansuy's group at the University of Paris found that when this compound was treated with carbon tetrachloride in benzene the iron(III) complex FeCl(TPP) was formed. However, when the same reaction was carried out in the presence of a reducing agent (sodium dithionite, H_2/Pd or iron powder), the product was the iron(II) carbene Fe(CCl_2)(TPP). Analogously, related complexes Fe[C(R)Cl](TPP) with R = Me, CH_2OH, CF_3, CN, CO_2Et, and SPh were obtained from Fe(TPP) and $RCCl_3$ under reducing conditions [146]. The generation of the carbene ligand proceeds possibly via a radical pathway involving the paramagnetic iron(III) species Fe[C(R)Cl_2](TPP) as an intermediate. The reaction of Fe(TPP) with carbon tetraiodide and iron powder or sodium dithionite gave the dinuclear complex (TPP)Fe=C=Fe(TPP), which was the first molecule with a "naked" carbon atom bridging two transition metal atoms in a linear fashion [147].

Warren Roper's first dichlorocarbene complex was the osmium(II) derivative **70** (Scheme 8.18), obtained from the chloro(hydrido) compound **69** and $Hg(CCl_3)_2$ [148]. The related octahedral complexes $RuCl_2(CCl_2)(CO)(PPh_3)_2$, $OsCl_2(CCl_2)(CS)(PPh_3)_2$ and $IrCl_3(CCl_2)(PPh_3)_2$ were prepared in a similar way

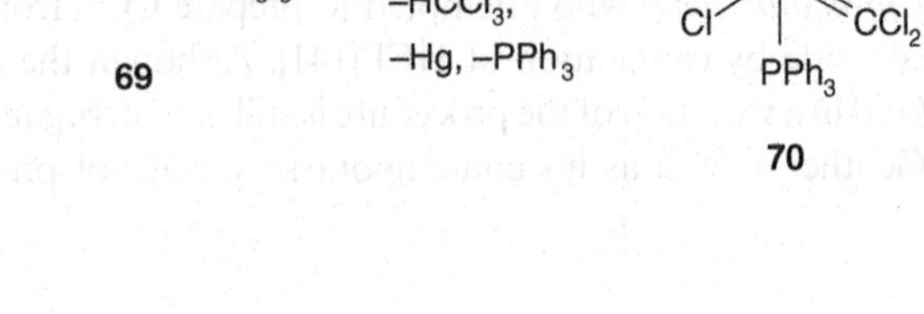

$$\underset{\mathbf{69}}{OsHCl(CO)(PPh_3)_3} \xrightarrow[-HCCl_3,\ -Hg,\ -PPh_3]{Hg(CCl_3)_2} \underset{\mathbf{70}}{OsCl_2(CO)(=CCl_2)(PPh_3)_2}$$

Scheme 8.18 Preparation of the dichlorocarbene osmium(II) complex **70** from the chloro(hydrido) compound **69** and $Hg(CCl_3)_2$ as the precursors

[140]. The reactions of **70** and of the ruthenium counterpart $RuCl_2(CCl_2)$-(CO)$(PPh_3)_2$ with $AgClO_4$ in acetonitrile afforded the cationic species $[MCl(CCl_2)(CO)(MeCN)(PPh_3)_2]^+$, the chloride abstraction occurring from the position *trans* to the carbene ligand [140].

For the difluorocarbene metal complexes prepared in Roper's group, both $Hg(CF_3)_2$ and $Cd(CF_3)_2$(dme) were used as CF_2 sources. The first reaction being studied was between the ruthenium(0) compound **71** and $Hg(CF_3)_2$, which gave the ruthenium(II) derivative **72** by oxidative addition (Scheme 8.19) [149]. In contrast to corresponding $M(CCl_3)$ species (M = Ru, Os, Ir), which in the analogous reactions with $Hg(CCl_3)_2$ could not be observed [140], the heteronuclear Ru—Hg compound **72** could be isolated and structurally characterized. The C—F distances of the CF_3 unit bound to ruthenium were found to be relatively long, thus explaining why one fluoride could be easily removed from intermediate **73** to yield the ruthenium difluorocarbene **74**. The reactions of **71** and of the osmium analogue $Os(CO)_2(PPh_3)_3$ with $Cd(CF_3)_2$(dme) did not lead to the expected $M(CF_3)(CdCF_3)$ derivatives, but gave the ruthenium(0) and osmium(0) complexes $M(CF_2)(CO)_2(PPh_3)_2$ (M = Ru, Os) in excellent yield [140]. X-ray diffraction studies confirmed that these $Ru(CF_2)$ and $Os(CF_2)$ compounds have a trigonal bipyramidal structure with the phosphine ligands in the apical positions and the carbene plane perpendicular to the equatorial plane of the trigonal bipyramid. Treatment of Vaska's complex *trans*-[IrCl-(CO)$(PPh_3)_2$] with $Cd(CF_3)_2$(dme) gave $Ir(CF_3)(CF_2)(CO)(PPh_3)_2$. This appears to be the only example, where both CF_3 and CF_2 were transferred from a $M(CF_3)_2$ derivative of group XII to a single metal center [140].

In summarizing his work on the chemistry of $M(CCl_2)$ and $M(CF_2)$ complexes, Roper pointed out that these metal carbenes were unique in bearing two excellent leaving groups as the carbene substituents [140]. Owing to this fact, nucleophilic substitution at the carbene carbon atom afforded a rich selection of mono- and disubstituted carbene ligands, the former added to a group of monohalocarbene metal compounds obtained by other routes [49, 60–64]. With

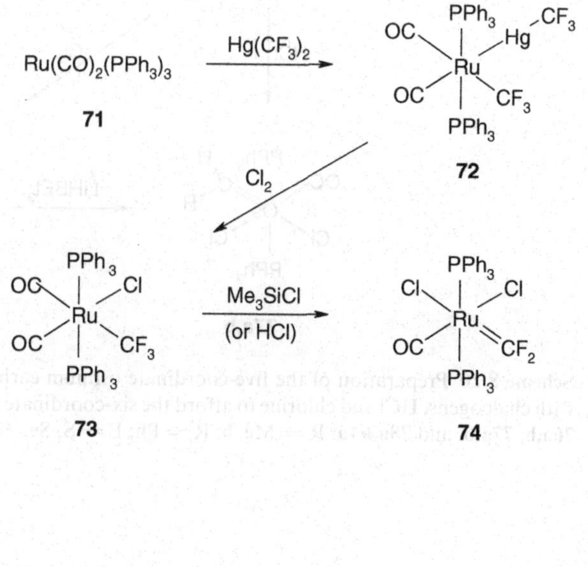

Ru(CO)₂(PPh₃)₃ (**71**) —[$Hg(CF_3)_2$]→ Ru(CO)₂(PPh₃)₂(CF₃)(HgCF₃) (**72**) —[Cl_2]→ Ru(CO)₂(PPh₃)₂Cl(CF₃) (**73**) —[Me_3SiCl (or HCl)]→ RuCl₂(CO)(PPh₃)₂(=CF₂) (**74**)

Scheme 8.19 Preparation of the heteronuclear RuHg compound **72** and the transformation of **72** into the difluorocarbene ruthenium(II) complex **74**

$OsCl_2(CCl_2)(CO)(PPh_3)_2$ as the starting material, the first complete series of chalcocarbonyl complexes $OsCl_2(CE)(CO)(PPh_3)_2$ (E = O, S, Se, Te), including the first M(CTe) derivative, was prepared using the EH^- anion as the nucleophile [150]. The most general reaction of the dichlorocarbene metal compounds took place with primary amines, resulting in products of the general composition $M(CNR)(L)_n$ [140]. The mechanism of the transformation of the CCl_2 ligand to coordinated isocyanides was inferred from a detailed kinetic study by our group on the reaction of the methoxycarbene complex $Cr[C(OMe)Ph](CO)_5$ with primary amines RNH_2, the rate of which was, surprisingly, third order in the amine concentration [151].

Roper's investigations into the substitution reactions of dihalocarbene ruthenium and osmium complexes also led to a highly unexpected result. Addition of two equivalents of lithium phenyl or other lithium aryls to the dichlorocarbene compounds **70** (see Scheme 8.18) and $RuCl_2(CCl_2)(CO)(PPh_3)_2$ did not lead to simple replacement of the carbon-bound chlorides and formation of $MCl_2(CR_2)(CO)(PPh_3)_2$ (R = Ph, *p*-Tol, etc.) but gave the ruthenium and osmium carbynes $MCl(CR)(CO)(PPh_3)_2$ in good yield [152–154]. The use of aryl lithium species as nucleophiles obviously led both to substitution of the chlorides at the carbene carbon atom and formal reduction of the metal center, which is different from the other examples of electrophilic reactivity of dihalocarbene metal complexes mentioned above. Another puzzling observation was that in contrast to Fischer's carbynes, such as **56a,b** and **57a,b** (with a d^6 configuration), the ruthenium and osmium compounds $MCl(CR)(CO)(PPh_3)_2$ (with a d^8 configuration) were nucleophilic at the carbyne carbon atom. Scheme 8.20 shows a few examples of the reactivity of the

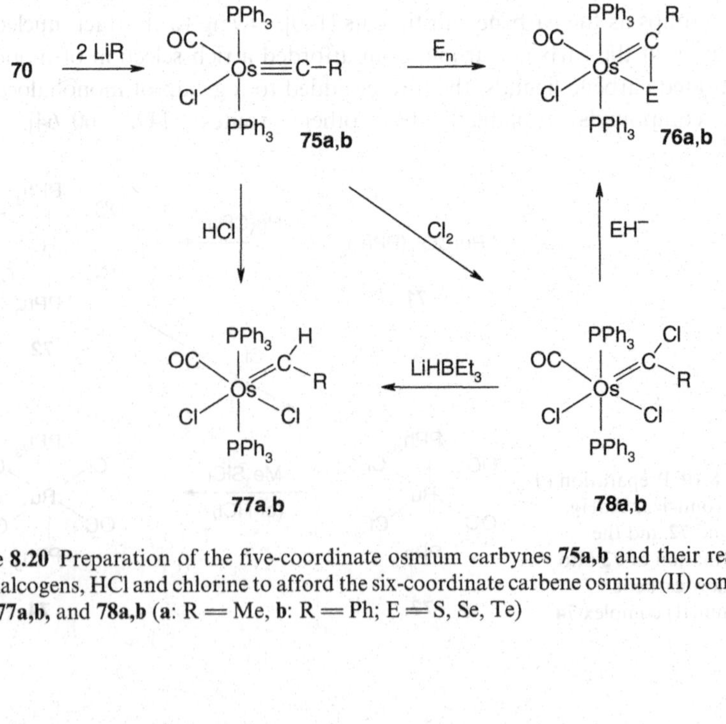

Scheme 8.20 Preparation of the five-coordinate osmium carbynes **75a,b** and their reactions with chalcogens, HCl and chlorine to afford the six-coordinate carbene osmium(II) complexes **76a,b**, **77a,b**, and **78a,b** (**a**: R = Me, **b**: R = Ph; E = S, Se, Te)

osmium carbynes **75a,b** [153]. It is interesting to note that the reactions of **75a,b** with electrophiles and those of the halocarbenes **78a,b** with nucleophiles are complementary and provide access to the three-membered MC(R)E ring carbene derivatives **76a,b** as well as to the secondary carbene complexes **77a,b**.

8.11 The Congeners of Metal Carbynes with M≡E Triple Bonds

The progress made in the chemistry of metal carbynes since Fischer reported the synthesis of the first example in 1973 (see Scheme 8.5) also initiated an intense search for compounds having triple bonds between a transition metal and the heavier elements of group XIV. The first compound with a M≡Ge triple bond was $(\eta^5\text{-}C_5H_5)Mo(GeR)(CO)_2$ (R = 2,6-bis(2,4,6-trimethylphenyl)phenyl), prepared by Phil Power at the University of California in Davis (for biography see Chap. 9) by a simple salt elimination reaction from $Na[(\eta^5\text{-}C_5H_5)Mo(CO)_3]$ and RGeCl in 1996 [155]. Somewhat later, Alexander Filippou (E. O. Fischer's last Ph.D. student, who is now Professor of Inorganic Chemistry at the University of Bonn) reacted the molybdenum and tungsten compounds *trans*-$M(N_2)_2(dppe)_2$ with $(\eta^1\text{-}C_5Me_5)GeX$ (X = Cl, Br, I) and obtained the six-coordinate germylyne complexes *trans*-$MX[Ge(\eta^1\text{-}C_5Me_5)](dppe)_2$ (M = Mo, W), along with *trans*-$MX_2(dppe)_2$, $Ge(C_5Me_5)_2$ and N_2 [156, 157]. In 2005/06, Filippou's group also reported the preparation of *trans*-$MCl(GeR)(PMe_3)_4$ (M = Mo, W; R = 2,6-bis[2,4,6-tris(2-propyl)methylphenyl]phenyl) [158] and of the remarkable dinuclear compounds *trans,trans*-$Cl(dppe)_2M{\equiv}Ge{-}Ge{\equiv}M(dppe)_2Cl$ (M = Mo, W), the latter being formed by thermolysis of *trans*-$MCl[Ge(\eta^1\text{-}C_5Me_5)](dppe)_2$ in the solid state [159]. The metal–tin and metal–lead complexes *trans*-$MCl(SnR)(dppe)_2$ and *trans*-$MCl(PbR)(PMe_3)_4$ (M = Mo, W; R = 2,6-bis[2,4,6-tris(2-propyl)methylphenyl]phenyl) were prepared in a similar way as their M≡GeR counterparts. Power's and Filippou's pioneering work was complemented by the synthesis of several compounds with triple bonds between a transition metal and an element of group XV such as phosphorus, arsenic, and antimony [160] and there is no doubt that activities will continue in this field.

8.12 The First and Second Generation of Grubbs' Ruthenium Carbenes

The most recent important step forward in developing the chemistry of metal carbenes and carbynes was carried out by Robert (Bob) Grubbs in the 1980s. Already at the beginning of his independent career in the late 1960s, he was interested in the mechanism of olefin metathesis and by using the Tebbe reagent $(C_5H_5)_2Ti(\mu\text{-}CH_2)(\mu\text{-}Cl)AlMe_2$ succeeded in isolating a bis(cyclopentadienyl) titanacyclobutane from the in situ generated $[(C_5H_5)_2TiCH_2]$ and $CH_2{=}CMe_2$ [161]. Although this metallacycle was an active metathesis catalyst, it did not promote the formation of ring-opened polymers from 7-oxo-norbornenes which Grubbs anticipated would act as model systems for ionophoric membranes [162, 163]. However, by examining the literature he discovered that in addition to the Ziegler catalysts and a series of carbene chromium carbonyls, extensively investigated by Katz [164], a few late-transition metal catalysts had previously also been used for the polymerization of strained olefins. Since he believed that Yves Chauvin's mechanism [165] for olefin metathesis was correct and that ruthenium could be the metal of choice, he felt that the synthesis of well-defined carbene ruthenium complexes was vital for catalyst optimization. After some unsuccessful attempts in the late 1980s, his graduate student Son-Binh Nguyen was able to prepare the air-stable ruthenium(II) complex **80** from **79** and 3,3-diphenylcyclopropene and found that this compound was an effective catalyst for the polymerization of norbornenes in protic media [166]. The catalytic activity of **80** could be significantly increased by replacing the PPh_3 ligands for tricyclohexylphosphine to give the related five-coordinative complex **81** (Scheme 8.21) [167]. Later, this complex was prepared in high yield from readily available starting materials such as $[(C_8H_{12})RuCl_2]_x$, propargylic chlorides, PCy_3 and H_2 [168]. At roughly the same time, a second one-pot procedure for ruthenium carbenes $RuCl_2(CHCH_2R)(PCy_3)_2$ (R = H, Ph) with $RuCl_3 \cdot aq$ as the ruthenium source was developed by us [169, 170] and is now, like the Grubbs method, routinely carried out on a kilogram scale in commercial and industrial laboratories. Another preparative route for $RuCl_2(CHCH{=}CPh_2)(PR_3)_2$ (R = isopropyl, cyclopentyl, cyclohexyl) also using $RuCl_3 \cdot aq$ as the starting material was disclosed in 2003 [171].

With the experimental tools of Grubbs' group, SonBinh Nguyen discovered in a short period of time that **81** did not only catalyze the metathesis of acyclic olefins but was also competitive with Schrock's molybdenum carbenes in ring-closing metathesis (RCM) [171]. Compared with Schrock's compounds, the ruthenium carbenes had the advantage that they were only moderately sensitive

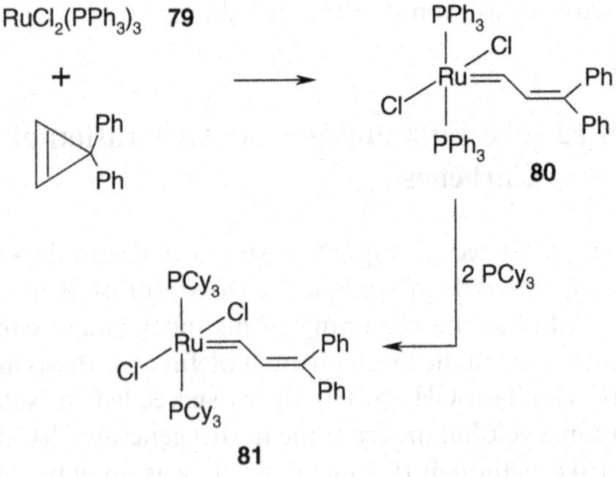

Scheme 8.21 Preparation of the carbene ruthenium(II) complex **81**, the first representative of the "first generation Grubbs catalysts", from **79** and 3,3-diphenylcyclopropene via the related ruthenium carbene **80** as the intermediate

to air and moisture and showed significant tolerance to functional groups. Given this fact and taking into account that the preparation of cyclopropenes on a larger scale was time-consuming and rather expensive, the challenge was to find a more convenient source for the carbene ligand. This was achieved by Peter Schwab (see Fig. 2.26), who had prepared a series of four-coordinate carbene rhodium(I) complexes *trans*-RhCl(CRR')$(L)_2$ (L = PR_3, AsR_3, SbR_3) from rhodium(I) precursors and diazoalkanes in our laboratory [172, 173]. Based on this experience, Peter reacted the starting material **79** with alkyl- and aryldiazoalkanes in dichloromethane at –78°C and, following the spontaneous evolution of N_2, isolated the desired ruthenium(II) compounds **82a,b** in 80–90% yield (Scheme 8.22) [174, 175]. Subsequent phosphine exchange led to the formation of the bis(tricyclohexylphosphine) complexes **83b** (X = H, Me, OMe, NMe_2, F, Cl, NO_2), which proved to be highly efficient catalysts for the ring-opening metathesis polymerization (ROMP) of cyclooctene and 1,5-cyclooctadiene. The parent ruthenium benzylidene $RuCl_2(CHPh)(PCy_3)_2$ (**83b** with X = H), considered to be the prototype of the "first generation Grubbs catalysts", was found to be 20–10^3 times more active than the previously prepared compounds **80** and **81** in several types of olefin metathesis. The reaction of $RuCl_2(CHPh)(PCy_3)_2$ with excess ethene gave the corresponding

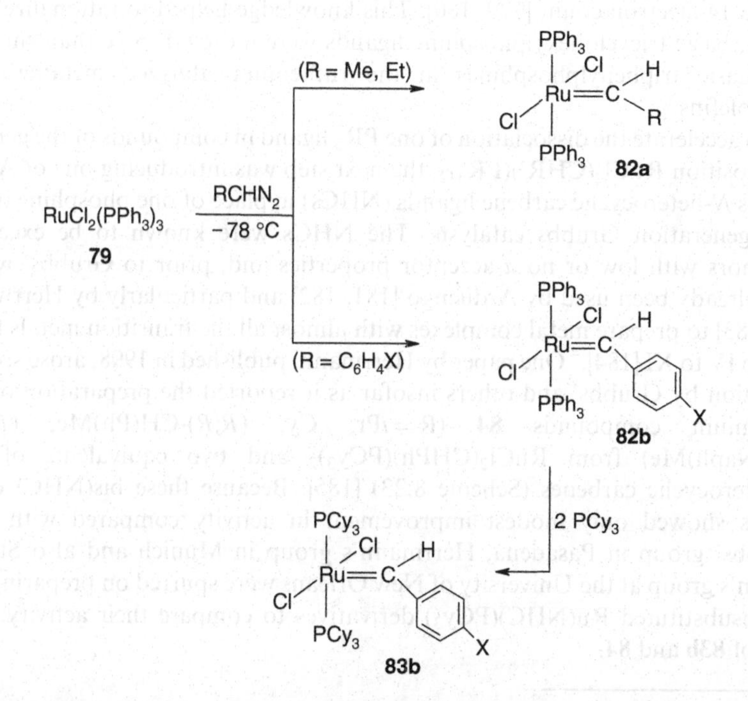

Scheme 8.22 Preparation of the carbene ruthenium(II) complexes **82a,b** from **79** and diazoalkanes, and of the highly active catalysts **83b** (X = H, Me, OMe, NMe_2, F, Cl, NO_2) by phosphine exchange

methylene derivative $RuCl_2(CH_2)(PCy_3)_2$ [174, 175], which was independently prepared by Montserrat Oliván and Kenneth G. Caulton from $RuH_2(H_2)_2(PCy_3)_2$ and dichloromethane by an unprecedented oxidative addition of both C—Cl bonds of CH_2Cl_2 to zerovalent $Ru(PCy_3)_2$, the latter being formed as an intermediate [176]. With regard to the structure of $RuCl_2(CHPh)(PCy_3)_2$ and its analogues, an X-ray diffraction study of **83b** (X = Cl) revealed that the molecule possesses a distorted square-pyramidal geometry with the two phosphine ligands *trans* to each other and the carbene unit lying in the Cl—Ru—Cl plane [174, 175].

To improve the activity and efficiency of $RuCl_2(CHPh)(PCy_3)_2$, Grubbs' group and others attempted to modify the coordination sphere of the five-coordinate ruthenium center. However, neither the exchange of the phosphine and the anionic ligands nor the modification of the carbene moiety itself led to a significant improvement [171]. Introducing chelating ligands, in particular bidentate phosphines [177, 178], led in part to an increase in the thermal stability of the ruthenium(II) complexes, but in most cases did not improve their overall olefin metathesis activity. Based on these results and several mechanistic studies, it was concluded that $RuCl_2(CHPh)(PCy_3)_2$ and its analogues are not catalysts but catalyst precursors which after loosing one of the neutral ligands generated the metathesis-active species $RuCl_2(CHR)(PCy_3)$ with a 14-electron count [179, 180]. This knowledge helped to rationalize why bulky, basic tricyclohexylphosphine ligands were more effective than smaller, less basic triphenylphosphines in the ruthenium-catalyzed metathesis of cycloolefins.

To accelerate the dissociation of one PR_3 ligand in compounds of the general composition $RuCl_2(CHR')(PR_3)_2$, the next step was introducing one of Arduengo's *N*-heterocyclic carbene ligands (NHCs) in place of one phosphine in the first generation Grubbs catalysts. The NHCs were known to be excellent σ-donors with low or no π-acceptor properties and, prior to Grubbs' work, had already been used by Arduengo [181, 182] and particularly by Herrmann [18, 183] to prepare metal complexes with almost all the transition metals from group IV to XI [184].[4] One paper by Herrmann, published in 1998, arose special attention by Grubbs' and others insofar as it reported the preparation of the ruthenium compounds **84** (R = *i*Pr, Cy, (*R,R*)-CH(Ph)Me, (*R,R*)-CH(Naph)Me) from $RuCl_2(CHPh)(PCy_3)_2$ and two equivalents of the *N*-heterocyclic carbenes (Scheme 8.23) [185]. Because these bis(NHC) complexes showed only modest improvement in activity compared with **83b**, Grubbs' group in Pasadena, Herrmann's group in Munich and also Steven Nolan's group at the University of New Orleans were spurred on preparing the monosubstituted Ru(NHC)(PCy_3) derivatives to compare their activity with that of **83b** and **84**.

[4] It is interesting to note, that apart from the impressive series of transition metal complexes with monodentate NHCs, tetracarbene palladium(II) and platinum(II) as well as hexacarbene iron(III) complexes with bi- and tridentate chelating carbene ligands were prepared.

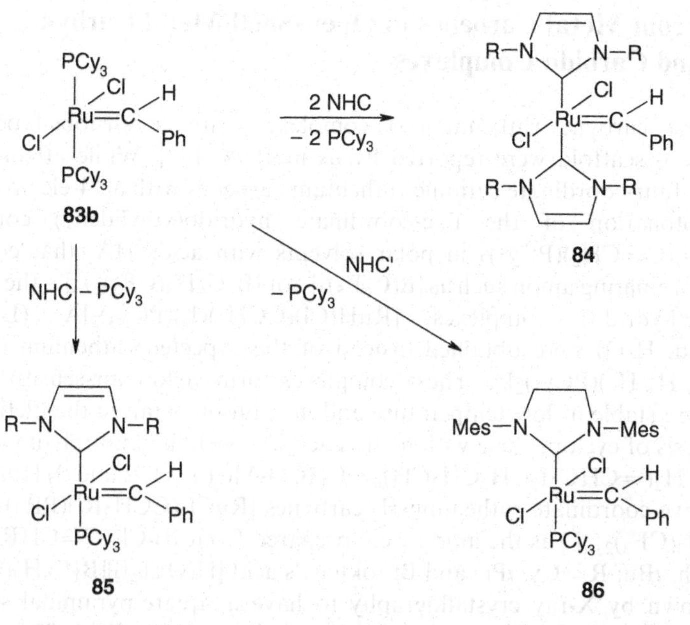

Scheme 8.23 Preparation of Herrmann's bis(NHC) ruthenium(II) complexes **84** and the mono(NHC) complexes **85** and **86** from the first generation Grubbs catalyst **83b** (X = H) as the precursor

As a result, at almost the same time three papers (received on the 1st of September 1998 [186], the 9th of December 1998 [187], and the 13th of December 1998 [188]) appeared reporting the synthesis of mono-NHC ruthenium carbenes **85** (R = *p*-Tol, Mes, Cy, (*R*)-CH(Ph)Me, (*R*)-CH(Naph)Me) from $RuCl_2(CHPh)(PCy_3)_2$ and equimolar amounts of NHCs. Somewhat later, the synthesis of the analogue **86** containing a five-membered *N*-heterocyclic carbene with a saturated backbone was described [189]. The three groups [186–188], who had prepared the mono-NHC complexes of types **85** and **86**, immediately recognized that these compounds were not only inert towards air and moisture but also displayed an excellent catalytic performance including the RCM of sterically demanding dienes to form tri- and tetrasubstituted olefins. Moreover, both **85** and **86** catalyzed the ROMP of 1,5-cyclooctadiene at rates approximately 100–1,000 times greater than those observed for $RuCl_2(CHPh)(PCy_3)_2$. The superior activity of compound **86** in particular, showing a productivity in ROMP comparable to that of industrially used Ziegler–Natta systems [190], gave rise to the name "second generation Grubbs catalysts" for the class of ruthenium(II) compounds represented by **85** and **86** [191]. Fine tuning of the substituents of the NHC ligand in **86** afforded chiral catalysts which showed excellent activity in asymmetric ring-opening cross-metathesis reactions (AROCM) with enantioselectivities ranging from 68 to 82%. An asymmetric cross-metathesis reaction (ACM) of substituted 1,4-, 1,5- and 1,6-dienes with *cis*-1,4-diacetoxy-2-butene was also achieved with catalysts similar to **86** [192].

8.13 From Metal Carbenes to Open-Shell Metal Carbyne and Carbido Complexes

The first carbyne ruthenium(II) complexes with a Grubbs-type *trans*-$Ru(PCy_3)_2$ scaffold were reported by us in 1998 [193]. While attempting to prepare four-coordinate cationic ruthenium carbenes with a 14-electron count by protonation of the five-coordinate hydrido(vinylidene) compound $RuH(Cl)(C{=}CH_2)(PCy_3)_2$ in polar solvents with acids HA, that contain a non-coordinating anion such as $[B(C_6F_5)_4]^-$ or $[B\{C_6H_3(CF_3)_2\}_4]^-$, the cationic carbyne(hydrido) complexes $[RuH(Cl)(CCH_3)(L)(PCy_3)_2]A$ ($L = OEt_2$, NMe_2Ph, H_2O) were obtained instead of the expected ruthenium carbenes $[RuCl(CHCH_3)(PCy_3)_2]A$. These complexes form yellow air-sensitive solids, which are stable at low temperature and in solution catalyze the olefin cross-metathesis of cyclopentene with methyl acrylate yielding multiply unsaturated esters $CH_2({=}CHCH_2CH_2CH_2CH)_n{=}CHCO_2Me$ ($n = 1$, 2 and 3) [193]. More stable five-coordinate ruthenium(II) carbynes $[RuCl_2(CCH_2R')(PR_3)_2]^+$ with $[B\{C_6H_3(CF_3)_2\}_4]^-$ as the anion were prepared from $RuCl_2(C{=}CHR')(PR_3)_2$ (R' = Ph, *t*Bu; R = Cy, *i*Pr) and Brookhart's acid $[H(OEt_2)_2][B\{C_6H_3(CF_3)_2\}_4]$ and shown by X-ray crystallography to have a square-pyramidal structure analogous to the first generation Grubbs carbenes [194]. With **87** instead of $RuCl_2(C{=}CHPh)(P\textit{i}Pr_3)_2$ as the precursor, the cationic carbene complexes **89** (Scheme 8.24) were obtained by intramolecular nucleophilic attack of the carboxylato ligand to the carbyne unit of the intermediate **88**. The chelate complexes **89** are reminiscent of the Hoveyda-type ruthenium carbenes in which the phenyl group at the carbene carbon atom contains an alkoxy substituent in the *ortho* position that coordinates via the oxygen to the metal. In contrast to the Grubbs systems, the Hoveyda-type complexes contain only one PCy_3 or NHC ligand bound to ruthenium [195, 196]. In some cases, the reactivity of these complexes is significantly higher than that of the Grubbs

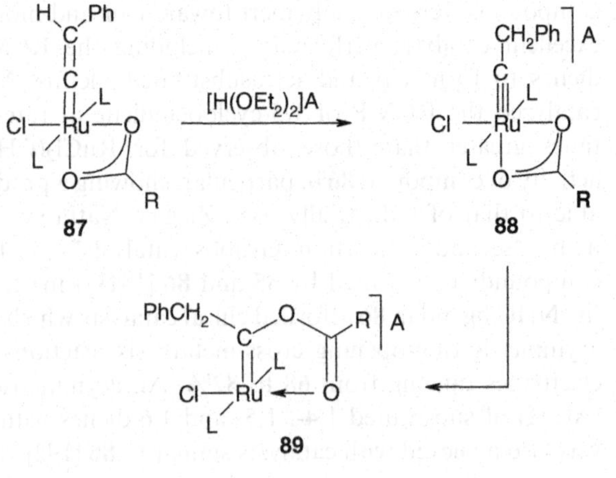

Scheme 8.24 Preparation of cationic chelating ruthenium(II) carbenes **89** by protonation of neutral ruthenium(II) vinylidenes **87** via ruthenium carbynes **88** as intermediates ($L = P\textit{i}Pr_3$; R = H, Me, Ph; $A = B[C_6H_3(CF_3)_2]_4$)

carbenes since the presence of the weak coordinating OR moiety supports the formation of a 14-electron intermediate and thus facilitates the metathesis.

In analogy to Grubbs' isolable 14-electron ruthenium(II) carbene complexes $Ru(OtBu)_2(CHPh)(PCy_3)$ [197], three groups led by Caulton, Fogg, and Johnson recently also succeeded in isolating stable four-coordinate ruthenium(0) carbynes [198–200]. The original aim of this work was to modify the coordination sphere of first generation Grubbs catalysts by replacing the two chlorides for non-halide anionic ligands. Thus, Caulton reacted the carbene precursors $RuCl_2(CHPh)(PR_3)_2$ (R = Cy, *i*Pr) with two equivalents of NaOPh and obtained the zerovalent ruthenium carbynes $Ru(OPh)(CPh)(PR_3)_2$ via the expected short-lived intermediates $Ru(OPh)_2(CHPh)(PR_3)_2$ [198]. In contrast, treatment of $RuCl_2(CHPh)(PR_3)_2$ with bulky alkoxides caused phosphine displacement apart from chloride substitution to give the four-coordinate carbenes $Ru(OR)_2(CHPh)(PR_3)$. Similar to Caulton's route, Deryn Fogg's group prepared the compound $Ru(OC_6F_5)(CPh)(PCy_3)_2$ from $RuCl_2(CHPh)(PCy_3)_2$ and $TlOC_6F_5$ [199]. Marc Johnson obtained the first carbyne(chloro) ruthenium(0) complexes of the composition $RuCl(CR)(PCy_3)_2$ (R = *n*Bu, *p*-Tol) either in one step by treatment of $RuCl_2(CHR)(PCy_3)_2$ with Lappert's bulky germylene $Ge[CH(SiMe_3)_2]_2$ or, more conveniently, in two steps with NaO-*p*-C_6H_4*t*Bu (leading first to Ru(O-*p*-C_6H_4*t*Bu)(CR)$(PCy_3)_2$) followed by treatment with $SnCl_2$ [200]. Oxidation of RuX(C-*p*-Tol)$(PCy_3)_2$ by, respectively, XeF_2, C_2Cl_6, Br_2 and I_2 gave either octahedral bis(phosphine) ruthenium(II) derivatives RuX_3(C-*p*-Tol)$(PCy_3)_2$ (X = F, Cl) or square-pyramidal mono(phosphine) complexes RuX_3(C-*p*-Tol)(PCy_3) (X = Br, I) depending on the size of the halide ligands [200]. In contrast to Roper's ruthenium and osmium carbynes $MCl(CR)(CO)(PPh_3)_2$ (M = Ru, Os) [137], Johnson's mono(phosphine) derivatives RuX_3(C-*p*-Tol)(PCy_3) are coordinatively unsaturated having a 16-electron count.

An exciting discovery, recently made by Joseph Heppert [201], has shed new light on the chemistry of first and second generation Grubbs catalysts. In the course of investigating ruthenium-mediated ring expansion reactions, Heppert's group found that the ruthenium carbenes **83b** (X = H) and **86** reacted with the methylene cyclopropane derivative **90**, known as Feist's ester, to give the diamagnetic, air-stable (!) carbido complexes **91a,b** via metathesis and elimination of diethyl fumarate (Scheme 8.25). An X-ray diffraction study of **91b** showed that the ruthenium atom possesses a distorted trigonal bipyramidal coordination sphere, the terminal carbido moiety being shielded by two adjacent cyclohexyl groups and one mesityl unit of the NHC ligand. The Ru—C bond length of 1.650 Å is among the shortest distances observed for ruthenium–carbon triple bonds [194, 201].

The knowledge of Heppert's results prompted Grubbs' group immediately to re-examine the reactivity of the cyclic carbene complex **92**, previously obtained from **82b** (X = H) and Feist's ester. Rather surprisingly, the addition of excess PCy_3 to **92** caused, apart from phosphine exchange, the instant release of dimethyl fumarate and gave **91b** in 70% yield [202]. It thus became obvious that the more electron-donating PCy_3 ligands accelerated the olefin elimination.

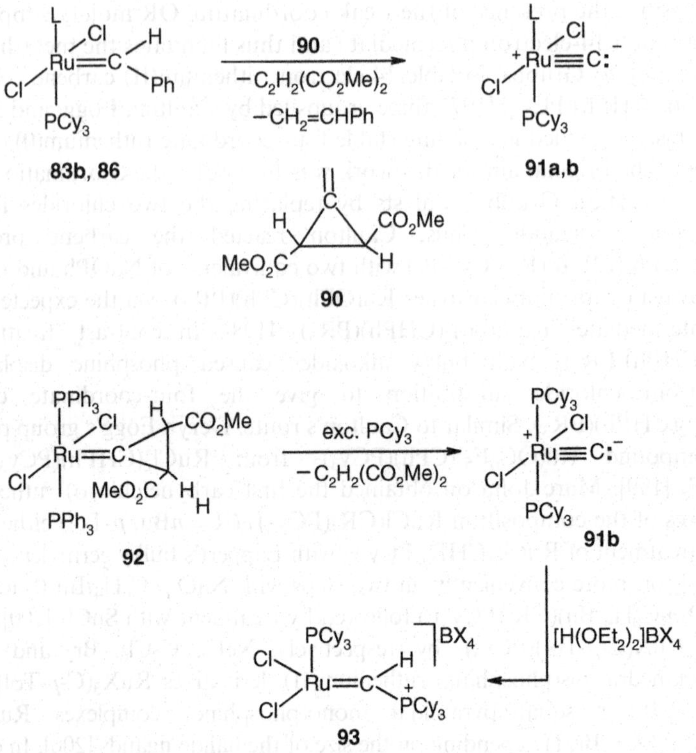

Scheme 8.25 Preparation of carbido ruthenium(II) complexes **91a,b** either from **83b**, **86** or from **92** as the precursors, and protonation of **91b** to afford the cationic four-coordinate ruthenium carbenes **93** (X = F or C_6F_5); (**91a**: L = NHC as in **86**; **91b**: L = PCy_3)

The reaction of **91b** with $PdCl_2(SMe_2)_2$ and $Mo(CO)_5(NMe_3)$ resulted in the formation of the corresponding heterometallic compounds with a Ru≡C—M bridge [202], thus illustrating that the terminal carbido ligand could function as a σ-donor. This observation was supplemented by Warren Piers' work on the protonation of **91b** with strong acids $[H(OEt_2)_2]BX_4$ affording the unusually stable four-coordinate complexes **93** (X = F or C_6F_5) [203]. The reaction of **91a** with $[H(OEt_2)_2]BX_4$ proceeded analogously and gave the corresponding products $[RuCl_2\{CH(PCy_3)\}(L)]BX_4$. These 14-electron PCy_3-substituted ruthenium carbenes are exceptionally active RCM catalysts, significantly exceeding the activity of **83b** and **86**. Piers believes that the reason for the high activity is that there is no need for the dissociation of a ligand from the catalyst precursor to enter the catalytic cycle and bind the C=C bond of the olefinic substrate [203].

Although Heppert's discovery and the subsequent work by Grubbs and others has received special attention, it must be emphasized that prior to the studies on ruthenium carbido complexes Christopher Cummins from MIT had reported the preparation of the anionic molybdenum carbide $[CMo\{N(R)Ar\}_3]^-$ (R =

$C(CD_3)_2CH_3$, Ar = 3,5-$C_6H_3Me_2$) in 1997. It was obtained from the methylidyne precursor $HCMo[N(R)Ar]_3$ by deprotonation with benzyl potassium [204, 205]. The four-coordinate methylidyne was first prepared via a stepwise synthesis by deoxygenation of bound carbon monoxide [204], and secondly by a CH transfer reaction from $Mo[N(R)Ar]_3$ and 7-chloronorbornadiene in the presence of $Ti[N(t\text{Bu})Ar]_3$ as a specific chlorine-atom abstractor [206]. In this case, benzene was formed as a by-product. As Cummins' group showed, the terminal carbide atom of the ion $[CMo\{N(R)Ar\}_3]^-$ is a strong nucleophile and readily undergoes addition reactions with several electrophilic substrates to furnish functionalized molybdenum carbynes. Particularly interesting is the intermediacy of the ionic species $[CMo\{N(t\text{Bu})Ar\}_3]^-$ in the construction of the phosphaisocyanide ligand CPPh in the $[(PhPC)Mo\{N(t\text{Bu})Ar\}_3]^-$ anion, which due to its spectroscopic and structural data can equally well be described as an anionic substituted carbyne complex with a Mo—C triple bond [206, 207]. With Cummins' starting material $Mo[N(R)Ar]_3$, the nitrido compound $NMo[N(t\text{Bu})Ar]_3$ and the phosphido analogue $PMo[N(t\text{Bu})Ar]_3$ were also prepared; the latter being used as the precursor of the novel transition metal complex $O{=}P{=}Mo[N(t\text{Bu})Ar]_3$ with phosphorus monoxide as ligand [207].

8.14 The Dötz Reaction and the Use of Metal Carbenes for Organic Synthesis

The importance of the Grubbs-type ruthenium carbenes as catalysts for the synthesis of organic compounds including a series of natural products not withstanding, it should be recalled, that about two decades prior to the isolation of $RuCl_2(CHPh)(PCy_3)_2$ [174, 175], and almost at the same time as Schrock's first paper on the preparation of the tantalum complex $(Me_3CCH_2)_3Ta(CHCMe_3)$ appeared [73], Fischer-type chromium carbenes were used in organic synthesis [208–214]. The breakthrough occurred in 1975, when Karl-Heinz Dötz (in the continuation of his work on cyclopropanation reactions with alkoxycarbene complexes [37]) reported, that treatment of $Cr[C(OMe)Ph](CO)_5$ (**16**) with tolane in di-*n*-butyl ether at 45°C resulted in the formation of compound **94a** (Scheme 8.26) [215]. Stirring a solution of **94a** in $(n\text{-}C_4H_9)_2O$ for 3 h at 60–90°C led to a complete conversion to the thermodynamically more stable isomer **94b**. Oxidative degradation of either **94a** or **94b** with nitric acid in glacial acetic acid gave the free naphthol molecule, which was built up in the coordination sphere of the metal from the carbene moiety, the alkyne and one CO ligand [209]. Subsequent studies, initially by Dötz and later by many others, revealed that various other carbene chromium pentacarbonyl complexes with polycyclic, heterocyclic, vinyl or alkynyl substituents at the carbene carbon atom can also be used as precursors for the benzannulation reaction. Moreover, since apart from alkyl- and aryl-substituted alkynes also non-conjugated enynes, vinylimines, alkynyl esters, alkynyl amines, and several other unsaturated molecules are useful building blocks for the ring system, it has become common to refer the new metal-mediated cycloaddition as the "Dötz reaction" [208, 216, 217].

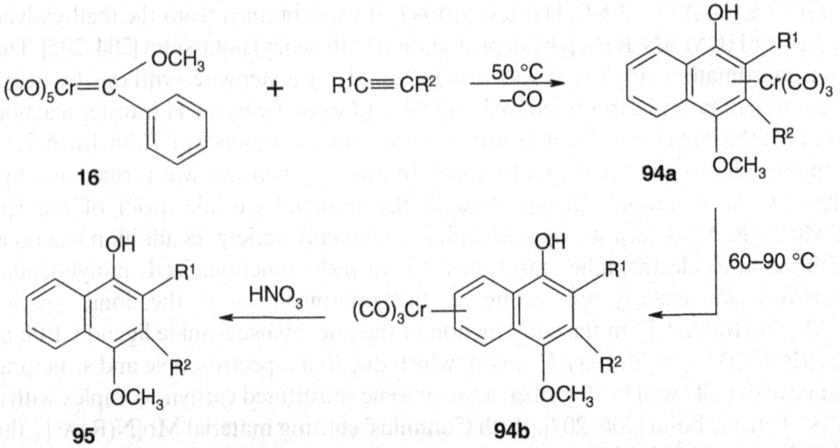

Scheme 8.26 Preparation of naphthol chromium(0) complexes **94a,b** via the Dötz reaction, and the displacement of the free naphthol molecule **95** by oxidative degradation

At present, the scope of using Fischer-type carbene complexes for synthetic purposes appears nearly unlimited. In Armin de Meijere's view, they are "chemical multitalents" [214], which almost from the beginning have not only improved the repertoire of cycloaddition and cocyclization reactions, but also exhibited new routes for the preparation of natural products. Apart from Weiss and Fischer's work to generate peptides with Cr[C(OMe)Ph](CO)$_5$ (**16**) as the starting material (see Scheme 8.4), Dötz showed in 1981 that **16** reacted with enynes $RCH_2C(CH_3){=}CHCH_2C{\equiv}CCH_3$ to yield the vitamins K$_1$ and K$_2$ in three steps. By a similar reaction sequence, he prepared vitamin E from the *E*-2-butenylcarbene complex Cr[C(OMe)C(CH$_3$)=CHCH$_3$](CO)$_5$ and a C$_{23}$ enyne via a hydroquinone intermediate [209]. Martin Semmelhack supplemented these studies by preparing antibiotics such as nanaomycin A and deoxyfrenolicin using the storable and easy to handle tetramethylammonium salt (NMe$_4$)[Cr{C(O)C$_6$H$_4$-2-OMe}(CO)$_5$] as the precursor [218, 219]. More recently, Dötz reported that the reactions of Na$_2$[Fe(CO)$_4$] or K$_2$[M(CO)$_5$ (M = Cr, Mo, W)) with an acetylated D-galactonic acid chloride gave carbene-complex functionalized sugars, which provide the chance for a controlled modification of the sugar framework [220, 221].

8.15 Olefin Metathesis: A Landmark in Applied Organometallic Chemistry

Beyond the use of the Fischer-type chromium carbenes as *stoichiometric reagents* and the Grubbs-type ruthenium carbenes as versatile catalysts for the preparation of organic compounds, Schrock's molybdenum and tungsten complexes of the general composition M(CHR)(NAr)(OR')$_2$ (and derivatives thereof) and

Grubbs' ruthenium compounds $Ru(CHR)(NHC)(PCy_3)X_2$ (with sterically protected *N*-heterocyclic carbenes) were also applied for the production of polymers since the mid 1990s. The impetus for this came mainly from petrochemical companies which were not only interested in new materials but also in a convenient supply of olefins as precursors for polymers with well-defined structures. The industrial interest in this field arose already in the 1950s when Herbert Eleuterio from DuPont's Research Department in Wilmington observed that passing a propene feed over a molybdenum-halide-on-aluminium-oxide catalyst gave a propene–ethene copolymer [222]. A decade later, Robert Banks and Grant Bailey of Phillips Petroleum reported the disproportionation of propene into ethene and butene using $Mo(CO)_6$ or $W(CO)_6$ supported on Al_2O_3. The unexpected products were obviously formed by cleavage and reformation of the olefinic double bonds. Based on these results, Nissim Calderon and his group at Goodyear in Akron, OH, developed a soluble catalyst system that could not only induce the polymerization of cycloolefins via ring-opening but also converted propene into ethene and 2-butene. The Goodyear researchers named the reaction "olefin metathesis", derived from the Greek words *meta* (change) and *thesis* (position). They were also the first to establish a mechanistic scheme involving a cyclobutane metal intermediate [223, 224]. However, since cyclobutanes were not found in the mixture of products, and olefins were not generated by putting cyclobutanes into olefin metathesis systems, Rowland Pettit (for biography, see Chap. 5) proposed the formation of a tetracarbene complex, in which four CHR units are coordinated to a central metal atom [225]. Not long afterwords, Grubbs modified this scheme, assuming that the redistribution of groups around the carbon–carbon double bonds could be due to the rearrangement of a metallacyclopentane intermediate [226].

Had Pettit and Grubbs been aware of an article, published by Yves Chauvin and Jean-Louis Hérisson in the journal *Makromolekulare Chemie* in January 1971 [165], their ideas might not have come up. The French scientists from the *Institut Francais du Pétrole* had studied the reaction of cyclopentene with 2-pentene in the molar ratio of 1:1 in the presence of a tungsten-centered homogeneous catalyst. They found that the main products were C_9, C_{10} and C_{11} dienes in a ratio if 1:2:1. This observation was rather unexpected because the mechanistic ideas, favored at that time, predicted that only the C_{10} product should be formed by metathesis. After Chauvin and Hérisson verified that other pairs of olefins, such as cyclooctene/2-pentene, cyclopentene/2-hexene, cyclopentene/3-heptene, etc., also gave a mixture of three main products, they concluded that the "most obvious hypothesis (to explain the results) is the formation of a metallocarbene sequestering an olefin to form a metallacyclobutane intermediate" [227]. In addition, Chauvin's group showed that a small amount of propene is generated from 2-butene in the presence of WCl_6 and LiMe or $SnMe_4$, presumably via a methyltungsten compound, which gave a $W{=}CH_2$ species by α-hydrogen elimination. This methylene metal intermediate could then react with 2-butene yielding ethene and propene.

However, as convincing Chauvin's proposal appeared, consensus regarding the involvement of metal carbenes in olefin metathesis did not emerge in the following years. The observation, among some others, that in agreement with the conventional mechanism the reaction of cyclooctene and 1-pentene gave almost entirely the C_{13} product, led Calderon to affirm (still in 1977) that "a clear-cut selection of a single scheme [among four proposed, including Chauvin's] over the rest" could not be made [222]. This scepticism mainly disappeared after Casey found a model system for Chauvin's mechanism, using a preformed tungsten carbene [228], and after Katz analyzed the kinetics of olefin metathesis [229, 230]. Moreover, Grubbs supported Chauvin's proposal by labeling studies on a ring-closing metathesis reaction [163], and Schrock showed that metal carbenes could, in fact, be generated under conditions similar to those used to prepare Ziegler-type catalysts [78]. Finally, the preparation of Schrock's molybdenum and tungsten complexes $M(CHR)(NAr)(OR')_2$ (M = Mo, W) as well as of Grubbs' ruthenium compounds $Ru(CHR)(PCy_3)_2X_2$ and $Ru(CHR)(NHC)(PCy_3)X_2$, and their reactions with olefins proved that the "recipe" offered by Chauvin was correct and that, based on his ideas, "olefin metathesis (became) one of the most useful synthetic techniques to be added to the chemist's toolbox in recent years".[5]

8.16 An Extension: Metal Complexes with Unsaturated Carbenes

The chapter on metal carbenes and carbynes should not come to an end, however, without mentioning the huge family of transition metal complexes with unsaturated carbenes as ligands. By far the majority of complexes belonging to this family are the metal vinylidenes $M(C{=}CRR')(L)_n$ and the metal allenylidenes $M(C{=}C{=}CRR')(L)_n$, which are most conveniently prepared from acetylene or substituted alkynes (for ligated $C{=}CH_2$, $C{=}CHR$ and $C{=}C(R)SiMe_3$) and from propargylic alcohols or propargylic halides (for ligated $C{=}C{=}CRR'$) as precursors [231–237]. Representatives with four and five carbon atoms in the $M({=}C)_n$ chain are rare and accessible from either enolized acylalkynes or substituted diynes [238–243]. Chromium and tungsten complexes with a linear $M({=}C)_7$ unit were recently prepared by Helmut Fischer (a former Ph.D. student of E. O. Fischer and now Professor of Inorganic Chemistry at the University of Konstanz, Germany) from the anionic species $[M(CO)_5(C{\equiv}C)_3C(NMe_2)_3]^-$ (M = Cr, W) by silica-induced elimination of dimethylamide [244]. Since the late 1980s, metal vinylidenes and metal allenylidenes received also attention in homogeneous catalysis, although in most cases the respective $M(C{=}CRR')(L)_n$ and $M(C{=}C{=}CRR')(L)_n$ derivatives were not characterized by chemical analysis or spectroscopic techniques but generated in situ and reacted with the corresponding substrate to the desired product [245–247].

[5] Editor's page of *Chemical & Engineering News*, October 10, 2005.

While the most straightforward routes to metal vinylidenes and metal allenylidenes used terminal and functionalized alkynes as starting materials, it is interesting to note retrospectively that the first mononuclear complexes of the general composition M(C=CRR')$(L)_n$ and M(C=C=CRR')$(L)_n$, were prepared from other precursors. Bruce King from the University of Georgia in Athens reported in 1972 that the reaction of the molybdenum vinyl compound (η^5-C_5H_5)Mo($CO)_3$[C(Cl)=$C(CN)_2$] with triphenylphosphine led to the displacement of all three carbonyl ligands and gave the dicyanovinylidene complex (η^5-C_5H_5)MoCl[C=$C(CN)_2$]$(PPh_3)_2$ as a mixture of two stereoisomers [248]. The first allenylidene complexes were independently obtained by E. O. Fischer and Heinz Berke (who at that time was working towards his *Habilitation* at the University of Konstanz and has been Professor of Inorganic Chemistry at the University of Zurich since 1988) in 1976. Fischer's group prepared the chromium and tungsten compounds **97a,b** from the corresponding vinyl-substituted carbenes **96a,b** by successive reactions with a Lewis acid such as BF_3 or $AlEt_3$ and a weak base such as THF (Scheme 8.27) [249]. Berke used (η^5-C_5H_5)Mn(HC≡CCO_2Me)$(CO)_2$ and *t*BuLi as starting materials and obtained the half-sandwich type complex (η^5-C_5H_5)Mn(C=C=C*t*Bu_2)$(CO)_2$ upon protonation of a proposed manganese alkynyl intermediate [250].

King's, Fischer's and Berke's reports initiated an immense activity in this field and in the following decades hundreds of vinylidene and allenylidene transition metal complexes were prepared and their reactivities studied. The general perception now is that in the vinylidenes M(C=CRR')$(L)_n$ the α-carbon atom is an electrophilic and the β-carbon atom a nucleophilic center. In the allenylidenes M(C=C=CRR')$(L)_n$ both the α- and the γ-carbon atoms are electron-poor and thus electrophilic, while the β-carbon atom is electron-rich and thus reacts as a nucleophile [231–237]. It is therefore possible to obtain metal carbynes from metal vinylidenes by protonation, as was shown for the first time by us with the formation of *trans*-[IrCl(CCH_2R)(P*i*$Pr_3)_2$]BF_4 from *trans*-[IrCl(C=CHR)(P*i*$Pr_3)_2$] (R = H, Me, Ph) and HBF_4 [251]. With regard to iridium as the metal center, a complete series of structurally analogous square-planar complexes from Ir=CRR' (iridium carbenes) to Ir=C=C=C=C=CRR' (iridium pentatetraenylidenes), sometimes referred to as "the Würzburg collection" (Scheme 8.28), had been described and their chemistry investigated [252]. Among the most surprising results was the fragmentation of CH_3OH into CO and four hydrogen atoms by the square-planar

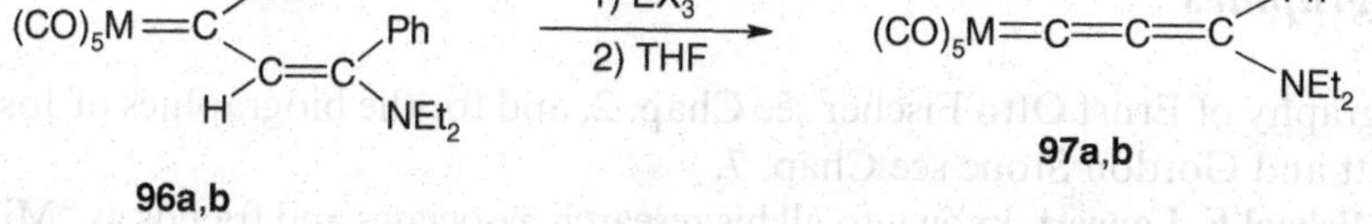

$(CO)_5$M=C(OEt)–CH=C(Ph)(NEt_2) **96a,b** —[1) EX_3; 2) THF]→ $(CO)_5$M=C=C=C(Ph)(NEt_2) **97a,b**

Scheme 8.27 Preparation of allenylidene metal pentacarbonyls **97a,b** from vinyl-substituted carbenes **96a,b** by successive reactions with a Lewis acid and THF (**a**: M = Cr, **b**: M = W)

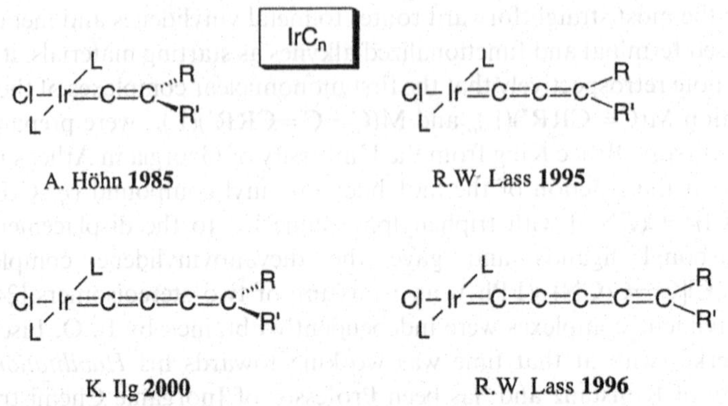

IrC_n

A. Höhn **1985** R.W. Lass **1995**

K. Ilg **2000** R.W. Lass **1996**

Scheme 8.28 A series of iridium(I) complexes containing iridium–carbon double bonds ("the Würzburg collection"), with the names of the coworkers and the year in which the result was published

iridium(I) complex *trans*-$[Ir(OH)({=}C{=}C{=}C{=}CPh_2)(PiPr_3)_2]$, affording apart from H_2O the octahedral dihydrido iridium(III) compound $IrH_2(CH{=}C{=}C{=}CPh_2)(CO)(PiPr_3)_2$ in almost quantitative yield.

Although more than four decades have now passed since the landmark paper by Fischer and Maasböl appeared in 1964 [25], the chemistry of metal carbenes and carbynes is still full of excitement. Moreover, this development is a compelling example of how the serendipitous discovery of two types of $M(CR_2)(L)_n$ and $M(CR)(L)_n$ complexes, the so-called Fischer-type and Schrock-type carbenes and carbynes, has been transformed into a set of well-defined tools with manyfold applications in synthetic chemistry, including organic synthesis and industrial processes. From today's perspective, it is not difficult to predict that the developments in this field will continue and other carbenes and carbynes beyond CH_2, CCl_2, CF_2, $CHNMe_2$, CHPh or CPh_2 and CH, CMe, $CNMe_2$ or CPh, respectively, will be "tamed" in the coordination sphere of either an electron-poor or an electron-rich transition metal. Recent work, in particular by Jean-Marie Basset at Lyon [253], may indicate that the highly developed molecular chemistry of metal carbenes and carbynes is going to be more strongly combined with surface chemistry, not only to provide catalysts for alkene metathesis but also for alkane activation.

Biographies

Biography of Ernst Otto Fischer see Chap. 2, and for the biographies of Joseph Chatt and Gordon Stone see Chap. 7.

Michael F. Lappert, known to all his research associates and friends as "Mike", is Research Professor of Chemistry at the University of Sussex in the UK. According to David Cardin, one of his former colleagues at Brighton, "his scientific output

Fig. 8.4 Michael Lappert at the time when he became Research Professor of Chemistry at the University of Sussex (photo by courtesy from M. F. L.)

can genuinely be described as seminal, in the strict sense that it has provided over many years a source, from which much subsequent advance has sprung" [254].

Lappert (Fig. 8.4) was born on the 31st of December 1928 at Brno in the then Czechoslovakia and came to England as a child together with his brother as a consequence of the occupation of his home territory by the German army. After attending Wilson's Grammar School in London, he moved to the Northern Polytechnic, where he graduated as an external London University student with first class honors in 1949. At the Northern Polytechnic he continued with his doctoral studies under the supervision of William Gerrard and investigated the reactions of boron trichloride with alcohols and ethers, including optically active derivatives. After obtaining the Ph.D. in 1951, he extended this work to reactions of boron trichloride with organic compounds in general. In 1956, he became both a Fellow of the Royal Institute of Chemistry and a recognized teacher of the University of London, being awarded a D.Sc. by the same university in 1960. He left Northern Polytechnic in 1959 to take up a lectureship at the University of Manchester Institute of Science and Technology (UMIST), where he was appointed as Senior Lecturer in 1962. Two years later, he was recruited by the then new University of Sussex as Reader and promoted to Professor in 1969, a post he held until 1996.

Throughout his whole career, Lappert has been a firm supporter of basic research driven purely by intellectual curiosity, rather than by current fashions of applying chemistry to its several sister disciplines. He mentored well over 100 Ph.D. students, and with his coworkers published nearly 800 papers and reviews on various aspects of inorganic and organometallic chemistry. The emphasis of his work has been on the synthesis and the characterization of compounds in novel or unusual oxidation states and/or geometries. Most of the results relate to molecular chemistry of the metals and metalloids, a characterization that allows inclusion of the extensive contributions to the chemistry of amide derivatives and related alkoxide species. During the Manchester period, his previous research on amido tin(IV) compounds led on to the discovery of the first molecular germanium(II), tin(II) and lead(II) amides and their alkyl analogues. At the end of the 1960s, he extended this work to alkyl compounds of the d- and f-block elements, which for decades were considered not to be isolable. Other important contributions, apart from those on metal carbenes, were on metal β-diketiminato compounds, transition metal complexes with metal to phosphorus double bonds, and methyl-bridged lanthanide species of the type $[(C_5R_5)_2Ln(\mu\text{-}CH_3)]_2$, which were the first compounds to catalyze olefin polymerization at a single metal site.

Lappert has received numerous awards including the first Chemical Society Award for Main Group Metal Chemistry, the Frank Stanley Kipping Award for Organosilicon Chemistry from the American Chemical Society, and the Organometallic Chemistry Award from the Royal Society of Chemistry. He has been a Tilden, Nyholm, and Sir Edward Frankland Medallist and was elected Fellow of the Royal Society in 1979. On the occasion of the 40th anniversary of the Royal Society of Chemistry's journal *Chemical Communication* in 2005, he was identified as its most cited author, in any area of chemistry and by a comfortable margin, over the years since 1965. Apart from his scientific merits, those who know him personally will certainly agree that he is an inspiring human being, equipped with fine humor, and admired by his friends, colleagues, and former students all over the world.

When **Yves Chauvin** was informed on the 5th of Octover 2005 that the Royal Swedish Academy awarded the Nobel Prize in Chemistry 2005 to three scientists, including him, he found it "embarrassing, above all" [255]. Although the decision was obviously a great surprise for him, it was not for the chemical community since, as Professor Per Ahlberg pointed out in the Presentation Speech at the Stockholm Concert Hall on the 10th of December 2005, the "achievements of this year's Laureates have given us one of the most important methods for building new organic molecules – metathesis". He also emphasized that "a number of researchers made proposals about how metathesis works. The breakthrough came in 1971 when Yves Chauvin presented new experiments and suggested that the catalyst is a carbene metal compound in which the metal is bonded to carbon with a double bond. Chauvin also presented a novel mechanism for how the catalyst works in metathesis" [256].

Yves Chauvin (Fig. 8.5) was born on the 10th of October 1930 in Menin in Western Flanders near to the border between Belgium and France. His father was an electrical engineer and both of his parents were French. He went first to the pre-school in Flanders and then to the primary school in France, which meant that he had to cross the border every day. He continued his higher education in various places but was – as he confessed in his autobiography – "not a very brilliant student, even at chemistry school" [257]. The decision to choose chemistry as the main discipline occurred "rather by chance" following his belief "that you can become passionately involved in your work whatever it is". In 1954, he received a

Fig. 8.5 Yves Chauvin at the time when he received the Nobel Prize (photo reproduced with permission of CPE Lyon, by courtesy of Professor Jean-Marie Basset)

diploma degree from Lyon's School of Chemistry, Physics and Electronics but due to special circumstances, he never obtained a Ph.D. After he finished his military service, he accepted an offer from industry where his main duty was in process development. This consisted "primarily of copying what already existed, with no possibility of exploring other fields". He was told to "change as little as possible .... (since) we know it will work" [257]. However, this was not what he wanted, and thus he left industry in 1960.

In that same year he joined the *Institut Francais du Pétrole* at Rueil-Malmaison in the western suburbs of Paris and became immediately acquainted with research. While at this institute (which was supported by oil companies) the main focus was on heterogeneous catalysis, Chauvin tried to avoid this area and was looking for new challenges. Due to his continuing interest in the chemical literature, he was fascinated by the work of Ziegler, Natta, Chatt, Wilke, and others and decided to follow their footsteps. Soon he became the "French specialist" [257] in the field of organometallic catalysis, and in the 1960s and 1970s developed two valuable homogeneous catalytic processes. The first is the "Dimersol Process", which is based on a nickel catalyst and exists in two versions. In the "gasoline" version, propene is dimerized to high-octane isohexenes, which are presently produced on an annual scale of approximately 3.5 million tonnes. In the "chemical" version, *n*-butenes are dimerized to isooctenes, which are converted to precursors for plastics via the oxo reaction. The second process, called the "Alphabutol Process", is based on a titanium catalyst and consists in the dimerization of ethene to 1-butene, used as co-monomer for low-density linear polyethene.

For Chauvin, 1964 was "an especially magical year", since it brought forth the isolation of the first carbene metal complex by Fischer, the catalytic polymerization of cyclopentene via ring opening by Natta, and the disproportionation of olefins catalyzed by a molybdenum- or tungsten-based heterogeneous catalyst, discovered by Banks and Bailey of Phillips Petroleum. Although apparently these results had nothing in common, Chauvin assumed that the "disproportionation and ring opening were part of the same reaction" [227]. With his coworker Jean-Louis Hérisson he designed a series of ingenious experiments, the results of which led to the conclusion that the olefin disproportionation (later called metathesis) is a chain reaction and, with a catalyst generated from WCl_6 and LiMe or $SnMe_4$, "the existence on the tungsten of a methylidene group" drives the propagation steps [227]. Given the fact that in January 1971, when the seminal paper by Chauvin and Hérisson appeared [165], Schrock's work on carbene complexes of tantalum, molybdenum and tungsten in higher oxidation states was unknown, part of the chemical community were rather sceptical about Chauvin's mechanistic scheme. However, based on preparative and kinetic investigations by Casey, Katz, Grubbs, and others, more and more evidence accumulated during the mid and late 1970s illustrating that the "non-pairwise" or "odd-carbon" mechanism proposed by Chauvin was correct. Final proof was provided by Schrock's and Grubbs' studies in the 1980s and 1990s and since then, Chauvin's ideas have been universally accepted [258]. He continued to work in the field of catalysis at the *Institut Francais du Pétrole* until his retirement in 1995 and not long afterwards became Emeritus Director of Research in the surface organometallic chemistry laboratory of CNRS/CPE in Lyon. Yves Chauvin is a member of the French Académie des Sciences and, according to Jean-Marie Basset, "one of the spiritual fathers in the world of the industrial homogeneous catalysis".

Richard Royce Schrock was the youngest of the three Nobel laureates, who received the prize in Chemistry 2005. In the press release from the Royal Swedish Academy of Sciences it was emphasized that, based on the ideas of Chauvin, he and Grubbs "made metathesis into one of organic chemistry's most important reactions". And it continued that "Schrock was the first to produce an efficient metal-compound catalyst for metathesis".

Schrock was born on the 4th of January 1945 in Berne, a northeast Indiana farming community proud of its Swiss heritage. His father was a carpenter and in his childhood Richard spent much time in the woodworking shop discovering "that it is not easy to drive nails in maple" [259]. When he received, from his older brother, a chemistry set on his eighth birthday, he created a small laboratory in a storage area of the family house and used his woodworking skills "to build shelves for the ever expanding collection of test tubes, beakers and flasks". While he recalled "there were no serious mishaps", his mother remembered that once the local fire brigade was called to the house. Fortunately, only a small rug was burning and Richard was "quick to think and act".

After moving with his parents to San Diego in 1959, he finished high school in 1963 and, after being accepted as an undergraduate student of the University of California, decided to attend the relatively new campus at Riverside. During a summer job, he began research in what could be called "atmospheric chemistry" and learned to measure low concentrations of photolysis products using "a temperamental, delicate, almost impossible to align, multi-pass Perkin-Elmer IR machine". In view of these problems, he was really surprised to find his name on a paper entitled "*Rate Constant Ratios During Nitrogen Dioxide*

Photolysis" 1 year later. In his last semester at Riverside, he took an inorganic chemistry course taught by Frederick Hawthorne and, though he had enjoyed organic chemistry, became fascinated by the idea of "exploring the chemistry of all elements in the periodic table" [259].

Following the advice of one of his teachers, he applied to Harvard University for graduate studies and was accepted. Before he arrived, he had no plan concerning the kind of chemistry he wanted to pursue, and it was just by chance that on one day he passed by John Osborn's office, who had recently been appointed as an assistant professor. Osborn had studied organorhodium chemistry with Geoffrey Wilkinson at Imperial College London and discovered the unusually high catalytic activity of $RhCl(PPh_3)_3$ ("Wilkinson's catalyst") in hydrogenation reactions. He told Schrock about the excitement of creating new transition metal compounds and the manifold of applications those species could have, particularly in catalysis. Schrock felt that this "sounded like what I wanted to do ... and so I signed on". In the following years, he prepared a series of cationic phosphine rhodium complexes, which turned out to be quite useful for asymmetric hydrogenation. He obtained a Ph.D. from Harvard University in 1971, and with a fellowship from the National Science Foundation he then moved to Cambridge University to work in the group of Jack Lewis for one year. There he met Earl Muetterties, who offered him a position in the Central Research Department of DuPont in Wilmington, which he accepted.

As was mentioned above, Schrock worked at DuPont in the group of George Parshall and shared a lab with Fred Tebbe, who is well known for the discovery of the Tebbe reagent [260]. Schrock's research was on organotantalum chemistry which early on led to the preparation of the tantalum carbene $(Me_3CCH_2)_3Ta(CHCMe_3)$. This compound marked the beginning of high oxidation state carbene or alkylidene chemistry and was soon followed by the synthesis of the first isolable transition metal methylene complex. In spring 1975, Richard accepted an offer from MIT which fulfilled his "dream of obtaining an academic position at a top institution" [259]. During the next decade, well-defined catalysts for both the alkene and alkyne metathesis were prepared which contained sterically protecting imido and/or alkoxide ligands. At the same time, the activation of dinitrogen became another hot topic in his group and more recent they achieved the reduction of N_2 to NH_3 with a molybdenum catalyst in an identified manner at room temperature and pressure.

In 1980, at age 35, Schrock (Fig. 8.6) was promoted to full professor at MIT and named the Frederick G. Keyes Professor of Chemistry in 1989. He was elected to the American Academy of Arts and Science and the National Academy of Science. Apart from the Nobel Prize, he received inter alia the ACS Award in Organometallic Chemistry, the ACS Award in Inorganic Chemistry, an ACS Arthur C. Cope Scholar Award, and the Sir Geoffrey Wilkinson Medal. In the concluding remarks of his Nobel Lecture [78], he emphasized that he and others "have come an enormous distance in the last 30 years, from

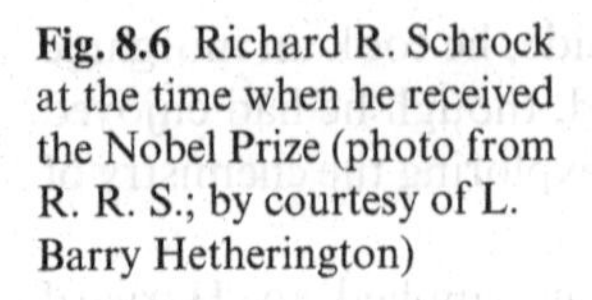

Fig. 8.6 Richard R. Schrock at the time when he received the Nobel Prize (photo from R. R. S.; by courtesy of L. Barry Hetherington)

'ill-defined' metathesis catalysts to those whose structure and reactivity in solution ... we can control with pinpoint precision. Fundamental problems with known catalysts remain ... (but) I expect these challenges to be met and hope that the consequences of the synthesis and fundamental study of new types of catalysts, and the application of them to a wide variety of problems, will continue to be felt in the coming years". It is not overoptimistic to predict that this hope will probably be fulfilled.

Robert H. Grubbs, the tallest of the three recipients of the 2005 Nobel Prize, came from "the other side", which means from organic and not from inorganic chemistry like Fischer, Wilkinson or Schrock, to discover the beauties and power of organo-transition metal compounds. In spite of this "black mark", he succeeded, as Professor Per Ahlberg complemented him, in "that they [the Grubbs catalysts] have become the first well-defined catalysts for general use in metathesis in ordinary laboratories, and new visions of the opportunities of organic synthesis have emerged from their use" [256].

Grubbs was born on the 27th of February 1942 in rural Kentucky near Possum Trott, where the families of both his parents lived. His father was a practical engineer and, after he served 2 years in the army during World War II, took night classes to become a diesel mechanic. Given this background, it was not surprising that in his childhood was, as he recalled, "always interested in building things. Instead of buying candy, I would purchase nails which I used to construct things out of scrap wood" [261].

Grubbs' interest in science started in junior high school, where an outstanding teacher introduced him "to the joys of science". Nevertheless, coming from a rural area and having spent many hours working on farms, in retrospective he considered it "natural" to start college as an agricultural chemistry major at the University of Florida in Gainesville. There he met Merle Battiste, a new faculty member, who started him "on the chemical journey" and saved him "from a life of analyzing animal matter". In Battiste's group, he worked on reactions of cyclopropenes, a class of compounds which later opened the way to the first generation Grubbs catalysts. Battiste also gave him E. S. Gould's textbook "*Mechanism and Structure in Organic Chemistry*" with which he was fascinated "by being able to do rather simple chemical transformations to learn about the details of how organic compounds reacted at the molecular level". As a result, he later conceded that "building new molecules was even more fun than building houses".

Although it would have been best for Battiste if Grubbs had stayed in his laboratory and finished his degree, he encouraged him to move to Columbia University and work with Ronald Breslow. However, before he left for New York he listened to a seminar by Rowland Pettit (for biography see Chap. 5), who was talking about the stabilization of unstable molecules such as cyclobutadiene by coordination to transition metals. It was an inspirational talk and thus, after he joined Breslow's group, he initially chose a project related to Pettit's work for his graduate studies. He finished with investigating the anti-aromaticity of cyclobutadiene and obtained his Ph.D. from Columbia University in 1968.

In following the desire to turn his research toward organo-transition metal chemistry, he accepted an offer by Jim Collman to work in his group at Stanford University. At that time, Collman had developed a systematic method of discussing reaction types that provided the basis for understanding the individual steps in catalytic cycles. Grubbs considered olefin metathesis as one of the most exciting of these catalytic processes and soon felt encouraged to study the mechanism of this reaction. He finished his fellowship at Stanford in 1969 and moved to Michigan State University which was the only school that offered him a position for starting an independent academic career. In the next "nine very productive years" [261] he began his work on olefin metathesis and "a number of other areas of catalysis".

In 1978, he was offered a chair at the California Institute of Technology (Caltech) which he accepted. The rich intellectual environment at Caltech allowed him to grow his research in many directions and also to return to organic synthesis and the development of new synthetic processes based on transition metals. Using the Tebbe reagent, his group not only prepared the first metallacyclobutane, which was active in metathesis, but also developed a general route for the conversion of esters to vinyl ethers. The quest for a more stable reagent for carrying out olefin metathesis finally led to the synthesis of the ruthenium-based catalysts and their history is recorded above. After early frustrations of attempts to get several major companies producing polymers involved in this research, Grubbs, with the aid of some coworkers, established Materia, Inc., where many of the commercial applications of the ruthenium technology and other metathesis-based products have since been developed.

Grubbs (Fig. 8.7) was named Victor and Elizabeth Atkins Professor of Chemistry at Caltech in 1990 and, apart from the Nobel Prize, has received a long list of awards including the ACS Award in Organometallic Chemistry, the ACS Award in Polymer Chemistry, the Benjamin Franklin Medal in Chemistry, the ACS Herbert C. Brown Award for Creative Research in Synthetic

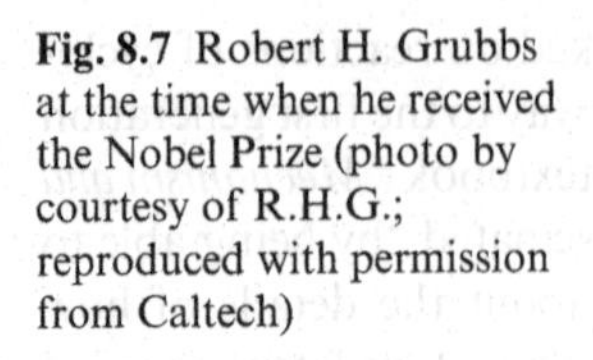

Fig. 8.7 Robert H. Grubbs at the time when he received the Nobel Prize (photo by courtesy of R.H.G.; reproduced with permission from Caltech)

Methods, the ACS Arthur C. Cope Award, and the Paul Karrer Gold Medal. He is a Fellow of the National Academy of Science, the American Academy of Arts and Science, and the Royal Society of Chemistry. In his Nobel Lecture [163], he conceded that starting "from seemingly unrelated work, our investigations into the fundamental chemistry of this transformation have been an exciting journey, with major advances often resulting from complete surprises, mistakes, and simple intuition. Ultimately, these efforts have contributed to olefin metathesis becoming the indispensable synthetic tool that it is today".

References

1. M. Regitz, Stable Carbenes – Illusion or Reality, *Angew. Chem. Int. Ed. Engl.* **30**, 674–676 (1991).
2. J. P. A. Dumas, and E. M. Peligot, Mémoire sur l'Esprit de bois et sur les divers Composés Éthérés qui en proviennent, *Ann. Chim. Phys.* **58**, 5–74 (1835).
3. J. R. Partington, *A History of Chemistry* (Macmillan, London, 1964, Vol. 4, p. 353).
4. W. Kirmse, *Carbene, Carbenoide und Carbenanaloge* (Verlag Chemie, Weinheim, 1969).
5. H. Staudinger, Über aliphatische Diazoverbindungen und Ketene, *Helv. Chim. Acta* **5**, 87–103 (1922).
6. Houben-Weyl, *Methoden der Organischen Chemie* (Ed. M. Regitz, Thieme Verlag, Stuttgart, 1989, Vol. E19b).
7. J. Hine, *Divalent Carbon* (The Ronald Press, New York, 1964).
8. W. Kirmse, *Carbene Chemistry, 2. Ed.* (Academic Press, New York, 1971).
9. M. Jones, and R. A. Moss, *Carbenes I* (Wiley, New York, 1971).
10. R. A. Moss, and M. Jones, *Carbenes II* (Wiley, New York, 1975).
11. H.-J. Schönherr, and H.-W. Wanzlick, 1.3.4.5-Tetraphenyl-imidazoliumperchlorat, *Liebigs Ann. Chem.* **731**, 176–179 (1970).
12. H.-W. Wanzlick, and H.-J. Schönherr, Direct Synthesis of a Mercury Salt-Carbene Complex, *Angew. Chem. Int. Ed. Engl.* **7**, 141–142 (1968).
13. K. Öfele, 1,3-Dimethyl-4-imidazolinyliden-(2)-pentacarbonylchrom – Ein neuer Übergangsmetall-Carben-Komplex, *J. Organomet. Chem.* **12**, P42–P43 (1968).
14. A. J. Arduengo, III, R. L. Harlow, and M. Kline, A Stable Crystalline Carbene, *J. Am. Chem. Soc.* **113**, 361–363 (1991).

15. A. J. Arduengo, III, and R. Krafczyk, Auf der Suche nach stabilen Carbenen, *Chemie in unserer Zeit* **32**, 6–14 (1998).
16. A. J. Arduengo, III, Looking for Stable Carbenes: The Difficulty in Starting Anew, *Acc. Chem. Res.* **32**, 913–921 (1999).
17. M. Regitz, Nucleophilic Carbenes: An Incredible Renaissance, *Angew. Chem. Int. Ed. Engl.* **35**, 725–728 (1996).
18. W. A. Herrmann, and C. Köcher, N-Heterocyclic Carbenes, *Angew. Chem. Int. Ed. Engl.* **36**, 2162–2187 (1997).
19. D. Bourissou, O. Guerret, F. P. Gabbai, and G. Bertrand, Stable Carbenes, *Chem. Rev.* **100**, 39–91 (2000).
20. G. Bertrand (Ed.) *Carbene Chemistry: From Fleeting Intermediates to Powerful Reagents* (Marcel Decker, New York, 2002).
21. G. W. Nyce, S. Csihony, R. M. Waymouth, and J. L. Hedrick, A General and Versatile Approach to Thermally Generated N-Heterocyclic Carbenes, *Chem. Eur. J.* **10**, 4073–4079 (2004).
22. F. E. Hahn, Heterocyclic Carbenes, *Angew. Chem. Int. Ed.* **45**, 1348–1352 (2006).
23. A. J. Arduengo, III, J. R. Goerlich, R. Krafczyk, and W. J. Marshall, 1,3,4,5-Tetraphenylimidazol-2-ylidene: The Realization of Wanzlick's Dream, *Angew. Chem. Int. Ed.* **37**, 1963–1965 (1998).
24. E. O. Fischer, Early Days of Transition-Metal Carbene Chemistry, in: *Advances in Metal Carbene Chemistry* (Ed. U. Schubert, NATO NSI Series Vol. 269, Kluwer Academic Publ., Dordrecht, 1989, pp. 1–9).
25. E. O. Fischer, and A. Maasböl, Tungsten Carbonyl-carbene Complex, *Angew. Chem. Int. Ed. Engl.* **3**, 580 (1964).
26. R. Aumann, and E. O. Fischer, Addition of Isocyanides to Transition Metal Carbene Complexes, *Angew. Chem. Int. Ed. Engl.* **6**, 879–880 (1967).
27. O. S. Mills, and A. D. Redhouse, Crystal and Molecular Structure of Phenylmethoxycarbenepentacarbonylchromium, *J. Chem. Soc. A* **1968**, 642–647.
28. E. O. Fischer, Structure, Bonding and Reactivity of (Stable) Transition Metal Carbonyl Carbene Complexes, *Pure Appl. Chem.* **24**, 407–423 (1970).
29. E. O. Fischer, Recent Aspects of Transition Metal Carbonyl Carbene Complexes, *Pure Appl. Chem.* **30**, 353–372 (1972).
30. E. O. Fischer, Auf dem Weg zu Carben- und Carbin-Komplexen, *Angew. Chem.* **86**, 651–663 (1974) [Nobel Lecture].
31. E. O. Fischer, On the Way to Carbene and Carbyne Complexes, *Adv. Organomet. Chem.* **14**, 1–32 (1976).
32. H. Fischer, The Synthesis of Carbene Complexes, in: *Transition Metal Carbene Complexes* (Eds. K. H. Dötz, H. Fischer, P. Hofmann, F. R. Kreissl, U. Schubert, and K. Weiss, Verlag Chemie, Weinheim, 1983, pp. 4–8).
33. H. Werner, Substitutionsreaktionen von Metallcarbonyl-Komplexen des Chroms, Molybdäns und Wolframs; einige neue kinetische und mechanistische Aspekte, *J. Organomet. Chem.* **94**, 285–302 (1975).
34. C. F. Bernasconi, Developing the Physical Organic Chemistry of Fischer Carbene Complexes, *Chem. Soc. Rev.* **26**, 299–307 (1997).
35. K. Weiss, and E. O. Fischer, Der Pentacarbonyl(phenylcarben)chrom(0)-Rest als Aminoschutzgruppe für Peptidsynthesen, *Chem. Ber.* **106**, 1277–1284 (1973).
36. K. Weiss, and E. O. Fischer, Pentacarbonyl(organylcarben)chrom– und –wolfram-Reste als Aminoschutzgruppen für Peptidsynthesen, *Chem. Ber.* **109**, 1868–1886 (1976).
37. E. O.Fischer, and K. H. Dötz, Synthese von Cyclopropanderivaten aus Vinyläthern mit Hilfe von Übergangsmetall-Carbonyl-Carben-Komplexen, *Chem. Ber.* **105**, 3966–3973 (1972).
38. M. D. Cooke, and E. O. Fischer, Proof of the Absence of a Free Carbene in the Preparation of a Cyclopropane from a Metal-Carbene Complex, *J. Organomet. Chem.* **56**, 279–284 (1973).

39. C. P. Casey, and R. L. Anderson, Thermolysis of (2-Oxacyclopentylidene)-pentacarbonylchromium(0): Evidence Against Free Carbenes in Thermal Decomposition of Metal-Carbene Complexes, *J. Chem Soc., Chem. Comm.* **1975**, 895–896.
40. E. O. Fischer, and H.-J. Beck, Transfer of Methoxy(phenyl)carbene Ligands from Molybdenum to Iron and Nickel, *Angew. Chem. Int. Ed. Engl.* **9**, 72–73 (1970).
41. A. De Renzi, and E. O. Fischer, 1,1-Dihalo-2-phenylvinyl Methyl Ethers from [Phenyl-(methoxy)carbene]pentacarbonylchromium(0) and Phenyl(trihalomethyl)mercurials, *Inorg. Chim. Acta* **8**, 185–189 (1974).
42. E. O. Fischer, G. Kreis, C. G. Kreiter, J. Müller, G. Huttner, and H. Lorenz, *trans*-Halo[alkyl(aryl)carbyne]tetracarbonyl Complexes of Chromium, Molybdenum and Tungsten. A New Complex Type with Transition Metal-Carbon Triple Bond, *Angew. Chem. Int. Ed. Engl.* **12**, 564–565 (1973).
43. G. Huttner, H. Lorenz, and W. Gartzke, Transition Metal-Carbon Triple Bonds, *Angew. Chem. Int. Ed. Engl.* **13**, 609–610 (1974).
44. E. O. Fischer, and U. Schubert, Darstellung, Eigenschaften und Reaktionsverhalten von Übergangsmetallcarbonyl-Carbin-Komplexen, *J. Organomet. Chem.* **100**, 59–81 (1975).
45. E. O. Fischer, U. Schubert, and H. Fischer, Selectivity and Specificity in Chemical Reactions of Carbene and Carbyne Metal Complexes, *Pure Appl. Chem.* **50**, 857–870 (1978).
46. H. P. Kim, and R. J. Angelici, Transition Metal Complexes with Terminal Carbyne Ligands, *Adv. Organomet. Chem.* **27**, 51–111 (1987).
47. A. Mayr, Comments on the Chemistry of Low-Valent Alkylidyne Complexes of the Group 6 Transition Metals, *Comments Inorg. Chem.* **10**, 227–266 (1990).
48. A. Mayr, and H. Hoffmeister, Recent Advances in the Chemistry of Metal-Carbon Triple Bonds, *Adv. Organomet. Chem.* **32**, 227–324 (1991).
49. H. Fischer, P. Hofmann, F. R. Kreissl, R. R. Schrock, U. Schubert, and K. Weiss, *Carbyne Complexes* (VCH Publishers, Weinheim-New York, 1988).
50. K. Öfele, Pentacarbonyl(2,3-diphenylcyclopropenylidene)chromium(0), *Angew. Chem. Int. Ed. Engl.* **7**, 950 (1968).
51. C. P.Casey, and T. J. Burkhardt, (Diphenylcarbene)pentacarbonyltungsten(0), *J. Am. Chem. Soc.* **95**, 5833–5834 (1973).
52. C. P.Casey, T. J. Burkhardt, C. A. Bunnell, and J. C. Calabrese, Synthesis and Crystal Structure of Diphenylcarbene(pentacarbonyl)tungsten(0), *J. Am. Chem. Soc.* **99**, 2127–2134 (1977).
53. E. O. Fischer, W. Held, and F. R. Kreissl, Synthese von Pentacarbonylalkylchromaten(—I) und -wolframaten(—I), *Chem. Ber.* **110**, 3842–3848 (1977).
54. E. M. Badley, J. Chatt, R. L. Richards, and G. A. Sim, The Reactions of Isocyanide Complexes Platinum(II): a Convenient Route to Carbene Complexes, *Chem. Comm.* **1969**, 1322–1323.
55. J. S. Miller, and A. L. Balch, Preparation and Reactions of Tetrakis(methyl isocyanide) Complexes of Divalent Nickel, Palladium, and Platinum, *Inorg. Chem.* **11**, 2069–2074 (1972).
56. L. Tschugajeff, M. Skanawy-Grigorjewa, and A. Posnjak, Über die Hydrazin-Carbylamin-Komplexe des Platins, *Z. anorg. allg. Chem.* **148**, 37–42 (1925). [preliminary communication: *J. Russ. Chem. Soc.* **47**, 776 (1915)].
57. G. Rouschias, and B. L. Shaw, The Chemistry and Structure of Chugaev's Salt and Related Compounds containing a Cyclic Carbene Ligand, *J. Chem. Soc. A* **1971**, 2097–2104.
58. A. Burke, A. L. Balch, and J. H.Enemark, Palladium and Platinum Complexes Resulting from the Addition of Hydrazine to Coordinated Isocyanide, *J. Am. Chem. Soc.* **92**, 2555–2557 (1970).
59. D. J. Cardin, B. Cetinkaya, M. F. Lappert, L. J. Manojlovic-Muir, and K. W. Muir, An Electron-rich Olefin as a Source of Co-ordinated Carbene; Synthesis of *trans*-$PtCl_2[C(NPhCH_2)_2]PEt_3$, *Chem. Comm.* **1971**, 400–401.
60. D. J. Cardin, B. Cetinkaya, and M. F. Lappert, Transition Metal-Carbene Complexes, *Chem. Rev.* **72**, 545–574 (1972).
61. D. J. Cardin, B. Cetinkaya, M. J. Doyle, and M. F. Lappert, Chemistry of Transition-Metal Carbene Complexes and Their Role as Reaction Intermediates, *Chem. Soc. Rev.* **2**, 99–144 (1972).

62. M. F. Lappert, The Coordination Chemistry of Bivalent Group IV Donors; Nucleophilic-Carbene and Dialkylstannylene Complexes, *J. Organomet. Chem.* **100**, 139–159 (1975).
63. M. F. Lappert, The Coordination Chemistry of Electron-rich Alkenes (Enetetramines), *J. Organomet. Chem.* **358**, 185–214 (1988).
64. M. F. Lappert, Contributions to the Chemistry of Carbenemetal Chemistry, *J. Organomet. Chem.* **690**, 5467–5473 (2005).
65. M. F. Lappert, and A. J. Oliver, A Three-Fragment Oxidative Addition Reaction as a Route to Transition Metal Carbene Complexes: Imidoyl Halides and Rhodium(I) Compounds as Precursors for Rhodium(III) Carbenes, *J. Chem Soc., Chem. Comm.* **1972**, 274–275.
66. P. B. Hitchcock, M. F. Lappert, G. M. McLaughlin, and A. J. Oliver, Complexes of Imidoyl Chloride and Rhodium(I) Precursors, and the Crystal and Molecular Structure of Carbonyltri-iodo-[α-(N-methyl-α-methyliminobenzylamino)benzylidene-N,C]rhodium, *J. Chem. Soc., Dalton Trans.* **1974**, 68–74.
67. B. Cetinkaya, M. F. Lappert, and K. Turner, Three-fragment Oxidative Addition of Chloroform-iminium or –amidinium Chlorides to Rh^{I} or Pt^{II} Substrates; Complexes of the Secondary Carbene $CHNMe_2$, *J. Chem Soc., Chem. Comm.* **1972**, 851–852.
68. B. Cetinkaya, M. F. Lappert, G. M. McLaughlin, and K. Turner, Chloromethylene-ammonium Chlorides. Electron-Rich Carbenoids as Precursors to Secondary Carbene Metal Complexes; Crystal and Molecular Structure of Trichloro(dimethylamino-methylene)bis(triethylphosphine)rhodium(III), *J. Chem. Soc., Dalton Trans.* **1974**, 1591–1599.
69. M. J. Doyle, M. F. Lappert, G. M. McLaughlin, and J. McMeeking, The Synthesis of Alkylideneamido(carbene)rhodium(I) Complexes and Related Chemistry; the Crystal and Molecular Structure of *trans*-$Rh[N{=}C(CF_3)_2][C(NMeCH_2)_2](PPh_3)_2$, *J. Chem. Soc., Dalton Trans.* **1974**, 1494–1501.
70. P. J. Davidson, and M. F. Lappert, Stabilisation of Metals in a Low Co-ordination Environment using the Bis(trimethylsilyl)methyl Ligand; Coloured Sn^{II} and Pb^{II} Alkyls, $M[CH(SiMe_3)_2]_2$, *J. Chem Soc., Chem. Comm.* **1973**, 317.
71. J. Chatt, A New Structure for Olefin Co-Ordination Compounds, *Research* **4**, 180–183 (1950).
72. F. G. A. Stone, *Leaving No Stone Unturned* (American Chemical Society, Washington, DC, 1993, p. 209).
73. R. R. Schrock, An "Alkylcarbene" Complex of Tantalum by Intramolecular α-Hydrogen Abstraction, *J. Am. Chem. Soc.* **96**, 6796–6797 (1974).
74. G. L. Juvinall, σ-Bonded Alkyl Compounds of Niobium and Tantalum. Trimethyldichloroniobium and Trimethyldichlorotantalum, *J. Am. Chem. Soc.* **86**, 4202–4203 (1964).
75. R. R.Schrock, and P. Meakin, Pentamethyl Complexes of Niobium and Tantalum, *J. Am. Chem. Soc.* **96**, 5288–5290 (1974).
76. R. R. Schrock, Preparation and Characterization of $M(CH_3)_5$ (M = Nb, Ta) and $Ta(CH_2C_6H_5)_5$ and Evidence for Decomposition by α-Hydrogen Atom Abstraction, *J. Organomet. Chem.* **122**, 209–225 (1976).
77. R. R.Schrock, The Reaction of Niobium and Tantalum Neopentylidene Complexes with the Carbonyl Function, *J. Am. Chem. Soc.* **98**, 5399–5400 (1976).
78. R. R. Schrock, Multiple Metal-Carbon Bonds for Catalytic Metathesis Reactions, *Angew. Chem. Int. Ed.* **45**, 3748–3759 (2006) [Nobel Lecture].
79. L. Li, M. Hung, and Z. Xue, Direct Observation of $(Me_3ECH_2)_5Ta$ (E = C, Si) as the Precursors to $(Me_3ECH_2)_3Ta{=}CHEMe_3$ and $(Me_3SiCH_2)_2Ta(\mu\text{-}CSiMe_3)_2Ta(CH_2SiMe_3)_2$. Kinetic and Mechanistic Studies of the Formation of Alkylidene and Alkylidyne Ligands, *J. Am. Chem. Soc.* **117**, 12746–12750 (1995).
80. L. J. Guggenberger, and R. R. Schrock, A Tantalum Carbyne Complex, *J. Am. Chem. Soc.* **97**, 2935 (1975).
81. R. R. Schrock, The First Isolable Transition Metal Methylene Complex and Analogs. Characterization, Mode of Decomposition, and Some Simple Reactions, *J. Am. Chem. Soc.* **97**, 6577–6578 (1975).

82. R. R. Schrock, and L. J. Guggenberger, Structure of Bis(cyclopentadienyl)methylmethylenetantalum and the Estimated Barrier to Rotation about the Tantalum–Methylene Bond, *J. Am. Chem. Soc.* **97**, 6578–6579 (1975).
83. R. R. Schrock, and P. R. Sharp, Preparation and Characterization of Ta(η^5-$C_5H_5)_2(CH_2)(CH_3)$, a Study of its Decomposition, and Some Simple Reactions, *J. Am. Chem. Soc.* **100**, 2389–2399 (1978).
84. R. R. Schrock, Alkylidene Complexes of Niobium and Tantalum, *Acc. Chem. Res.* **12**, 98–104 (1979).
85. R. R. Schrock, L. W. Messerle, C. D. Wood, and L. J. Guggenberger, Preparation and Characterization of Several Alkylidene Complexes, M(η^5-$(C_5H_5)_2$(alkylidene)X (M = Ta or Nb), and the X-Ray Structure of Ta(η^5-$(C_5H_5)_2(CHC_6H_5)(CH_2C_6H_5)$. An Investigation of Alkylidene Ligand Rotation, *J. Am. Chem. Soc.* **100**, 3793–3800 (1978).
86. P. W. Jolly, and R. Pettit, Evidence for a Novel Metal-Carbene System, *J. Am. Chem. Soc.* **88**, 5044–5045 (1966).
87. W. A. Herrmann, B. Reiter, and H. Biersack, Einbau von Methylene in Mangan-Komplexe, *J. Organomet. Chem.* **97**, 245–251 (1975).
88. W. A. Herrmann, The Methylene Bridge, *Adv. Organomet. Chem.* **20**, 159–263 (1982).
89. W. A. Herrmann, Reaktionen aliphatischer Diazoverbindungen mit thermolabilen Mangan-Komplexen, *Chem. Ber.* **108**, 486–499 (1975).
90. W. A. Herrmann, Organometallic Syntheses with Diazoalkanes, *Angew. Chem. Int. Ed. Engl.* **17**, 800–813 (1978).
91. J. D. Fellmann, G. A. Rupprecht, C. D. Wood, and R. R. Schrock, Bisneopentylidene Complexes of Niobium and Tantalum, *J. Am. Chem. Soc.* **100**, 5964–5966 (1978).
92. M. Brookhart, and M. L. H. Green, Carbon–Hydrogen–Transition Metal Bonds, *J. Organomet. Chem.* **250**, 395–408 (1983).
93. M. Brookhart, M. L. H. Green, and L.-L. Wong, Carbon–Hydrogen–Transition Metal Bonds, *Progr. Inorg. Chem.* **36**, 1–124 (1988).
94. G. A. Rupprecht, L. W. Messerle, J. D. Fellmann, and R. R. Schrock, Octahedral Alkylidene Complexes of Niobium and Tantalum by Ligand-Promoted α Abstraction, *J. Am. Chem. Soc.* **102**, 6236–6244 (1980).
95. D. N. Clark, and R. R. Schrock, Tungsten and Molybdenum Neopentylidyne and Some Tungsten Neopentylidene Complexes, *J. Am. Chem. Soc.* **100**, 6774–6776 (1978).
96. R. R. Schrock, High-Oxidation-State Molybdenum and Tungsten Alkylidyne Complexes, *Acc. Chem. Res.* **19**, 342–348 (1986).
97. J. S. Murdzek, and R. R. Schrock, High Oxidation State Alkylidyne Complexes, in: *Carbyne Complexes* (Eds. H. Fischer, P. Hofmann, F. R. Kreissl, R. R. Schrock, U. Schubert, and K. Weiss, VCH Publishers, Weinheim-New York, 1988, p. 147–203).
98. L. G. McCullough, R. R. Schrock, J. C. Dewan, and J. S. Murdzek, Preparation of Trialkoxymolybdenum(VI) Alkylidyne Complexes. Their Reactions with Acetylenes, and the X-Ray Structure of Mo$[C_3(CMe_3)_2][OCH(CF_3)_2]_2(C_5H_5N)_2$, *J. Am. Chem. Soc.* **107**, 5987–5998 (1985).
99. R. R. Schrock, High Oxidation State Multiple Metal–Carbon Bonds, *Chem. Rev.* **102**, 145–179 (2002).
100. R. R. Schrock, High Oxidation State Alkylidene and Alkylidyne Complexes, *Chem. Comm.* **2005**, 2773–2777.
101. R. A. Andersen, M. H. Chisholm, J. F. Gibson, W. W. Reichert, I. P.Rothwell, and G. Wilkinson, (Trimethylsilyl)methylidene and (Trimethylsilyl)methylidyne Compounds of Molydenum and Tungsten: $(Me_3SiCH_2)_3M{\equiv}CSiMe_3$ (M = Mo, W) and $(Me_3SiCH_2)_3Mo{=}CHSiMe_3$, *Inorg. Chem.* **20**, 3934–3936 (1981).
102. T. J. Katz, and S. J. Lee, Initiation of Acetylene Polymerization by Metal Carbenes, *J. Am. Chem. Soc.* **102**, 422–424 (1980).

103. J. Kress, M. Wesolek, J.-P. Le Ny, and J. A. Osborn, Molecular Complexes for Efficient Metathesis of Olefins. The Oxo-Ligand as a Catalyst-Cocatalyst Bridge and the Nature of the Active Species, *J. Chem. Soc., Chem. Comm.* **1981**, 1039–1040.
104. J. Kress, M. Wesolek, and J. A. Osborn, Tungsten(IV) Carbenes for the Metathesis of Olefins. Direct Observation and Identification of the Chain Carrying Carbene Complexes in a Highly Active Catalyst System, *J. Chem. Soc., Chem. Comm.* **1982**, 514–516.
105. R. R. Schrock, M. L. Listemann, and L. G. Sturgeoff, Metathesis of Tungsten-Tungsten Triple Bonds with Acetylenes and Nitriles To Give Alkylidyne and Nitrido Complexes, *J. Am. Chem. Soc.* **104**, 4291–4293 (1982).
106. M. L. Listemann, and R. R. Schrock, A General Route to Tri-*tert*-butoxytungsten Alkylidyne Complexes. Scission of Acetylenes by Tungsten Hexa-*tert*-butoxide, *Organometallics* **4**, 74–83 (1985).
107. F. Huq, W. Mowat, A. C. Skapski, and G. Wilkinson, Crystal Structure of Bis-μ-(trimethylsilylmethylidyne)tetrakis(trimethylsilylmethyl)diniobium(V). A New Type of Carbon Bridging Group, *Chem. Comm.* **1971**, 1477–1478.
108. M. H. Chisholm, J. C. Huffman, and I. P. Rothwell, Addition of Alkynes to Hexaalkoxydimolybdenum (M≡M) Compounds and Structure of μ-Ethyne-hexaisopropoxydipyridinodimolybdenum, *J. Am. Chem. Soc.* **103**, 4245–4246 (1981).
109. M. R. Churchill, J. W. Ziller, L. McCullough, S. F. Pedersen, and R. R. Schrock, $W(\eta^5\text{-}C_5H_5)[C(Ph)C(CMe_3)C(Ph)]Cl_2$: A Molecule Having a Localized, Nonplanar, Fluxional Metallacyclobutadiene Ring, *Organometallics* **2**, 1046–1048 (1983).
110. M. R. Churchill, and W. J. Youngs, X-Ray Crystal Structure of $W(\equiv CCMe_3)(=CHCMe_3)(CH_2CMe_3)(Me_2PCH_2CH_2PMe_2)$, an Ordered Five-coordinate Tungsten(VI) Complex with Metal–Alkyl, Metal–Alkylidene, and Metal–Alkylidyne Linkages, *J. Chem. Soc., Chem. Comm.* **1979**, 321–322.
111. S. J. Holmes, D. N. Clark, H. W. Turner, and R. R. Schrock, α-Hydride Elimination from Methylene and Neopentylidene Ligands. Preparation and Protonation of Tungsten(IV) Methylidyne and Neopentylidyne Complexes, *J. Am. Chem. Soc.* **104**, 6322–6329 (1982).
112. P. R. Sharp, S. J. Holmes, R. R. Schrock, M. R. Churchill, and H. J. Wasserman, Tungsten Methylidyne Complexes, *J. Am. Chem. Soc.* **103**, 965–966 (1981).
113. K.-Y. Shih, K. Totland, S. W. Seidel, and R. R. Schrock, Spontaneous Loss of Molecular Hydrogen from Tungsten(IV) Alkyl Complexes to Give Alkylidyne Complexes, *J. Am. Chem. Soc.* **116**, 12103–12104 (1994).
114. H. Fischer, The Synthesis of Fischer-Type Carbyne Complexes, in: *Carbyne Complexes* (Eds. H. Fischer, P. Hofmann, F. R. Kreissl, R. R. Schrock, U. Schubert, and K. Weiss, VCH Publishers, Weinheim-New York, 1988, pp. 1–38).
115. F. R. Kreissl, Selected Reactions of Carbyne Complexes, in: *Carbyne Complexes* (Eds. H. Fischer, P. Hofmann, F. R. Kreissl, R. R. Schrock, U. Schubert, and K. Weiss, VCH Publishers, Weinheim-New York, 1988, pp. 99–146).
116. A. J. Hartshorn, and M. F. Lappert, The Role of C-Chlorocarbenemetal Complexes in Carbene– and Carbyne–Metal Complex Chemistry; Experiments with $[Cr(CO)_5\{C(Cl)NMe_2\}]$ and $[Cr(CO)_5(\equiv CNMe_2)]^+$, *J. Chem. Soc., Chem. Comm.* **1976**, 761–762.
117. E. O. Fischer, W. Kleine, and F. R. Kreissl, Chlordiäthylaminocarben-pentacarbonylchrom, *J. Organomet. Chem.* **107**, C23–C25 (1976).
118. H. Fischer, A. Mosch, and W. Kleine, Unusual Selectivity in the Thermal Rearrangement-Elimination Reaction of a Carbene Complex in Solution and in the Solid State, *Angew. Chem. Int. Ed. Engl.* **17**, 842–843 (1978).
119. B. D. Dombeck, and R. J. Angelici, Electrophilic and Oxidative Addition Reactions of Tungsten Thiocarbonyl Complexes, *Inorg. Chem.* **15**, 2397–2402 (1976).
120. A. J. L. Pombeiro, and R. L. Richards, Reactivity of Carbyne and Carbene Complexes of Molybdenum and Tungsten, *Transition Met. Chem.* **5**, 55–59 (1980).
121. A. J. L. Pombeiro, D. L. Hughes, C. J. Pickett, and R. L. Richards, The Aminocarbyne Ligand CNH_2: Metal-centred Synthesis from a Cyanosilane, Preparation and X-Ray

Structure of *trans*-[ReCl(CNH_2)($Ph_2PCH_2CH_2PPh_2$)$_2$]BF_4, *J. Chem. Soc., Chem. Comm.* **1986**, 246–247.
122. H. Fischer, and E. O. Fischer, Umsetzung von Lithiumbenzoyl-pentacarbonyl-wolframat mit Triphenylphosphindihalogeniden; ein neuer Weg zu Carbin-Komplexen, *J. Organomet. Chem.* **69**, C1–C3 (1974).
123. A. Mayr, G. A. McDermott, and A. M. Dorries, Synthesis of (Carbyne)metal Complexes by Oxide Abstraction from Acyl Ligands, *Organometallics* **4**, 608–610 (1985).
124. G. A. McDermott, A. M. Dorries, and A. Mayr, Synthesis of Carbyne Complexes of Chromium, Molybdenum, and Tungsten by Formal Oxide Abstraction from Acyl Ligands, *Organometallics* **6**, 925–931 (1987).
125. E. O. Fischer, A. Ruhs, and F. R. Kreissl, Carbonylsubstitutionen an *trans*-Bromotetracarbonyl(phenylcarbin)-Komplexen von Chrom und Wolfram, *Chem. Ber.* **110**, 805–815 (1977).
126. A. C. Filippou, and E. O. Fischer, Auf dem Weg zu den ersten, carbonylfreien, neutralen und kationischen Diethylaminocarbin-Komplexen des Wolframs, *J. Organomet. Chem.* **365**, 317–323 (1989).
127. R. Hoffmann, Building Bridges Between Inorganic and Organic Chemistry, *Angew. Chem. Int. Ed. Engl.* **21**, 711–725 (1982) [Nobel Lecture].
128. F. G. A. Stone, Metal–Carbon and Metal–Metal Multiple Bonds as Ligands in Transition-Metal Chemistry: The Isolobal Connection, *Angew. Chem. Int. Ed. Engl.* **23**, 89–99 (1984).
129. M. Green, J. C. Jeffery, S. J. Porter, H. Razay, and F. G. A. Stone, Triangulo-Metal Complexes Containing Tungsten with Iron, Cobalt, Rhodium, or Nickel and a Capping Tolylidyne Ligand; Crystal Structure of the Complex [RhFeW(μ_3-CC_6H_4Me-4)(μ-CO)(CO)$_5$(η-C_5H_5)(η-C_9H_7)], *J. Chem Soc., Dalton Trans.* **1982**, 2475–2483.
130. G. A. Carriedo, J. A. K. Howard, and F. G. A. Stone, Complexes of the Pentamethylcyclopentadienylcopper Group and the Crystal Structures of the Compounds [CuPtW(μ_3-CC_6H_4Me-4)(CO)$_2$(PMe_3)$_2$(η-C_5H_5)(η-C_5Me_5)] and [$CuRh_2$(μ-CO)$_2$(η-C_5Me_5)$_3$], *J. Chem Soc., Dalton Trans.* **1984**, 1555–1561.
131. T. V. Ashworth, M. J. Chetcuti, J. A. K. Howard, F. G. A. Stone, S. J. Wisbey, and P. Woodward, Synthesis of the Trimetal Compounds [M{W(μ-CC_6H_4Me-4)(CO)$_2$(η-C_5H_5)}$_2$] (M = Ni, Pd, or Pt) and the Crystal Structures of the Platinum and Nickel Complexes, *J. Chem Soc., Dalton Trans.* **1981**, 763–770.
132. W. R. Roper, Platinum Group Metals in the Formation of Metal–Carbon Multiple Bonds, *J. Organomet. Chem.* **300**, 167–190 (1986).
133. D. F. Christian, and W. R. Roper, Neutral, Cationic, and Dicationic Ruthenium(II) Complexes of the Secondary Carbenes CHNH(*p*-$C_6H_4CH_3$) and CHN(CH_3)(*p*-$C_6H_4CH_3$), *J. Organomet. Chem.* **80**, C35–C38 (1974).
134. T. J. Collins, and W. R. Roper, Thioformyl and Formyl Complexes of Osmium(II), *J. Chem Soc., Chem. Comm.* **1976**, 1044–1045.
135. T. J. Collins, and W. R. Roper, Co-Ordinated Thioformaldehyde Monomer. Synthesis and Reactions of [Os(η^2-CH_2S)(CO)$_2$(PPh_3)$_2$], *J. Chem Soc., Chem. Comm.* **1977**, 901–902.
136. A. F. Hill, W. R. Roper, J. M. Waters, and A. H. Wright, A Mononuclear, Low-Valent, Electron-Rich Osmium Methylene Complex, *J. Am. Chem. Soc.* **105**, 5939–5940 (1983).
137. M. A. Gallop, and W. R. Roper, Carbene and Carbyne Complexes of Ruthenium, Osmium, and Iridium, *Adv. Organomet. Chem.* **25**, 121–198 (1986).
138. H. Werner, R. Flügel, B. Windmüller, A. Michenfelder, and J. Wolf, Synthesis and Reactions of Stable 16-Electron Osmium(0) Complexes [OsCl(NO)(PR_3)$_2$] Including the X-ray Crystal Structure of [$OsCl_2$(NO)(η^1-CH=C=CPh_2)(P-*i*-Pr_3)$_2$], *Organometallics* **14**, 612–618 (1995).
139. R. Flügel, B. Windmüller, O. Gevert, and H. Werner, Synthesis, Molecular Structure and Reactions of Stable Square-Planar 16-Electron Ruthenium(0) Complexes: *trans*-[RuCl(NO)(PR_3)$_2$], *Chem. Ber.* **129**, 1007–1013 (1996).
140. P. J. Brothers, and W. R. Roper, Transition-Metal Dihalocarbene Complexes, *Chem. Rev.* **88**, 1293–1326 (1988).

141. A. Geuther, Ueber die Zersetzung des Chloroforms durch alkoholische Kalilösung, *Ann. Chem. Pharm.* **123**, 121–122 (1862).
142. M. Schmeisser, and H. Schröter, Darstellung des Dichlorcarbens CCl_2 aus CCl_4 und Kohlenstoff, *Angew. Chem.* **72**, 349–350 (1960).
143. M. Schmeisser, H. Schröter, H. Schilder, J. Massone, and F. Rosskopf, Zur Frage der Isolierung von Dichlorcarben CCl_2. Die Pyrolyse von Tetrachlorkohlenstoff und Tetrachloeäthylen bei 1200–1250° im Hochvakuum in Gegenwart von Aktivkohle, *Chem. Ber.* **95**, 1648–1656 (1962).
144. G. Chu, R. A. Moss, and R. R. Sauers, Dichlorodiazirine: A Nitrogenous Precursor of Dichlorocarbene, *J. Am. Chem. Soc.* **127**, 14206–14207 (2005).
145. D. Mansuy, M. Lange, J.-C. Chottard, P. G, P. Morliere, D. Brault, and M. Rougee, Reaction of Carbon Tetrachloride with 5,10,15,20-Tetraphenyl-porphinatoiron(II) [(TPP)FeII]: Evidence for the Formation of the Carbene Complex [(TPP)$Fe^{II}(CCl_2)$], *J. Chem. Soc., Chem. Comm.* **1977**, 648–649.
146. D. Mansuy, New Iron-Porphyrin Complexes with Metal–Carbon Bond – Biological Implications, *Pure Appl. Chem.* **52**, 681–690 (1980).
147. D. Mansuy, J.-P. Lecomte, J.-C. Chottard, and J.-F. Bartoli, Formation of a Complex with a Carbide Bridge Between Two Iron Atoms from the Reaction of (Tetraphenylporphyrin)iron(II) with Carbon Tetraiodide, *Inorg. Chem.* **20**, 3119–3121 (1981).
148. G. R. Clark, K. Marsden, W. R. Roper, and L. J. Wright, Carbonyl, Thiocarbonyl, Selenocarbonyl, and Tellurocarbonyl Complexes Derived from a Dihalocarbene Complex of Osmium, *J. Am. Chem. Soc.* **102**, 1206–1207 (1980).
149. G. R. Clark, S. V. Hoskins, and W. R. Roper, Difluorocarbene Complexes of Ruthenium Derived from Trifluoromethyl Compounds. $RuCl_2(CF_2)(CO)(PPh_3)_2$, $RuCl_2(CFNMe_2)(CO)(PPh_3)_2$, $RuCl_2(CFOMe)(CO)(PPh_3)_2$, and the Structure of $Ru(CF_3)(HgCF_3)(CO)_2(PPh_3)_2$, *J. Organomet. Chem.* **234**, C9–C12 (1982).
150. G. R. Clark, K. Marsden, C. E. F. Rickard, W. R. Roper, and L. J. Wright, Syntheses and Structures of the Chalcocarbonyl Complexes $OsCl_2(CO)(CE)(PPh_3)_2$, *J. Organomet. Chem.* **338**, 393–410 (1988).
151. H. Werner, E. O. Fischer, B. Heckl, and C. G. Kreiter, Kinetik und Mechanismus der Aminolyse von (Methoxyphenylcarben)pentacarbonylchrom(0) – eine Reaktion 4. Ordnung mit negativer Arrhenius-Aktivierungsenergie, *J. Organomet. Chem.* **28**, 367–389 (1971).
152. G. R. Clark, K. Marsden, W. R. Roper, and L. J. Wright, An Osmium Carbyne Complex, *J. Am. Chem. Soc.* **102**, 6570–6571 (1980).
153. G. R. Clark, C. M. Cochrane, K. Marsden, W. R. Roper, and L. J. Wright, Synthesis and some Reactions of a Terminal Carbyne Complex of Osmium. Crystal Structures of $Os(\equiv CR)Cl(CO)(PPh_3)_2$ and $Os(=C[AgCl]R)Cl(CO)(PPh_3)_2$, *J. Organomet. Chem.* **315**, 211–230 (1986).
154. L.-J. Baker, G. R. Clark, C. E. F. Rickard, W. R. Roper, S. D. Woodgate and L. J. Wright, Syntheses and Reactions of the Carbyne Complexes $M(\equiv CR)Cl(CO)(PPh_3)_2$ (M = Ru, Os; R = 1-Naphthyl, 2-Naphthyl. The Crystal Structures of [Os(≡C-1-naphthyl)-$(CO)_2(PPh_3)_2$]ClO_4, Os(= CH-2-naphthyl))$Cl_2(CO)(PPh_3)_2$, and Os(2-naphthyl)-$Cl(CO)_2(PPh_3)_2$, *J. Organomet. Chem.* **551**, 247–259 (1998).
155. R. S. Simons, and P. P. Power, (η^5-C_5H_5)$(CO)_2MoGeC_6H_3$-2,6-Mes_2: A Transition-Metal Germylyne Complex, *J. Am. Chem. Soc.* **118**, 11966–11967 (1996).
156. A. C. Filippou, A. I. Philippopoulos, P. Portius, and D. U. Neumann, Synthesis and Structure of the Germylyne Complexes *trans*-[X$(dppe)_2$W≡Ge(η^1-Cp*)] (X = Cl, Br, I) and Comparison of the W≡E Bonds (E = C, Ge) with DFT Calculations, *Angew. Chem. Int. Ed.* **39**, 2778–2781 (2000).
157. A. C. Filippou, P. Portius, and A. I. Philippopoulos, Molybdenum and Tungsten Germylyne Complexes of the General Formula *trans*-[X$(dppe)_2$W≡Ge(η^1-Cp*)] (X = Cl, Br, I; dppe = $Ph_2PCH_2CH_2PPh_2$; Cp* = C_5Me_5): Synthesis, Molecular Structures, and Bonding Features of the Germylyne Ligand, *Organometallics* **21**, 653–661 (2002).

158. A. C. Filippou, N. Weidemann, A. I. Philippopoulos, and G. Schnakenburg, Activation of Aryl Germanium(II) Chlorides by $[Mo(PMe_3)_6]$ and $[W(\eta^2\text{-}CH_2PMe_2)H(PMe_3)_4]$: A New Route to Metal-Germanium Triple Bonds, *Angew. Chem. Int. Ed.* **45**, 5987–5991 (2006).
159. A. C. Filippou, G. Schnakenburg, A. I. Philippopoulos, and N. Weidemann, Ge_2 Trapped by Triple Bonds Between Two Metal Centers: The Germylidyne Complexes *trans, trans*-$[Cl(depe)_2M{\equiv}Ge{-}Ge{\equiv}M(depe)_2Cl]$ (M = Mo, W) and Bonding Analyses of the M≡Ge—Ge≡M Chain, *Angew. Chem. Int. Ed.* **44**, 5979–5985 (2005).
160. G. Balázs, L. J. Gregoriaded, and M. Scheer, Triple Bonds Between Transition Metals and the Heavier Elements of Groups 14 and 15, *Organometallics* **26**, 3058–3075 (2007).
161. T. R. Howard, J. B. Lee, and R. H. Grubbs, Titanium Metallacarbene–Metallacyclobutane Reactions: Stepwise Metathesis, *J. Am. Chem. Soc.* **102**, 6876–6878 (1980).
162. T. M. Trnka, and R. H. Grubbs, The Development of $L_2X_2Ru{=}CHR$ Olefin Metathesis Catalysts: An Organometallic Success Story, *Acc. Chem. Res.* **34**, 18–29 (2001).
163. R.H. Grubbs, Olefin-Metathesis Catalysts for the Preparation of Molecules and Materials, *Angew. Chem. Int. Ed.* **45**, 3760–3765 (2006) [Nobel Lecture].
164. T. J. Katz, Metal Carbenes in Low Oxidation States as Initiators for Olefin Metathesis and Related Reactions, *Angew. Chem. Int. Ed.* **44**, 3010–3019 (2005).
165. J.-L. Hérisson, and Y. Chauvin, Catalysis of Olefin Transformations by Tungsten Complexes. Telomerization of Cyclic Olefins in the Presence of Acyclic Olefins, *Makromol. Chem.* **141**, 161–167 (1971).
166. S. T. Nguyen, L. K. Johnson, R. H. Grubbs, and J. W. Ziller, Ring-Opening Metathesis Polymerization (ROMP) of Norbornene by a Group VIII Carbene Complex in Protic Media, *J. Am. Chem. Soc.* **114**, 3974–3975 (1992).
167. S. T. Nguyen, R. H. Grubbs, and J. W. Ziller, Synthesis and Activities of New Single-Component, Ruthenium-Based Olefin Metathesis Catalysts, *J. Am. Chem. Soc.* **115**, 9858–9859 (1993).
168. T. E. Wilhelm, T. R. Belderrain, S. N. Brown, and R. H. Grubbs, Reactivity of $Ru(H)(H_2)Cl(PCy_3)_2$ with Propargylic and Vinyl Chlorides: New Methodology to Give Metathesis-Active Ruthenium Carbenes, *Organometallics* **16**, 3867–3869 (1997).
169. J. Wolf, W. Stüer, C. Grünwald, H. Werner, P. Schwab, and M. Schulz, Ruthenium Trichloride, Tricyclohexylphosphane, 1-Alkynes, Magnesium, Hydrogen, and Water – Ingredients of an Efficient One-Pot Synthesis of Ruthenium Catalysts for Olefin Metathesis, *Angew. Chem. Int. Ed.* **37**, 1124–1126 (1998).
170. H. Werner, and J. Wolf, Synthesis of Rhodium and Ruthenium Carbene Complexes with a 16-Electron Count, in: *Handbook of Metathesis, Vol. 1* (Ed. R. H. Grubbs, Wiley-VCH, Weinheim, 2003, Chap. 1.8).
171. S. T. Nguyen, and T. M. Trnka, The Discovery and Development of Well-Defined, Ruthenium-Based Olefin Metathesis Catalysts, in: *Handbook of Metathesis, Vol. 1* (Ed. R. H. Grubbs, Wiley-VCH, Weinheim, 2003, Chap. 1.6).
172. P. Schwab, N. Mahr, J. Wolf, and H. Werner, Carbenerhodium Complexes with Diaryl- and Aryl(alkyl)carbenes as Ligands: The Missing Link in the Series of the Double Bond Systems *trans*-$[RhCl\{{=}C({=}C)_nRR'\}(L)_2]$ Where n = 0, 1, and 2, *Angew. Chem. Int. Ed. Engl.* **32**, 1480–1482 (1993).
173. H. Werner, Success and Serendipity During Studies Aimed at Preparing Carbenerhodium(I) Complexes, *J. Organomet. Chem.* **500**, 331–336 (1995).
174. P. Schwab, M. B. France, J. W. Ziller, and R. H. Grubbs, A Series of Well-Defined Metathesis Catalysts – Synthesis of $[RuCl_2({=}CHR')(PR_3)_2]$ and Their Reactions, *Angew. Chem. Int. Ed. Engl.* **34**, 2039–2041 (1995).
175. P. Schwab, R. H. Grubbs, and J. W. Ziller, Synthesis and Applications of $RuCl_2({=}CHR')(PR_3)_2$: The Influence of the Alkylidene Moiety on Metathesis Activity, *J. Am. Chem. Soc.* **118**, 100–110 (1996).
176. M. Oliván, and K. G. Caulton, The First Double Oxidative Addition of CH_2Cl_2 to a Metal Complex: Facile Synthesis of $[Ru(CH_2)Cl_2\{P(C_6H_{11})_3\}_2]$, *Chem. Commun.* **1997**, 1733–1734.

177. S. M. Hansen, F. Rominger, M. Metz, and P. Hofmann, The First Grubbs-Type Metathesis Catalyst with *cis* Stereochemistry: Synthesis of $[(\eta^2\text{-dtbpm})Cl_2Ru{=}CH{-}CH{=}CMe_2]$ from a Novel, Coordinatively Unsaturated Dinuclear Ruthenium Dihydride, *Chem. Eur. J.* **5**, 557–566 (1999).
178. H. Werner, S. Jung, P. González-Herrero, K. Ilg, and J. Wolf, Vinylidene, Vinyl, and Carbene Ruthenium Complexes with Chelating Diphosphanes as Ligands, *Eur. J. Inorg. Chem.* **2001**, 1957–1961.
179. E. L. Dias, S. T. Nguyen, and R. H. Grubbs, Well-Defined Ruthenium Olefin Metathesis Catalysts: Mechanism and Activity, *J. Am. Chem. Soc.* **119**, 3887–3897 (1997).
180. M. S. Sanford, J. A. Love, and R. H. Grubbs, Mechanism and Activity of Ruthenium Olefin Metathesis Catalysts, *J. Am. Chem. Soc.* **123**, 6543–6554 (2001).
181. A. J. Arduengo, III, H. V. R. Dias, J. C. Calabrese, and F. Davidson, Homoleptic Carbene-Silver(I) and Carbene-Copper(I) Complexes, *Organometallics* **12**, 3405–3409 (1993).
182. A. J. Arduengo, III, S. F. Gamper, J. C. Calabrese, and F. Davidson, Low-Coordinate Carbene Complexes of Nickel(0) and Platinum(0), *J. Am. Chem. Soc.* **116**, 4391–4394 (1994).
183. T. Weskamp, V. P. W. Böhm, and W. A. Herrmann, *N*-Heterocyclic Carbenes: State of the Art in Transition-Metal Complex Synthesis, *J. Organomet. Chem.* **600**, 12–22 (2000).
184. U. Kernbach, M. Ramm, P. Luger, and W. P. Fehlhammer, *Angew. Chem. Int. Ed. Engl.* **35**, 310–312 (1996).
185. T. Weskamp, W. C. Schattenmann, M. Spiegler, and W. A. Herrmann, A Novel Class of Ruthenium Catalysts for Olefin Metathesis, *Angew. Chem. Int. Ed.* **37**, 2490–2493 (1998).
186. J. Huang, E. D. Stevens, S. P. Nolan, and J. L. Petersen, Olefin Metathesis-Active Ruthenium Complexes Bearing a Nucleophilic Carbene Ligand, *J. Am. Chem. Soc.* **121**, 2674–2678 (1999).
187. T. Weskamp, F. J. Kohl, W. Hieringer, D. Gleich, and W. A. Herrmann, Highly Active Ruthenium Catalysts for Olefin Metathesis: The Synergy of N-Heterocyclic Carbenes and Coordinatively Labile Ligands, *Angew. Chem. Int. Ed.* **38**, 2416–2419 (1999).
188. M. Scholl, T. M. Trnka, J. P. Morgan, and R. H. Grubbs, Increased Ring Closing Metathesis Activity of Ruthenium-Based Olefin Metathesis Catalysts Coordinated with Imidazolin-2-ylidene Ligands, *Tetrahedron Lett.* **1999**, 2247–2250.
189. M. Scholl, S. Ding, C. W. Lee, and R. H. Grubbs, Synthesis and Activity of a New Generation of Ruthenium-Based Olefin Metathesis Catalysts Coordinated with 1,3-Dimesityl-4,5-dihydroimidazol-2-ylidene Ligands, *Org. Lett.* **1**, 953–956 (1999).
190. C. W. Bielawski, and R. H. Grubbs, Highly Efficient Ring-Opening Metathesis Polymerization (ROMP) Using New Ruthenium Catalysts Containing *N*-Heterocyclic Carbene Ligands, *Angew. Chem. Int. Ed.* **39**, 2903–2906 (2000).
191. S. Stinson, *Chem. Eng. News* **78**(No. 35), 6–7 (2000).
192. J. M. Berlin, S. D. Goldberg, and R. H. Grubbs, Highly Active Chiral Ruthenium Catalysts for Asymmetric Cross- and Ring-Opening Cross-Metathesis, *Angew. Chem. Int. Ed.* **45**, 7591–7595 (2006).
193. W. Stüer, J. Wolf, H. Werner, P. Schwab, and M. Schulz, Carbynehydridoruthenium Complexes as Catalysts for the Selective, Ring-Opening Metathesis of Cyclopentene with Methyl Acrylate, *Angew. Chem. Int. Ed.* **37**, 3421–3423 (1998).
194. P. Gonzàlez-Herrero, B. Weberndörfer, K. Ilg, J. Wolf, and H. Werner, The Sensitive Balance Between Five-Coordinate Carbene Ruthenium Complexes and Six-Coordinate Carbyne Ruthenium Complexes Formed from Ruthenium Vinylidene Precursors, *Organometallics* **20**, 3672–3685 (2001).
195. J. S. Kingsbury, J. P. A. Harrity, P. J. Bonitatebus, Jr., and A. H. Hoveyda, A Recyclable Ru-Based Metathesis Catalyst, *J. Am. Chem. Soc.* **121**, 791–799 (1999).
196. S. B. Garber, J. S. Kingsbury, B. L. Gray, and A. H. Hoveyda, Efficient and Recyclable Monomeric and Dendritic Ru-Based Metathesis Catalysts, *J. Am. Chem. Soc.* **122**, 8168–8179 (2000).

197. M. S. Sanford, L. M. Henling, M. W. Day, and R. H. Grubbs, Ruthenium-Based Four-Coordinate Olefin Metathesis Catalysts, *Angew. Chem. Int. Ed.* **39**, 3451–3453 (2000).
198. J. N. Coalter,III, J. C. Bollinger, O. Eisenstein, and K. G. Caulton, R-Group Reversal of Isomer Stability for $RuH(X)L_2(CCHR)$ vs. $Ru(X)L_2(CCH_2R)$: Access to Four-Coordinate Ruthenium Carbenes and Carbynes, *New. J. Chem.* **24**, 925–927 (2000).
199. J. C. Conrad, D. Amoroso, P. Czechura, G. P. A. Yap, and D. E. Fogg, The First Highly Active, Halide-Free Ruthenium Catalyst for Olefin Metathesis, *Organometallics* **22**, 3634–3636 (2003).
200. S. R. Caskey, M. H. Stewart, Y. J. Ahn, M. J. A. Johnson, J. L. C. Rowsell, and J. W. Kampf, Synthesis, Structure, and Reactivity of Four-, Five-, and Six-Coordinate Ruthenium Carbyne Complexes, *Organometallics* **26**, 1912–1923 (2007).
201. R. G. Carlson, M. A. Gile, J. A. Heppert, M. H. Mason, D. R. Powell, D. Vander Velde, and J. M. Vilain, The Metathesis-Facilitated Synthesis of Terminal Ruthenium Carbide Complexes: A Unique Carbon Atom Transfer Reaction, *J. Am. Chem. Soc.* **124**, 1580–1581 (2002).
202. A. Hejl, T. M. Trnka, M. W. Day, and R. H. Grubbs, Terminal Ruthenium Carbido Complexes as σ-Donor Ligands, *Chem. Commun.* **2002**, 2524–2525.
203. P. E. Romero, W. E. Piers, and R. McDonald, Rapidly Initiating Ruthenium Olefin-Metathesis Catalysts, *Angew. Chem. Int. Ed.* **43**, 6161–6165 (2004).
204. J. C. Peters, A. L. Odom, and C. C. Cummins, A Terminal Molybdenum Carbide Prepared by Methylidyne Deprotonation, *Chem. Commun.* **1997**, 1995–1996.
205. C. C. Cummins, Reductive Cleavage and Related Reactions Leading to Molybdenum-Element Multiple Bonds: New Pathways Offered by Three-Coordinate Molybdenum(III), *Chem. Commun.* **1998**, 1777–1786.
206. T. Agapie, P. L. Diaconescu, and C. C Cummins, Methine (CH) Transfer via a Chlorine Atom Abstraction/Benzene-Elimination Strategy: Molybdenum Methylidyne Synthesis and Elaboration to a Phosphaisocyanide Complex, *Angew. Chem. Int. Ed.* **45**, 862–870 (2006).
207. C. C. Cummins, Terminal, Anionic Carbide, Nitride, and Phosphide Transition-Metal Complexes as Synthetic Entries to Low-Coordinate Phosphorus Derivatives, *Angew. Chem. Int. Ed.* **45**, 862–870 (2006).
208. F. Z. Dörwald, *Metal Carbenes in Organic Synthesis* (Wiley-VCH, Weinheim, 1999).
209. K. H. Dötz, Carbene Complexes in Organic Synthesis, *Angew. Chem. Int. Ed. Engl.* **23**, 587–608 (1984).
210. K. H. Dötz, Carbene Complexes in Stereoselective Cycloaddition Reactions, *New J. Chem.* **14**, 433–445 (1990).
211. L. S. Hegedus, Synthesis of Amino Acids and Peptides Using Chromium Carbene Complex Photochemistry, *Acc. Chem. Res.* **28**, 299–305 (1995).
212. D. F. Harvey, and D. M. Sigano, Carbene-Alkyne-Alkene Cyclization Reactions, *Chem. Rev.* **96**, 271–288 (1996).
213. R. Aumann, and H. Nienaber, (1-Alkynyl)carbene Complexes: Tools for Synthesis, *Adv. Organomet. Chem.* **41**, 163–242 (1997).
214. A. De Meijere, H. Schirmer, and M. Duetsch, Fischer Carbene Complexes as Chemical Multitalents: The Incredible Range of Products from Carbenepentacarbonyl Metal α,β-Unsaturated Complexes, *Angew. Chem. Int. Ed.* **39**, 3964–4002 (2000).
215. K. H. Dötz, Synthesis of the Naphthol Skeleton from Pentacarbonyl[methoxy(phenyl)-carbene]chromium(0) and Tolane, *Angew. Chem. Int. Ed. Engl.* **14**, 644–645 (1975).
216. P. Hofmann, and M. Hämmerle, Mechanism of Dötz Reaction: Alkyne-Carbene Linkage to Chromacyclobutenes?, *Angew. Chem. Int. Ed. Engl.* **28**, 908–910 (1989).
217. M. Torrent, M. Duran, and M. Solà, Density Functional Study on the Preactivation Scenario of the Dötz Reaction: Carbon Monoxide Dissociation versus Alkyne Addition as the First Reaction Step, *Organometallics* **17**, 1492–1501 (1998).

218. M. F. Semmelhack, J. J. Bozell, T. Sato, W. Wulff, E. Spiess, and A. Zask, Synthesis of Nanaomycin A and Deoxyfrenolicin by Alkyne Cycloaddition to Chromium-Carbene Complexes, *J. Am. Chem. Soc.* **104**, 5850–5852 (1982).
219. M. F. Semmelhack, J. J. Bozell, L. Keller, T. Sato, E. Spiess, W. Wulff, and A. Zask, Synthesis of Naphthoquinone Antibiotics by Intramolecular Alkyne Cycloaddition to Carbene-Chromium Complexes, *Tetrahedron* **41**, 5803–5812 (1985).
220. K. H. Dötz, W. Straub, R. Ehlenz, K. Peseke, and R. Meisel, Carbene Complex Functionalized Sugars, *Angew. Chem. Int. Ed. Engl.* **34**, 1856–1858 (1995).
221. K. H. Dötz, C. Jäkel, and W.-C. Haase, Organotransition Metal Modified Sugars, *J. Organomet. Chem.* **617–618**, 119–132 (2001).
222. A. M. Rouhi, Olefin Metathesis: The Early Days, *Chem. Eng. News* **80**(51), 34–38 (2002).
223. N. Calderon, H. Y. Chen, and K. W. Scott, Olefin Metathesis – A Novel Reaction for Skeletal Transformations of Unsaturated Hydrocarbons, *Tetrahedron Lett.*, **1967**, 3327–3329.
224. E. A. Ofstead, J. P. Ward, W. A. Judy, K. W. Scott, and N. Calderon, Olefin Metathesis I. Acyclic Vinylenic Hydrocarbons, *J. Am. Chem. Soc.* **90**, 4133–4140 (1968).
225. G. S. Lewandos, and R. Pettit, On the Mechanism of the Metal-Catalyzed Disproportionation of Olefins, *J. Am. Chem. Soc.* **93**, 7087–7088 (1971).
226. R. H. Grubbs, and T. K. Brunck, A Possible Intermediate in the Tungsten-Catalyzed Olefin Metathesis Reaction, *J. Am. Chem. Soc.* **94**, 2538–2540 (1972).
227. Y. Chauvin, Olefin Metathesis: The Early Days, *Angew. Chem. Int. Ed.* **45**, 3741–3747 (2006) [Nobel Lecture].
228. C. P. Casey, and T. J. Burkhardt, Reactions of (Diphenylcarbene)pentacarbonyl-tungsten(0) with Alkenes. Role of Metal-Carbene Complexes in Cyclopropanation and Olefin Metathesis Reactions, *J. Am. Chem. Soc.* **96**, 7808–7809 (1974).
229. T. J. Katz, and J. McGinnis, The Mechanism of the Olefin Metathesis Reaction, *J. Am. Chem. Soc.* **97**, 1592–1594 (1975).
230. T. K. Katz, The Olefin Metathesis Reaction, *Adv. Organomet. Chem.* **16**, 283–318 (1977).
231. M. I. Bruce, Organometallic Chemistry of Vinylidene and Related Unsaturated Carbenes, *Chem. Rev.* **91**, 197–257 (1991).
232. H. Werner, Organometallic Chemistry of Alkenes and Alkynes, *J. Organomet. Chem.* **475**, 45–55 (1994).
233. H. Werner, Allenylidenes: Their Multifaceted Chemistry at Rhodium, *Chem. Comm.* **1997**, 903–910.
234. M. I. Bruce, Transition Metal Complexes Containing Allenylidene, Cumulenylidene, and Related Ligands, *Chem. Rev.* **98**, 2797–2858 (1998).
235. V. Cadierno, M. P. Gamasa, and J. Gimeno, Recent Developments in the Reactivity of Allenylidene and Cumulenylidene Complexes, *Eur. J. Inorg. Chem.* **2001**, 571–591.
236. M. I. Bruce, Metal Complexes Containing Cumulenylidene Ligands, $\{L_mM\}{=}C({=}C)_n{=}CRR'$, *Coord. Chem. Rev.* **248**, 1603–1625 (2004).
237. H. Fischer, and N. Szesni, π-Donor-Substituted Metallacumulenes of Chromium and Tungsten, *Coord. Chem. Rev.* **248**, 1659–1677 (2004).
238. K. Ilg, and H. Werner, The First Structurally Characterized Metal Complex with the Molecular Unit $M{=}C{=}C{=}C{=}CR_2$, *Angew. Chem. Int. Ed.* **39**, 1632–1634 (2000) ($M({=}C)_4$).
239. K. Ilg, and H. Werner, Closing the Gap between MC_3 and MC_5 Metallacumulenes: The Chemistry of the First Structurally Characterized Transition-Metal Complex with $M{=}C{=}C{=}C{=}CR_2$ as the Molecular Unit, *Chem. Eur. J.* **8**, 2812–2820 (2002) ($M({=}C)_4$).
240. K. Venkatesan, F. J. Fernandez, O. Blacque, T. Fox, M. Alfonso, H. W. Schmalle, and H. Berke, A Facile and Novel Route to Unprecedented Manganese C_4 Cumulenic Complexes, *Chem. Comm.* **2003**, 2006–2008 ($M({=}C)_4$).
241. D. Touchard, P. Haquette, A. Daridor, L. Toupet, and P. H. Dixneuf, First Isolable Pentatetraenylidene Metal Complex Containing the $Ru{=}C{=}C{=}C{=}C{=}CPh_2$ Assembly. A Key Intermediate to Provide Functional Allenylidene Complexes, *J. Am. Chem. Soc.* **116**, 11157–11158 (1994) ($M({=}C)_5$).

242. R. W. Lass, P. Steinert, J. Wolf, and H. Werner, Synthesis and Molecular Structure of the First Neutral Transition-Metal Complex Containing a Linear M=C=C=C=C=CR$_2$ Chain, *Chem. Eur. J.* **2**, 19–23 (1996) (M(=C)$_5$).
243. G. Roth, and H. Fischer, Complexes with Diamino-Substituted Unsaturated C_3 and C_5 Ligands: First Group 6 Pentatetraenylidenes and New Allenylidene Complexes, *Organometallics* **15**, 1139–1145 (1996) (M(=C)$_5$).
244. M. Dede, M. Drexler, and H. Fischer, Heptahexaenylidene Complexes: Synthesis and Characterization of the First Complexes with an M=C=C=C=C=C=C=CR$_2$ Moiety (M = Cr, W), *Organometallics* **26**, 4294–4299 (2007).
245. C. Bruneau, and P. H. Dixneuf, Metal Vinylidenes in Catalysis, *Acc. Chem. Res.* **32**, 311–323 (1999).
246. H. Katayama, and F. Ozawa, Vinylideneruthenium Complexes in Catalysis, *Coord. Chem. Rev.* **248**, 1703–1715 (2004).
247. C. Bruneau, and P. H. Dixneuf, Metal Vinylidenes and Allenylidenes in Catalysis: Applications in Anti-Markovnikov Additions to Terminal Alkynes and Alkene Metathesis, *Angew. Chem. Int. Ed.* **45**, 2176–2203 (2006).
248. R. B. King, and M. S. Saran, Metal Complexes with Terminal Dicyanomethylenecarbene Ligands formed by Chlorine Migration Reactions, *J. Chem. Soc., Chem. Comm.* **1972**, 1053–1053.
249. E. O. Fischer, H. J. Kalder, A. Frank, F. H. Köhler, and G. Huttner, 3-Dimethylamino-3-phenylallenylidene, a New Ligand on the Pentacarbonylchromium and –tungsten Framework, *Angew. Chem. Int. Ed. Engl.* **15**, 623–624 (1976).
250. H. Berke, Simple Synthesis of Dicarbonyl(η-cyclopentadienyl)(3,3-di-*tert*-butylallenylidene)manganese, *Angew. Chem. Int. Ed. Engl.* **15**, 624 (1976).
251. A. Höhn, and H. Werner, Carbyne-Iridium Complexes: Evidence for an Equilibrium Between Alkenylidene(hydrido) and Carbyne Isomers, *Angew. Chem. Int. Ed. Engl.* **25**, 737–738 (1986).
252. H. Werner, K. Ilg, R. Lass, and J. Wolf, Iridium-Containing Cumulenes: How to Prepare and How to Use, *J. Organomet. Chem.* **661**, 137–147 (2002).
253. C. Copéret, and J.-M. Basset, Strategies to Immobilize Well-Defined Olefin Metathesis Catalysts: Supported Homogeneous Catalysis vs. Surface Organometallic Chemistry, *Adv. Synth. Catal.* **349**, 78–92 (2007).
254. D. J. Cardin, Protagonists in Chemistry, *Inorg. Chim. Acta* **360**, 1245–1247 (2007).
255. Y. Chauvin, Telephone Interview, 5th of October 2005; Published on the Official Web Site of the Nobel Foundation (Copyright Nobel Web AB 2008).
256. P. Ahlberg, Presentation Speech, Stockholm, 10th of December 2005 (Copyright Nobel Web AB 2005).
257. Y. Chauvin, Autobiography, in: *Les Prix Nobel, The Nobel Prizes 2005* (Ed. K. Grandin, [Nobel Foundation], Stockholm, 2006).
258. D. Astruc, The Metathesis Reactions: From a Historical Perspective to Recent Developments, *New. J. Chem.* **29**, 42–56 (2005).
259. R. R. Schrock, Autobiography, in: *Les Prix Nobel, The Nobel Prizes 2005* (Ed. K. Grandin, [Nobel Foundation], Stockholm, 2006).
260. F. N. Tebbe, G. W. Parshall, and G. S. Reddy, Olefin Homologation with Titanium Methylene Compounds, *J. Am. Chem. Soc.* **100**, 3611–3613 (1978).
261. R. H. Grubbs, Autobiography, in: *Les Prix Nobel, The Nobel Prizes 2005* (Ed. K. Grandin, [Nobel Foundation], Stockholm, 2006).

Chapter 9
Metal Alkyls and Metal Aryls: The "True" Transition Organometallics

Only those who do not seek are safe from error.
Albert Einstein, Nobel Laureate 1921 (1879–1955)

9.1 The Extensions of Frankland's Pioneering Work

The quest for stable alkyls and aryls of the transition metals, which can be handled under normal conditions, dates back to the second half of the nineteenth century. Following Frankland's seminal studies on $Zn(CH_3)_2$ and $Zn(C_2H_5)_2$ in the late 1840s and the 1850s, analogous metal alkyls such as $Cd(C_2H_5)_2$, $Hg(CH_3)_2$, $Mg(C_2H_5)_2$, $Al(CH_3)_3$ and $Pb(C_2H_5)_4$ were prepared (see Chap. 3) and attempts were made to use them for the synthesis of related organometallic compounds of the transition elements. While those experiments failed, the discovery of Grignard reagents at the turn to the twentieth century gave a new impetus to the field and only a few years later William Pope and Stanley Peachey at the Municipal School of Technology in Manchester prepared stable $Pt(CH_3)_3I$ from $PtCl_4$ and excess CH_3MgI [1, 2]. At the meeting of the Chemical Society in London in March 1907, at which the discovery was announced, the chairman, Sir Henry Roscoe, complimented the authors for their results "which might indeed be said to be a wonderful find" [3].

In those days, the preparation and characterization of $Pt(CH_3)_3I$ was probably not easy and, as Pope and Peachey indicated, required "attention to a number of details". The analysis of the "bright yellow crystalline powder" also presented "some difficulties, owing to the tendency of explosive decomposition on heating" [1, 2]. Nevertheless, a correct elemental analysis for platinum and iodine was obtained, and some derivatives of the general composition $Pt(CH_3)_3X$ with X = OH, Cl, CN and NO_3 as well as a few adducts with donor ligands such as $Pt(CH_3)_3I(NH_3)_2$ could be prepared [1, 2]. Three decades later, in 1938, Henry Gilman and Myrl Lichtenwalter from Iowa State University succeeded in isolating $Pt(CH_3)_4$, which was formed upon treatment of $Pt(CH_3)_3I$ with sodium methyl and isolated in 46% yield [4]. As part of this work it was recognized that tetramethylplatinum was "one of several by-products of the Pope and Peachey reaction of platinic chloride and methylmagnesium iodide". In addition

H. Werner, *Landmarks in Organo-Transition Metal Chemistry*,
Profiles in Inorganic Chemistry, DOI 10.1007/978-0-387-09848-7_9,

to $Pt(CH_3)_4$, Gilman and Lichtenwalter equally obtained $Pt_2(CH_3)_6$ by reduction of $Pt(CH_3)_3I$ with potassium in benzene, and reconverted the dinuclear species to the starting material with an equimolar amount of iodine. Convinced of the importance of their results, the authors emphasized that the syntheses of $Pt(CH_3)_4$ and $Pt_2(CH_3)_6$ "demonstrate for the first time that true organoplatinum compounds not having acid radicals can be prepared" [4]. While both Pope and Gilman assumed that the tri- and tetramethylplatinum(IV) compounds were monomers, an X-ray diffraction study of $Pt(CH_3)_3Cl$, carried out by Rundle and Sturdivant in 1947, revealed that in the solid the molecule is a tetramer with platinum atoms and chlorine atoms at alternate corners of a distorted cube (Fig. 9.1) [5]. The structure of $Pt(CH_3)_4$ was reported to be similar [5], with methyl units replacing the chlorine atoms. Later it was shown, however, that the crystalline material used by Rundle and Sturdivant was in fact $Pt(CH_3)_3OH$ instead of $Pt(CH_3)_4$, with oxygen atoms acting as the bridges [6].

Sir William Pope's group not only reacted $PtCl_4$ but also several other transition metal halides such $CuCl_2$, $FeCl_2$, $FeCl_3$, $CoBr_2$, $NiBr_2$, $PdCl_2$, PdI_2, $RuCl_3$, $RhCl_3$, and $MoCl_5$ with Grignard reagents, mainly with phenyl magnesium iodide, but only obtained biphenyl in most cases instead of phenyl metal compounds [7]. The reactions of $AuCl_3$ or $HAuBr_4$ with ethyl magnesium bromide gave $Au(C_2H_5)_2Br$, which due to its lability could be isolated in not more than 10–15% yield. $Au(C_2H_5)_2Cl$ was also found to be labile but gave a stable 1:1 adduct with NH_3 [8].

The first attempts to prepare phenylchromium compounds were disclosed by Julius Sand and Fritz Singer as early as in 1903 [9]. They observed *eine günstige Reaktion* (meaning a favorable reaction) between CrO_2Cl_2 and PhMgBr but failed to isolate a pure product. A few years later, similar observations were reported by Bennett and Turner [10] and Kondyrew and Fomin [11], who used either $CrCl_3$ or $Cr(SCN)_3$ and PhMgBr as the starting materials. The general conclusion by both groups was that "organochromium compounds are unstable and cannot be isolated".

Despite this pessimistic prediction, Franz Hein began his pioneering work on the reactivity of $CrCl_3$ and other chromium(III) and chromium(II) derivatives

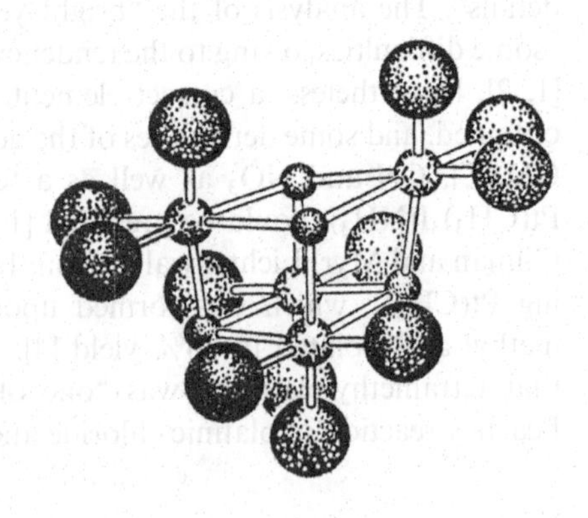

Fig. 9.1 Molecular structure of $Pt(CH_3)_3Cl$; the *smallest spheres* representing the chlorine atoms, the *medium-sized spheres* the platinum atoms and the *largest spheres* the methyl groups (from ref. 5; reproduced with permission of the American Chemical Society)

towards aryl magnesium halides in 1919, and he continued this for almost four decades. The results of his investigations, which received world-wide attention [12, 13], were described in detail in Chap. 5. It was not before the mid 1950s, when Ernst Otto Fischer and Harold Zeiss proved that the whole series of Hein's "polyphenylchromium compounds" were indeed π-arene chromium(I) complexes, which did not contain Cr—C_6H_5 σ-bonds. The fact that, nevertheless, phenylchromium(III) compounds existed and could be isolated in pure form, was shown independently by Herwig and Zeiss [14–16], and Hein and Weiss [17, 18] in 1957/58. At that time the chemistry of aryl and alkyl transition metal compounds began to flourish and soon disproved Cotton's pessimistic view, expressed in 1955, that "the often heard generalization that they [the alkyls and aryls of transition metals] are much less stable. . .than those of non-transition metals is quite true" [12].

9.2 Heteroleptic Complexes with Metal–Alkyl and Metal–Aryl Bonds

For the preparation of stable compounds with transition metal–alkyl and –aryl σ-bonds, the striking developments in the field of carbonyl, cyclopentadienyl, and phosphine metal complexes was very supportive. In the late 1950s, Thomas Coffield [19] as well as Walter Hieber [20–22] succeeded in obtaining the methyl compounds $CH_3Co(CO)_4$ and $CH_3M(CO)_5$ (**2a**: M = Mn; **2b**: M = Re) from the metal carbonylates $[Co(CO)_4]^-$ and $[M(CO)_5]^-$ (**1a,b**) and CH_3I. They could be converted into the corresponding acyl metal derivatives **3** on treatment with CO, amines or tertiary phosphines (Scheme 9.1) [23–26]. Gordon Stone, Walter Hieber and others also prepared the related fluoroalkyl and fluoroaryl metal complexes $R_FCo(CO)_4$ and $R_FM(CO)_5$ (M = Mn, Re; R_F = CF_3, C_2F_5, C_3F_7, C_6F_5, etc.), which were significantly more stable than their non-fluorinated counterparts [27].

The first cyclopentadienyl metal compounds containing σ-bonded carbon ligands were the titanocene derivatives $(C_5H_5)_2TiR_2$ (R = Ph, *p*-Tol, *p*-$C_6H_4NMe_2$), obtained by Lawrence Summers and Robert Uloth from $(C_5H_5)_2TiCl_2$ and two equivalents of the corresponding aryl lithium reagent in 1954 [28]. The bis(phenyl) and bis(*p*-tolyl) complexes were quite stable and

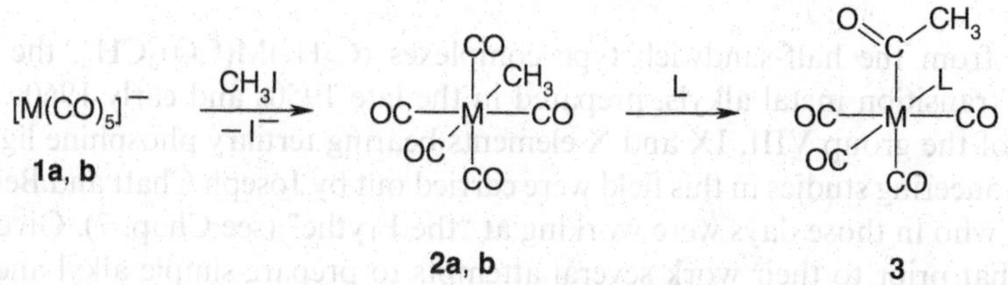

$[M(CO)_5]^-$ (**1a, b**) $\xrightarrow[-I^-]{CH_3I}$ $CH_3M(CO)_5$ (**2a, b**) $\xrightarrow{L}$ $CH_3C(O)M(CO)_4L$ (**3**)

Scheme 9.1 Preparation of the methylmanganese and –rhenium complexes **2a,b** from the pentacarbonyl metallates **1a,b** and their conversion into the acyl metal derivatives **3** (L = CO, PR_3, NH_2R) via migratory insertion of a carbonyl ligand into the M—CH_3 bond (**a**: M = Mn; **b**: M = Re) decomposed only slowly at room temperature. The rather unstable dimethyltitanocene $(C_5H_5)_2Ti(CH_3)_2$ was prepared by Geoffrey Wilkinson and Stanley Piper from $(C_5H_5)_2TiCl_2$ and CH_3MgI in 1956, but at that time could be isolated in only 1% yield [29]. An efficient synthesis of $(C_5H_5)_2Ti(CH_3)_2$ was reported by Carl Clauss and Herbert Bestian from the Hoechst Company in 1962, who used methyl lithium instead of CH_3MgI as the alkylating reagent. With this procedure, the yield of isolated $(C_5H_5)_2Ti(CH_3)_2$ was 95% [30]. Even less stable than the $(C_5H_5)_2TiR_2$ species were the monocyclopentadienyl compounds $(C_5H_5)TiR_3$, of which the trimethyl derivative was obtained by Umberto Giannini and Sebastiano Cesca in 1960 [31]. Bis(organyl)zirconocenes and -hafnocenes $(C_5H_5)_2MR_2$ with M = Zr and Hf, and R = Me, Ph, CH_2Ph and CH_2SiMe_3 were prepared in the early 1970s and found to be significantly more stable than the titanium analogues [32]. At nearly the same time, Lappert's, de Liefde Meijer's and Parshall's groups reported the preparation of the paramagnetic dialkyl and diaryl complexes $(C_5H_5)_2MR_2$ of niobium and tantalum, the stability of which surprisingly was higher than those of the diamagnetic $(C_5H_5)_2ZrR_2$ and $(C_5H_5)_2HfR_2$ analogues [32].

Together with the synthesis of $(C_5H_5)_2Ti(CH_3)_2$, Piper and Wilkinson also reported the preparation of the first cyclopentadienyl(carbonyl) and cyclopentadienyl(nitrosyl) metal complexes $(C_5H_5)M(CO)_3R$ (M = Mo, W), $(C_5H_5)Fe(CO)_2R$ and $(C_5H_5)Cr(NO)_2R$ containing a metal–alkyl bond [29, 33, 34]. These complexes were formed either by metathetical reactions of the metallate anions $[(C_5H_5)M(CO)_3]^-$ and $[(C_5H_5)Fe(CO)_2]^-$ with alkyl halides or by nucleophilic substitution of the chloro or iodo ligands of $(C_5H_5)Cr(NO)_2Cl$ and $(C_5H_5)Fe(CO)_2I$ with the corresponding Grignard reagent. In contrast to the general belief at that time [12], these alkyl metal compounds were remarkably stable and most of them could be handled in air. The molybdenum and the tungsten complexes $(C_5H_5)Mo(CO)_3CH_3$ and $(C_5H_5)W(CO)_3CH_3$ were also accessible via an "exotic preparative scheme" [32], namely from the hydrides $(C_5H_5)M(CO)_3H$ and diazomethane by inserting the CH_2 unit into the metal–hydrogen bond [33–35]. In 1958, $CH_3Mn(CO)_5$ was prepared from $HMn(CO)_5$ and CH_2N_2 in the same way [22].

9.3 Chatt and His Contemporaries

Apart from the half-sandwich type complexes $(C_5H_5)M(CO)_3CH_3$, the most stable transition metal alkyls, prepared in the late 1950s and early 1960s, were those of the group VIII, IX and X elements bearing tertiary phosphine ligands. The pioneering studies in this field were carried out by Joseph Chatt and Bernard Shaw, who in those days were working at "the Frythe" (see Chap. 7). Given the fact, that prior to their work several attempts to prepare simple alkyl and aryl derivatives of the transition metals of the general composition MR_2 (with M in the oxidation state two) and MR_3 (with M in the oxidation state three) had failed, Chatt and Shaw concluded on the basis of the ligand field theory "that ligands causing large splittings of the d-energy levels should stabilise planar and octahedral [alkyl and aryl metal] complexes of d^3–d^8 types" [36]. To achieve this goal and prepare compounds of the general composition $MR_2(L)_n$ and $MR_3(L)_n$, they chose tertiary phosphines as stabilizing ligands L, because these "appear to be the most strongly double-bonding to the metal of all ligands except those containing carbon as ligand atoms, e.g., CO and C_2H_4". They also indicated that tertiary phosphines "form very stable complexes of the type $[(PR_3)_2MX_2]$ (M = Ni, Pd, Pt; X = halogen) of square planar configuration" [36].

Based on the above mentioned propositions, Chatt and Shaw prepared at first a series of methylplatinum(II) compounds of the types *trans*-$Pt(CH_3)X(PR_3)_2$ and *cis*-$Pt(CH_3)_2(PR_3)_2$ (R = Et, *n*Pr, Ph) from *cis*- or *trans*-$PtX_2(PR_3)_2$ and methyl lithium or methyl Grignard reagents. These compounds were "remarkably inert" and were "not hydrolysed by dilute acids or oxidised in moist air". The dimethyl derivative *cis*-$Pt(CH_3)_2(PEt_3)_2$ could be distilled without decomposition at 85°C in vacuo, and a sample of *trans*-$Pt(CH_3)I(PEt_3)_2$ was "unchanged after a year" [36]. The reaction of the monomethyl compounds *trans*-$Pt(CH_3)I(PR_3)_2$ (**4**) with methyl iodide afforded the platinum(IV) complexes **6** (Scheme 9.2), for which either the configuration **6a** or **6b** was proposed on the basis of the dipole moments. The related dichloroplatinum(IV) derivatives **7a** or **7b** were obtained by treatment of *cis*-$Pt(CH_3)_2(PR_3)_2$ (**5**) with chlorine [36]. In the following paper, Chatt and Shaw reported the preparation of several analogues of **4** and **5** of the general composition $Pt(R)X(L)_2$ and $PtR_2(L)_2$, where R was not only an alkyl but also an aryl or ethynyl group [37]. The stabilities of these compounds increased in the order R = C≡CH < alkyl < phenyl ~ *para*-substituted phenyl < C≡CPh < *ortho*-substituted phenyl, and the same order was subsequently found for the corresponding nickel(II) derivatives [38].

Apart from the platinum and nickel compounds with σ-bonded alkyl or aryl ligands, Chatt's group reported between 1961 and 1966 the synthesis of an impressive series of related cobalt(II), iron(II), rhodium(III), iridium(III), rhenium(III) and rhenium(V) complexes [39–42]. The paramagnetic cobalt(II) and iron(II) derivatives $MR_2(PR'_3)_2$ were stable when R was an aryl group with

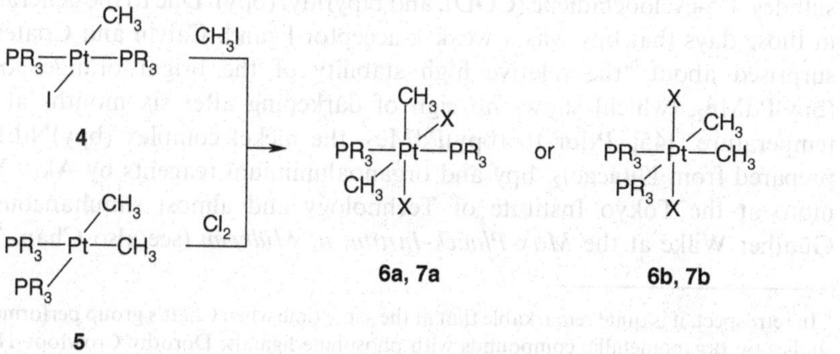

Scheme 9.2 Preparation of dimethylplatinum(IV) complexes **6a** or **6b** (X = I) and **7a** or **7b** (X = Cl) from the platinum(II) compounds **4** (R = Et, *n*Pr) and **5** (R = Et) by oxidative addition

"somewhat bulky" substituents in the *ortho* positions. Based on the magnetic moments, a square–planar configuration for $CoR_2(PR'_3)_2$ and $FeR_2(PR'_3)_2$ was proposed, with the aryl groups and the phosphine ligands in *trans* disposition [39]. This was confirmed by an X-ray crystal structure analysis of the bis(mesityl) derivative $Co(2,4,6\text{-}C_6H_2Me_3)_2(PEt_2Ph)_2$ [43]. With rhodium as the metal atom, the five-coordinate compounds $Rh(1\text{-}Naph)_2Br(PR_3)_2$ (PR_3 = $PnPr_3$, PEt_2Ph) were obtained from $RhX_3(PR_3)_3$ and 1-naphthyl magnesium bromide in low yield. They were diamagnetic and possessed a "surprisingly high stability" [40]. The analogous iridium(III) precursors $IrX_3(PR_3)_3$ reacted with excess methyl magnesium chloride in refluxing tetrahydrofuran to give the octahedral trimethyl derivatives $Ir(CH_3)_3(PR_3)_3$, in which the two sets of monodentate ligands adopted facial arrangements [42]. Ruthenium(II) and osmium(II) formed octahedral alkyl and aryl complexes of the types *cis*- and *trans*-$M(R)X(L\text{–}L)_2$ and *cis*-$MR_2(L\text{–}L)_2$, where L–L was a chelating ligand such as dmpe, dppe or dppm [44]. The monoalkyl and monoaryl compounds $M(R)X(L\text{–}L)_2$ with X = H received special attention because they were the first transition metal complexes having both a hydrogen atom and an organic group attached to the same metal center by σ-bonds. [1]

As often occurs in science, Chatt's idea about the stabilizing effect of π-accepting ligands on metal–alkyl and metal–aryl σ-bonds, published in 1959 [36], almost simultaneously emerged elsewhere. Calvin and Coates at the University of Durham in the UK reported the preparation of alkyl- and arylpalladium(II) compounds of the general composition *trans*-$Pd(R)X(PEt_3)_2$ and *trans*-$PdR_2(PEt_3)_2$ in 1960 [45]. The monomethyl and monoaryl complexes were obtained from $PdX_2(PEt_3)_2$ and RMgX, while the dimethyl and diaryl derivatives were synthesized by reacting $PdX_2(PEt_3)_2$ with methyl or aryl lithium reagents. Analogous compounds with ethyl and *n*-propyl instead of methyl appeared to be formed but could not be isolated on account of their instability. Apart from *trans*-$Pd(R)X(PEt_3)_2$ and *trans*-$PdR_2(PEt_3)_2$, similar alkyl- and arylpalladium(II) complexes were obtained from palladium dihalides coordinated to tertiary arsines, sulfides, 1,5-cyclooctadiene (COD), and bipyridyl (bpy). Due to the general belief in those days that bpy was a weak π-acceptor ligand, Calvin and Coates were surprised about "the relative high stability of the bright orange complex $(bpy)PdMe_2$, [which] shows no sign of darkening after six months at room temperature" [45]. Prior to $(bpy)PdMe_2$, the nickel complex $(bpy)NiEt_2$ was prepared from $Ni(acac)_2$, bpy and organoaluminium reagents by Akio Yamamoto at the Tokyo Institute of Technology and almost simultaneously by Günther Wilke at the *Max-Planck-Institut at Mülheim* (see also Chap. 7) [46,

[1] In retrospect, it is quite remarkable that at the same time when Chatt's group performed their studies on organometallic compounds with phosphine ligands, Dorothy Crowfoot–Hodgkin completed her work on the structure of vitamine B_{12}, for which she was awarded the Nobel Prize for Chemistry in 1964. As far as we presently know, vitamine B_{12} is the only natural product containing a transition metal–carbon σ-bond [see: P. G. Lenhert, and D. Crowfoot-Hodgkin, Structure of the 5,6-Dimethylbenzimidazolylcobamide Coenzyme, *Nature* **192**, 937–938 (1961)].

47]. The diethyliron compound (bpy)$FeEt_2$ was obtained in a similar way and found to be an active catalyst for butadiene oligomerization [48]. Later, Yamamoto's group also reported an alternate method for preparing the palladium(II) complexes $PdR_2(PR'_3)_2$, where R was not only methyl but ethyl and *n*-propyl as well. The reactions of these compounds with carbon monoxide gave the corresponding dialkyl ketones, possibly via the acyl(alkyl) derivatives $PdR(COR)(PR'_3)_2$ as intermediates [49].

Parallel to Calvin and Coates' work [45], the stabilizing effect that olefins such as COD appear to have on metal–carbon σ-bonds was equally used by John Doyle's group at the University of Iowa. In 1961, they reported the synthesis of a series of platinum(II) complexes of the general composition Pt(R)I(olefin) and PtR_2(olefin), where R was an alkyl or aryl group and the olefinic ligand norbornadiene, dicyclopentadiene, COD, or cyclooctatetraene (COT) [50, 51]. The mononuclear complexes **9a,b** could be converted to the corresponding binuclear species **10a,b**, which contained the tub-shaped COT as a bridging ligand between the two PtR_2 units (Scheme 9.3). The reactions of the dimethyl and diaryl derivatives PtR_2(COD) with PPh_3 and pyridine led to the displacement of the diolefin and afforded the bis(phosphine) and bis(pyridine) compounds $PtR_2(PPh_3)_2$ and $PtR_2(py)_2$ in excellent yield [50, 51].

Although the work on alkyl and aryl transition metal complexes containing carbonyl, phosphine and cyclopentadienyl ligands received worldwide attention, in retrospect it must be emphasized that one year before the preparation of diaryltitanocenes by Summers and Uloth was reported in 1954, Herman and Nelson isolated a phenyltitanium compound with an alkoxide instead of CO, PR_3 or C_5H_5 as the ancillary ligand [52, 53]. The initial product of the composition PhTi(O*i*Pr)$_3$–LiO*i*Pr–LiBr–OEt_2 was obtained from the reaction of Ti(O*i*Pr)$_4$ with a solution of LiPh and LiBr in diethyl ether, and was converted to solvent-free PhTi(O*i*Pr)$_3$ by reaction with $TiCl_4$. Herman and Nelson indicated that after "the preparation of compounds with covalent titanium-carbon bonds... has been attempted repeatedly and unsuccessfully in the past since the first study of Cahours in 1861" (see Chap. 3), they were "quite fortunate to isolate a stable compound with a titanium–carbon σ-bond". The white crystalline substance "decomposed rapidly with oxygen and water...but shows no darkening after storage at 10°C for over a year" [52, 53].

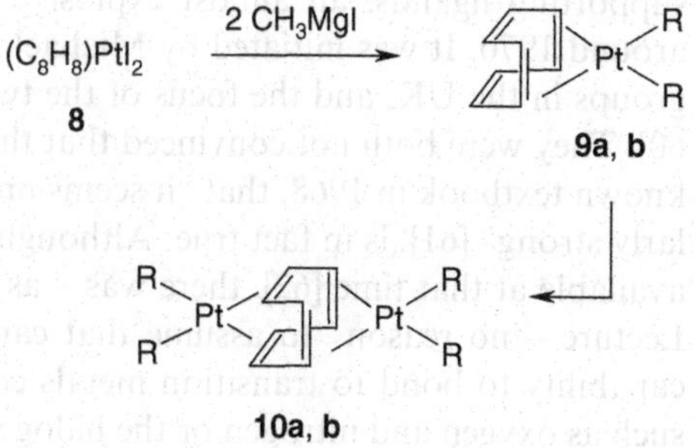

2 CH_3MgI

$(C_8H_8)PtI_2$

8

9a, b

10a, b

Scheme 9.3 Preparation of mono- and dinuclear dimethyl- and diphenylplatinum(II) complexes **9a,b** and **10a,b** from the dihalide **8** as the precursor (**a**: R = Me; **b** = Ph)

Following this report, phenyl titanium trichloride and tribromide as well as a series of highly labile alkyl titanium trihalides $RTiX_3$ (R = Me, Et, *i*Pr, *n*Bu; X = Cl, Br) were prepared from $TiCl_4$ or $TiBr_4$ and various arylating or alkylation reagents, generally at low temperatures [54]. Moreover, Me_2TiCl_2 was obtained from $TiCl_4$ and trimethylaluminium at −80°C and found to be less stable than the monomethyl derivative. Attempts to alkylate titanium tetrahalides with organo lithium or Grignard reagents under normal conditions generally resulted in reduction of titanium(IV) to lower valency states. Thermally more stable compounds than $MeTiCl_3$ were prepared from phenyl or alkyl titanium trihalides by replacing the halide by alkoxide or dialkylamino groups or by forming 1:1 and 1:2 adducts with Lewis bases such as py, en or dme [54]. In the course of this work, also the long-sought $TiMe_4$ was prepared by Clauss and Beermann from the Hoechst company in 1959 [55]. After the synthesis of $PtMe_4$, it was the second binary (or homoleptic) transition metal methyl compound to be described in the literature. Tetramethyltitanium was obtained from $TiCl_4$ and a solution of methyl lithium in diethyl ether at −78°C and codistilled with OEt_2 from the reaction mixture at −30°C. In 1963, Berthold and Groh from the University of Mainz in Germany isolated crystalline $TiMe_4$ from a diethyl ether/hexane solution after removing the ether in vacuo at −78°C. The crystals began to decompose upon warming to a few degrees above −78°C [56]. The reactions of $TiMe_4$ with several Lewis bases afforded five-coordinate and six-coordinate adducts, $TiMe_4L$ and $TiMe_4L_2$, of which those with L = pyridine and PMe_3, and those with L_2 = bpy and dmpe were the most stable. In 1964, tetraphenyltitanium was prepared by Razuvaev's group and characterized by treatment with $HgCl_2$ yielding PhHgCl [57]. $TiPh_4$ decomposed at −10°C to give a presumably polymeric $TiPh_2$ and biphenyl. Tetrabenzyltitanium, obtained by Giannini and Zucchini in 1968, was significantly more stable and melted at 70–71°C [58].

9.4 Lappert, Wilkinson and the Isolation of Stable Metal Alkyls und Aryls

After the limited success of the aforementioned work aimed to prepare isolable transition metal alkyls and aryls of the general composition MR_n without supporting ligands, an almost explosive development in this field began at around 1970. It was initiated by Michael Lappert's and Geoffrey Wilkinson's groups in the UK, and the focus of the two main actors was quite similar [59, 60]. They were both not convinced that the statement, still expressed in a well-known textbook in 1968, that "it seems unlikely that M—C bonds are particularly strong" [61], is in fact true. Although only few thermodynamic data were available at that time [62], there was – as Wilkinson pointed out in his Nobel Lecture – no reason "to assume that carbon would differ appreciably in its capability to bond to transition metals compared to other first row elements such as oxygen and nitrogen or the halogens" [3]. Moreover, he [63, 64] as well as Lappert [65, 66] indicated that the known bond energies, e. g., for Pt—C_6H_5 in *trans*-$PtPh_2(PEt_3)_2$ or for Ti—CH_3 in $(C_5H_5)_2TiMe_2$, were quite comparable to those between non-transition metals and carbon. Therefore, they both concluded that the reason for the instability of simple binary transition metal alkyls and aryls was not thermodynamic but kinetic in nature.

In those days it was already known, that several pathways existed by which a metal alkyl could decompose, and for metal methyls the most obvious was homolytic cleavage of the M—CH_3 bond. This had been established for main group compounds such as $SnMe_4$ and $PbMe_4$, and there were good reasons to believe that the decomposition of $TiMe_4$ occurred similarly. In contrast, for metal alkyls bearing hydrogen atoms in the β-position of the alkyl group, such as ethyl, *n*-propyl or *n*-butyl, the preferred mode of decomposition seemed to be the so-called β-hydride elimination. In this mechanism, it was assumed that the initial step was the transfer of a hydrogen from the β-carbon atom of the alkyl chain to the metal via a four-center transition state. The hydrido(olefin) metal intermediate could then eliminate the olefin, and the resulting metal hydrido species decomposed further, for instance to metal and dihydrogen. An early example (though not for a simple binary metal alkyl) supporting the mechanistic pathway shown in Scheme 9.4, was reported in 1970 by George Whitesides, who found that Cu(*n*Bu)(P*n*Bu_3) decomposed to 1-butene, P*n*Bu_3 and CuH. The latter then reacted with another molecule of Cu(*n*Bu)(P*n*Bu_3) to afford *n*-butane, 1-butene, and dihydrogen, together with P*n*Bu_3 and metallic copper [67]. In contrast, the analogous neophyl compound $Cu(CH_2CMe_2Ph)$(P*n*Bu_3), with no hydrogen atoms at the β-carbon atom, decomposed primarily by a radical mechanism [68].

Taking the proposals about the mechanisms of decomposition of metal alkyls into account, Lappert made it quite clear, from the beginning of his work, which alkyl groups should be used for the preparation of stable compounds of the general composition MR_n. Already in his first paper [59], he drew attention "to ligands R^- ... in which one or more α-H atom(s) have been replaced by an organometallic or t-alkyl fragment." He was convinced that "the groups $Me_2PhSiCH_2$, $(Me_3Si)_2CH$, $(Me_3Si)_3C$, $(Me_3Si)Ph_2C$, Me_3SnCH_2, Me_3CCH_2, and Me_2PhCH_2 ... appeared to be of interest", because in this case "a β-hydrogen abstraction pathway is [not] available", and "groups such as Me_3SiCH_2 may be electronically related to $PhCH_2$ from the standpoint of radical carbanion stabilization". In less than 4 years, Lappert's group prepared

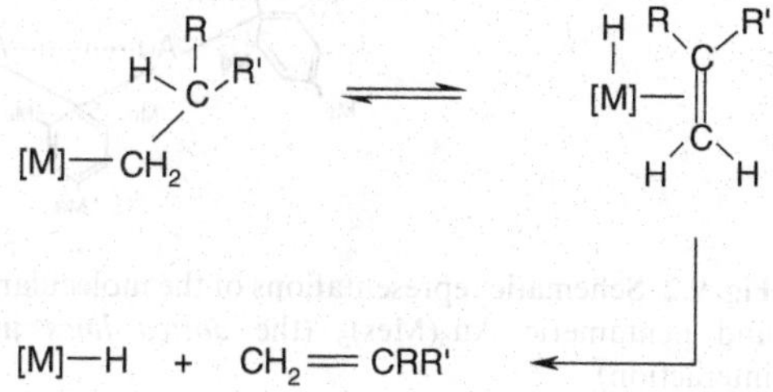

Scheme 9.4 Proposed mechanistic pathway for the decomposition of metal alkyls bearing hydrogen atoms at the β-carbon atom of the alkyl chain ([M] represents a non-specified metal–ligand fragment)

[M]—CH_2—C(H)(R)(R') ⇌ H–[M](η²-CH₂=CRR') → [M]—H + CH_2═CRR' a whole range of stable binary alkyls of the transition metals and the lanthanides including those of low as well as of high-oxidation states [65, 66]. While in most cases organolithium compounds were the most versatile reagents for the preparation of metal alkyls MR_n from metal halides MX_n, the use of RLi sometimes led to undesirable reduction. When this was significant, e.g., with $TiCl_4$ as the starting material, magnesium dialkyls and aluminium trialkyls as well as the scandium and yttrium derivatives $M(CH_2EMe_3)_3(THF)_2$ (M = Sc and Y; E = C and Si) were the reagents of choice [65, 66]. Apart from mononuclear dialkyl, trialkyl and tetraalkyl compounds MR_n, the tetranuclear copper complex $Cu_4(CH_2SiMe_3)_4$ was prepared from CuI and Me_3SiCH_2Li by Lappert's group. The X-ray crystal structure analysis revealed, that the Cu_4C_4 core of the tetramer is strictly planar with each alkyl ligand forming a single bridge between two copper atoms (Fig. 9.2) [69]. Single alkyl bridges were also identified in manganese dialkyls $[MnR_2]_n$, in which the degree of molecular aggregation is strongly dependent on the nature of the alkyl unit. While the compound with R = CH_2CMe_2Ph is a dimer, the analogue with R = CH_2CMe_3 is a tetramer (see Fig. 9.3), and that with R = CH_2SiMe_3 a polymer [70]. Corresponding aryl bridges were found in dimeric bis(mesityl)cobalt(II) [71], trimeric bis(mesityl)manganese(II) [72] and pentameric mesitylcopper(I) and mesitylgold as well. The ten-membered rings of alternate metal and carbon

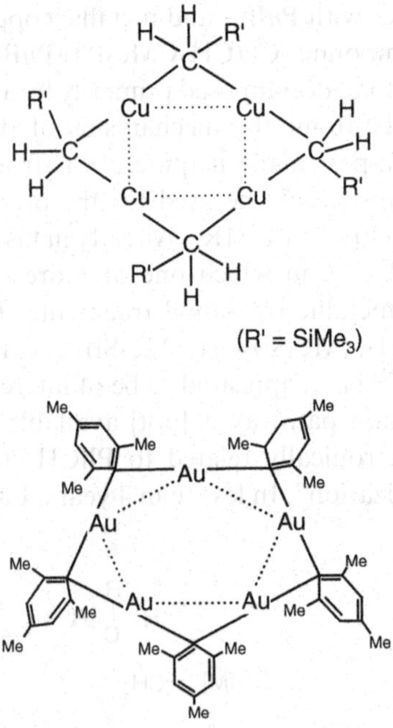

(R' = $SiMe_3$)

Fig. 9.2 Schematic representations of the molecular structures of tetrameric $Cu_4(CH_2SiMe_3)_4$ and pentameric $Au_5(Mes)_5$ (the *dotted lines* indicate a presumed weak metal–metal interaction)

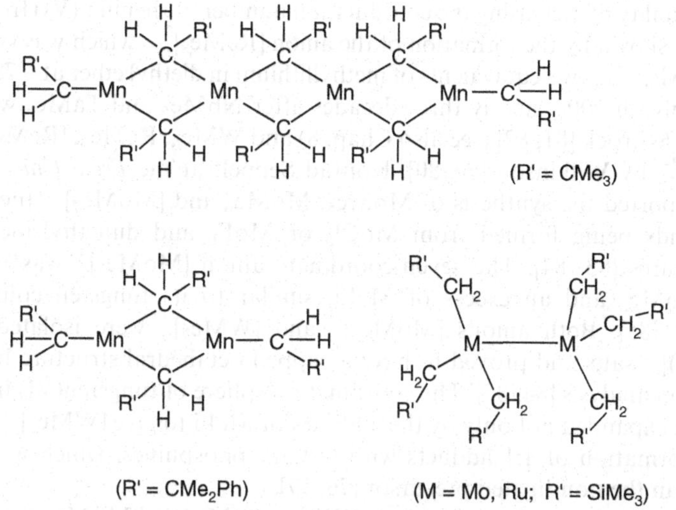

(R' = CMe_3)

(R' = CMe_2Ph) (M = Mo, Ru; R' = $SiMe_3$)

Fig. 9.3 Schematic representations of the molecular structures of tetranuclear and dinuclear manganese dialkyls $[Mn(CH_2R')_2]_n$, and of dinuclear $Mo_2(CH_2SiMe_3)_6$ and $Ru_2(CH_2CMe_3)_6$

atoms in $Cu_5(Mes)_5$ and $Au_5(Mes)_5$ form a five-pointed star-shaped skeleton with the copper and gold atoms bridged by mesityl groups (Fig. 9.2) [73]. $Au_5(Mes)_5$ is thermally quite stable, in contrast to the known binary alkylgold(I) derivatives [74, 75].

Wilkinson's approach to obtaining stable transition metal alkyls was in some respects similar to that of Lappert's, but in addition to CH_2SiMe_3, CH_2CMe_3 and CH_2CMe_2Ph he also used methyl as an alkyl group having no β-hydrogen atoms. Following the idea that the aforementioned thermal lability of $TiMe_4$ was due to its coordinative unsaturation, which allowed decomposition pathways through a dinuclear intermediate or transition state with one or two bridging methyl groups, he assumed that for a six-coordinated transition metal compound MMe_6 "there should be a good chance for thermal stability" [63, 64]. An obvious candidate seemed to be hexamethyltungsten, since in contrast to chromium and molybdenum the hexachloride WCl_6 was "readily available, and the obvious route [was] the interaction of WCl_6 with methyl lithium in ether". Moreover, Wilkinson argued that possibly "this reaction has been studied previously without success, since for success it is necessary that only half of the stoichiometric quantity of CH_3Li be used" [63, 64]. Under those conditions, $W(CH_3)_6$ was obtained as a relatively stable, red crystalline compound which is extremely volatile and melts at ca. 30°C. It only slowly decomposed in the presence of water and could be kept indefinitely at −78°C [76, 77]. Subsequent to this work, Wilkinson reported an alternative synthesis of $W(CH_3)_6$ from WCl_6 and $AlMe_3$ as the alkylating reagent [78]. Trimethylaluminium had also allowed to prepare $ReMe_6$ from $ReOMe_4$ [78, 79], the latter being obtained either from $ReOCl_4$ and MeLi or from $ReOCl_3(PPh_3)_2$ and MeLi in the presence of traces of oxygen [80].

The possibility of increasing the coordination number of rhenium(VI) from six to eight was shown by the formation of the anion $[ReMe_8]^{2-}$, which was generated from $ReMe_6$ and two equivalents of methyllithium in diethyl ether at –78°C [79].

Notably, in 2000, nearly three decades after $NbMe_5$ and $TaMe_5$ were prepared by Schrock [81, 82] (see also Chap. 8) and WMe_6, $ReMe_6$, $[ReMe_7]^-$ and $[ReMe_8]^{2-}$ by Wilkinson [76–80], Konrad Seppelt at the *Freie Universität* in Berlin reported the synthesis of $MoMe_5$, $MoMe_6$ and $[MoMe_7]^-$, the neutral compounds being formed from $MoCl_5$ or MoF_6 and dimethylzinc at low temperatures [83, 84]. The seven-coordinate anion $[MoMe_7]^-$ was prepared from $MoMe_6$ and an excess of MeLi, similar to its tungsten counterpart $[WMe_7]^-$ [85]. Both anions $[MoMe_7]^-$ and $[WMe_7]^-$ were isolated as the $[Li(OEt_2)]^+$ salts and proved to have a capped octahedral structure by X-ray diffraction analysis [84, 85]. The coordination sphere of tungsten(VI) in WMe_6 had been expanded not only by the addition of MeLi to give $[WMe_7]^-$ but also by the formation of 1:1 adducts with tertiary phosphines, which were more stable than the non-ligated precursor [76, 77].

The elucidation of the structures of WMe_6, $ReMe_6$ and $MoMe_6$ provides an interesting insight into the interplay between intuition and experiment including computational work. Geoffrey Wilkinson, having been trained in inorganic chemistry in the days of the valence bond theory and the VSEPR model, was convinced that hexamethyltungsten was octahedral. In the paper describing in detail the preparation and properties of WMe_6 [77], he expressed his view that "the i. r. spectrum is consistent with an octahedral structure, as is the photoelectron spectrum ... on the basis of which a simple and obvious molecular orbital scheme has been constructed". The ESR spectrum of $ReMe_6$ in petroleum or toluene seemed to be consistent with a distorted octahedral symmetry [86]. Regarding the attempts to grow single crystals, it was mentioned that compounds like WMe_6, $ReMe_6$ and $TaMe_5$ occasionally decomposed explosively, when isolated in a reasonably pure state, and therefore had to be handled with extreme caution even in the absence of air [78, 87]. It is worth mentioning that, unlike the highly labile WMe_6, the related *o*-xylidene tungsten complex $W[(CH_2)_2C_6H_4]_3$, which also contains six W—C(sp^3) σ-bonds, is thermally stable [88].

Nevertheless, despite the thermal lability the molecular structures of WMe_6 [89–91], $ReMe_6$ [90, 91], $MoMe_6$ [85], $MoMe_5$ [84], and $TaMe_5$ [92] have been determined, either in the gas phase or in the crystalline state at low temperatures. In all cases, they were in agreement with theoretical work using state-of-the-art quantum chemistry [93–95]. In contrast to Wilkinson's proposal, hexamethyltungsten is not octahedral but has a distorted trigonal prismatic structure with C_{3v} symmetry, which is not the result of intermolecular forces but rather represents a true minimum on the energy surface. Hexamethylrhenium possesses an almost perfect trigonal prismatic configuration, while the d^0 metal monoanions $[MMe_6]^-$ (M = Nb, Ta) [90, 91] as well as the dianion $[ZrMe_6]^{2-}$ [96] show a slight distortion from ideal trigonal prismatic symmetry. The structure of $MoMe_6$ is quite similar to that of WMe_6, although the bond lengths and bond angles differ somewhat. This has been attributed to relativistic effects in the case of the tungsten compound [95]. For $MoMe_5$ and $TaMe_5$, the square pyramid is favored over the trigonal bipyramid, which was found for the tetra- and trimethyltitanium(IV) etherates $TiMe_4$–OEt_2 and $TiMe_3Cl$–OEt_2 [97]. Solvent-free $TiMe_3Cl$ is a heterocubane tetramer with the chloro ligands in triply bridging positions, quite similar to the above mentioned $PtMe_3Cl$. In the dinuclear transition metal alkyls $Mo_2(CH_2SiMe_3)_6$, $W_2(CH_2SiMe_3)_6$ and $Ru_2(CH_2CMe_3)_6$, the two MR_3 halves adopt a mutually staggered configuration (see Fig. 9.3) [98, 99]. The short M—M distances have been explained by a bond order of three, in analogy to the dinuclear hexa(dimethylamides) $M_2(NMe_2)_6$ and hexakis(alkoxides) $M_2(OR)_6$ (M = Mo, W) [100].

In addition to $TiMe_4$–OEt_2, $TiMe_3Cl$–OEt_2 and $[TiMe_3Cl]_4$, Kleinhenz and Seppelt prepared and structurally characterized salts of the anions $[TiMe_5]^-$ and $[Ti_2Me_9]^-$ [97], and thus extended the family of ate complexes, the history of which dates back to Georg Wittig (Nobel laureate for chemistry 1979) and his school [101]. Up to now, ate complexes of the general composition $[MR_n]^{x-}$ (R = alkyl, aryl, alkynyl, etc.) are known for almost all the transition metals including the lanthanides, with n ranging from 2 to 8 and x from 1 to 4 [65, 66, 102–104]. Dinuclear ate complexes $[M_2R_n]^{x-}$ have also been described, not only in the case of titanium(IV) [97] but also in that of vanadium(II), chromium(II) and molybdenum(II) [65, 66]. With acetylides as anionic ligands, ate complexes such as $[Ni(C{\equiv}CR)_4]^{2-}$ and $[Fe(C{\equiv}CR)_6]^{4-}$ with nickel and iron in the oxidation state two as well as $[Ni(C{\equiv}CH)_4]^{4-}$ and $[Ni_2(C{\equiv}CR)_8]^{6-}$ (R = H, Me, Ph) with nickel in the oxidation states zero and one, had been reported by Reinhard Nast's group from the University of Hamburg early on in 1955 [105, 106]. A particularly noteworthy example was the synthesis of pyrophoric, diamagnetic $K_4[Ni(C{\equiv}CH)_4]$, which was obtained from $K_2[Ni(C{\equiv}CH)_4]$ and potassium in liquid ammonia and unexpectedly found not to be explosive [107].

In contrast to the anionic complexes, cationic transition metal alkyls or aryls are rare and have mainly been obtained for copper, silver and gold [65, 66]. The first cation with a non-d^{10} metal center was $[Mo(Mes)_4]^+$, which was prepared from tetramesitylmolybdenum(IV) and Br_2 or I_2 by Wolfgang Seidel (a former Ph.D. student of Hein and later Professor of Inorganic Chemistry at the University of Jena) [108]. Seidel's group also published the synthesis of Fe(2,4,6-$C_6H_2t Bu_3)_2$, being the first monomeric diaryliron(II) compound [109], and of the ate complexes $[Ti(Mes)_4]^-$, $[Mo(Mes)_4]^-$ and $[Cr(Mes)_4]^-$ [110]. The latter were easily oxidized to $Ti(Mes)_4$, $Mo(Mes)_4$ and $Cr(Mes)_4$, respectively. Apart from several other compounds of the general composition MR_4, which for R = CH_2CMe_3, CH_2CPh_3, CH_2SiMe_3 and CH_2SnMe_3 were prepared by Wilkinson [63, 64] and Lappert [65, 66], those with R = 1-norbornyl received special attention. With this ligand, which is bonded to the metal via one of the bridge-head tertiary carbon atoms, even the corresponding iron(IV) and cobalt (IV) complexes FeR_4 and CoR_4 could be obtained [111].

9.5 An Apparent Conflict: Metal Alkyls and Aryls Containing σ- and π-Donor Ligands

Stable compounds with transition metal–alkyl bonds, however, were not only prepared with bulky alkyl groups such as 1-norbornyl, 1-adamantyl and neopentyl, or with so-called stabilizing ligands such as CO, PR_3 and η^5-C_5H_5, but also if *σ-donor* or *π-donor* groups are linked to the metal. Already in one of his first papers on metal alkyls, Wilkinson pointed out that "even ligands such as ammonia and water can give stable alkyls provided the species are substitution-inert, as those of Cr^{III}, Co^{III}, and Rh^{III}" [60]. Typical examples are the dications $[(PhCH_2)Cr(OH_2)_5]^{2+}$, $[(CH_3)Cr(OH_2)_5]^{2+}$ [112, 113] and $[(C_2H_5)Rh(NH_3)_5]^{2+}$ [114], the latter of which does not decompose by β-hydride elimination in contrast to several ethylrhodium(I) derivatives. Zeiss and Herwig's $CrPh_3(THF)_3$ [14–16], Kurras' $Cr(p\text{-Tol})Cl_2(THF)_3$ [32], and Floriani's $V(Mes)_3(THF)$ [115] also belong to this class of compounds.

The first stable transition metal alkyls containing the strongly π-donating oxo ligand O^{2-} were the vanadium(V) and rhenium(VI) oxides $VO(CH_2SiMe_3)_3$ [60] and $ReOMe_4$ [80], which were reported by Wilkinson in the early 1970s. The exceptional methyltrioxorhenium(VII), which more recently gained significant importance as a versatile catalyst for olefin oxidation, metathesis and aldehyde olefination [116], was first obtained accidentally by Ian Beattie and Peter Jones in 1979 [117]. When they repeated the preparation of ReO_2Me_3 as described by Wilkinson in 1976 [79], they observed "the growth of needle crystals in tubes containing residues [of ReO_2Me_3] and which had been left open to the atmosphere". Based on this observation, they then prepared air-stable ReO_3Me (**11**) by exposure of either $ReOMe_4$ or ReO_2Me_3 to dry air in a closed system and isolated the product in "yields in excess of 50%" [117]. About 10 years later, a more efficient synthesis for **11** and other alkyltrioxorhenium(VII) compounds as well was published by Wolfgang Herrmann (Fig. 9.4), who used Re_2O_7 as the starting material and treated this with organotin or organozinc reagents [118]. However, the drawback of this method was the formation of an equimolar amount of tin or zinc perrhenates, which consumed half of the dirhenium heptoxide used. Subsequently, Herrmann's group found that the trioxorhenium(VII) derivatives **12** and **13** were more convenient precursors (see Scheme 9.5), which allowed the synthesis of alkyl- and aryltrioxorhenium(VII) complexes in multigram quantities without loss of rhenium [119–121]. Somewhat later, Wilkinson's group reported the preparation of the mesityl derivative $ReO_3(Mes)$ by a different route via the oxidation of $ReO_2(Mes)_2$ with NO_2 [122].

Apart from the wealth of monoalkyl- and monoarylrhenium complexes ReO_3R (type I), other mono- and dinuclear organorhenium oxides in the oxidation states seven (types II and III) and six (types IV to VII) have been reported (see Scheme 9.6) [119–121]. The first representative of the type II family was $ReO_2(CH_2SiMe_3)_3$, which was obtained by Wilkinson's group in 1975, albeit in low yield [123]. Mertis and Wilkinson prepared also the

Fig. 9.4 Wolfgang Anton Herrmann was born in 1948 in Kelheim, Bavaria, and studied chemistry at the *Technische Universität* in München. He received the Diploma degree with E. O. Fischer in 1971, and then moved to the newly founded University of Regensburg, where he worked for his Ph.D. under the supervision of Professor Henri Brunner. After postdoctoral work in the research group of Professor Philip Skell at the Pennsylvania State University, he returned to Regensburg and finished his *Habilitation* in 1978. He was appointed to Full Professor at the University of Frankfurt/Main in 1982 and, as successor of E. O. Fischer, held the Chair of Inorganic Chemistry at the *Technische Universität* in München since 1985. He is the recipient of several major awards, including the Gottfried Wilhelm Leibniz Award of the German Research Council, the Max Planck Research Prize, the Alexander von Humboldt Award, the Wilhelm Klemm Prize of the German Chemical Society, and the ACS Award in Organometallic Chemistry. Since October 1995, Wolfgang Herrmann has been the President of *Technische Universität München*. His research interests comprise the chemistry of metal carbenes and metal alkyls, the preparation of organometallic compounds with metal–metal multiple bonds, the development of new catalysts for the hydroformylation in biphase systems, the chemistry of organolanthanide complexes, and olefin metathesis. He is the author of more than 750 papers in peer-reviewed journals and the co-author (with B. Cornils) of the well-known handbook "*Applied Homogeneous Catalysis with Organometallic Complexes*" (photo by courtesy from W. A. H.)

corresponding methyl derivative ReO_2Me_3 by oxidation of $ReOMe_4$ with NO in diethyl ether at −78°C [79]. A more general preparative procedure for dioxorhenium(VII) compounds ReO_2R_3 was developed by David Hoffman at the University of Houston, who used the pyridine adducts of the dioxorhenium-(VII) halides $R_2ReO_2X(py)$ and zinc dialkyls as starting materials [124]. Besides the complexes ReO_2R_3, those with two different alkyl groups or two alkyl groups and one aryl group were obtained by this route. In 1989, Hoffman also reported the preparation of $Re_2O_5(CH_2CMe_3)_4$, being as yet the only example of an unsolvated type III compound, and confirmed the presence of a slightly bent ReORe bridge by X-ray diffraction analysis [125].

Mononuclear paramagnetic complexes of type IV with rhenium in the oxidation VI have been reported for R = Mes and Xyl. They were prepared either from $(Me_3NH)ReO_4$ or $Me_3SiOReO_3$ and RMgBr in THF at −78°C, followed by oxidation of the isolated rhenium(V) intermediates $[ReO_2R_2]_2Mg(THF)_2$

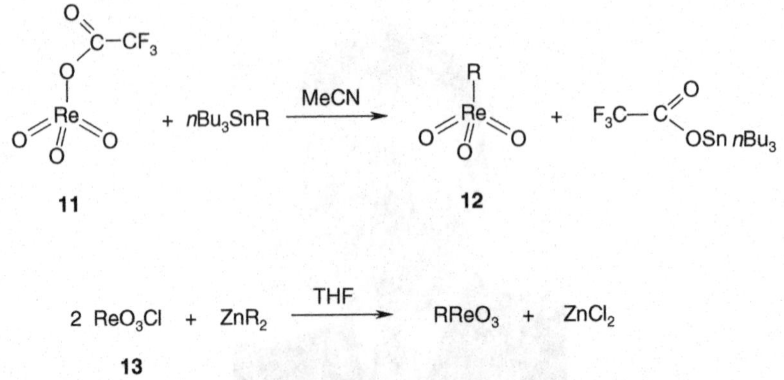

Scheme 9.5 Herrmann's preparative routes to alkyl and aryl rhenium(VII) trioxides **12** [R = Me, Et, *n*Pr, *i*Bu, CH_2SiMe_3, CH_2CHMe_2, $CH_2CH—CH_2$, C(Me)—CH_2, Ph, *p*-Tol, *o*-Tol, Mes, C_6F_5, C_5H_5, C_5Me_5, etc.] with **11** and **13** as the precursors

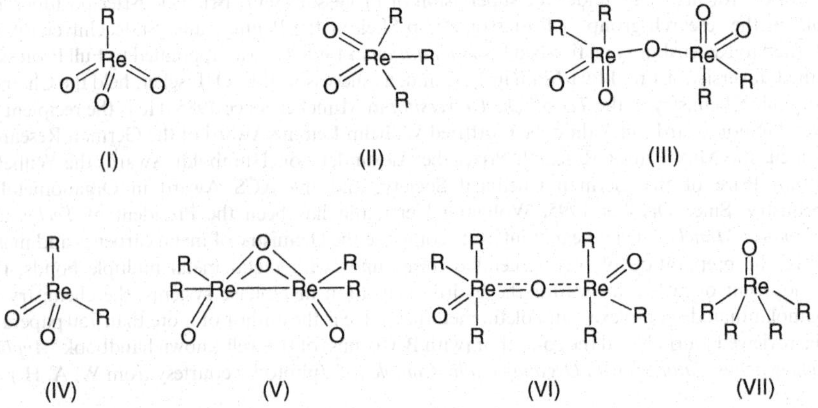

Scheme 9.6 The series of the known mono- and dinuclear organorhenium oxides in the oxidation states seven (types I–III) and six (types IV–VII)

with oxygen or aqueous H_2O_2 [126, 127]. If under the same conditions the analogous *o*-tolyl or *o*-methoxyphenyl Grignard reagents were used, tetraarylrhenium(VI) compounds $ReOR_4$ of type VII were obtained [127]. The neopentyl species $Re_2O_4(CH_2CMe_3)_4$, the first complex of type V, was prepared by treatment of $Me_3SiOReO_3$ with $Al(CH_2CMe_3)_3(THF)$ and characterized crystallographically [128]. Other dinuclear rhenium(VI) oxo complexes of type V were obtained from Re_2O_7 and $RTi(OiPr)_3$ or, more conveniently, by treatment of ReO_3R with ZnR_2 at low temperature [120, 129–131]. The doubly bridged compound $Me_4Re_2O_2(\mu\text{-}O)_2$ reacted upon treatment with excess $ZnMe_2$ in THF yielding the singly bridged complex $Me_6Re_2O_2(\mu\text{-}O)$ of type VI, which could be reconverted to its precursor by oxidation with O_2 in the presence of water [120].

The Re—O distances of the central Re—O—Re unit of $Me_6Re_2O_2(\mu\text{-}O)$ were in agreement with some double bond character [132].

An unusual reaction of methylrhenium trioxide, which illustrates the relation between the structure of an organometallic oxo complex and that of the corresponding inorganic oxide ReO_3, is particularly interesting. In the course of investigating the properties of ReO_3Me, it was observed that heating this compound in aqueous solutions at 70°C for several hours produced a gold-colored polymer of the empirical formula $H_{0.5}(Me_{0.92}ReO_3)$, which contained crystalline domains with double layers of corner-sharing $MeReO_5$ octahedra and layers of intercalated water molecules [120, 133]. The oxorhenium double layers are interconnected by van der Waals forces between the methyl groups, which are oriented inside the double layer. The observed stoichiometric ratio of CH_3:Re = 4.6:5 and partial reduction by additional hydrogen equivalents are responsible for the electric conductivity of the polymer. With regard to the above mentioned application of ReO_3Me as a catalyst for the oxidation of olefins and aromatic compounds with H_2O_2, it is interesting to note that Herrmann's group succeeded in isolating crystals of the 1:1 adducts **16a,b** of the explosive bis(peroxo) complex **15** with σ-donor ligands such as H_2O and $OP(NMe_2)_3$ (Scheme 9.7) [134]. In solution, these adducts (which have a distorted trigonal bipyramidal structure if each peroxo group is considered to occupy one coordination site) are in equilibrium with **15**. Experiments carried out in NMR tubes revealed that the starting material ReO_3Me reacted with an equimolar amount of H_2O_2 to the mono(peroxo) derivative **14**, which could not be isolated but was shown to be catalytically active in some oxidation processes [119–121].

Apart from the rhenium(VI) compounds $ReOR_4$, ReO_2R_2 and $Re(NR')_2R_2$, Wilkinson also reported the synthesis of the first tetra- and diorganylosmium(VI) oxo complexes $OsOR_4$ and OsO_2R_2. While the monooxo compound $OsO(CH_2SiMe_3)_4$ was prepared from OsO_4 and $Mg(CH_2SiMe_3)_2$ as the alkylating (and reducing) reagent, albeit in low quantities [135], the dioxo derivatives $OsO_2(Mes)_2$ and $OsO_2(Xyl)_2$ were obtained from OsO_4 and (Mes)MgBr or (Xyl)MgBr in diethyl ether/THF at low temperature in reasonable yield [126, 127]. Later, Herrmann developed a more efficient route to tetraalkylosmium(VI) oxo complexes such as **18** and **19**, using the osmium(VI) glycolates **17a,b** as the precursors instead of OsO_4 (Scheme 9.8) [136]. At the same time, Patricia Shapley from the University of Illinois in Urbana reported a high-yield synthesis of Wilkinson's $OsO(CH_2SiMe_3)_4$ by alkylation of the osmium(VI) dianions $[OsO_2X_4]^{2-}$ (X = Cl, $OSiMe_3$)

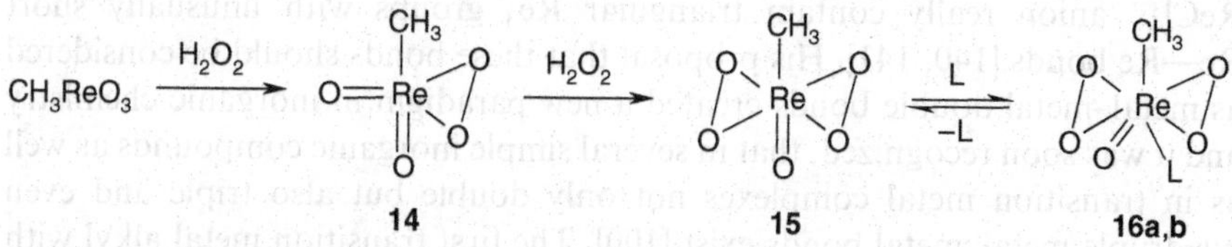

$CH_3ReO_3 \xrightarrow{H_2O_2}$ **14** $\xrightarrow{H_2O_2}$ **15** $\underset{-L}{\overset{L}{\rightleftharpoons}}$ **16a,b**

Scheme 9.7 Preparation of the labile rhenium(VII) peroxo complexes **14** and **15** and the stable 1:1 adducts **16a,b** (**a**: L = H_2O; **b**: L = $OP(NMe_2)_3$) from ReO_3Me as the precursor

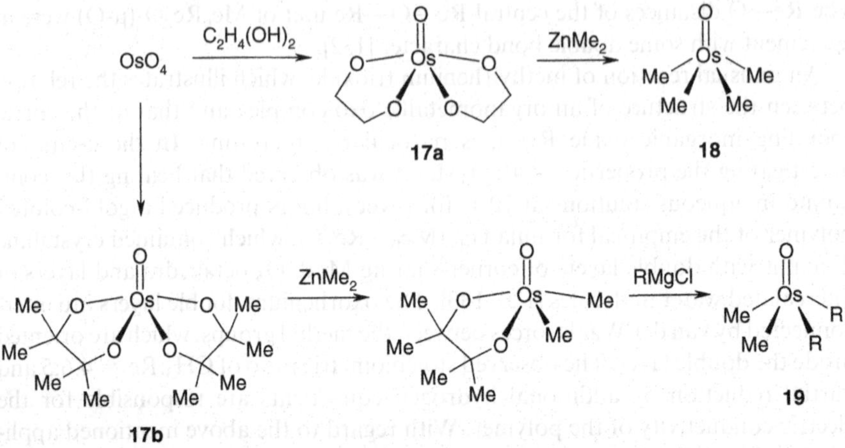

Scheme 9.8 Preparation of the tetraalkyloxoosmium(VI) complexes **18** and **19** ($R = CH_2SiMe_3$) from the osmium(VI) glycolates **17a,b** as the precursors

with either $Mg(CH_2SiMe_3)_2$ or $(Me_3SiCH_2)MgCl$ [137]. The ruthenium counterpart $RuO(CH_2SiMe_3)_4$ was prepared analogously. The X-ray diffraction analyses of $OsOMe_4$ and $OsO(CH_2SiMe_3)_4$ revealed that both compounds have a square pyramidal geometry, with the four carbon atoms forming the base of the pyramid and the osmium atom lying slightly above the C_4 plane [136, 137]. Although the monooxo- and the dioxoosmium(VI) alkyls and aryls have a d^2 electronic configuration, they are diamagnetic. In addition to the oxo complexes $MO(CH_2SiMe_3)_4$ (M = Ru, Os), Patricia Shapley's group also prepared the first imidoosmium(VI) analogues $Os(NMe)R_4$ ($R = Me, CH_2Ph, CH_2SiMe_3$) by alkylation of the nitridometal anions $[OsNR_4]^-$ with Meerwein's oxonium salt $[OMe_3]BF_4$ [138].

9.6 Binary Metal Alkyls with M—M Multiple Bonds

The most exciting development in the area of "true" transition organometallics in recent years was the preparation of binary alkyl and aryl compounds with metal–metal *multiple* bonds. The general field of compounds with multiply bonded metal atoms was pioneered by Cotton (see Fig. 4.4) who discovered, simultaneously with a group in the UK [139], that salts with the assumed $ReCl_4^-$ anion really contain triangular Re_3 groups with unusually short Re—Re bonds [140, 141]. His proposal that these bonds should be considered as metal–metal double bonds created a new paradigm in inorganic chemistry and it was soon recognized, that in several simple inorganic compounds as well as in transition metal complexes not only double but also triple and even quadruple metal–metal bonds exist [100]. The first transition metal alkyl with a M—M bond order of three was Wilkinson's dimolybdenum complex $Mo_2(CH_2SiMe_3)_6$ [98], the structure of which had already been shown in

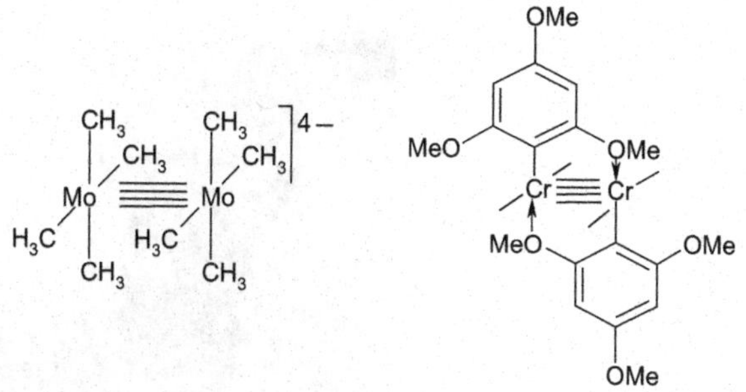

Fig. 9.5 Schematic representations of the structures of $[Mo_2(CH_3)_8]^{4-}$ and $Cr_2[2,4,6\text{-}C_6H_2(OMe)_3]_4$. In the case of the chromium complex the two other tris(methoxy)phenyl ligands lie in a plane perpendicular to the skeleton shown, and were omitted for clarity

Fig. 9.3. A M—M bond order of four probably exists in the isostructural anions $[Cr_2Me_8]^{4-}$ [142, 143] and $[Mo_2Me_8]^{4-}$ [144], where the two MMe_4 halves adopt an eclipsed conformation (Fig. 9.5). Aryl derivatives with a M—M multiple bond were first prepared by Seidel [145] and somewhat later by Cotton [146 ,147], and structurally characterized by Cotton's group. In both $V_2[2,6\text{-}C_6H_3(OMe)_2]_4$ and $Cr_2[2,4,6\text{-}C_6H_2(OMe)_3]_4$, the *ortho* substituted methoxyphenyl groups behave as bridging ligands, similary to acetate or benzoate moieties in $Mo_2(O_2CMe)_4$, $Re_2(O_2CPh)_4Cl_2$, etc. [100].

9.7 The Recent Highlight: Power's RCrCrR and the Fivefold Cr—Cr Bonding

A milestone in the rich chemistry of transition metal compounds with metal–metal multiple bonds has been Philip Power's report about the isolation of a dinuclear chromium(I) aryl RCrCrR [R = $2,6(2,6\text{-}i\mathrm{Pr}_2C_6H_3)_2C_6H_3$] **(20)** with a "fivefold chromium–chromium bonding" [148]. The compound was obtained by reduction of the chloro-bridged chromium(II) dimer $R_2Cr_2(\mu\text{-}Cl)_2$ with a slight excess of potassium graphite and isolated at room temperature in 41% yield. The dark red crystals were very air- and moisture-sensitive but thermally stable up to 200°C. The X-ray crystal structure analysis revealed a planar *trans*-bent core symmetry with a Cr—Cr bond length of 1.835 Å (Fig. 9.6). Although this short distance appeared to agree with the value expected for a compound of the general composition RCrCrR, Power (Fig. 9.7) was originally cautious about using the word

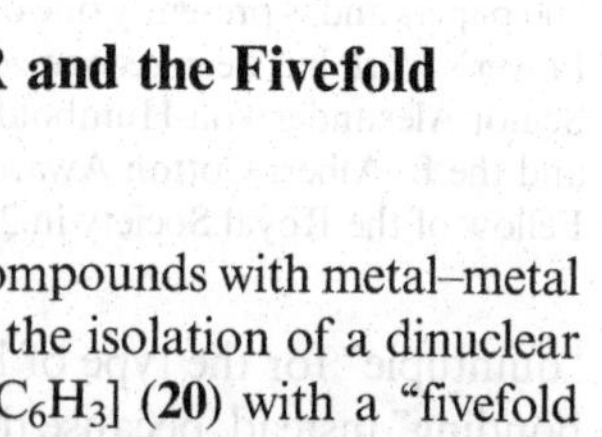

Fig. 9.6 Structural representation of Power's diaryldichromium(I) compound **20** with "fivefold bonding" between the two chromium atoms

Fig. 9.7 Philip P. Power (born in 1953) obtained his B.A. from Trinity College of Dublin in 1974 and his Ph.D. degree, under the supervision of Michael Lappert, from the University of Sussex in 1977. After postdoctoral studies with Richard H. Holm at Stanford University, he joined the faculty of the University of California, Davis, where he is currently Professor of Chemistry. His research interests comprise low-coordinate main group and transition metal chemistry, multiply bonded compounds of the heavier main group and transition metal elements, stable free radicals and bi-radicaloids, the synthesis of clusters of silicon, germanium and tin, and the development of new sterically crowded ligands to stabilize unusual geometries. He has published more than 300 papers and is presently one of the most creative researchers in synthetic inorganic chemistry. Power's work has been recognized by several distinguished lectureships and awards including a Senior Alexander-von-Humboldt Award, the Mond Medal of the Royal Society of Chemistry, and the F. Albert Cotton Award in Synthetic Inorganic Chemistry. Moreover, he was elected Fellow of the Royal Society in 2005 (photo by courtesy from P. P. P.)

"quintuple" for the type of bonding present in **20**. He preferred to call it "fivefold bonding" instead, because the two aryl ligands were in the *trans*-bent rather than in the linear arrangement anticipated for a traditional multiply bonded molecule such as acetylene or another alkyne. Nevertheless, he emphasized that apart from the short metal–metal distance the temperature-independent paramagnetism as well as theoretical studies "support the sharing of five electron pairs in five bonding molecular orbitals between the two $3d^5$ chromium(I) ions" [148].

The enthusiastic comments not only by Cotton ("This is beautiful and fascinating chemistry") but also by theoretical chemists such as Frank Weinhold and Clark Landis from the University of Wisconsin (Power's work "is an empirical landmark in the evolution of our understanding of metal–metal bonding in particular and bonding across the periodic table in general") and Björn Roos of Lund University in Sweden ("This new compound fills a hole in our understanding of the bonding between transition metal atoms") immediately initiated a more extensive computational analysis of **20** [149]. Using PhCrCrPh as the model compound, this study revealed that "fivefold bonding with filled bonding orbitals $(\sigma_g)^2(\pi_u)^4(\delta_g)^4$ is the predominant configuration in the wave function" [150]. Moreover, it showed that the *trans*-bent planar structure for PhCrCrPh is only 1 kcal mol^{-1} higher in energy than the linear structure. The preference for the *trans*-bent configuration in the isolated compound **20** was interpreted as the result "of the secondary interaction between the chromium and the flanking ring of the [bulky disubstituted aryl] ligand". Though the calculated Cr—Cr bond order was less than five, the theoreticians led by Brynda and Gagliardi concluded that "the bond is formally quintuple, because five orbitals and five electrons on each atom are involved in the bonding" [150]. An independent computational study by Landis and Weinhold was completely in agreement with this result [151].

The appreciation of the theoretical work notwithstanding, the greatest achievement, however, was undoubtedly the isolation and precise characterization of the novel dinuclear compound **20**. Not too long ago, probably nobody would have assumed that simple molecules of the general composition A_2B_2 with a fivefold bonding between the two atoms A could exist as stable entities under normal conditions. It is a relatively safe bet to predict that Phil Power's chromium dimer was just a start and, as he proposed [152], "could be the forerunner of other quintuple bonds".

9.8 Novel Perspectives: Metal Alkyls and Aryls Formed by C—H and C—C Bond Activation

The well established routes for preparing alkyl transition metal compounds, using in particular organo alkali, organo zinc and Grignard reagents, were supplemented in 1982 by the seminal discovery that metal complexes, in particular those with electron-rich metal centers, can directly react with unsubstituted alkanes to form metal–alkyl bonds. Almost simultaneously, but independently, Robert Bergman at the University of California in Berkeley (Fig. 9.8), and William Graham at the University of Alberta in Edmonton, Canada (Fig. 9.9), observed that the 18-electron pentamethylcyclopentadienyl iridium(I) and iridium(III) compounds **21** and **22** generated the short-lived 16-electron fragments **23a,b** under UV irradiation, which in the presence of saturated hydrocarbons such as cyclohexane or neopentane afforded the cycloalkyl(hydrido) and alkyl(hydrido) complexes **24a,b** and **25a,b** by intermolecular C—H bond activation (Scheme 9.9) [153–155]. The air-sensitive products were converted with CCl_4 or $CHBr_3$ to the corresponding cycloalkyl- and alkyl(halogeno) derivatives $(C_5Me_5)Ir(R)X(L)$ (X = Cl, Br; L = PMe_3, CO), which were exceedingly stable and could be analytically and spectroscopically characterized. Shortly thereafter, William Jones at the University of Rochester as well as Bergman's group found that $(C_5Me_5)RhH_2(PMe_3)$ behaved similarly to the iridium counterpart **21** and reacted with propane, hexane, cyclopropane, cyclopentane and cyclohexane to furnish the alkyl(hydrido) and cycloalkyl(hydrido) species $(C_5Me_5)RhH(R)(PMe_3)$ by photochemical activation in high yield [156–159]. Prior to Bergman's, Graham's and Jones' work, Joseph Chatt [160], Malcolm Green [161, 162] and Hans Brintzinger [163] had observed that the in situ generated 16-electron fragments $Ru(dmpe)_2$ and $W(C_5H_5)_2$ reacted not only by *intra*molecular but, in the presence of arenes, also by *inter*molecular C—H

Fig. 9.8 Robert G. Bergman (born in 1942) completed his undergraduate studies at Carleton College in 1963 and received his Ph.D. degree, under the direction of Jerome A. Berson, from the University of Wisconsin in 1966. After postdoctoral studies with Ronald Breslow at Columbia University, he went to the California Institute of Technology in Pasadena as a Noyes Research Instructor. He was promoted to Assistant Professor in 1969, Associate Professor in 1971, and Full Professor 2 years later. In 1977 he accepted an appointment as Professor of Chemistry at the University of California, Berkeley, where he has been Gerald E. K. Branch Distinguished Professor since 2002. His research interests focus on the discovery of new chemical reactions, on studies how those reactions work, and on using this knowledge in catalysis and organic synthesis. The thermal cyclization of *cis*-1,5-hexadiyne-3-enes to 1,4-dehydrobenzene diradicals, reported in 1972, became a "name reaction" and is presently well known as the "Bergman cyclization". In the field of organometallic chemistry, Bergman investigated the mechanism of migratory insertion and oxidative addition reactions, the reactivity of compounds containing metal–oxygen and metal–nitrogen bonds, the activation of carbon–hydrogen bonds of alkanes and cycloalkanes by electron-rich transition metals, and more recently the application of C—H activation reactions to problems in organic synthesis. His work has been recognized by several major awards including the ACS Award in Organometallic Chemistry, the ACS James Flack Norris Award in Physical Organic Chemistry, the ACS Arthur C. Cope Award, and the Award in Chemical Sciences by the National Academy of Sciences. He was elected to membership in the National Academy of Sciences and became a Fellow of the National Academy of Arts and Sciences (photo by courtesy from R. G. B.)

bond activation to afford the 18-electron aryl(hydrido) ruthenium(II) and tungsten(IV) complexes $RuH(R)(dmpe)_2$ (R = Ph, Naph) and $(C_5H_5)_2WH(R)$ (R = Ph, Tol, C_6H_4F), respectively. Subsequent studies by Robert Crabtree [164] and Hugh Felkin [165] provided good indirect evidence that intermediates formed from dihydrido iridium(III) and heptahydrido rhenium(VII) precursors interacted with cycloalkanes such as cyclopentane and cyclooctane by sp^3-C—H bond cleavage to give η^5-cyclopentadienyl and η^4-cyclooctadiene metal derivatives. However, both of these systems involve multiple hydrogen-atom loss in the cycloalkane and require an added olefin as hydrogen acceptor.

Given the state of the art around 1980, Bergman's and Graham's findings were indeed a landmark since they proved for the first time that transition metal compounds were "capable of intermolecular oxidative addition to single C—H

Fig. 9.9 William A. G. Graham (born in 1930), a Canadian citizen, obtained his B.A. (1952) and M.A. (1953) from the University of Saskatchewan and his Ph.D. (1956) from Harvard University, where he studied the chemistry of boron hydrides under the supervision of Gordon Stone. Following postdoctoral work with Anton Burg at the University of Southern California and employment with a Boston research firm, he moved in 1962 to the University of Alberta in Edmonton, where he is now Professor Emeritus. His research interests centered on the synthesis and study of compounds where main group elements below carbon were bonded to transition metal moieties. A particular highlight was the development of novel synthetic methods involving insertions into Si—H bonds, leading to isolable silyl–transition metal hydrides. In some way, they presaged and modeled aliphatic C—H bond activation, which he and Bergman reported independently in 1982. Graham's work has been recognized by a number of awards and lectureships including the Palladium Medal of the Chemical Institute of Canada, the Centenary Lectureship of the Royal Society of Chemistry, and the election as a Fellow of the Royal Society of Canada. A special issue of the *Canadian Journal of Chemistry* was published to mark his 65th birthday in 1995 (photo by courtesy from W.A.G.G.)

bonds in saturated hydrocarbons, leading to hydridoalkylmetal complexes in high yield at room temperature in homogeneous solution" [153, 154]. However, even more spectacular than the formation of **24a,b** and **25a,b** was the result, again disclosed independently by Graham [166] and Bergman [167], that methane (according to Robert Crabtree being "notoriously unreactive" [168]) interacted with the in situ generated fragments **23a,b** giving the methyl(hydrido) iridium complexes $(C_5Me_5)IrH(CH_3)(CO)$ and $(C_5Me_5)IrH(CH_3)(PMe_3)$ in excellent yield. The crucial point in preparing these species (and of the analogous cyclopentadienyl compound $(C_5H_5)IrH(CH_3)(CO)$ as well [166]) was the use of an inert solvent which did not react with the labile 16-electron intermediates by oxidative addition. Graham [166] and Bergman [169] found that perfluoroalkanes and noble gases such as Kr and Xe served for this purpose. In collaboration with Anthony Rest of the University of Southampton, Graham showed that the iridium compound **22** as well as the analogous rhodium dicarbonyl $(C_5Me_5)Rh(CO)_2$ reacted upon irradiation in a dilute methane matrix at 12 K to form exactly the same methyl(hydrido) derivatives

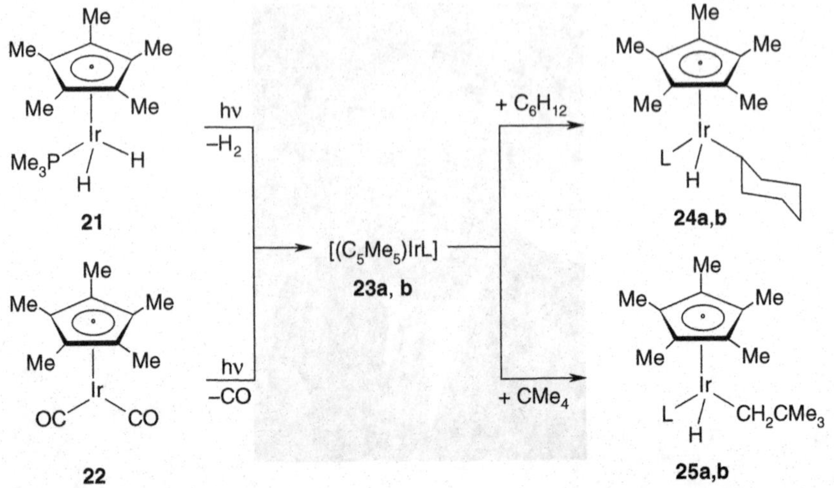

Scheme 9.9 Preparation of the cycloalkyl(hydrido) and alkyl(hydrido) iridium(III) complexes **24a,b** and **25a,b** by intermolecular C—H bond activation from the dihydrido iridium(III) and the dicarbonyl iridium(I) compounds **21** and **22** as the precursors (**a**: L = PMe_3; **b**: L = CO)

$(C_5Me_5)MH(CH_3)(CO)$ (M = Ir, Rh) which had been prepared at room temperature in perfluorohexane [170]. This observation implied that the kinetic barrier for oxidative addition of CH_4 to the initially produced 16-electron fragments $(C_5Me_5)M(CO)$ must be low. Continuing efforts by Rest and Graham provided good evidence that monocarbonyl(methane) species $(C_5Me_5)M(CO)(CH_4)$ are formed in low concentrations in methane matrices and should be considered as labile intermediates along the reaction coordinate [171]. Experimental studies by Bergman's group [172, 173] on alkane rhodium compounds $(C_5Me_5)Rh(CO)(RH)$ (in analogy to Kubas' $M(H_2)$ species sometimes described as "σ-complexes", because of the interaction of the R—H σ-bond with the electrophilic metal center) as well as theoretical work by Roald Hoffmann, Tom Ziegler and Michael Hall were in agreement with this proposal [174–176]. In this context it is worth mentioning that Chatt assumed already in the early 1960s [160] that the formation of $RuH(Ph)(dmpe)_2$ and RuH-(Naph)$(dmpe)_2$ from the labile 16-electron fragment $Ru(dmpe)_2$ and benzene or naphthalene was preceded by the precoordination of the arene to the metal center before oxidative addition occurred. The proposed η^2-arene–metal type of bonding has been confirmed more recently by the isolation and structural characterization of $(C_5H_5)Rh(PMe_3)(\eta^2\text{-}C_6F_6)$ and $(C_5Me_5)Ru(NO)(\eta^2\text{-}$NaphH) [177, 178] as well as by the identification of intermediates such as $(C_5H_5)Rh(PMe_3)(\eta^2\text{-}C_6H_6)$ and $(C_5Me_5)Rh(CO)(\eta^2\text{-}C_6H_4R_2)$ [179]. Upon warming, the latter are converted to the corresponding aryl(hydrido) isomers.

The stirring discovery, that transition metal alkyls and aryls were accessible via oxidative addition of C—H bonds to electrophilic metal centers, soon brought up the question whether dialkyl, diaryl or alkyl(aryl) transition metal compounds could be formed by a similar breaking of C—C bonds. Earlier work by Jack Tipper and Chatt in the 1950s and 1960s [180, 181] as well as subsequent studies by Jack Halpern [182] and others revealed that electron-deficient species such as $PtCl_2$ or $Rh(CO)_2Cl$ (generated from the dimer) interacted with strained hydrocarbons such as cyclopropane or quadricyclene by insertion of the metal into one of the cyclic C—C bonds yielding substituted metallacyclobutanes. [2] However, it was assumed that a similar metal insertion into C—C bonds of unstrained alkanes and alkylarenes to be both thermodynamically and kinetically less favorable than the competing insertion into C—H bonds. Despite these predictions, David Milstein (see Fig. 9.10) at the Weizmann Institute of

Fig. 9.10 David Milstein (born in 1947) completed his undergraduate studies at the Hebrew University of Jerusalem in 1973 and received his Ph.D. degree, under the direction of Professor Jochanan Blum, from the same university in 1976. He carried out postdoctoral studies at the University of Iowa and Colorado State University, where he discovered, together with his advisor John Stille, the *Stille Reaction.* In 1979, he joined the Central Research and Development Department of DuPont in Wilmington, Delaware, where he became a Group Leader in the area of homogeneous catalysis. In 1987, he accepted an appointment as Associate Professor of Chemistry at the Weizmann Institute of Science in Rehovot, Israel. He was promoted to Full Professor in 1993, became Department Head 3 years later, and Head of the Kimmel Center for Molecular Design at the Weizmann Institute in 2000. Since 1996, he is the Israel Matz Professor of Organic Chemistry. His research interests focus on the development of fundamental organometallic chemistry and its application to the design of new processes catalyzed by transition metal complexes. He has studied metal-mediated activation and functionalization of strong chemical bonds, including C—C, C—H, C—F, N—H and O—H bonds. Among his recent contributions is a new method for preparing amides by coupling of alcohols and amines with a pincer-type catalyst. Milstein's work has been recognized by several major awards including the Kolthoff Prize, the Israel Chemical Society Award and the ACS Award in Organometallic Chemistry. He was elected to membership in the German Academy of Sciences Leopoldina in 2006 (photo by courtesy from D. M.)

---

[2] For a review on previous work in this field see [168].

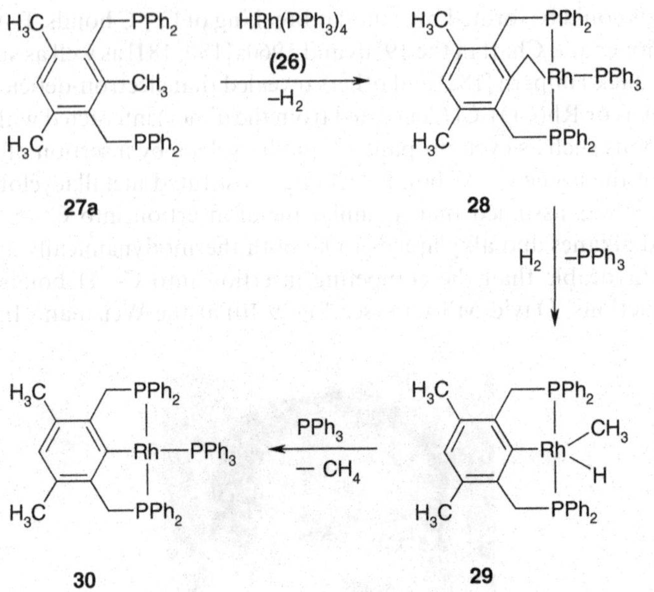

Scheme 9.10 Preparation of the chelating aryl rhodium(I) complex **30** from **26** and **27a** via the stable intermediate **28** by C—C bond activation

Science in Rehovot, Israel, showed in a series of ingenious experiments that this obstacle could be overcome. In the initial part of his work, published in 1993 [183], he demonstrated that the reaction of the rhodium(I) complex $HRh(PPh_3)_4$ (**26**) with the pincer-type ligand **27a** at room temperature led to the elimination of H_2 and formation of the kinetically favored benzylic compound 28 (Scheme 9.10). The C—H activation process could be reversed by heating of **28** under a mild hydrogen pressure, resulting in complete C—C bond breaking and methane elimination. Since the arylated product **30** did not react with methane, the overall C—C bond cleavage was irreversible and thermodynamically more favorable than the overall C—H activation sequence. Regarding the mechanism of formation of **30**, it is conceivable that a rhodium(III) intermediate **29** is formed which reacts with free triphenylphosphine by reductive elimination leading to **30** and methane [183].

However, the initial work, spectacular as it was, did not answer the question whether oxidative addition of a strong C—C bond to a soluble metal compound could be thermodynamically favorable *in the absence* of H_2 or other added reagents such as $HSiX_3$ or $Si_2(OMe)_6$ [184]. By changing the pincer-type ligand and replacing the PPh_2 for PMe_2 units, Milstein's group proved in 1995 that a direct C—C insertion could occur upon heating a mixture of the chelating ligand **27b** and the chlororhodium(I) complex $Rh(PEt_3)_3Cl$ (**31a**) at 150°C [185–187]. The alkyl(aryl) rhodium(III) derivative **32**, which was characterized crystallographically, was obtained in almost quantitative yield. The corresponding C—H activation product **33** was prepared from **27b** and $Rh(PEt_3)_3Ph$ (**31b**) and converted to

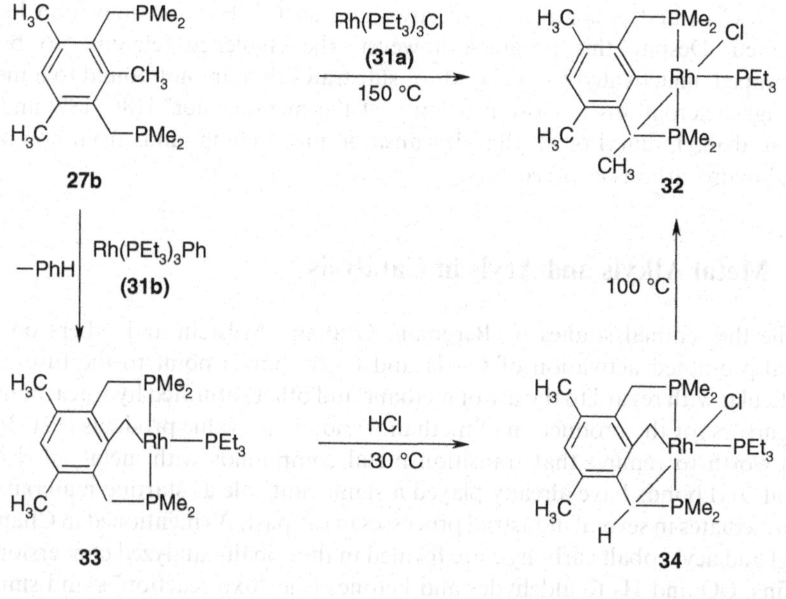

Scheme 9.11 Preparation of the alkyl(aryl) rhodium(I) complex **32** from **27b** and the conversion of the C—H activation product **33** to the C—C bond activation product **32** via **34** as the intermediate

32 via the hydrido rhodium(III) compound **34** as the intermediate. Although the aryl—CH_3 bond is significantly stronger than the arylCH_2—H bonds in the pincer-type ligands **27a,b** [186, 187], the sequence of reactions shown in Scheme 9.11 confirmed that the metal insertion into the aryl—CH_3 bond of **27b** is thermodynamically favored over the insertion into the arylCH_2—H bond. Increasing the steric bulk at the phosphorus atoms of the pincer-type ligands by substituting the phenyl or methyl groups in **27a,b** for *t*-butyl and using $[Rh(C_8H_{14})_2Cl]_2$ or $[Ir(C_8H_{14})_2Cl]_2$ as complex precursors led to the concomitant formation of the isomeric C—H and C—C activation products [188]. In the case of iridium, two isomers were formed in a constant ratio within the temperature range of 20–60°C, indicating that they were generated in two independent, concurrent processes. The surprising fact, established in this study, was that the barrier for C—C oxidative addition is slightly lower than for C—H oxidative addition which could be due to the specific orientation of the relevant metal orbitals toward the carbon–carbon bond. By taking the steric requirements and the influence of the ligands in the transition metal substrate into consideration, Milstein also accomplished a catalytic cleavage of the strong aryl—CH_3 bond in a pincer-type ligand by $[Rh(C_8H_{14})_2Cl]_2$ in dioxane under mild H_2 pressure or with an excess of $HSi(OEt)_3$ [189]. More recently, he reported the first example of an apparent *single step metal insertion* into a carbon–carbon bond in solution and determined the activation parameters for this process [190]. The particular reaction (with the pincer-type ligand that contains two bulky P*t*Bu_2 units) takes place at −70°C (!) and is completely selective, i.e., no C—H bond activation being observed. Despite this progress, however, the challenge remains "to bring unstrained, unactivated C—C bonds of substrates that are not bound to a metal through heteroatoms in close proximity of the metal center" [186, 187] and to obtain dialkyl, diaryl or alkyl(aryl) transition metal compounds from non-activated hydrocarbons as precursors.

9.9 Metal Alkyls and Aryls in Catalysis

While the seminal studies by Bergman, Graham, Milstein and others on the metal-promoted activation of C—H and C—C bonds point to the future, in particular with regard to the use of methane and other saturated hydrocarbons as precursors for the production of methanol and higher value products [191–194], it is worth to remind that transition metal compounds with metal–alkyl and metal–aryl bonds have already played a significant role as starting materials or intermediates in several industrial processes in the past. As mentioned in Chap. 4, alkyl and acyl cobalt carbonyls are formed in the cobalt-catalyzed conversion of olefins, CO and H_2 to aldehydes and ketones (the "oxo reaction"), and similar compounds are probably involved in the cobalt-catalyzed carbonylation of methanol to acetic acid [195]. In the related rhodium–iodide and iridium–iodide catalyzed carbonylation of methanol (the "Monsanto Process" in the case of rhodium and the "Cativa Process" in the case of iridium), intermediates with Rh—CH_3 and Ir—CH_3 bonds have been fully characterized by spectroscopic techniques [195–197]. Stable aryl and allyl nickel(II) chelate complexes are employed as catalysts for the "Shell Higher Olefin Process" (SHOP), which uses ethene as the main feedstock and yields a mixture of α-olefins and internal olefins [198, 199]. Similar four-coordinate aryl and alkyl palladium(II) compounds are involved as intermediates in the arylation of olefins and alkynes via the Heck Reaction and the Sonogashira Reaction, in the coupling of aryl and alkyl halides via the Suzuki Reaction, and in several other important processes leading to the formation of carbon–carbon bonds [200–203].

In summary, it is obvious that the field of transition metal alkyls and aryls, which in the 1950s and 1960s was sidelined by the worldwide attention on the chemistry of sandwich complexes, has matured during the last four decades. The statement made in a review published in 1968 that "by any criterion, simple transition metal alkyls are very unstable" [204] has been convincingly disproved, and there is now no d^1–d^{10} element of which stable and isolable compounds with M—C single bonds were unknown. Based on the pioneering work by Lappert [65, 66] and Wilkinson [3, 63, 64], the introduction of ligands such CH_2CMe_3, CH_2SiMe_3, etc., significantly broadened the scope of the field, increasing the range of available stable compounds and the possibility of studying new types of chemical reactions. Moreover, the work carried out after 1970 proved that the previous assessment, that the stability of transition metal alkyls is favored by a low-oxidation state as well as by the presence of π-acceptor ligands such as CO, PR_3, or C_5H_5, was premature. Recent studies also confirmed that coordinatively unsaturated compounds with M—C_2H_5 or M—C_4H_9 bonds are thermally stable and resistant to β-hydride elimination provided the metal center is shielded by sterically demanding groups [205, 206]. There is still good reason for the optimism expressed by Wilkinson at the end of his Nobel lecture that we "can expect other types of transition metal alkyls to be made in due course and can hope that in addition to their own intrinsic interest some of them may find uses in catalytic or other syntheses" [3].

Biography of Geoffrey Wilkinson see Chap. 5, for that of Joseph Chatt see Chap. 7, and for the biographies of Michael Lappert and Richard Schrock see Chap. 8.

References

1. W. J. Pope, and S. J. Peachey, *Proc. Chem. Soc.* **23**, 86–87 (1907).
2. W. J. Pope, and S. J. Peachey, The Alkyl Compounds of Platinum, *J. Chem. Soc.* **95**, 571–576 (1909).
3. G. Wilkinson, The Long Search for Stable Transition Metal Alkyls, in: *Les Prix Nobel en 1973*, (Ed. W. Odelberg, [Nobel Foundation], Stockholm, 1974).
4. H. Gilman, and M. Lichtenwalter, Tetramethylplatinum and Hexamethyldiplatinum, *J. Am. Chem. Soc.* **60**, 3085–3086 (1938).
5. R. E. Rundle, and J. H. Sturdivant, The Crystal Structures of Trimethylplatinum Chloride and Tetramethylplatinum, *J. Am. Chem. Soc.* **69**, 1561–1567 (1947).
6. T. G. Spiro, D. H. Templeton, and A. Zalkin, The Crystal Structure of Trimethylplatinum Hydroxide, *Inorg. Chem.* **7**, 2165–2167 (1968).
7. F. Runge, *Organo-Metallverbindungen, 2. Aufl.* (Wissenschaftliche Verlagsgesellschaft m. b.h., Stuttgart, 1944, pp. 246–249 and 667–668).
8. W. J. Pope, and C. S. Gibson, The Alkyl Compounds of Gold, *J. Chem. Soc.* **91**, 2061–2066 (1907).
9. J. Sand, and F. Singer, Stickoxyd und das Grignard'sche Reagens, *Liebigs Ann. Chem.* **329**, 190–194 (1903).
10. G. M. Bennett, and E. E. Turner, The Action of Chromic Chloride on the Grignard Reagent, *J. Chem. Soc.* **105**, 1057–1062 (1914).
11. E. Krause, and A. von Grosse, *Die Chemie der metall-organischen Verbindungen* (Gebrüder Bornträger, Berlin, 1937, p. 768).
12. F. A. Cotton, Alkyls and Aryls of Transition Metals, *Chem. Rev.* **55**, 551–594 (1955).
13. D. Seyferth, Bis(benzene)chromium. 1. Franz Hein at the University of Leipzig and Harold Zeiss and Minoru Tsutsui at Yale, *Organometallics* **21**, 1520–1530 (2002).
14. W. Herwig, and H. Zeiss, Triphenylchromium, *J. Am. Chem. Soc.* **79**, 6561 (1957).
15. W. Herwig, and H. Zeiss, The Preparation and Reactions of Triphenylchromium(III), *J. Am. Chem. Soc.* **81**, 4798–4801 (1959).
16. S. I. Khan, and R. Bau, Crystal and Molecular Structure of $Cr(C_6H_5)_3(OC_4H_8)_3$, *Organometallics* **2**, 1896–1897 (1983).
17. F. Hein, and R. Weiss, Zur Existenz echter Organochromverbindungen, *Z. anorg. allg. Chem.* **295**, 145–152 (1958).
18. M. M. Olmstead, P. Power, and S. C. Shoner, Isolation and X-Ray Crystal Structures of Homoleptic, σ-Bonded, Transition-Metal Aryl Complexes $[(LiEt_2O)_4VPh_6]$ and $[(LiEt_2O)_3CrPh_6]$, *Organometallics* **7**, 1380–1385 (1988).

19. R. D. Closson, J. Kozikowski, and T. H. Coffield, Alkyl Derivatives of Manganese Carbonyl, *J. Org. Chem.* **22**, 598 (1957).
20. W. Hieber, O. Vohler, and G. Braun, Über Methylkobalttetracarbonyl, *Z. Naturforsch., Part B,* **13**, 192–193 (1958).
21. W. Hieber, and G. Wagner, Über Organomanganpentacarbonyle, *Liebigs Ann. Chem.* **618**, 24–30 (1958).
22. W. Hieber, G. Braun, and W. Beck, Organorheniumpentacarbonyle, *Chem. Ber.* **93**, 901–908 (1960).
23. T. H. Coffield, J. Kozikowski, and R. D. Closson, Acyl Manganese Pentacarbonyl Compounds, *J. Org. Chem.* **22**, 598 (1957).
24. K. A. Keblys, and A. H. Filbey, Amine Acyl Manganese Tetracarbonyl Complexes, *J. Am. Chem. Soc.* **82**, 4204–4206 (1960).
25. R. F. Heck, and D. S. Breslow, Reactions of Alkyl- and Acyl-Cobalt Carbonyls with Triphenylphosphine, *J. Am. Chem. Soc.* **82**, 4438–4439 (1960).
26. F. Calderazzo, and F. A. Cotton, The Carbonylation of Methyl Manganese Pentacarbonyl and Decarbonylation of Acetyl Manganese Pentacarbonyl, *Inorg. Chem.* **1**, 30–36 (1962).
27. P. M. Treichel, and F. G. A. Stone, Fluorocarbon Derivatives of Metals, *Adv. Organomet. Chem.* **1**, 143–220 (1964).
28. L. Summers, and R. H. Uloth, Reactions of Bis-(cyclopentadienyl)-titanium Dichloride with Aryllithium Compounds, *J. Am. Chem. Soc.* **76**, 2278–2279 (1954).
29. G. Wilkinson, and T. S. Piper, Alkyl and Aryl Derivatives of π-Cyclopentadienyl Compounds of Chromium, Molydenum, Tungsten, and Iron, *J. Inorg. Nucl. Chem.* **3**, 104–124 (1956).
30. K. Clauss, and H. Bestian, Über die Einwirkung von Wasserstoff auf einige metallorganische Verbindungen und Komplexe, *Liebigs Ann. Chem.* **654**, 8–19 (1962).
31. U. Giannini, and S. Cesca, Cyclopentadienyltrimethyltitanium, *Tetrahedron Lett.* **1960**, 19–20.
32. R. R. Schrock, and G. W. Parshall, σ-Alkyl and -Aryl Complexes of the Group 4–7 Transition Metals, *Chem. Rev.* **76**, 243–268 (1976).
33. T. S. Piper, and G. Wilkinson, Chromium Methyl Compound, *Chem. & Ind.* **1955**, 1296.
34. T. S. Piper, and G. Wilkinson, Iron Alkyl and Aryl Compounds, *Naturwissenschaften* **43**, 15–16 (1956).
35. E. O. Fischer, W. Hafner, and H. O. Stahl, Über Cyclopentadienyl-metall-carbonyl-wasserstoffe des Chroms, Molybdäns und Wolframs, *Z. anorg. allg. Chem.* **282**, 47–62 (1955).
36. J. Chatt, and B. L. Shaw, Alkyls and Aryls of Transition Metals: Part I. Complex Methylplatinum(II) Derivatives, *J. Chem. Soc.* **1959**, 705–716.
37. J. Chatt, and B. L. Shaw, Alkyls and Aryls of Transition Metals: Part II. Platinum(II) Derivatives, *J. Chem. Soc.* **1959**, 4020–4032.
38. J. Chatt, and B. L. Shaw, Alkyls and Aryls of Transition Metals: Part III. Nickel(II) Derivatives, *J. Chem. Soc.* **1960**, 1718–1729.
39. J. Chatt, and B. L. Shaw, Alkyls and Aryls of Transition Metals: Part IV. Cobalt(II) and Iron(II) Derivatives, *J. Chem. Soc.* **1961**, 285–290.
40. J. Chatt, and A. E. Underhill, Alkyls and Aryls of Transition Metals: Part V. Rhodium Derivatives, *J. Chem. Soc.* **1963**, 2088–2089.
41. J. Chatt, J. D. Garforth, and G. A. Rowe Alkyls and Aryls of Transition Metals: Part VI. Rhenium, *J. Chem. Soc. A.* **1966**, 1834–1836.
42. J. Chatt, and B. L. Shaw, Alkyls and Aryls of Transition Metals: Part VII. Trimethyliridium Derivatives, *J. Chem. Soc. A* **1966**, 1836–1837.
43. P. G. Owston, and J. M. Rowe, The Crystal Structure of *trans*-Dimesitylbis(diethylphenylphosphine)cobalt(II), *J. Chem. Soc.* **1963**, 3411–3419.
44. J. Chatt, and R. G. Hayter, Some Halido- and Hydrido-alkyl and –aryl Complexes of Ruthenium(II) and Osmium(II), *J. Chem. Soc.* **1963**, 6017–6027.
45. G. Calvin, and G. E. Coates, Organopalladium Compounds, *J. Chem. Soc.* **1960**, 2008–2016.

46. A. Yamamoto, K. Morifuji, S. Ikeda, T. Saito, Y. Uchida, and A. Misono, Butadiene Polymerization Catalysts. Diethylbis(dipyridyl)iron and Diethyldipyridylnickel, *J. Am. Chem. Soc.* **87**, 4652–4653 (1965).
47. G. Wilke, and G. Herrmann, Stabilized Dialkylnickel Compounds, *Angew. Chem. Int. Ed. Engl.* **5**, 581–582 (1966).
48. A. Yamamoto, K. Morifuji, S. Ikeda, T. Saito, Y. Uchida, and A. Misono, Diethylbis-(dipyridyl)iron. A Butadiene Cyclodimerization Catalyst, *J. Am. Chem. Soc.* **90**, 1878–1883 (1968).
49. T. Ito, H. Tsuchiya, and A. Yamamoto, Dialkylpalladium(II) Complexes. Synthesis, Characterizations, and Reactions with CO, I_2, and CH_3I, *Bull. Chem. Soc. Jpn.* **50**, 1319–1327 (1977).
50. J. R. Doyle, J. H. Hutchinson, N. C. Baenziger, and L. W. Tresselt, Reactions of *cis*-Diiodo-(cyclooctatetraene)-platinum(II) with Grignard Reagents, *J. Am. Chem. Soc.* **83**, 2768–2769 (1961).
51. C. R. Kistner, J. H. Hutchinson, J. R. Doyle, and J. C. Storlie, The Preparation and Properties of Some Aryl and Alkyl Platinum(II)-Olefin Compounds, *Inorg. Chem.* **2**, 1255–1261 (1963).
52. D. F. Herman, and W. K. Nelson, Isolation of a Compound Containing the Titanium—Carbon Bond, *J. Am. Chem. Soc.* **75**, 3877–3882 (1953).
53. D. F. Herman, and W. K. Nelson, Stability of the Titanium–Carbon Bond, *J. Am. Chem. Soc.* **75**, 3882–3887 (1953).
54. M. Bottrill, P. D. Gavens, J. W. Kelland, and J. McMeeking, σ-Bonded Hydrocarbyl Complexes of Titanium(IV), in: *Comprehensive Organometallic Chemistry* (Eds. G. Wilkinson, F. G. A. Stone, and E. W. Abel, Pergamon Press, Oxford, 1982, Vol. 3, Chap. 22.4).
55. K. Clauss, and C. Beermann, Halogenfreie Methyl-Verbindungen des Titans und Chroms, *Angew. Chem.* **71**, 627 (1959).
56. H. J. Berthold, and G. Groh, Über die Isolierung von Tetramethyltitan, *Z. anorg. allg. Chem.* **319**, 230–235 (1963).
57. V. N. Latjaeva, G. A. Razuvaev, A. V. Malisheva, and G. A. Kiljakova, Phenyl Derivatives of Organotitanium Compounds, *J. Organomet. Chem.* **2**, 388–397 (1964).
58. U. Giannini, and U. Zucchini, Benzyltitanium Compounds, *Chem. Comm.* **1968**, 940.
59. M. R. Collier, M. F. Lappert, and M. M. Truelock, μ-Methylene Transition Metal Binuclear Compounds: Complexes with Me_3Si- and Related Ligands, *J. Organomet. Chem.* **25**, C36–C38 (1970).
60. G. Yagupsky, W. Mowat, A. Shortland, and G. Wilkinson, Trimethylsilylmethyl Compounds of Transition Metals, *Chem. Comm.* **1970**, 1369–1370.
61. M. L. H. Green, *Organometallic Compounds* (Eds. G. E. Coates, M. L. H. Green, and K. Wade, Methuen, London, 1968, Vol. 2, p. 221).
62. H. A. Skinner, The Strengths of Metal-to-Carbon Bonds, *Adv. Organomet. Chem.* **2**, 49–114 (1964).
63. G. Wilkinson, Aspects of the Transition Metal to Carbon σ-bond; Stable Binary Alkyls, *Chimia* **27**, 165–169 (1973).
64. G. Wilkinson, Long Search for Stable Transition Metal Alkyls, *Science* **185**, 109–112 (1974).
65. P. J. Davidson, M. F. Lappert, and R. Pearce, Stable Homoleptic Metal Alkyls, *Acc. Chem. Res.* **7**, 209–217 (1974).
66. P. J. Davidson, M. F. Lappert, and R. Pearce, Metal σ-Hydrocarbyls, MR_n. Stoichiometry, Structures, Stabilities, and Thermal Decomposition Pathways, *Chem. Rev.* **76**, 219–242 (1976).
67. G. M. Whitesides, E. R. Stedronsky, C. P. Casey, and J. S. Filippo, Jr., Mechanism of Thermal Decomposition of *n*-Butyl(tri-*n*-butylphosphine)copper(I), *J. Am. Chem. Soc.* **92**, 1426–1427 (1970).
68. G. M. Whitesides, E. J. Panek, and E. R. Stedronsky, Radical Intermediates in the Thermal Decomposition of Neophyl(tri-*n*-butylphosphine)copper(I) and Neophyl(tri-*n*-butylphosphine)silver(I), *J. Am. Chem. Soc.* **94**, 232–239 (1972).

69. J. A. J. Jarvis, B. T. Kilbourn, R. Pearce, and M. F. Lappert, Crystal Structure (at –40°) of Tetrakistrimethylsilylmethylcopper(I)], an Alkyl Bridged, Square Planar, Tetranuclear Copper(I) Cluster, *J. Chem. Soc., Chem. Comm.* **1973**, 475–476.
70. R. A. Andersen, E. Carmona-Guzman, J. F. Gibson, and G. Wilkinson, Neopentyl, Neophyl, and Trimethylsilylmethyl Compounds of Manganese. Manganese(II) Dialkyls; Manganese(II) Dialkyl Amine Adducts; Tetraalkylmanganate(II) Ions and Lithium Salts; Manganese(IV) Tetraalkyls, *J. Chem. Soc., Dalton Trans.* **1976**, 2204–2211.
71. K. H. Theobold, J. Silvestre, E. K. Byrne, and D. S. Richeson, Homoleptic Mesityl Complexes of Cobalt(II). Synthesis, Crystal Structure, and Theoretical Description of Bis(μ-mesityl)dimesityldicobalt, *Organometallics* **8**, 2001–2009 (1989).
72. S. Gambarotta, C. Floriani, A. Chiesi-Villa, and C. Guastini, A Homoleptic Arylmanganese(II) Complex: Synthesis and Structure of a Thermally Stable Trinuclear Mesitylmanganese(II) Complex, *J. Chem. Soc., Chem. Comm.* **1983**, 1128–1129.
73. E. M. Meyer, S. Gambarotta, C. Floriani, A. Chiesi-Villa, and C. Guastini, Polynuclear Aryl Derivatives of Group 11 Metals: Synthesis, Solid State-Solution Structural Relationship, and Reactivity with Phosphines, *Organometallics* **8**, 1067–1079 (1989).
74. R. J. Puddephatt, Gold, in: *Comprehensive Organometallic Chemistry* (Eds. G. Wilkinson, F. G. A. Stone, and E. W. Abel, Pergamon Press, Oxford, 1982, Vol. 2, Chap. 15).
75. A. Grohmann, and H. Schmidbaur, Gold, in: *Comprehensive Organometallic Chemistry II* (Eds. E. W. Abel, F. G. A. Stone, and G. Wilkinson, Pergamon Press, Oxford, 1995, Vol. 3, Chap. 1).
76. A. Shortland, and G. Wilkinson, Hexamethyltungsten, *J. Chem. Soc., Chem. Comm.* **1972**, 318.
77. A. J. Shortland, and G. Wilkinson, Preparation and Properties of Hexamethyltungsten, *J. Chem. Soc., Dalton Trans.* **1973**, 872–876.
78. L. Gayler, K. Mertis, and G. Wilkinson, New Synthesis of Hexamethyltungsten(VI), Hexamethylrhenium(VI) and Dioxotrimethylrhenium(VII), *J. Organomet. Chem.* **85**, C37–C38 (1975).
79. K. Mertis, and G. Wilkinson, The Synthesis and Reactions of Hexamethylrhenium(VI), *cis*-Trimethyldioxorhenium(VII), and the Octamethylrhenate(VI) Ion, *J. Chem. Soc., Dalton Trans.* **1976**, 1488–1492.
80. K. Mertis, J. F. Gibson, and G. Wilkinson, Tetramethyloxorhenium(VI), a Paramagnetic d^1 Alkyl, *J. Chem. Soc., Chem. Comm.* **1974**, 93.
81. R. R. Schrock, and P. Meakin, Pentamethyl Complexes of Niobium and Tantalum, *J. Am. Chem. Soc.* **96**, 5288–5290 (1974).
82. R. R. Schrock, Preparation and Characterization of $M(CH_3)_5$ (M = Nb, Ta) and $Ta(CH_2C_6H_5)_5$ and Evidence for Decomposition by α-Hydrogen Atom Abstraction, *J. Organomet. Chem.* **122**, 209–225 (1976).
83. B. Roessler, S. Kleinhenz, and K. Seppelt, Pentamethylmolybdenum, *Chem. Comm.* **2000**, 1039–1040.
84. B. Roessler, and K. Seppelt, $[Mo(CH_3)_6]$ and $[Mo(CH_3)_7]^-$, *Angew. Chem. Int. Ed.* **39**, 1259–1261 (2000).
85. V. Pfennig, N. Robertson, and K. Seppelt, The Structures of the Anions $[W(CH_3)_7]^-$ and $[Re(CH_3)_8]^{2-}$, *Angew. Chem. Int. Ed. Engl.* **36**, 1350–1352 (1997).
86. J. F. Gibson, G. M. Lack, K. Mertis, and G. Wilkinson, Electron Spin Resonance Spectra of Hexamethylrhenium(VI) and the Octamethylrhenate(VI) Ion, *J. Chem. Soc., Dalton Trans.* **1976**, 1492–1495.
87. K. Mertis, L. Gayler, and G. Wilkinson, Permethyls of Tantalum, Tungsten and Rhenium: A Warning, *J. Organomet. Chem.* **97**, C65 (1975).
88. M. F. Lappert, C. L. Raston, G. L. Rowbottom, B. W. Skelton, and A. H. White, Tungsten *o*-Xylidene Complexes Derived from Tetrachloro(oxo)tungsten(VI), *J. Chem. Soc., Dalton Trans.* **1984**, 883–891.

89. A. Haaland, A. Hammel, K. Rypdal, and H. V. Volden, Coordination Geometry of Gaseous Hexamethyltungsten: Not Octahedral, *J. Am. Chem. Soc.* **112**, 4547–4549 (1990).
90. V. Pfennig, and K. Seppelt, Crystal and Molecular Structures of Hexamethyltungsten and Hexamethylrhenium, *Science* **271**, 626–628 (1996).
91. S. Kleinhenz, V. Pfennig, and K. Seppelt, Preparation and Structures of $[W(CH_3)_6]$, $[Re(CH_3)_6]$, $[Nb(CH_3)_6]^-$, and $[Ta(CH_3)_6]^-$, *Chem. Eur. J.* **4**, 1687–1691 (1998).
92. C. Pulham, A. Haaland, A. Hammel, K. Rypdal, H. P. Verne, and H. V. Volden, The Structures of Pentamethyltantalum and -Antimony: One Square Pyramid and one Trigonal Bipyramid, *Angew. Chem. Int. Ed. Engl.* **31**, 1464–1466 (1992).
93. S. K. Kang, T. A. Albright, and O. Eisenstein, The Structure of d^0 ML_6 Complexes, *Inorg. Chem.* **28**, 1611–1613 (1989).
94. T. A. Albright, and H. Tang, The Structure of Pentamethyltantalum, *Angew. Chem. Int. Ed. Engl.* **31**, 1462–1464 (1992).
95. M. Kaupp, The Nonoctahedral Structures of d^0, d^1, and d^2 Hexamethyl Complexes, *Chem. Eur. J.* **4**, 1678–1686 (1998).
96. P. M. Morse, and G. S. Girolami, Are d^0 ML_6 Complexes Always Octahedral? The X-Ray Structure of Trigonal Prismatic $[Li(tmed)]_2[ZrMe_6]$, *J. Am. Chem. Soc.* **111**, 4114–4116 (1989).
97. S. Kleinhenz, and K. Seppelt, Preparation and Structures of Methyltitanium Compounds, *Chem. Eur. J.* **5**, 3573–3580 (1999).
98. F. Huq, W. Mowat, A. Shortland, A. C. Skapski, and G. Wilkinson, Crystal Structure of Hexakis(trimethylsilylmethyl)dimolybdenum, *Chem. Comm.* **1971**, 1079–1080.
99. R. P. Tooze, M. Motevalli, M. B. Hursthouse, and G. Wilkinson, Alkyl Compounds of Diruthenium(III) and Diosmium(III). X-Ray Crystal Structure of the First Ruthenium Peralkyl, Hexakis(neopentyl)diruthenium(III), *J. Chem. Soc., Chem. Comm.* **1984**, 799–800.
100. F. A. Cotton, C. A. Murillo, and R. A. Walton, *Multiple Bonds between Metal Atoms, 3. Ed.* (Springer Science and Business Media, New York, 2005).
101. G. Wittig, Komplexbildung und Reaktivität in der metallorganischen Chemie, *Angew. Chem.* **70**, 65–71 (1958).
102. R. Taube, New Results in the Coordination Chemistry of σ-Bonded Carbanionic Ligands, *Pure Appl. Chem.* **55**, 165–176 (1983).
103. R. Taube, D. Steinborn, H. Drevs, P. N. Chuong, N. Stransky, and J. Langlotz, Redoxstabilität in der σ-Organoübergangsmetallchemie, *Z. Chem.* **28**, 381–396 (1988).
104. S. U. Koschmieder, and G. Wilkinson, Homoleptic and Related Aryls of Transition Metals, *Polyhedron* **10**, 135–173 (1991).
105. R. Nast, Komplexe Acetylide von Übergangsmetallen, *Angew. Chem.* **72**, 26–31 (1960).
106. R. Nast, Coordination Chemistry of Metal Alkynyl Compounds, *Coord. Chem. Rev.* **47**, 89–124 (1982).
107. R. Nast, and K. Vesper, Alkinylokomplexe von Nickel, *Z. anorg. allg. Chem.* **279**, 146–156 (1955).
108. W. Seidel, and I. Bürger, Bildung von Tetramesitylmolybdän(V); ein homoleptisches σ-Arylübergangsmetallkation, *J. Organomet. Chem.* **177**, C19–C21 (1979).
109. H. Müller, W. Seidel, and H. Görls, $Fe(2,4,6\text{-}t Bu_3C_6H_2)_2$, A Monomeric Diaryl Complex with Two-Coordinated Iron(II), *Angew. Chem. Int. Ed. Engl.* **34**, 325–327 (1995).
110. W. Seidel, and I. Bürger, Tetramesitylchrom – eine stabile Arylchrom(IV)-Verbindung, *Z. anorg. allg. Chem.* **426**, 155–158 (1976).
111. B. K. Bower, and H. G. Tennent, Transition Metal Bicyclo[2.2.1]hept-1-yls, *J. Am. Chem. Soc.* **94**, 2512–2514 (1972).
112. F. A. L. Anet, and E. Leblanc, A Novel Organochromium Compound, *J. Am. Chem. Soc.* **79**, 2649–2650 (1957).
113. M. Ardon, K. Woolmington, and A. Pernick, The Methylpentaaquochromium(III) Ion, *Inorg. Chem.* **10**, 2812 (1971).

114. K. Thomas, J. A. Osborn, A. R. Powell, and G. Wilkinson, The Preparation of Hydridopentammine- and Hydridoaquotetraammine-rhodium(III) Sulfates and Other Salts; the Formation of Alkyl and Fluoroalkyl Derivatives, *J. Chem. Soc. A* **1968**, 1801–1806.
115. S. Gambarotta, C. Floriani, A. Chiesi-Villa, and C. Guastini, A Tri-σ-aryl Vanadium(III) Derivative: Structural Determination of Trimesitylvanadium(III)-Tetrahydrofuran, *J. Chem. Soc., Chem. Comm.* **1984**, 886–887.
116. F. E. Kühn, A. Scherbaum, and W. A. Herrmann, Methyltrioxorhenium and Its Applications in Olefin Oxidation, Metathesis and Aldehyde Olefination, *J. Organomet. Chem.* **689**, 4149–4164 (2004).
117. R. I. Beattie, and P. J. Jones, Methyltrioxorhenium. An Air-Stable Compound Containing a Carbon–Rhenium Bond, *Inorg. Chem.* **18**, 2318–2319 (1979).
118. W. A. Herrmann, J. G. Kuchler, J. K. Felixberger, E. Herdtweck, and W. Wagner, Methylrhenium Oxides: Synthesis from Dirhenium Heptoxide and Catalytic Activity in Olefin Metathesis, *Angew. Chem. Int. Ed. Engl.* **27**, 394–396 (1988).
119. W. A. Herrmann, Laboratory Curiosities of Yesterday, Catalysts of Tomorrow: Organometallic Oxides, *J. Organomet. Chem.* **500**, 149–174 (1995).
120. W. A. Herrmann, and F. E. Kühn, Organorhenium Oxides, *Acc. Chem. Res.* **30**, 169–180 (1997).
121. C. C. Romao, F. E. Kühn, and W. A. Herrmann, Rhenium(VII) Oxo and Imido Complexes: Synthesis, Structures, and Applications, *Chem. Rev.* **97**, 3197–3246 (1997).
122. B. S. McGilligan, J. Arnold, G. Wilkinson, B. Hussain-Bates, and M. B. Hursthouse, Reactions of Dimesityldioxo-osmium(VI) with Donor Ligands; Reactions of $MO_2(2,4,6\text{-}Me_3C_6H_2)_2$, M = Os or Re, with Nitrogen Oxides. X-Ray Crystal Structures of $[2,4,6\text{-}Me_3C_6H_2N_2]^+[OsO_2(ONO_2)(2,4,,6\text{-}Me_3C_6H_2)]^-$, $OsO(NBu\textit{t})(2,4,6\text{-}Me_3C_6H_2)_2$, $OsO_3(NBu\textit{t})$, and $ReO_3[N(2,4,6\text{-}Me_3C_6H_2)_2]$, *J. Chem. Soc., Dalton Trans.* **1990**, 2465–2475.
123. K. Mertis, D. H. Williamson, and G. Wilkinson, Synthesis and Properties of Oxorhenium(VI) Methyl and Trimethylsilylmethyl Compounds, *J. Chem. Soc., Dalton Trans.* **1975**, 607–611.
124. S. Cai, D. M. Hoffman, and D. A. Wierda, Rhenium(VII) Oxo–Alkyl Complexes: Reductive and α-Elimination Reactions, *Organometallics* **15**, 1023–1032 (1996).
125. S. Cai, D. M. Hoffman, and D. A. Wierda, Alkoxide and Thiolate Rhenium(VII) Oxo–Alkyl Complexes and $Re_2O_5(CH_2CMe_3)_4$, a Compound with a $[O_2Re—O—ReO_2]^{4+}$ Core, *Inorg. Chem.* **28**, 3784–3786 (1989).
126. C. J. Longley, P. D. Savage, G. Wilkinson, B. Hussain, and M. B. Hursthouse, Alkylimido and Oxo Aryls of Rhenium. X-Ray Structures of $(Bu\textit{t}N)_2ReCl_2(\textit{o}\text{-}MeC_6H_4)$ and $MO_2(2,6\text{-}Me_2C_6H_3)_2$, M = Re and Os, *Polyhedron* **7**, 1079–1088 (1988).
127. P. Stavropoulos, P. G. Edwards, T. Behling, G. Wilkinson, M. Motevalli, and M. B. Hursthouse, Oxoaryls of Rhenium-(V) and -(VI) and Osmium(VI). X-Ray Crystal Structures of Dimesityldioxorhenium(VI), Tetramesityloxorhenium(VI), and Dimesityldioxoosmium(VI), *J. Chem. Soc., Dalton Trans.* **1987**, 169–176.
128. S. Cai, D. M. Hoffman, J. C. Huffman, D. A. Wierda, and H.-G. Woo, Synthesis and Structural Characterization of $[Re_2(\mu\text{-}O)_2O_2(CH_2CMe_3)_4]$ and the Li^+ and $[NEt_4]^+$ Salts of the d^2 $[ReO_2(CH_2CMe_3)_2]^-$ Anion, *Inorg. Chem.* **26**, 3693–3700 (1987).
129. J. M. Huggins, R. R. Whitt, and L. Lebioda, Synthesis and Structure of a Novel Bis-μ-oxo Rhenium(VI) Dimer: $Re_2O_2(\mu\text{-}O)_2(CH_2CMe_2Ph)_4$, *J. Organomet. Chem.* **312**, C15–C19 (1986).
130. F. E. Kühn, J. Mink, and W. A. Herrmann, Alkylrhenium(VI) Oxides – Synthesis and Characterization, *Chem. Ber.* **130**, 295–298 (1987).
131. W. A. Herrmann, C. C. Romao, P. Kiprof, J. Behm, M. R. Cook, and M. Taillefer, Methyl- und Ethylrheniumoxide: Präparative und strukturchemische Aspekte, *J. Organomet. Chem.* **413**, 11–25 (1991).

132. P. Stavropoulos, P. G. Edwards, G. Wilkinson, M. Motevalli, K. M. A. Malik, and M. B. Hursthouse, Oxoalkyls of Rhenium-(V) and -(VI). X-Ray Crystal Structures of $(Me_4ReO_2)_2Mg(thf)_4$, $[(Me_3SiCH_2)_4ReO_2]Mg(thf)_2$, $Re_2O_3Me_6$, and $Re_2O_3(CH_2SiMe_3)_6$, *J. Chem. Soc., Dalton Trans.* **1985**, 2167–2175.
133. W. A. Herrmann, R. W. Fischer, and W. Scherer, Polymerization of Methyltrioxorhenium(VII): Organometallic Oxides as Precursors for Oxide Ceramics, *Adv. Mater.* **4**, 653–658 (1992).
134. W. A. Herrmann, R. W. Fischer, W. Scherer, and M. U. Rauch, Methyltrioxorhenium(VII) as an Epoxidation Catalyst: Structure of the Active Species and the Mechanism of Catalysis, *Angew. Chem. Int. Ed. Engl.* **32**, 1157–1160 (1993).
135. A. S. Alves, D. S. Moore, R. A. Andersen, and G. Wilkinson, Triphenylphosphonium and Trimethylphosphonium Hexachloroosmates(IV) and Their Reactions with Alkylating Agents; Methyl and Trimethylsilylmethyl Osmium Compounds; *trans*-Dichlorotetrakis(trimethylphosphine)osmium(II); Tetrakis(trimethylsilylmethyl)oxoosmium(VI), *Polyhedron* **1**, 83–87 (1982).
136. W. A. Herrmann, S. J. Eder, P. Kiprof, K. Rypdal, and P. Watzlowik, Organoosmium Oxides: Efficient Syntheses and Structures, *Angew. Chem. Int. Ed. Engl.* **29**, 1445–1448 (1990).
137. R. W. Marshman, W. S. Bigham, S. R. Wilson, and P. A. Shapley, Synthesis and Characterization of Monooxo Alkyl Complexes of Ruthenium(VI) and Osmium(VI), *Organometallics* **9**, 1341–1343 (1990).
138. P. A. (Belmonte) Shapley, and Z.-Y. Own, Alkylation of Osmium(VI) Nitrido Complexes: Reaction of $[Os(N)R_4][N\text{-}n\text{-}Bu_4]$ with $[Me_3O][BF_4]$ and the X-Ray Crystal Structures of $[Os(N)(CH_2SiMe_3)_4][N\text{-}n\text{-}Bu_4]$ and $[Os(NMe)(CH_2SiMe_3)_4]$, *Organometallics* **5**, 1269–1271 (1986).
139. W. T. Robinson, J. E. Fergusson, and B. R. Penfold, The Configuration of the Anion in $CsReCl_4$, *Proc. Chem. Soc.* **1963**, 116.
140. J. A. Bertrand, F. A. Cotton, and W. A. Dollase, The Metal–Metal Bonded, Polynuclear Complex Anion in $CsReCl_4$, *J. Am. Chem. Soc.* **85**, 1349–1350 (1963).
141. F. A. Cotton's Centenary Lecture "Quadruple Bonds and other Multiple Metal to Metal Bonds", *Chem. Soc. Rev.* **4**, 27–53 (1975) (The detailed story of elucidating the structure of the rhenium anions is given).
142. E. Kurras, and J. Otto, Komplexe Methylchromverbindungen, *J. Organomet. Chem.* **4**, 114–118 (1965).
143. J. Krausse, G. Marx, and G. Schödl, IR- and Röntgenstrukturanalyse des $Li_4Cr_2(CH_3)_8 \cdot 4C_4H_8O$, *J. Organomet. Chem.* **21**, 159–168 (1970).
144. F. A. Cotton, J. M. Troup, T. R. Webb, D. H. Williamson, and G. Wilkinson, The Preparation, Chemistry, and Structure of the Lithium Salt of the Octamethyldimolybdate(II) Ion, *J. Am. Chem. Soc.* **96**, 3824–3828 (1974).
145. W. Seidel, G. Kreisel, and H. Mennenga, Bis(2,6-dimethoxyphenyl)-vanadin – eine diamagnetische Vanadin(II)-Verbindung, *Z. Chem.* **16**, 492–493 (1976).
146. F. A. Cotton, and M. Millar, The Probable Existence of a Triple Bond Between Two Vanadium Atoms, *J. Am. Chem. Soc.* **99**, 7886–7891 (1977).
147. F. A. Cotton, and M. Millar, Tetrakis-(2,4,6-trimethoxyphenyl)dichromium. A Homologous New Compound with an Exceedingly Short Bond, *Inorg. Chim. Acta* **25**, L105-L106 (1977).
148. T. Nguyen, A. D. Sutton, M. Brynda, J. C. Fettinger, G. J. Long, and P. P. Power, Synthesis of a Stable Compound with Fivefold Bonding Between Two Chromium(I) Centers, *Science* **310**, 844–846 (2005).
149. Short review: U. Radius, and F. Breher, To Boldly Pass the Metal–Metal Quadruple Bond, *Angew. Chem. Int. Ed.* **45**, 3006–3010 (2006).
150. M. Brynda, L. Gagliardi, P.-O. Widmark, P. P. Power, and B. O. Roos, A Quantum Chemical Study of the Quintuple Bond Between Two Chromium Centers in

[PhCrCrPh]: *trans*-Bent versus Linear Geometry, *Angew. Chem. Int. Ed.* **45**, 3804–3807 (2006).
151. C. R. Landis, and F. Weinhold, Origin of Trans-Bent Geometries in Maximally Bonded Transition Metal and Main Group Molecules, *J. Am. Chem. Soc.* **128**, 7335–7345 (2006).
152. S. Ritter, "Quintuple" Bond Makes its Debut, *Chem. Eng. News* **83**(39), 9 (2005).
153. A. H. Janowicz, and R. G. Bergman, C—H Activation in Completely Saturated Hydrocarbons: Direct Observation of M + R—H → M(R)H, *J. Am. Chem. Soc.* **104**, 352–354 (1982).
154. A. H. Janowicz, and R. G. Bergman, Activation of C—H Bonds in Saturated Hydrocarbons on Photolysis of (η^5-C_5Me_5)(PMe_3)IrH_2. Relative Rates of Reaction of the Intermediate with Different Types of C—H Bonds and Functionalization of the Metal-Bound Alkyl Groups, *J. Am. Chem. Soc.* **105**, 3929–3939 (1983).
155. J. K. Hoyano, and W. A. G. Graham, Oxidative Addition of the Carbon–Hydrogen Bonds of Neopentane and Cyclohexane to a Photochemically Generated Iridium(I) Complex, *J. Am. Chem. Soc.* **104**, 3723–3725 (1982).
156. W. D. Jones, and F. J. Feher, Alkane Carbon–Hydrogen Bond Activation by Homogeneous Rhodium(I) Compounds, *Organometallics* **2**, 562–563 (1983).
157. W. D. Jones, and F. J. Feher, The Mechanism and Thermodynamics of Alkane and Arene Carbon–Hydrogen Bond Activation in (C_5Me_5)Rh(PMe_3)(R)H, *J. Am. Chem. Soc.* **106**, 1650–1663 (1984).
158. R. A. Periana, and R. G. Bergman, Oxidative Addition of Rhodium to Alkane C—H Bonds: Enhancement in Selectivity and Alkyl Group Functionalization, *Organometallics* **3**, 508–510 (1984).
159. R. A. Periana, and R. G. Bergman, Isomerization of the Hydridoalkylrhodium Complexes Formed on Oxidative Addition of Rhodium to Alkane C—H Bonds. Evidence for the Intermediacy of η^2-Alkane Complexes, *J. Am. Chem. Soc.* **108**, 7332–7346 (1986).
160. J. Chatt, and J. M. Davidson, The Tautomerism of Arene and Ditertiary Phosphine Complexes of Ruthenium(0), and the Preparation of New Types of Hydrido-Complexes of Ruthenium(II), *J. Chem. Soc.* **1965**, 843–855.
161. C. Giannotti, and M. L. H. Green, Photo-induced Insertion of Bis-π-cyclopentadienyltungsten into Aromatic Carbon–Hydrogen Bonds, *J. Chem. Soc., Chem. Comm.* **1972**, 1114–1115.
162. M. L. H. Green, Studies on Synthesis, Mechanism and Reactivity of Some Organo-–Molybdenum and –Tungsten Compounds, *Pure Appl. Chem.* **50**, 27–35 (1978).
163. K. L. T. Wong, J. L. Thomas, and H. H. Brintzinger, Photogeneration of Coordinatively Unsaturated Sandwich Compounds of Molybdenum and Tungsten from Their Carbonyl Complexes, *J. Am. Chem. Soc.* **96**, 3694–3695 (1974).
164. R. H. Crabtree, J. M. Mihelcic, and J. M. Quirk, Iridium Complexes in Alkane Dehydrogenation, *J. Am. Chem. Soc.* **101**, 7738–7740 (1979).
165. D. Baudry, M. Ephritikine, and H. Felkin, The Activation of C—H Bonds in Cyclopentane by Bis(phosphine)rhenium Heptahydrides, *J. Chem. Soc., Chem. Comm.* **1980**, 1243–1244.
166. J. K. Hoyano, A. D. McMaster, and W. A. G. Graham, Activation of Methane by Iridium Complexes, *J. Am. Chem. Soc.* **105**, 7190–7191 (1983).
167. M. J. Wax, J. M. Stryker, J. M. Buchanan, C. A. Kovac, and R. G. Bergman, Reversible C—H Insertion/Reductive Elimination in (η^5-Pentamethylcyclopentadienyl)(trimethylphosphine)iridium Complexes. Use in Determining Relative Metal–Carbon Bond Energies and Thermally Activating Methane, *J. Am. Chem. Soc.* **106**, 1121–1122 (1984).
168. R. H. Crabtree, The Organometallic Chemistry of Alkanes, *Chem. Rev.* **85**, 245–269 (1985).
169. M. B. Sponsler, B. H. Weiller, P. O. Stoutland, and R. G. Bergman, Liquid Xenon: An Effective Inert Solvent for C—H Oxidative Addition Reactions, *J. Am. Chem. Soc.* **111**, 6841–6843 (1989).

170. A. J. Rest, I. Whitwell, W. A. G. Graham, J. K. Hoyano, and A. D. McMaster, Photo-activation of Methane at 12 K by (η^5-Cyclopentadienyl)- and (η^5-Pentamethylcyclopentadienyl)-dicarbonyl-rhodium and –iridium Complexes, *J. Chem. Soc., Chem. Comm.* **1984**, 624–626.
171. A. J. Rest, I. Whitwell, W. A. G. Graham, J. K. Hoyano, and A. D. McMaster, Photoactivation of Methane by η^5-Cyclopentadienyl and Substituted η^5-Cyclopentadienyl Group 8 Metal Complexes, [M(η^5-C_5R_5)$(CO)_2$] (M = Rh or Ir, R = H or Me), and Dicarbonyl(η^5-indenyl)iridium: A Matrix Isolation Study, *J. Chem. Soc., Dalton Trans.* **1987**, 1181–1190.
172. R. G. Bergman, Activation of Carbon-Hydrogen Bonds in Alkanes and Other Organic Molecules Using Organotransition Metal Complexes, *Adv. Chem. Ser.* **230**, 211–20 (1992).
173. T. A. Mobley, C. Schade, and R. G. Bergman, Diastereomeric and Isotopic Scrambling in (Hydrido)alkyliridium Complexes. Evidence for the Presence of a Common "Alkane Complex" Intermediate, *J. Am. Chem. Soc.* **117**, 7822–7823 (1995).
174. J.-Y. Saillard, and R. Hoffmann, C—H and H—H Activation in Transition Metal Complexes and on Surfaces, *J. Am. Chem. Soc.* **106**, 2006–2026 (1984).
175. T. Ziegler, V. Tschinke, L. Fan, and A. D. Becke, Theoretical Study on the Electronic and Molecular Structures of (C_5H_5)M(L) (M = Rh, Ir; L = CO, PH_3) and $M(CO)_4$ (M = Ru, Os) and Their Ability to Activate the C—H Bond in Methanes, *J. Am. Chem. Soc.* **111**, 9177–9185 (1989).
176. J. Song, and M. B. Hall, Methane Activation on Transient Cyclopentadienylcarbonylrhodium, *Organometallics* **12**, 3118–3126 (1993).
177. S. T. Belt, S. B. Duckett, M. Helliwell, and R. N. Perutz, Activation and η^2-Co-ordination of Arenes: Crystal and Molecular Structure of an (η^2-Hexafluorobenzene)rhodium Complex, *J. Chem. Soc., Chem.Comm.* **1989**, 928–930.
178. C. D. Tagge, and R. G. Bergman, Synthesis, X-Ray Structure Determination, and Reactions of (Pentamethylcyclopentadienyl)(nitrosyl)ruthenium η^2-Arene Complexes, *J. Am. Chem. Soc.* **118**, 6908–6915 (1996).
179. W. D. Jones, and F. J. Feher, Comparative Reactivities of Hydrocarbon C—H Bonds with a Transition-Metal Complex, *Acc. Chem. Res.* **22**, 91–100 (1989) (references therein).
180. C. F. H. Tipper, Some Platinous-*cyclo*Propane Complexes, *J. Chem. Soc.* **1955**, 2045–2046.
181. D. M. Adams, J. Chatt, R. G. Guy, and N. Sheppard, The Structure of "Cyclopropane Platinous Chloride", *J. Chem. Soc.* **1961**, 738–742.
182. L. Cassar, and J. Halpern, Oxidative Addition of Quadricyclene to Di-μ-chlorotetracarbonyldirhodium(I) and the Mechanism of the Rhodium(I)-catalyzed Isomerization of Quadricyclene to Norbornadiene, *J. Chem. Soc. D* **1970**, 1082–1083.
183. M. Gozin, A. Weisman, Y. Ben-David, and D. Milstein, Activation of a Carbon–Carbon Bond in Solution by Transition-Metal Insertion, *Nature* **364**, 699–701 (1993).
184. M. Gozin, M. Aizenberg, S.-Y. Liou, A. Weisman, Y. Ben-David, and D. Milstein, Transfer of Methylene Groups Promoted by Metal Complexation, *Nature* **370**, 42–44 (1994).
185. S.-Y. Liou, M. Gozin, and D. Milstein, Directly Observed Oxidative Addition of a Strong Carbon–Carbon Bond to a Soluble Metal Complex, *J. Am. Chem. Soc.* **117**, 9774–9775 (1995).
186. D. Milstein, Frontiers in Bond Activation by Electron-Rich Metal Complexes, in: *Stereoselective Reactions of Metal-Activated Molecules* (Eds. H. Werner, and J. Sundermeyer, Vieweg, Braunschweig/Wiesbaden, 1995, pp. 107–115).
187. B. Rybtchinski, and D. Milstein, Metal Insertion into C—C Bonds in Solution, *Angew. Chem. Int. Ed.* **38**, 870–883 (1999).

188. B. Rybtchinski, A. Vigalok, Y. Ben-David, and D. Milstein, A Room Temperature Direct Metal Insertion into a Nonstrained Carbon–Carbon Bond in Solution. C—C vs C—H Bond Activation, *J. Am. Chem. Soc.* **118**, 12406–12415 (1996).
189. S.-Y. Liou, M. E. van der Boom, and D. Milstein, Catalytic Selective Cleavage of a Strong C—C Single Bond by Rhodium in Solution, *Chem. Comm.* **1998**, 687–688.
190. M. Gandelman, A. Vigalok, L. Konstantinovski, and D. Milstein, The First Observation and Kinetic Evaluation of a Single Step Metal Insertion into a C—C Bond, *J. Am. Chem. Soc.* **122**, 9848–9849 (2000).
191. A. E. Shilov, *Activation of Saturated Hydrocarbons by Transition Metal Complexes* (D. Reidel Publishing Company, Dordrecht, 1984).
192. J. H. Lunsford, The Catalytic Oxidative Coupling of Methane, *Angew. Chem. Int. Ed. Engl.* **34**, 970–981 (1995).
193. S. S. Stahl, J. A. Labinger, and J. E. Bercaw, Homogeneous Oxidation of Alkanes by Electrophilic Late Transition Metals, *Angew. Chem. Int. Ed.* **37**, 2180–2192 (1998).
194. C. Jia, T. Kitamura, and Y. Fujiwara, Catalytic Functionalization of Arenes and Alkanes via C—H Bond Activation, *Acc. Chem. Res.* **34**, 633–639 (2001).
195. P. M. Maitlis, and A. Haynes, Carbonylation Reactions of Alcohols and Esters, in: *Metal-catalysis in Industrial Organic Processes* (Eds. G. P. Chiusoli, and P. M. Maitlis, RSC Publishing, Cambridge, 2006, Chap. 4.2).
196. P. M. Maitlis, A. Haynes, G. J. Sunley, and M. J. Howard, Methanol Carbonylation Revisited. Thirty Years on, *J. Chem. Soc., Dalton Trans.* **1996**, 2187–2196.
197. A. Haynes, P. M. Maitlis, G. E. Morris, G. J. Sunley, H. Adams, P. W. Badger, C. M. Bowers, D. B. Cook, P. I. P. Elliot, T. Ghaffar, H. Green, T. R. Griffin, M. Payne, J. M. Pearson, M. J. Taylor, P. W. Vickers, and R. J. Watt, Promotion of Iridium-Catalyzed Methanol Carbonylation: Mechanistic Studies of the Cativa Process, *J. Am. Chem. Soc.* **126**, 2847–2861 (2004).
198. W. Keim, B. Hoffmann, R. Lodewick, M. Peuckert, G. Schmitt, J. Fleischhauer, and U. Meier, Linear Oligomerization of Olefins via Nickel Chelate Complexes and Mechanistic Considerations Based on Semiempirical Calculations, *J. Mol. Catal.* **6**, 79–97 (1979).
199. W. Keim, Nickel: An Element with Wide Application in Industrial Homogeneous Catalysis, *Angew. Chem. Int. Ed. Engl.* **29**, 235–244 (1990).
200. M. Beller, A. Zapf, and T. H. Riermeier, Palladium-Catalyzed Olefinations of Aryl Halides (Heck Reaction) and Related Transformations, in: *Transition Metals for Organic Synthesis, Second Ed.* (Eds. M. Beller and C. Bolm, Wiley-VCH, Weinheim, 2004, Vol. 1, Chap. 2.13).
201. K. Sonogashira, Development of Pd—Cu Catalyzed Cross-Coupling of Terminal Acetylenes with sp^2-Carbon Halides, *J. Organomet. Chem.* **653**, 46–49 (2002).
202. A. Zapf, Coupling of Aryl and Alkyl Halides with Organoboron Reagents (Suzuki Reaction), in: *Transition Metals for Organic Synthesis, Second Ed.* (Eds. M. Beller and C. Bolm, Wiley-VCH, Weinheim, 2004, Vol. 1, Chap. 2.10).
203. F. Calderazzo, M. Catellani, and G. P. Chiusoli, Carbon–Carbon Bond Formation through Activation of Aryl– and Vinyl–Halide Bonds: Fine Chemicals, in: *Metal-Catalysis in Industrial Organic Processes* (Eds. G. P. Chiusoli, and P. M. Maitlis, RSC Publishing, Cambridge, 2006, Chap. 5.3).
204. G. W. Parshall, and J. J. Mrowca, σ-Alkyl and –Aryl Derivatives of Transition Metals, *Adv. Organomet. Chem.* **7**, 157–209 (1968).
205. R. R. Schrock, S. W. Seidel, N. C. Mösch-Zanetti, K.-Y. Shih, M. B. O'Donoghue, W. M Davis, and W. M. Reiff, Synthesis and Decomposition of Alkyl Complexes of Molybdenum(IV) That Contain a $[(Me_3SiNCH_2CH_2)_3N]^{3-}$ Ligand. Direct Detection

of α-Elimination Processes That Are More than Six Orders of Magnitude Faster than β-Elimination Processes, *J. Am. Chem. Soc.* **119**, 11876–11893 (1997).

206. N. Shirasawa, M. Akita, S. Hikichi, and Y. Moro-oka, Thermally Stable Coordinatively Unsaturated Alkyl Complexes Resistant to β-Hydride Elimination: $Tp^{iPr}M—CH_2CH_3$ ($M = Co, Fe$), *Chem. Comm.* **1999**, 417–418.

Chapter 10
Epilogue

Die Chymie ist noch immer meine heimliche Geliebte.[1]
Johann Wolfgang von Goethe, German Poet (1749–1832)

Now, in 2008, it is 50 years ago that I started the work for my doctorate thesis and began to discover the beauties of organo-transition metal chemistry. About 40 years ago, I reported with Richard Prinz, my first Ph.D. student, the synthesis of hexamethylborazine chromium tricarbonyl, which was the first transition metal π-complex of an "inorganic benzene". Thirty five years ago, when I was at the University of Zürich, Albrecht Salzer, one of the Ph.D. students of my "Swiss family", prepared the first triple-decker sandwich, which initiated extensive research in this field in Germany, Russia, the United States and elsewhere. Thirty years ago, the chemistry of "metal bases" began to flourish in our laboratory and remained in the focus of our interest for nearly one decade. Twenty five years ago, Justin Wolf, the long-time senior research associate of my research group at Würzburg, isolated the first vinylidene rhodium complex, becoming the flagship of a series of "metalla-cumulenes" with linear M=C=C to M=C=C=C=C=C chains. Twenty years ago, Peter Schwab, who later prepared the ruthenium phenylcarbene $RuCl_2(CHPh)(PCy_3)_2$ in Grubbs' group, joined me and in the course of the studies for his Ph.D. obtained the first transition metal complex with a bridging trialkylstibine ligand. Fifteen years ago, Ruth Flügel succeeded with the synthesis of stable four-coordinate ruthenium(0) and osmium(0) compounds with a 16-electron count, which were used as starting materials for various $Ru(CH_2)$ and $Os(CH_2)$ derivatives. Ten years ago, in the continuation of Peter Schwab's work, Thomas Pechmann, my last Ph.D. student, prepared the first complexes with bridging trialkylphosphine and triarylphosphine ligands and thus provided new insights into the coordinative capability of tertiary phosphines.

What has changed in the past 50 years? When I moved from Jena to München across the "iron curtain" in 1958, the political and social circumstances on the Western side of the internal German border were consolidated

[1] In English: "Chemistry is still my secret love".

H. Werner, *Landmarks in Organo-Transition Metal Chemistry*,
Profiles in Inorganic Chemistry, DOI 10.1007/978-0-387-09848-7_10,

and the economy was in sustained growth. Based on this progress, the universities expanded and the Federal State governments as well as some restored funding institutions spent a considerable amount of money to support basic research. In those days, organo-transition metal chemistry had already left the state of its infancy and began to develop at an impressive rate. In the fresh spirit, reawakened in academic life after World War II, a young generation of scientists entered the scene and developed new areas of research. In transition organometallics, there were in particular two key players, Ernst Otto Fischer and Geoffrey Wilkinson, who took up the chance to explore the chemistry of a novel type of metal compounds, the "sandwich complexes", and they dedicated all their efforts to this goal. They also attracted a great number of students and postdoctoral fellows, and within less than 10 years the foundation of a new scientific field had been established. Fischer and Wilkinson's work was completely curiosity-driven and not influenced by public demands or economic needs. At that time, the authorities in the governments and funding institutions understood quite well that progress in science is unpredictable and that basic research in particular needed strong support.

Unfortunately, during the last decade the situation has changed. When scientists, even if young and highly motivated, apply for a new research grant they will be frequently asked "what is it good for" or "what is the relevance of this work". Apparently, politicians as well as people directing the funding agencies completely neglect what the prerequisites of innovations are and what some of the scientific pioneers have said after being asked about the reason of their success: August Kekulé welcomed "the time to dream", Louis Pasteur appreciated "having benefitted from serendipidity", and Justus von Liebig was grateful for the "germs of innovation, whizzing in the air", which hit him and matured in his mind.

Has all this been forgotten? And, thus, should we be pessimistic about the future of our science? The answer is No! My optimistic view is based on one hand on statements from leaders in research such as George Whitesides ("We need more curiosity") or Nobel laureate Sir Harold Kroto ("Fundamental advances in the chemical sciences are today more vital than in any other area for survival of the human race"), and on the other – maybe more important – on those from younger scientists who presently are taking the lead. Barbara Albert, a Professor of Chemistry at the University of Darmstadt, Germany, and a member of the board of the German Chemical Society, recently raised the question: "Where will we stand in research, teaching and engineering in twenty years time, if all of us are limited to those fields which in the short term appear particularly relevant?" And she continued: "In such a way we would produce a generation of 'me-too'-students, -researchers and -engineers with whom no modern science could be kept alive". Ursula Keller, a young professor of physics at ETH Zürich and winner of the highly prestigious Philip Morris Prize 2005, went a step further and in an interview following the award ceremony said that in a civilized society "we must definitely afford the luxury of basic research. That distinguishes human beings from animals. This is culture".

There are now some signs on the political scene that this is going to be understood.

In conclusion, I am very grateful for having had the chance to spend more than 50 years in academia. At the beginning, I was blessed with two outstanding supervisors, Franz Hein and Ernst Otto Fischer, both of whom were great scientists and strong personalities as well. I learned from Hein that above all it is crucial to maintain your convictions, and I learned from Fischer that success in research results from careful experimentation and especially hard work. Both Hein and Fischer gave me the motivation to study organo-transition metal chemistry, and I am glad having followed their advice. I was fortunate to collaborate with a superb group of about 150 graduate students and postdocs who suggested many original projects of research that I would not have conceived myself. The work carried out in our group has contributed to some of the main fields in transition organometallics, and I am pleased to say that several results of our research have made it into a number of monographs, textbooks and encyclopedias. Apart from research, I felt also privileged to teach, and it was probably no mere accident that I could attract as many of the best qualified science undergraduates as my colleagues in the seemingly more "popular" fields.

In the final paragraph of his Nobel Lecture, Alan MacDiarmid raised the question: "Of what use is a beautiful poem"? And he answered: "It gives intellectual stimulation and enjoyment. Similarly with research. If it has some practical use, that is merely 'icing on the cake'!" In this spirit, I close with a sentence written by the great German poet Johann Wolfgang von Goethe in his play *Iphigenie auf Tauris*:

> How blest is he who his progenitors
> With pride remembers, to the list'ner tells
> The story of their greatness, of their deeds,
> And, silently rejoicing, sees himself
> Link'd to this goodly chain!

Index

Note: Names of the family and personal friends were omitted

A
Abel, Edward W., 6, 82, 227
Adamson, Arthur, 62
Addison, Clifford, 40, 117
Ahlberg, Per, 278, 282
Ahrland, Sten, 37
Albano, Vincenzo G., 117
Albert, Barbara, 338
Anderson, John Stuart, 96
Anderson, J. S., 132, 200
Angelici, Robert J., 36, 255
Anschütz, Richard, 85
Arduengo, Anthony J., 237, 238, 239, 266
Ashe, Arthur J., III, 156
Ashford, T. A., 195, 196
Astruc, Didier, 146, 147, 148
Aubke, Friedhelm, 101, 102
Aumann, Rudolf, 239
Ausländer, Rose, 1

B
Bach, Johann Sebastian, 15
Bacon, Francis, 177
Bähr, Georg, 67
Baeyer, Adolf von, 70
Bailar, John C., 30, 38
Bailey, Grant C., 273, 279
Bailey, R. W., 88
Banks, Robert L., 273, 279
Barbier, Philippe Antoine, 74, 80, 81
Barton, Derek H. R., 168
Basolo, Fred, 35, 36, 38, 48
Basset, Jean-Marie, 276, 280
Battiste, Merle, 282, 283
Beattie, Ian R., 310
Beck, Wolfgang, 36, 37, 48, 99, 115
Becker, Ewald, 86, 89
Beermann, Claus, 304
Behrens, Helmut, 98, 113, 115
Bennett, George M., 298
Bennett, Martin A., 6, 57, 60, 62
Bercaw, John E., 32
Bergman, Robert G., 32, 317, 318, 319, 320, 324
Berke, Heinz, 47, 50, 275
Berson, Jerome A., 318
Berthelot, Marcellin, 43, 77, 198
Berthet, J.-C., 158
Berthold, Hans J., 304
Berzelius, Jöns Jakob, 69, 70, 71
Bestian, Herbert, 300
Biilmann, Einar, 197
Birnbaum, Karl, 71
Bjerrum, Jannik X., 167
Blaise, Emile Edmond, 81
Blanchard, Arthur A., 88
Blum, Jochanan, 321
Bock, Hans, 26
Bohr, Nils, 69
Bouveault, Louis, 80
Braunschweig, Holger, 63
Brentano, Clemens, 19
Breslow, David S., 111, 159
Breslow, Ronald, 283, 318
Brill, Rudolf, 88, 89, 98
Brimm, Eugene O., 94, 130
Brintzinger, Hans Herbert, 159, 160, 317
Briscoe, Henry Vincent Aired, 166–167
Bröring, Martin, 55, 59
Broglie, Louis-Victor de, 69
Brookhart, Maurice, 251, 268
Brown, David A., 39–40
Brown, Herbert C., 30

Brown, Robert D., 129
Bruce, Michael I., 57, 227
Brügel, Walter, 3–4
Brunner, Henri, 65, 97, 311
Brunner, John T., 82
Brynda, Marcin, 317
Buchner, Eduard, 51
Buckton, George Bowdler, 73
Bunsen, Robert Wilhelm, 69, 70, 71, 72, 79, 81
Burg, Anton B., 30, 227, 319

C
Cadet de Gassicourt, Louis-Claude, 69
Cahours, Auguste, 74, 77, 303
Calderazzo, Fausto, 36, 93, 94, 116
Calderon, Nissim, 273–274
Calvin, G., 302, 303
Cardin, David J., 276
Carmona, Ernesto, 100, 157, 158
Casey, Charles P., 241, 242, 246, 252, 274, 280
Caulton, Kenneth G., 266, 269
Cesca, Sebastiano, 300
Chatt, Joseph, 30, 117, 200, 202, 203, 204, 221, 222, 223, 224, 226, 242, 243, 244, 246, 276, 279, 300, 301, 302, 317, 320, 321, 325
Chauvin, Yves, 264, 273–274, 278, 279, 280
Chini, Paolo, 105–108, 116, 117
Chisholm, Malcolm H., 252
Churchill, Winston, 13
Ciani, Gianfranco, 107, 117
Clauss, Karl, 300, 304
Clemens, John, 47
Closson, Rex D., 94
Coates, Geoffrey E., 302–303
Cockcroft, John Douglas, 165
Coffield, Thomas H., 146, 150, 299
Cohen, John Michael, 76
Collman, James P., 258, 283
Confucius, 69
Connor, Joseph A., 157, 239
Cornils, Boy, 311
Cotton, Frank Albert, 94, 95, 96, 98, 140, 145, 161, 168, 206, 299, 314, 315, 316
Court, Trevor L., 47, 181
Cowley, Alan H., 184
Crabtree, Robert H., 318, 319
Criegee, Rudolf, 2, 149, 150, 151
Crowfoot-Hodgkin, Dorothy, 302
Cummins, Christopher C., 270–271

D
Dabard, Rene, 147
Dahl, Larry F., 93, 94, 107, 108, 117
Dalton, John, 69
Dauben, Hyp J., jr., 206
Davies, Stephen G., 165
Davison, Alan, 99
Deckelmann, Edith, 41–42
Deckelmann, Karl, 41–42
Denning, Robert G., 197
Dewar, James, 77, 86, 89
Dewar, Michael J. S., 168, 200, 201, 202, 203, 222, 246
Dirac, Paul Adrien Maurice, 69
Dodge, Richard P., 150
Döbereiner, Johann Wolfgang, 19
Doering, William von E., 134, 137, 235, 260
Dötz, Karl Heinz, 65, 271, 272
Dove, Michael, 181
Doyle, John R., 197, 303
Drefahl, Günther, 20
Dubler, Erich, 46
Dumas, Jean Baptiste André, 235
Duncanson, L. A., 200, 202–203
Dunitz, Jack D., 133–134
Dunken, Horst, 21

E
Eady, Colin R., 108
Ebsworth, Evelyn A. V., 46
Echter, Julius (Prince Bishop), 51
Edelmann, Frank, 191
Eichler, Theodor, 51
Eiland, Philip Frank, 133
Einstein, Albert, 69, 297
Eisfeld, Karl, 25, 142
Elguero, José, 61
Ellis, John Emmitt, 97, 98, 99–100, 105, 154
Elschenbroich, Christoph, 46, 144, 154
Eleuterio, Herbert S., 273
Emeléus, Harry J., 30, 46, 132, 200, 226
Emerson, George F., 151
Emerson, Ralph Waldo, 85
Ephritikhine, Michel, 150, 154, 157, 158
Ercoli, Raffaele, 116
Erker, Gerhard, 54, 55
Erlenmeyer, Emil, 31
Erlenmeyer, Hans, 160
Eschenmoser, Albert, 47
Esteruelas, Miguel Angel, 60
Ewens, R. V. G., 98

F
Fackler, John P., 6, 96
Fahr, Egon, 51
Fajans, Kasimir, 37
Fallab, Silvio, 160
Faraday, Michael, 69, 80
Felkin, Hugh, 318
Fermi, Enrico, 69
Fichte, Johann Gottlieb, 19
Figgis, Brian N., 117
Filippou, Alexander, 65, 263
Fischer, Ernst Otto, 1, 3, 4, 5, 6, 14, 25, 26, 27, 28, 29, 31, 32, 35, 36, 37, 45, 48, 49, 52, 64, 65, 97, 100, 114, 115, 131, 132, 133, 134, 135, 136, 137, 138, 139, 141, 142, 143, 144, 145, 146, 149, 150, 156, 157, 161, 162, 163, 165, 167, 207, 208, 216, 217, 221, 238, 239, 240, 241, 242, 244, 247, 248, 249, 253, 254, 255, 256, 257, 262, 263, 272, 274, 275, 276, 279, 282, 299, 311, 338, 339
Fischer, Franz, 111, 224
Fischer, Hans, 137
Fischer, Helmut, 65, 274
Fischer, Hermann Emil, 51
Fischer, Karl Tobias, 64
Fischer, Rainer Dieter, 65
Floriani, Carlo, 310
Flügel, Ruth, 337
Fogg, Deryn E., 269
Fomin, D., 298
Fowles, Gerald W. A., 46
Frankland, Edward, 71, 72, 73, 74, 75, 79, 80, 81, 87, 167, 297
Fredga, Arne, 163
Freni, Maria, 100
Freudenberg, Karl, 137, 224
Freyer, Walter, 100
Fritz, Heinz Peter, 3, 65

G
Gade, Lutz H., 55, 56, 57, 59
Gagliardi, Laura, 317
Garcia Alonso, Francisco Javier, 52
Gates, Marshall, 133
Gay-Lussac, Joseph Louis, 69, 146
Gerrard, William, 277
Geuther, Anton, 259
Giannini, Umberto, 300, 304
Gillard, Robert D., 196
Gilliland, William L., 88
Gilman, Henry, 226, 297–298
Gil-Rubio, Juan, 61
Ginsburg, David, 58, 59
Göser, Peter, 214, 217
Goethe, Johann Wolfgang von, 19, 71, 337, 339
González, Francisco, 158
González-Herrero, Pablo, 61
Gould, Edwin S., 283
Graham, Thomas, 164
Graham, William A. G., 227, 317, 318, 319, 320, 324
Green, Malcolm L. H., 144, 165, 185, 216, 217, 218, 249, 251, 317
Green, Michael, 227
Griess, Peter, 71
Grignard, Victor, 74, 80, 81
Grimes, Russel N., 156, 186
Groh, Gerda, 304
Grosse, Aristid von, 85, 200
Grove, John, 76
Grubbs, Robert H., 32, 53, 263, 264, 265, 266, 267, 268, 269, 270, 271, 272, 273, 274, 280, 282, 283, 284, 337

H
Hafner, Klaus, 58
Hafner, Walter, 3, 136, 137, 138, 141, 142, 143, 144
Hall, Michael B., 320
Hallam, B. F., 204
Halpern, Jack, 321
Hammond, George S., 32, 33
Hansen, Hans-Jürgen, 47
Hantzsch, Arthur, 43, 49, 51, 66
Harder, Viktor, 42, 46
Haworth, Robert D., 163
Hawthorne, M. Frederick, 281
Heaton, Brian T., 106, 117
Heck, Richard F., 111, 324
Hegel, Georg Friedrich Wilhelm, 19
Hein, Franz, 6, 19, 20, 21, 22, 23, 25, 27, 29, 32, 41, 44, 45, 65, 66, 67, 91, 92, 114, 136, 138, 139, 140, 141, 142, 143, 163, 199, 298, 299, 339
Heisenberg, Werner, 69
Hel'man, Anna D., 197
Heppert, Joseph A., 269, 270
Herberhold, Max, 65, 97
Herberich, Gerhard E., 65, 156, 182, 183
Hérisson, Jean-Louis, 273, 280
Herman, Daniel F., 303

Herrmann, Wolfgang A., 165, 213, 249, 266, 267, 310, 311, 312, 313
Herwig, Walter, 87, 143, 299, 310
Herzog, Siegfried, 21
Hesse, Herrmann, 9
Hieber, Walter, 5, 31, 35, 36, 48, 51, 52, 64, 65, 86, 89–91, 92, 93, 95, 96, 97, 98, 100, 110, 113, 114, 115, 132, 136, 137, 149, 199, 219, 299
Hill, Anthony, 227
Hine, Jack, 235, 260
Hoberg, Heinz, 150, 152
Hock, Heinrich, 92
Höhn, Arthur, 276
Hoffman, David M., 311
Hoffmann, Roald, 182, 191, 227, 256, 320
Hofmann, August Wilhelm von, 80
Hofmann, Hermann P., 157
Hofmann, Karl Andreas, 196, 197
Holm, Richard H., 316
Honecker, Erich, 61
Hopf, Henning, 51
Hoveyda, Amir H., 268
Howard, Judith A. K., 227
Hübel, Walter, 149, 150
Hünig, Siegfried, 48, 51
Hummel, Georg, 4
Huttner, Gottfried, 65, 241

I
Ibers, James A., 38, 92
Ilg, Kerstin, 276
Ingold, Christopher K., 246
Issleib, Kurt, 67

J
Jemmis, Eluvathingal D., 191
Jira, Reinhard, 64, 131, 132
Job, André, 86
Jørgensen, Christian Klixbüll, 37
Johnson, Brian F. G., 106, 108, 109
Johnson, Marc J. A., 269
Johnson, R. C., 38
Jonas, Klaus, 185, 219, 220, 221
Jonassen, Hans B., 197
Jones, Humphrey Owen, 77, 86, 89
Jones, Peter J., 310
Jones, William D., 317
Jutzi, Peter, 51
Juvinall, Gordon L., 247

K
Kaesz, Herbert D., 94, 227
Kaminsky, Walter, 159
Karrer, Paul, 42, 43, 47, 49
Katz, Thomas J., 58, 153, 252, 264, 274, 280
Kealy, Thomas J., 64, 129, 130, 131, 132, 134, 162
Kekulé, Friedrich August, 69, 338
Keller, R. N., 222
Keller, Ursula, 338
Kelvin, Lord (Thomson, William), 77
Kennedy, John F., 34
Kepler, Johannes, 195, 235
Khand, Ihsan, 164
Kharasch, Morris Selig, 163, 195, 196, 198
King, R. Bruce, 149, 155, 275
Kirmse, Wolfgang, 235
Kläui, Wolfgang, 42, 55
Kleinhenz, Sven, 309
Klemm, Wilhelm, 114, 139, 140, 141
Knox, Selby A. R., 227
Kögler, Hubert, 29, 146
Koerner von Gustorf, Ernst, 102
Kohlschütter, Hans W., 45
Kolbe, Hermann, 71, 81
Kondyrew, N. W., 298
Krause, Erich, 85, 200
Kreis, Gerhard, 241, 242
Kreiter, Cornelius G., 65
Kroto, Harold W., 338
Kruck, Thomas, 100
Kuball, Hans-Georg, 51
Kubas, Gregory J., 203, 320
Kudinov, Alexander R., 187
Kuhn, Norbert, 155
Kündig, Ernst Peter, 47, 145
Kurras, Erhard, 23, 310
Küster, William, 111

L
Lagowski, Joseph J., 39, 154
Lamola, Angelo A., 32
Landis, Clark R., 316–317
Langer, Carl, 6, 76–77
Lappert, Michael F.,46, 165, 244, 245, 246, 255, 269, 276, 277, 278, 304, 305, 306, 307, 309, 316, 324, 325
Lass, Raimund W., 276
Le Bel, Joseph Achille, 69
Le Blanc, Max, 66
Leermakers, Peter A., 32
Leigh, G. Jeffery, 36, 46

Leonhard, Konrad, 50
Leutert, Fritz, 90
Lewis, Gilbert N., 43
Lewis, Jack, 42, 46, 58, 59, 97, 105, 106, 108, 109, 117, 118, 165, 281
Lichtenwalter, Myrl, 297–298
Liebermann, Carl Theodore, 197
Liebig, Justus von, 6, 69, 70, 71, 79, 338
Liefde Meijer, H. J. de, 150, 300
Lippard, Stephen J., 96
Lipscomb, William N., 205
Löwig, Carl Jacob, 73
Longoni, Giuliano, 107, 116, 117
Longuett-Higgins, H. Christopher, 104, 150
Lopéz, Ana, 60
Lotz, Simon, 57
Lucas, Howard J., 198, 199

M
Maasböl, Alfred, 239, 276
MacDiarmid, Alan G., 339
Märkl, Gottfried, 51
Magnus, Gustav, 70
Maitlis, Peter M., 150, 169, 228
Malatesta, Lamberto, 117
Malisch, Wolfgang, 54
Manchot, Wilhelm, 31, 51, 87, 198, 199
Manchot, J. Wilhelm, 87
Mann, Frederick G., 222
Mansuy, Daniel, 260
Martín, Marta, 60
Martinengo, Secondo, 107, 117
Martius, C. A., 71
Mason, Ron, 108
Mathey, Francois, 154, 155, 165
Mayr, Andreas, 255
Meerwein, Hans, 30, 235
Meijere, Armin de, 272
Mendeleev, Dmitrii Ivanovich, 69
Mertis, Konstantinos, 310
Meyer, K. H., 112
Miller, Samuel A., 1, 129, 130, 131, 133, 150, 161
Miller, Wilhelm von, 31
Millikan, Robert Andrews, 69
Mills, Owen S., 204, 239
Milstein, David, 321, 322, 323, 324
Mingos, D. Michael P., 203
Mittasch, Alwin, 89, 160
Moffitt, William, 138
Moissan, Henri, 69, 77
Mond, Ludwig, 6, 75, 76–78, 81, 82, 86, 87, 88, 93, 97, 110, 114, 199
Mond, Robert L., 86, 87, 95, 96, 97
Muetterties, Earl L., 281
Müller, Jörn, 65, 214, 217
Müller-Westerhoff, Ullrich, 58, 152, 153
Müntzer, Thomas, 9
Murrell, John, 201

N
Nast, Reinhard, 115, 309
Natta, Giulio, 94, 116, 159, 160, 161, 267, 279
Nelson, Walter K., 303
Nesmeyanov, Aleksander N., 146
Neuber, Anna, 139, 140
Neukomm, Heinrich, 42, 53
Nguyen, SonBinh, 264
Niedenzu, Kurt, 39
Nixon, John F., 154
Nolan, Steven P., 266
Nöth, Heinrich, 26, 62
Novalis (Friedrich von Hardenberg), 19
Nyholm, Ronald Sydney, 40, 129

O
Odling, William, 73
Öfele, Karl, 29, 36, 38, 100, 144, 236, 237, 242, 246
Oersted, Hans Christian, 78–79
Ogden, J. Steven, 103
Oliván, Montserrat, 266
Onsager, Lars, 141
Orgel, Leslie E., 134, 150
Oro, Luis A., 59, 60
Osborn, John A., 252, 281
Oswald, Hans Rudolf, 40, 41, 43, 46
Ozin, Geoffrey A., 104, 216

P
Paal, C., 197
Pätzold, Peter, 26
Paneth, F. A., 166
Parker, Graham, 47
Parshall, George W., 247, 281, 300
Pasteur, Louis, 69, 338
Pauli, Wolfgang, 69
Pauling, Linus, 32, 198, 248
Pauson, Peter L., 1, 46, 64, 129, 130, 131, 132, 134, 145, 150, 154, 155, 161, 162, 163–165, 204, 206, 207, 221, 226

Peachey, Stanley J., 297
Pearson, Ralph G., 37, 38, 223,
Pechmann, Thomas, 54, 337
Pepinsky, Ray, 133
Pettit, Rowland, 151, 152, 168, 169, 249, 273, 283
Pfab, Wolfgang, 132
Pfeiffer, Paul, 20, 195, 196, 197
Piers, Warren E., 270
Pimentel, George, 104
Piper, T. Stanley, 300
Pitzer, Kenneth S., 198
Planck, Max Karl Ernst Ludwig, 69
Playfair, Lyon, 79
Podall, Harold E., 94
Poliakoff, Martyn, 103, 104, 207
Pombeiro, Armando J. L., 255
Pope, William J., 297, 298
Postgate, John R., 224
Powell, Herbert M., 98
Power, Philip P., 263, 315, 316, 317
Prandtl, Wilhelm, 196
Pregosin, Paul S., 47
Prelog, Vladimir, 47
Prinz, Richard, 39, 337

Q
Quincke, Friedrich, 76, 77

R
Rackow, Bogislav, 21
Raithby, Paul R., 118
Ramsay, William, 69
Rausch, Marvin D., 150, 151
Raymond, Kenneth N., 153
Razuvaev, Grigorii Alekseievich, 304
Reed, H. W. B., 210
Reichstein, Tadeus, 47
Reihlen, Hans, 87, 199, 200, 204, 207
Reppe, Walter, 110, 112, 113, 200, 208
Rest, Anthony J., 319, 320
Richards, John H., 31, 32, 33, 34, 57
Richards, Raymond L., 255
Roberts, John D., 32
Robertson, J. Monteath, 134
Robinson, Robert, 40, 201, 222
Rochow, Eugene R., 5, 30, 31, 227
Roelen, Otto, 110, 111, 112
Roos, Björn O., 316
Roosevelt, Franklin D., 13
Roper, Warren R., 47, 257, 258, 260, 261, 262, 269
Roscoe, Henry, 297
Rosenblum, Myron, 133, 150, 151, 165, 177
Rubinstein, A. M., 197
Rüdorf, Walter, 25
Rundle, Robert E., 94, 298
Rutherford, Ernest, 69
Ruzicka, Leopold, 47
Rybinskaya, Margarita, 182, 183

S
Sabatier, Paul, 81, 213
Sacco, Adriano, 100
Sacconi, Luigi, 94
Saltiel, Jack, 32
Salzer, Albrecht, 46, 178, 179, 180, 337
Sand, Julius, 298
Scotti, Mario, 57
Seaborg, Glenn Theodore, 166
Seel, Friedrich, 96, 115
Segikuchi, Akira, 151
Seidel, Wolfgang, 309, 315
Semmelhack, Martin F., 145, 272
Seppelt, Konrad, 308, 309
Seus, Dietlinde, 142
Seyferth, Dietmar, 70, 136, 138, 150
Shapley, Patricia A., 313, 314
Shaw, Bernard L., 3, 46, 300, 301
Sheline, Raymond K., 102
Sheppard, Norman, 104
Sidgwick, Nevil Vincent, 88, 90
Siebert, Walter, 51, 156, 187, 189, 191
Sieverts, Adolf, 19
Singer, Fritz, 298
Sinn, Hansjörg, 159
Skanawy-Grigorjewa, M., 243
Skell, Philip S., 144, 211, 216, 217, 235, 311
Slade, Philip E., jr., 197
Sola, Eduardo, 60
Sonogashira, Kenkichi, 324
Solvay, Ernest, 82
Spencer, John L., 214, 227
Summers, Lawrence, 299, 303
Sundermeyer, Jörg, 55, 56, 57, 59
Suzuki, Akira, 324
Schäfer, Konrad, 66
Scheer, Manfred, 155
Scheibe, Günter, 5
Schelling, Friedrich Wilhelm Joseph von, 19
Schenck, Günther O., 32
Schenk, Wolfdieter, 54

Scherer, Otto J., 51, 154, 155, 188, 189
Schiller, Friedrich, 19, 20, 129
Schlegel, August Wilhelm von, 19, 80
Schlegel, Friedrich von, 19
Schlenk, Wilhelm, 74
Schlenk, Wilhelm jr., 74
Schlögl, Karl, 55
Schmeisser, Martin, 260
Schmid, Hans, 40, 49
Schmidbaur, Hubert, 47, 51
Schmidpeter, Alfred, 26
Schmidt, Max, 26, 47–48, 51, 52, 65
Schneider, Walter, 95
Schomaker, Verner, 150
Schrauzer, Gerhard N., 208, 213
Schrock, Richard Royce, 147, 247–250, 251, 252, 253, 264, 271, 272, 274, 280, 281, 282, 308, 325
Schröder, Gerhard, 2
Schrödinger, Erwin, 69
Schröter, Heinz, 260
Schubert, Alfred, 67
Schubert, Maxwell, 91
Schubert, Ulrich, 54, 65
Schützenberger, Paul, 75, 85, 200
Schulten, H., 90
Schumacher, Ernst, 43, 177, 178, 182
Schumann, Herbert, 51
Schwab, Georg Maria, 30
Schwab, Peter, 53, 265, 337
Schwarzenbach, Gerold, 37, 40, 41, 43, 44
Stalin, Josef W., 13
Staudinger, Hermann, 30, 235
Steinhardt, Walter, 55
Stevens, Thomas S., 163
Stock, Alfred, 39, 115
Stone, F. Gordon A., 47, 142, 149, 150, 205, 214–216, 226–228, 246, 256, 257, 258, 276, 299, 319
Strauss, Steven H., 97, 101, 102
Strecker, Adolf Friedrich Ludwig, 51
Streitwieser, Andrew, 58, 150, 152, 153
Strohmeier, Walter, 102
Stuhlmann, H., 92
Sturdevant, J. H., 298

T
Taube, Rudolf, 45
Taubenest, Richard, 177, 182
Taufen, Harvey J. 202
Tebbe, Fred, 248, 281
Tebboth, John A., 133
Tieck, Ludwig, 19
Thiele, Johannes, 135
Timms, Peter L., 144, 154, 216, 217, 218, 227
Tipper, C. F. H., 321
Togni, Antonio, 47
Treadwell, William D., 40
Tremaine, John F., 133
Troup, Jan M., 98
Tschugajeff, Leo Alexandrovich, 243, 244
Tsutsui, Minoru, 136, 140, 141, 142, 143
Tune, Dave J., 47
Turner, Eustace E., 298
Turner, James J., 103, 104
Turner, Kevin, 47
Turro, Nicholas J., 32
Tyndall, John, 79

U
Uhlig, Egon, 67
Ulbricht, Walter, 19
Uloth, Robert H., 299, 303

V
Valero, Cristina, 60
Van't Hoff, Jacobus Henricus, 69
Vaska, Lauri, 93, 261
Venanzi, Luigi M., 47, 129, 197
Vetter, H., 90
Vicente, José, 61
Vögtle, Fritz, 51
Vogler, Arnd, 38
Vollhardt, K. Peter C., 148, 149

W
Wallis, Albert E., 95, 96
Walsh, Arthur Donald, 200
Wanklyn, J. Alfred, 73, 74
Wanzlick, Hans-Werner, 235, 236, 237, 238, 242
Watts, William E., 165
Weaver, D. L., 150
Weinhold, Frank, 316, 317
Weinland, Rudolf, 31, 114
Weiss, Erwin, 88, 115, 137
Weiss, Karin, 272
Weiss, Richard, 22, 143, 299
Wender, Irving, 100
Werner, Alfred, 20, 31, 37, 41, 42, 43, 49, 51, 66, 69, 85, 114, 132, 195
Weygand, Friedrich, 5

Whiting, Marc C., 133
Whitesides, George M., 305, 338
Wiberg, Egon, 26, 29, 30, 39, 115
Wiberg, Nils, 26
Wilke, Günther, 36, 55, 116, 209, 210, 211, 212, 213, 215, 216, 219, 221, 224, 225, 226, 279, 302
Wilkins, Cuthbert J., 258
Wilkins, Ralph G., 117
Wilkinson, Geoffrey, 1, 6, 30, 48, 62, 64, 96, 112, 131, 132, 133, 134, 135, 136, 138, 145–146, 158, 161–162, 163, 164, 165–168, 204, 218, 221, 226, 227, 247, 251, 252, 281, 282, 300, 304–309, 310, 313, 314, 324, 325, 338
Willner, Helge, 101, 102
Willstätter, Richard, 113
Winkler, Clemens, 74
Winstein, Saul, 198, 199, 200
Winter, R. M., 222
Wirzmüller, Anton, 238–239
Wislicenus, Johannes, 51
Wittig, Georg, 25, 29, 143, 309
Wöhler, Friedrich, 69, 259
Wolf, Justin, 52, 59, 337
Woodward, Peter, 227
Woodward, Robert Burns, 131, 132, 133, 134, 161, 162, 163, 164

Y
Yamamoto, Akio, 57, 212, 302–303

Z
Zalkin, Allen, 153
Zeise, Wilhelm Christoph, 6, 70–71, 78–79, 195–198, 221
Zeiss, Harold H., 87, 136, 137, 139, 140, 141, 142, 143, 299, 310
Ziegler, Karl, 114, 116, 152, 159, 160, 161, 209, 210, 224, 225, 264, 267, 274, 279
Ziegler, Tom, 320
Zinner, Helmut, 67
Zucchini, Umberto, 304
Zydowsky, Thomas M., 163